The Froehlich / Kent

ENCYCLOPEDIA OF
TELECOMMUNICATIONS

VOLUME 13

The Froehlich/Kent
ENCYCLOPEDIA OF TELECOMMUNICATIONS

Editor-in-Chief

Fritz E. Froehlich, Ph.D.

Professor of Telecommunications
University of Pittsburgh
Pittsburgh, Pennsylvania

Co-Editor

Allen Kent

Distinguished Service Professor of Information Science
University of Pittsburgh
Pittsburgh, Pennsylvania

Administrative Editor

Carolyn M. Hall

Pittsburgh, Pennsylvania

VOLUME 13

NETWORK-MANAGEMENT TECHNOLOGIES TO NYNEX

 **CRC Press**
Taylor & Francis Group
Boca Raton London New York

CRC Press is an imprint of the
Taylor & Francis Group, an **informa** business

First published 1997 by Marcel Dekker, Inc.

Published 2021 by CRC Press
Taylor & Francis Group
6000 Broken Sound Parkway NW, Suite 300
Boca Raton, FL 33487-2742

© 1997 by Taylor & Francis Group, LLC
CRC Press is an imprint of Taylor & Francis Group, an Informa business

No claim to original U.S. Government works

ISBN 13: 978-0-8247-2911-0 (hbk)
ISBN 13: 978-1-00-320989-8 (ebk)

DOI: 10.1201/9781003209898

Visit the Taylor & Francis Web site at
http://www.taylorandfrancis.com

and the CRC Press Web site at
http://www.crcpress.com

LIBRARY OF CONGRESS CATALOG CARD NUMBER: 90-3966

Library of Congress Cataloging-in-Publication Data

The Froehlich/Kent Encyclopedia of Telecommunications / editor-in-chief, Fritz E. Froehlich ; co-editor, Allen Kent.
    p. cm.
  Includes bibliographical references and indexes.
  ISBN 0-8247-2902-1 (v. 1 : alk. paper)
    1. Telecommunication—Encyclopedias. I. Froehlich, Fritz E.,
Kent, Allen.
TK5102.E646   1990   90-3966
384′.03—dc20   CIP

CONTENTS OF VOLUME 13

CONTRIBUTORS TO VOLUME 13

Jens Christian Arnbak, Dr.   Professor, Faculty of Electrical Engineering, Telecommunications and Traffic-Control Group, Delft University of Technology, Delft, The Netherlands: *Networks for Wireless Digital Communications*

Quirino Balzano, Ph.D.   Corporate Vice President and Director, Motorola Corporate Electromagnetic Research Laboratory, Motorola, Plantation, Florida: *Nonionizing Electromagnetic Wave Energy Exposures of Portable Cellular Telephones*

Joseph Bannister, Ph.D.   Director, Computer Systems Research Department, The Aerospace Corporation, El Segundo, California: *Networks for Local-Area Communications—An Overview of LANs*

Michael Botein, J.D, L.L.M., J.S.D.   Professor of Law and Founding Director, Communications Media Center, New York Law School, New York, New York: *The 1992 Cable Television Act*

L. Fredrik Cederqvist, J.D.   Policy Analyst, Teleport Communications Group; Senior Research Associate, Communications Media Center, New York Law School, New York, New York: *The 1992 Cable Television Act*

Frederick B. Cohen, Ph.D.   Management Analytics, Hudson, Ohio: *Network Protection*

David Frail   Director, Media Relations, NYNEX, New York, New York: *NYNEX*

Robert A. Gable   Senior Telecommunications Manager, Kennametal Inc., Latrobe, Pennsylvania: *North American 800 Services*

Oscar Garay   Manager of Motorola Corporate Electromagnetic Research Laboratory, Motorola, Plantation, Florida: *Nonionizing Electromagnetic Wave Energy Exposures of Portable Cellular Telephones*

Mario Gerla, Ph.D.   Professor, Computer Science Department, School of Engineering, University of California, Los Angeles, California: *Networks for Local-Area Communications—An Overview of LANs*

Allen S. Hammond, IV, J.D.   Professor of Law and Director of the Communications Media Center, New York Law School, New York, New York: *Network Regulation in the United States*

John P. Hudepohl, M.S.   Manager, Software Reliability Engineering and Tool Development, Northern Telecom, Research Triangle Park, North Carolina: *Network Software Reliability and Quality*

Asad Khamisy, Ph.D.   Member of Technical Staff, Sun Microsystems Labs, Mountain View, California: *Network Packet Delay Analysis Using Single-Server Queueing Models*

Roger Leo Kosak, M.S.   Senior Programmer, IBM Corporation, Raleigh, North Carolina: *Network-Management Technologies and Requirements in an Asynchronous Transfer Mode Environment*

Jean-Paul M. G. Linnartz, Dr.   Philips Research Labs, Eindhoven, The Netherlands: *Networks for Wireless Digital Communications*

Carl Eugene Loeffler, Ph.D.   SimLab, NASA Robotics Engineering Consortium, Carnegie Mellon University, Pittsburgh, Pennsylvania: *Networked Distributed Virtual Reality: Applications for Education, Entertainment, and Industry*

Tom Manning   Motorola, Plantation, Florida: *Nonionizing Electromagnetic Wave Energy Exposures of Portable Cellular Telephones*

Mahmood R. Noorchashm, Ph.D.   Member of the Technical Staff, AT&T Network and Computing Systems, Holmdel, New Jersey: *Neural Networks and Their Application in Communications*

Charles M. Preston, B.S.   Partner, Information Integrity, Anchorage, Alaska: *Network Protection*

Javad Salahi, Ph.D.   Member of the Technical Staff, AT&T Bell Laboratories, Holmdel, New Jersey: *Network Protocol and Performance Measurements for Packet Switching*

Moshe Sidi, D.Sc.   Professor, Electrical Engineering Department, Technion — Israel Institute of Technology, Haifa, Israel: *Network Packet Delay Analysis Using Single-Server Queueing Models*

Kazem Sohraby, Ph.D.   Member of the Technical Staff, AT&T Bell Laboratories, Holmdel, New Jersey: *Network Protocol and Performance Measurements for Packet Switching*

Roger Taylor, Ph.D.   Executive Director, ONet Networking Inc., Mississauga, Ontario, Canada: *NSFNET*

Yechiam Yemini, Ph.D.   Professor, Computer Science Department, Columbia University; Co-Founder and Director, Comverse Technology, Inc., Woodbury, New York; Co-Founder and Chief Scientific Advisor, System Management Arts, White Plains, New York; and Director, NYSERNet.Com, New York: *Network-Management Technologies*

Taieb F. Znati, Ph.D.   Associate Professor, Department of Computer Science and Telecommunications, University of Pittsburgh, Pittsburgh, Pennsylvania: *Network Protocols for Packet Switching (X.25) and Service Characteristics*

The Froehlich / Kent

ENCYCLOPEDIA OF TELECOMMUNICATIONS

VOLUME 13

Network-Management Technologies

Overview

This article reviews network-management problems, technologies and standards. The first section below uses a detailed example to introduce the problems and challenges of the field. The second section overviews the functions and architectures of various components of network-management systems (NMSs). The third section below provides an overview of network-management standards, and is followed by a section that describes key network-management application areas. The fifth section reviews research challenges and current directions. Bibliographic references are provided for further reading.

Introduction Through Examples

This section uses detailed examples to introduce the central problems of network management. Consider a typical networked system scenario depicted in Fig. 1. A client application, executing at a workstation on the right of the figure, wishes to exchange data with a server executing at a workstation on the left side. Data are exchanged using a Transmission Control Protocol (TCP) connection between the client and server.

The TCP connection uses routers A, B, C, and D to move Internet Protocol (IP) datagrams from the client domain I to the server domain V. Router A is connected to the client via an Ethernet link of the local-area network (LAN) domain I. Router D, similarly, is connected to the server via an Ethernet switch of the LAN domain V. The routers A and D are connected to a wide-area router backbone domain IV. Router A is connected via an asynchronous transfer mode (ATM) permanent virtual circuit (PVC) to the backbone router B. Router D is connected to the backbone router C via a T3 link. The router backbone network uses wide-area network (WAN) services provided by multiple carriers. The figure shows a Synchronous Optical Network (SONET) backbone domain III that offers various T3 link services used by the router backbone, as well as an ATM backbone II service that is layered over a different SONET backbone (not shown). Readers can consult Refs. 1 and 2 for an extensive introduction and detailed descriptions of the variety of network technologies involved in this scenario.

Consider now the flow of data from the client to the server. The client application transfers data to a local TCP transport entity via an application program interface (API) such as a socket. The TCP entity packages the data in IP packets with the appropriate IP address of the server. A local IP entity at the client workstation uses an address translation table to identify the interface address of Router A. It encapsulates the IP packet in a respective Ethernet frame that it

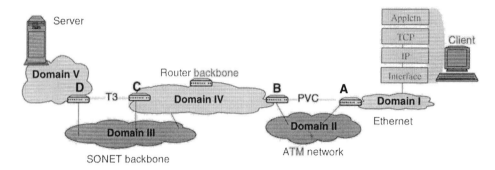

FIG. 1   A sample communication scenario involving federated domains (TCP = Transmission Control Protocol; IP = Internet Protocol).

transfers to the client interface hardware. The client interface broadcasts the Ethernet frame over the physical layer of Domain I to the interface of Router A. This flow of packets through the client workstation stack is depicted on the right side of Fig. 1.

Router A analyzes the IP address and relays the packet through the ATM PVC to Router B. The ATM interface at Router A fragments the IP packet into ATM cells and relays these cells through ATM switches of II to the ATM interface at Router B. There, the cells are reassembled into an IP packet that is relayed through the router backbone to Router C. Router C sends the packet through the T3 interface to Router D. Router D encapsulates the IP packet as an Ethernet frame and forwards it through the Ethernet switch to the server interface. The server interface passes the packet to the IP entity, which delivers it to a TCP entity, which passes it to the server application.

This entire flow of traffic through the network is accomplished by built-in protocol software and physical-layer mechanisms. Each of the domains in the figure (the LANs, WANs, routers, and applications) is typically administered by different organizations responsible for operations management. Below, we consider various operational management functions and problems associated with this scenario.

Configuration Management Problems

One of the first challenges of network management is that of establishing and maintaining a consistent configuration of various network components. Each of the domains of Fig. 1 must handle a range of such configuration management functions.

Consider first configuration management problems arising at the application layers of Domains I and V. How would the client application at Domain I determine the IP address of the server at Domain V? The client uses a directory service provided by a Domain Naming Service (DNS) (1). It issues a DNS query to a local DNS server, requesting resolution of the IP address corresponding to the symbolic name of the server. The DNS server of Domain I consults its directory database for the answer. The server may have to propagate the query further to DNS servers at the backbone domain IV and from there to Domain V.

For this name resolution process to succeed, the various DNS databases at different domains must have been configured properly. Furthermore, the client must be configured to recognize the address of a local DNS server, and the local DNS server must be configured to access DNS servers of the backbone domain IV that it uses for global name resolution. Directory databases are presently configured and maintained through manual processing. For example, the administrators of Domain I must configure each client to recognize the local DNS server and configure the local DNS server directory to include name-address mapping for all domain systems. A domain of only several hundred users typically involves directories with several thousand entries and a dozen configuration changes daily. These manual changes can lead to inconsistent entries and corrupt the directory databases. A corrupted directory configuration database can cause serious application-layer failures. Complexity of manual configuration management of directory services is thus rapidly stretching current network operations to their limits.

The administration of the router backbone domain IV, unlike those of I and V, is primarily concerned with ensuring correct and efficient packet flow among the backbone routers. Routing services involve rapidly growing complexity. Routers must be configured to support traffic by multiple protocols over multiple types of interfaces and ensure various policy constraints and security functions.

The complexity of router configuration data has been increasing rapidly. In under five years, the configuration data associated with a typical router has increased from a few score variables to several thousand. Routing configuration data, furthermore, are maintained by both manual configuration management and automatic updates by a variety of routing information protocols. This leads to routing configuration inconsistencies, resulting in complex forms of failures.

A configuration inconsistency of the routers will typically result in reachability problems. Operations staff typically use a PING tool to detect reachability problems of router backbones (1,3). The PING tool at a management workstation of the network operations center (NOC) generates packets periodically and sends them to echo servers at the routers. The echo servers reflect the packets back to the management workstation. Operations staff can monitor these PING packets to detect reachability problems. A lost PING indicates a broad spectrum of possible problems, from lost links to configuration inconsistencies. The router backbone administrators must correlate lost PING events with other symptoms to determine the problem. They must monitor configuration and operations data from different routers to detect and isolate configuration inconsistency problems.

The operations, administration, and maintenance (OAM) staff of the WAN backbones of Domains I and III face very different configuration management problems. The administration of Domain III is concerned with providing T1, T3, and other wide-area link services. To provide a T3 service connecting Routers C and D, a carrier must configure its multiplexing and interconnection equipment to allocate and route the respective T3 frames between the interfaces of the two routers. This configuration process, called *provisioning*, can be quite complex to accomplish. The T3 link may stretch over a number of physical links through a number of central offices. Each of these switching centers must allocate physical bandwidth over the respective links, set configuration tables of the multiplexing and interconnection equipment to reflect the allocation and route the link, set backup mechanisms to provide adequate service levels through failures, and configure monitoring of the link and provisioning of other service components associated with a T3 lease. With the rapid growth and changes in the underlying physical network and the variety of services that it supports, effective provisioning has become a problem of enormous complexity.

The administration of the ATM domain II, linking routers A and B, must establish and maintain a permanent virtual circuit to support this connection. The PVC must be established through multiple ATM switches, and respective quality of service (QoS) parameters must be configured and properly maintained. The assurance of QoS delivery renders the provisioning of PVC service of substantially greater complexity than providing fixed-bandwidth allocations such as T3 services. The ATM equipment shares bandwidth among multiple competing circuits. It must, therefore, regulate their access to the network by policing and shaping traffic. The provisioning process must configure these admission and flow control activities. Furthermore, ATM circuits use a SONET backbone to physically interconnect the ATM switches. The ATM domain must thus establish and maintain configuration of the mapping (multiplexing and routing) of ATM links to SONET links. The administration of the ATM domain must thus manage both the provisioning of PVC services offered to customers and provisioning of underlying SONET resources required by the ATM service. This gives rise to configuration management problems of substantial complexity.

In summary, the operations of resources at different layers and domains and the relations among these resources have parameters set by configuration data of increasing complexity. This configuration data must be set and maintained through changes by the operations staffs of the respective domains. Inconsistencies or corrupt configuration data can lead to significant operation failures, inefficient resource utilization, and low quality of services. At present, the tasks of managing configuration data and ensuring its consistency and quality require a significant level of complex, ad hoc manual processes. A central challenge of network-management technologies is to reduce this complexity and improve the quality of configuration management through software that substantially automates these functions.

Fault and Performance Management Problems

Consider now a typical fault scenario occurring in the network of Fig. 1. Suppose a clock at the interface hardware of the T3 link, used to connect Routers C and D, loses synchronization. Suppose the clock loss is intermittent and occurs for a brief period of 0.25 milliseconds (ms) approximately four times a second. This will result in an intermittent burst noise causing a loss of 0.1% in the T3 link capacity. This loss rate may be practically invisible to the WAN administration of Domain III that provides the T3 link. However, it can propagate to result in a very substantial impact on other domains. It is useful to consider the details of this propagation to best understand the challenges of problem management.

Propagation of Problems

Intermittent loss of T3 frames is typically amplified by the network layers above. A 0.1% loss over a T3 link will result in bit loss/error rates of approximately 45 kilobits per second (kb/s). Depending on the multiplexing techniques used, these bit errors will be distributed over a large number of packets. Loss or corruption of a single byte leads to corruption of an entire packet. The faulty link will generate erroneous packets at rates of scores or hundreds per second. Routers C and D, on the two sides of the T3 link, will discover and drop packets with headers that are damaged. The header could be only a small part of a packet (under 5%), depending on traffic characteristics. Therefore, only a small fraction of the erroneous packets is discovered and dropped locally at the router. Most damaged packets will only be discovered and discarded by error-detection mechanisms at end nodes. In the example of Fig. 1, the TCP connections at the client and server will identify bit errors and lost packets and will trigger respective retransmissions. Furthermore, the TCP mechanisms interpret packet loss as an indication of network congestion. They reduce the transmission window size dramatically to clear the congestion. The result of loss, retransmissions, and reduction in window size will be a significant slowdown of communications over any TCP link crossing the damaged link CD. The client-server application will see substantial degradation of throughput and a substantial increase in end-to-end delays.

Suppose the application of Fig. 1 involves a database server and clients performing transactions. The server uses locking mechanisms to protect its data. Remote clients accessing the server over the link CD will hold the locks for a substantially longer time due to their TCP connections' performance. In addition, the server's central processing unit (CPU) will be significantly taxed by the retransmissions and time-outs of TCP connections, processed as priority tasks of its operating system's kernel. This will cause a substantial degradation of the server's performance and the respective response time seen by all clients. Furthermore, database servers typically limit the time that clients are permitted to hold locks and abort transactions that exceed this time. Clients with transactions that are aborted will typically restart them from scratch, maintaining a high load on the server. Both the server and its clients will see significant performance problems.

In summary, this example illustrates the typical propagation of problems. Propagation pursues a vertical direction through network layers and a centrifugal direction from intermediate network nodes toward the end nodes. Once the problems reach end-node server applications, they may propagate horizontally to multiple clients. There are several elements of this example that are typical of emerging networked systems. It is useful to summarize them:

1. Network problems can propagate among multiple layers and domains. The manifestations of a problem may be observed in very different domains and layers than the ones in which it occurred.
2. Faults can cause performance problems and performance problems can cause faults. An interface clock problem can cause a TCP performance

problem, which in return causes failures of transactions at the applications layer. The traditional division of fault management versus performance management may need to be replaced with unified problem management paradigms.

3. There is a strong coupling between problems occurring at the network and those occurring at end-node systems and applications. Traditional organizational boundaries among network, system, and applications management may increase the complexity of effective problem management.

4. As transmission speeds increase, even a minor physical layer problem can be propagated by the multiplexing layers to cause substantial loss at layers above it. Current network architectures often amplify the impact of problems as they propagate through the layers.

5. Protocol architectures such as TCP/IP or fast packet switching have removed error detection and handling from network links to end-to-end connections. The impact of network failures may only be partially observable at intermediate nodes and will typically propagate to end nodes and be manifested at the transport and applications layers.

6. End-to-end transport mechanisms translate network failures to performance degradation. The problems of one group of connections can have a significant impact on the entire range of client-server interactions.

How can networks be managed to monitor, detect, and handle such problems? Below, we describe various technologies that are currently used to accomplish these tasks.

Monitoring Events

The first issue that must be addressed by problem management is how to detect the occurrence of a problem. A network-management system must be able to monitor operational behavior of various network elements and detect events that indicate a problem.

Figure 2 depicts the organization of a typical monitoring process. The managed element on the right is instrumented to collect various operational statistics. For example, the T3 equipment may be instrumented with counters that collect statistics of the number of lost T3 frames. Agent software, embedded at

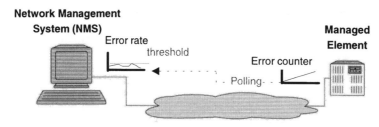

FIG. 2 Management systems monitor operations statistics of managed elements.

the element, permits management applications to access these counters. An NMS—also known as a management platform—polls the agent periodically to retrieve the error counters. The platform computes error rates and compares these to the threshold. If a threshold is exceeded, the platform will signal a problem to the operations staff (e.g., by changing the color of an icon on the screen). An alternative scheme is to compute threshold events at the agent and send event notifications to the central workstation. The protocol for exchanging management information between the platform and the agent is known as a *management protocol*.

Traditionally, the managed element, the agent software, the NMS, and the management protocol consisted of proprietary components delivered by element vendors. These systems are known as *element management systems*. The various domains of Fig. 1 can involve hundreds of different types of elements provided by different vendors. An OAM architecture that requires operations staff to use hundreds of separate element managers to monitor domain components is clearly unworkable. Creating an integrated architecture to manage heterogeneous multivendor elements thus emerged as a central driving force of the field through the late 1980s.

In the late 1980s, a temporary solution to the problem of integrated element management emerged in the form of manager of managers platforms. A number of vendors created an NMS that supported a large variety of proprietary protocols used by element managers. A manager of managers platform uses proprietary protocols and APIs to collect operations data from the element managers and present it on a single display. It unifies the display of data of heterogeneous element NMS. The difficulty with this approach is that the manager of managers platform has to support the large variety of proprietary protocols and APIs of different element vendors. This involves costly development and maintenance. The resulting graphical user interface (GUI) level integration is insufficient to handle the complexity of network management.

An alternative approach to integration, pursued concurrently, focused on standardization of the management protocol used by element agents. Two efforts sought to establish such management protocol standards. By 1990, the Internet Engineering Task Force (IETF) produced a first version of the Simple Network Management Protocol (SNMP). The Open System Interconnection (OSI) committee developed the Common Management Information Protocol (CMIP). We return to these protocols in the sections on functions and operations of management protocols. The reader is referred to Ref. 4 for a comprehensive description of both protocols.

A management protocol standard consists of two components: an architecture to organize and name various OAM instrumentation of elements and a protocol to query the data and retrieve it from the element agent to the NMS. By standardizing these components, element vendors can support syntactically unified access to OAM instrumentation of elements.

The protocol standard framework also permits simple integration of legacy systems and proprietary protocols. A proxy agent software can be constructed by an element vendor to provide standardized access to a legacy product. The proxy agent can be accessed by the platform using a protocol standard. The proxy translates these standard protocol requests to proprietary commands of the element protocol. The Telecommunications Management Network (TMN) reference framework, for example, uses proxy agents—known as q3 points—to convert proprietary element protocols to CMIP.

In summary, the operational behavior of an element is acquired by various instrumentations built into the element. Embedded agents permit remote management applications to access this data using a management protocol and to retrieve it to a management platform. Standardization of this management protocol in recent years has resulted in unification of the access and retrieval of element data.

Both SNMP and CMIP permit events to be monitored and detected at network-management platforms. Event notifications may be generated by embedded agents and dispatched to the NMS. Events may also be detected by polling operational data from agents and detecting threshold behaviors at the NMS. In the example of Fig. 1, the administration of the router domain monitors various events associated with the operations of routers and interface equipment (e.g., CSU/DSU) to the WAN links. An NMS is configured to monitor event notifications—"traps" in the language of SNMP—from agents embedded in the routers and CSU/DSU equipment. In addition, the NMS may poll various error and traffic counters to detect threshold events. These various events are reported to operations staff responsible for monitoring, analyzing, and handling them.

Correlating Events to Isolate Problems

Manual monitoring, analysis, and handling of events make up a useful paradigm to manage simple systems. However, with the growing scale and complexity of a networked system, this is no longer feasible. For example, the NMS of the router backbone domain may be responsible for managing several hundred routers and several thousand interfaces. Each router has several thousand variables collecting various operational statistics. How will operations staff know what should be monitored, how often it should be monitored, and what it means? How will they correlate this massive amount of operational data of different elements? How can they use the data to detect faults, isolate their causes, and handle them?

Consider again the T3 link failure of Fig. 1. The NMS of the router backbone could detect an excessive rate of frame errors at CSU/DSU equipment, loss of IP frames at Routers C and D, performance problems of TCP connections used to pass routing information to Router D, and respective routing problems. Furthermore, the operations staff of the router backbone may receive complaints from the administrations of the LAN domains I and V concerning connectivity and performance problems. They must correlate these various symptoms to isolate the problem to a T3 link problem and then pursue corrective actions to resolve the problem; these actions can include rerouting traffic through alternate links, interacting with the WAN provider to resolve the T3 service problems, and interactions with the end-user administrators of Domain V to alleviate the impact of the problem.

This problem management paradigm involves complex, labor-intensive processes and requires significant expertise and time. The operations staff must decide what operational data and events should be monitored. They must correlate these events to determine their cause. They must interact with operations of other domains to detect and isolate the problem and then resolve it. They must develop substantial expertise in the operations details and interactions among various components of the network. With the growing scale and complexity of networked systems, such labor-intensive problem management is no longer adequate. A single problem can result in hundreds of symptom events, generated at rates that cannot be manually processed. Operations staff cannot possibly track the complex patterns of causal propagation of problems. Technologies are needed to automate monitoring and correlation of events to isolate network problems and invoke routines to handle them.

Problem Handling

Once problems are detected, they must be handled. Often, this handling requires a number of activities by different individuals. To track and manage the problem resolution process, one uses trouble tickets. On detection of a problem, operators of an NOC open a trouble ticket and use it to record the problem's details. Trouble tickets are maintained in a database and are dispatched to the staffs responsible for handling them. In the example above, operators of a help desk of the router backbone domain IV will have created a trouble ticket in response to complaints by end-user domains I and V. Additional trouble tickets would have been opened in response to threshold events at Routers D and C and the CSU/DSU elements would have been attached to them. These different trouble tickets are assigned to operations staff, who pursue problem resolution. If the events that triggered the tickets are not correlated to isolate the problem that caused them, this process results in duplicated problem resolution activities. This can be both costly and error prone. Suppose the router backbone staff isolates the problem to the T3 link. They must then communicate the problem to the WAN backbone domain III. This triggers a similar problem resolution process by the WAN backbone administrators. The WAN NOC will generate trouble tickets to handle the T3 failure and pursue the tickets' handling.

Problem management currently involves complex ad hoc manual processes and interactions among multiple operations centers of different domains. This process typically reverses the direction of problem propagation. End users notice application-layer problems and report these to their administrations. These administrations pursue isolation of problems, resolution, and notifications to administrations of underlying domains (e.g., router backbone and WAN) providing network-layer services. These administrations perform further isolation, handling, and notifications to their underlying domains. The growing scale complexity and interdependency of domains are stretching this manual problem management to its limits. Technologies are needed to simplify and automate problem management.

Functions and Architecture of a Management System

The previous section illustrated some of the central challenges of network management. This section studies in greater detail the architecture and functions of network-management systems.

The task of network management can be loosely defined as that of monitoring, analyzing, and controlling the operations of a networked system. For the purpose of this article, the terms *network* or *networked system* are used in the broadest sense possible. They describe wide-area telecommunication networks, local-area networks, data communication networks, voice communications networks, and cellular or satellite networks. They include all operating components of a network, from the lower physical layers to the applications and services layers, from intermediate nodes that multiplex, switch, and route traffic, to end nodes that generate and consume this traffic.

The architecture of a typical network-management system is depicted in Fig. 3. The right side depicts an element managed by the NMS on the left side. The NMS platform includes manager software, which supports management protocol interactions with remote agents embedded at elements. The management protocol provides a standardized mechanism to query agents. A management application at the NMS can use the manager services to poll information from remote agents. The manager uses a transport protocol to exchange queries and responses with the agents.

The managed element includes components (e.g., port interface cards) instrumented for monitoring and configuration control. This proprietary instrumentation is organized in a tree-structured directory database, the management information base (MIB). Agent software provides a management protocol interface used to query MIB data and generate event notifications to the management platform.

In this section, we pursue a detailed description of the organization and operations of management software at the elements and then at the platform.

Organization and Function of Element Management Software

Management of elements starts with instrumentation of resources to monitor and control their behaviors. This instrumentation consists of routines to monitor operational statistics of the resource, as well as routines to configure it. Consider, for example, the management software of a typical LAN hub. The

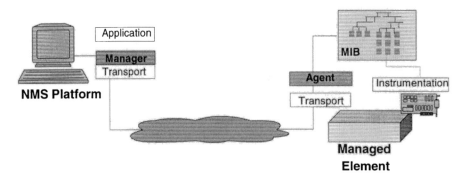

FIG. 3 Architecture of a network-management system (NMS) (MIB = management information base).

hub includes a variety of media interface resources such as thin- or thick-cable-based Ethernet, 10Base-T Ethernet, token ring, fiber distributed data interface (FDDI), 100-Mb/s (megabits per second) Fast Ethernet, bridge cards, or router cards. Each interface is a managed resource that is instrumented to monitor operational statistics such as traffic and errors. Interfaces also include instrumentation to configure their operations. For example, instrumentation of the Ethernet cards may support assignment of ports to different LAN segments (port-switching hubs).

Management instrumentation of an element can consist of thousands of variables. It is necessary to provide standardized organization of this data to enable external access to this information. This is the function of the MIB. An MIB provides a scheme to assign a unique identifier to each instrumented variable. It is essentially a tree-structured directory of instrumented managed variables. The unique identifier of a managed variable is formed by labels of a path along the MIB tree. For example, the MIB tree of SNMP labels each edge with a positive integer. A leaf node is uniquely identified by a sequence of the form 1.3.6.1.4.12.3.1, describing the sequence of edges leading from the root to the leaf. Managed data are conceptually stored at leaf nodes of the tree, much like the organization of files at the leaf nodes of a file directory. The management instrumentation of a T3 interface, for example, can maintain counters of T3 frame errors, as described in the section above, "Monitoring Events." This counter is located at a leaf and it is uniquely identified by a path descriptor as above. A remote manager can poll this counter by issuing a GET query to the respective identifier.

The agent software, embedded in elements, provides standardized access by a remote manager to managed information organized in the MIB. There are two forms of access that must be supported, synchronous and asynchronous. Synchronous access is conducted through query-response interactions. A manager can issue a GET or a SET request to read or write managed variables at the agent. The agent processes the request and dispatches a response to the manager synchronously. Asynchronous access consists of event notifications spontaneously generated by the agent and dispatched to the manager. Notifications are generated asynchronously with the manager's processing.

Traditionally, agents supported a primarily asynchronous interaction paradigm focused on event notifications. SNMP initially reversed this paradigm to focus primarily on synchronous interactions. A limited set of event notifications (traps) was provided by SNMPv1 (Version 1). The management paradigm recommended has been trap-directed polling. That is, a management center would monitor a limited set of traps, indicating significant failure conditions. On detecting such traps, the management platform would pursue polling of MIB variables using a synchronous interaction paradigm. Experience has shown, however, that more extensive support of event notifications is required and SNMPv2 has thus incorporated such support.

There are subtle differences between synchronous and asynchronous interaction paradigms that require some elaboration. On the one hand, a significant share of management is concerned with exceptional behaviors of managed elements. These behaviors are indicated by events occurring at the managed element and monitored best through local instrumentation. Indeed, elements are typically built to monitor operational events and invoke local handlers. For example, various network protocols are constructed to detect errors and failures and correct them. Therefore, it is only natural to extend this paradigm from local management of events by elements to global management of events by the NMS. An asynchronous interaction paradigm can be of great importance in supporting notifications of exceptional conditions at elements to management software that monitors and handles global conditions. It provides a natural extension of the organization of element operations.

An exception-focused management paradigm has additional advantages of requiring minimal communications bandwidth and affording a high degree of scalability. The management platform does not have to query elements periodically to find that they are operating correctly. Instead, elements notify the platform when they experience problems. In contrast, a synchronous polling-based paradigm requires constant polling of all elements in the network and thus produces significant traffic loads and leads to scalability problems.

On the other hand, an asynchronous paradigm can be highly unreliable. If an element is experiencing operational problems, it may be unable to generate event notifications. For example, suppose an element crashes and reboots. On rebooting, it may have lost the information about events that need to be notified. Even if an element generates an event notification, it is difficult to ascertain if the platform will actually receive it. Therefore, a platform could not depend on an element to notify it of its own failures. The design of a management protocol must thus accomplish a subtle balance between synchronous and asynchronous interaction models.

Management Communications

The manager software at the NMS must communicate with agents at the elements. The manager sends queries to the agent and receives responses and event notifications from the agent. The manager-agent communication model gives rise to several subtle issues considered in this section.

Consider first the transport protocol used to support manager-agent exchange. On the one hand, it is important to assure reliable transport of management information. The operations of management are typically the most highly protected component of the network. Therefore, a connection-oriented transport that assures reliable transfer may seem a useful choice. A connection-oriented transport enables effective exchange of both short messages and bulk transfer of data. A connection-oriented transport also simplifies the problem of securing manager-agent interactions since security can be established for each connection versus for each packet. For example, agents can authenticate managers and their access to element management functions at connection establishment time. Therefore, CMIP designers choose to utilize connection-oriented transport between a manager and an agent.

On the other hand, when an element experiences trouble, it may have great difficulty maintaining its connections. For example, if an element crashes and reboots, the connection is lost. The manager may waste time and scarce packets to try to reestablish the connection to the agent rather than manage it. A datagram transport would permit the manager to use more effectively whatever transport is available. The manager can use a datagram transport to accomplish directly reliable exchange or transport bulk data when it needs these services. A datagram transport protocol does not require maintenance of states at either the manager or the agent and is thus more resilient to failures. For example, if the agent crashes and reboots, the manager is not required to pursue any interactions to reestablish communications. A datagram transport protocol permits greater interoperability among heterogeneous communications architectures. A datagram transport protocol is also much simpler to implement and requires fewer resources at the managed element. SNMP designers thus select datagram transport, primarily User Datagram Protocol (UDP), to support manager-agent exchanges.

A second choice for manager-agent communications is between in-band and out-of-band models. An in-band model is one in which manager-agent communications are conducted over the very network to be managed. In-band management is typical of data networks. The manager-agent communications are conducted over the same links and use the same mechanisms as any other network application. An out-of-band model is one in which manager-agent communications are carried by a network separate from the managed network. An out-of-band model is often used to manage WAN telecommunications networks. Out-of-band communications offer the advantage of potentially greater resilience to failures. In-band management can lead to situations in which a manager is unable to communicate with the agent at a managed element because these communications depend entirely on the very functions of the managed element. On the other hand, out-of-band management can involve a substantial increase in the overall complexity of management. The network carrying management communications—the management network—requires its own management, leading to a recursive problem. This complexity has led in recent years to rapid convergence to in-band management. Mission-critical networks often use in-band management, as well as out-of-band backup links to critical resources.

Organization and Functions of Management Platform Software

The left side of Fig. 3 depicts a simplified organization of management platform software. A manager software supports exchanges between the platform and the agent. This software is typically split between a component that supports synchronous interactions (polling) and a component that supports asynchronous interactions (event notifications). The component supporting synchronous interactions is typically organized as a query client of the agent at an element, which is viewed as an MIB query server. Consider, for example, the operations of an MIB browser application. An end user utilizes a GUI to browse an MIB of a remote agent. End-user actions are translated by the browser to respective polling queries that are passed to the manager through a local API. The manager acts as a client of the agent. It issues query requests to the agent and passes the responses to the browser.

The asynchronous component of the manager is typically implemented as a background process—a daemon—that monitors the arrival of event notifications from agents. Applications subscribe to this event monitoring process to obtain notifications. Consider, for example, a graphical network display application at the platform. The application subscribes to the event-monitoring services for notifications of critical events. When any of these events arrives, the application gets a notification and invokes a routine to change the color of an icon corresponding to the element that generated the event.

The NMS platform includes components in addition to those described above. First, it typically provides a set of generic application tools. These applications include such tools as reachability tests using PING or a discovery tool that searches through the network to discover elements and incorporate them in the platform databases and display. Second, it provides vendor-specific tools to manage their elements. For example, an element vendor may provide a platform application to configure its products. Third, it may include support for applications development and for proxy agent development. Fourth, it may include various generic tools for problem and configuration management, such as diagnostics or trouble ticketing tools. Finally, the platform may include support for distribution of functions. For example, it may be necessary for multiple workstations to share access to a common network map. The platform provides map services to remote display clients.

Functions and Operations of Management Protocols

This section provides a comparative overview of SNMP and CMIP management protocols. It is important to note at the outset that the space allocated to discuss these protocols reflects their relative technical complexity rather then their relative impact. SNMP has had a significantly greater impact than CMIP, which is still struggling to be accepted by more than a minor subset of the market.

The distinguishing features of SNMP and CMIP derive primarily from their very different design goals. SNMP aims to support effective organization and exchange of managed information between managers and agents while minimizing the complexity and resources required at embedded elements. In contrast, CMIP aims to support a comprehensive range of agent services without regard to complexity or resource requirements. The enormous impact enjoyed by SNMP can be directly attributed to its simplicity. The early 1990s did not produce strong evidence that manageability requires the richer functionality of CMIP. Indeed, despite wide availability, only a minor fraction of SNMP agents and of the growing MIB data that they provide are ever used. The main challenges in accomplishing manageability are apparently less dependent on extensive ability to access operational data at agents. Rather, the primary problems are automation of the analysis of this data and the respective operations management procedures.

Protocol Operations Models

The overall operations of SNMP are depicted in Fig. 4. A manager entity can issue queries to an agent to GET or SET a specific MIB variable or to retrieve

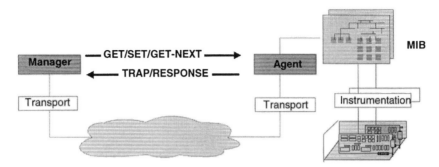

FIG. 4 Management protocol operations.

rows of an MIB table using GET-NEXT. The agent responds to these queries or emits trap messages notifying the manager of exceptional events. These queries and responses are transported by datagrams. CMIP operations are similar except the query language provides a richer set of operations and the transport layer is connection based.

Consider first the transport environment supporting manager-agent communications. SNMP requires a minimal datagram transport service to exchange queries, responses, and trap packets. While the primary datagram protocol used by SNMP is the Internet's UDP, implementations exist over other datagram transport protocols such as IPX, AppleTalk, and connectionless network service (CLNS). A datagram transport leaves the manager and agents the burden of handling bulk data transfer (e.g., large configuration tables) and recovering from unreliable transport behaviors (e.g., packet loss and reordering).

In contrast, CMIP requires an OSI connection-based transport to support manager-agent interactions. The manager must establish and maintain an association with each agent that it manages. A connection-based transport simplifies the transfer of bulk data and leaves recovery from packet loss or reordering to the transport layer. CMIP managers and agents, in contrast to SNMP, thus assume that the transport layer provides reliable stream communications. This, somewhat paradoxically, reduces the overall reliability of management. When an element experiences stress and requires management attention, it will typically tend to lose its connections. The few packets that it succeeds in transporting are used to maintain and reestablish connections rather than deliver critical management information. Thus, a connection-based transport operates at its worst just when manager-agent interactions require its operation most.

Managed Information Models

The fundamental data unit used by SNMP to describe managed information is that of a variable. SNMP uses a small number of variable data types such as counters, gauges, integers, or bit strings. These variables model instrumentation associated with managed resources. For example, a port of a managed communications system may have counters to measure cumulative traffic statistics such as packet arrivals, errors, and so on. SNMP uses a simple extension of Abstract Syntax Notation No. 1 (ASN.1) to specify the syntax of a variable. It limits the number of data types in order to minimize the complexity of converting data to or from internal representation at the agent to a common network representation (transfer syntax).

SNMP MIB variables are typically used for a few functions. Most variables are utilized to access operational statistics or configuration data. Some variables are used to provide configuration control through side effects of their changes. For example, a manager can partition a port by invoking a SET of a respective variable to a value 1. Still other variables are used by SNMP to coordinate concurrent accesses by multiple managers to agent resources. For example, multiple managers may wish to configure the variables of a remote monitoring (RMON) MIB to monitor packet statistics. If managers access configuration variables concurrently, they may cause inconsistencies. Special MIB variables are designated as locks to prevent the interference of such concurrent accesses.

SNMP provides means to group variables associated with the same resource. The fundamental structure used to accomplish such grouping is that of a table. For example, the rows of an interface table may describe various operational statistics and configuration of different interfaces. Each column of the table describes a specific attribute of an interface (e.g., a traffic counter or configuration state). Each row of the table describes the values of these attributes associated with a specific interface. Tables are used extensively by SNMP MIBs to capture data of managed objects with similar attributes. SNMP, however, does not view tables as part of its managed information model structure but, rather, views them as part of the conceptual organization and access of an MIB. In SNMPv1 table structures were entirely external to the specification of an MIB. This led to nonuniformity and difficulties in organizing and maintaining tabular data through changes. SNMPv2 thus includes extensions of the structure of managed information to specify roles of variables in indexing tables.

The fundamental data unit used by CMIP is that of a managed object with a type that is specified by a class. For example, a communications port object may include data attributes to describe operational statistics of the port such as error and packet arrival counters, as well as operations (methods, procedures) to control the port configuration (e.g., partition the port or run loopback tests).

CMIP, in contrast to SNMP, thus aggregates data attributes and operations as essential components of its managed information model. It uses an extended class model, commonly known as the Guidelines for the Development of Managed Objects (GDMO), to specify the syntax of managed object classes. GDMO classes, like those commonly used in object-oriented (OO) software, support specifications of data attributes and methods of a managed object and use inheritance to abstract common properties of similar objects. For example, one may define a class "router node" to describe generic data (e.g., routing tables, packet traffic statistics) and methods (e.g., warm and cold reboot, partition links) associated with a router node. Specific attributes and methods of routers for IP, IPX, or AppleTalk packets can then be incorporated in subclasses of the router node class that inherit the generic attributes and methods.

The data attributes of a managed object are similar to SNMP's variables and offer similar expressive power. They can usefully capture operational statistics and configuration data of a managed resource. The methods of a managed object provide explicit invocation of operations that contrasts with SNMP's implicit invocation of operations as side effects of a SET action. Furthermore, in contrast to SNMP, CMIP methods can be used to pass parameters to and from an operation. For example, a configuration change operation may require passing of multiple parameters to a resource and returning a value to the invoking manager routine. SNMP does not support this directly and requires complex manager-agent interactions to accomplish the same effect.

Class inheritance could be of great value in simplifying the modeling of the vast resources that compose a network. Inheritance enables one to model common features of diverse resources and add details incrementally to capture special features. For example, the interface table of MIB-II provides a collection of generic attributes of various interfaces common in internetworks. This interface table has been broadly used with variations and specializations by other MIBs. For example, the AtomMIB describing resources of an ATM switch uses variants of the interface table to describe a variety of resources, including virtual circuits. An OO modeling approach would have permitted one to recognize and take advantage of this similarity systematically. A class "interface" could have captured generic attributes and methods of an interface resource. This then could be specialized to model the large variety of interfaces occurring in various network nodes.

An OO approach to building software has been very useful in many fields, from computer-aided design/computer-aided manufacturing (CAD/CAM) systems to user interfaces. Class abstractions permit one to handle effectively computational tasks involving a large number of different objects. For example, CAD problems often involve thousands of different objects that must be accessed and manipulated by applications software. Class abstractions permit applications to accomplish these processing tasks in terms of a small number of common abstractions.

An OO model as pursued by CMIP could thus be of great value in simplifying the task of developing applications to process managed information. Application designers could focus on processing high-level common abstractions of managed information rather than low-level representations of this information. For example, a connectivity analysis application could focus on processing generic attributes of nodes and links regardless of their specific details. In the absence of such generic abstractions, an application code must reflect knowledge of specific MIB data that describe such connectivity information in each element. The complexity of coding this specific knowledge is perhaps one of the more serious obstacles in the development of management applications software. It could explain the increasing gap between a vast and growing range of MIBs available in networks and the minimal number of management applications that can use these MIBs.

At this time, however, the use of OO data models as advocated by CMIP has not produced significant evidence of the potential value. A number of GDMO specifications have been constructed by various committees. These typically make only minimal use of inheritance. Often, a managed object is specified as a single-level deep inheritance hierarchy. This means that every object is essentially defined as a special case of a single "top" class. Common generic features are not abstracted. Thus, no significant simplification is accomplished in developing management applications to manipulate these managed objects. Furthermore, CMIP implementations have focused on the agent side rather than demonstrating the value of an OO model in simplifying the creation of powerful applications on the manager side.

At present, the potential promise of OO management software organization as advocated by CMIP is yet to be proven by actual CMIP implementation and use. Interestingly, however, OO management software organization has been increasingly pursued in the design of management platforms. OO models of managed resources are used by several platform products to simplify the development of management applications. These OO models view SNMP data as instrumentation of objects at the platform. Thus, an interface object maintained by a platform repository may acquire its data attributes by polling respective SNMP MIB variables at agents. This OO approach to the design of management applications has accomplished important successes. It has demonstrated that OO design of management applications can be entirely accomplished and usefully applied without requiring the features and complexity of CMIP agents.

Organization and Access of Management Information Bases

Both SNMP and CMIP use hierarchical directory trees to organize access to managed information. A directory tree provides a unique identification of each node by concatenating the labels on the path to this node from the root. SNMP organizes MIB variables at the leaves of a static directory tree (see Fig. 5). Internal nodes of the tree represent grouping of variables along resources. For example, Group 1 can represent resources associated with a protocol layer, such as the cell layer of an ATM switch, and Subgroup 1 can represent a table of virtual circuits. The data associated with each virtual circuit is recorded as a row entry in the table.

Branches are labeled with numbers and the path to a leaf node is described as a string of numerical labels. For example, the path from the root to Attribute 4 is described by the string 1.1.3.1.1. These path descriptors provide unique identifiers of leaf nodes. MIB variables are conceptually stored at the leaves of

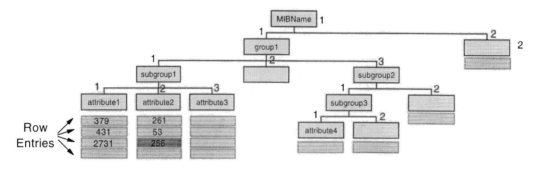

FIG. 5   The structure of a management information base tree.

the MIB tree. They are depicted as gray cells at the bottom of the figure. How can the leaf identification scheme be extended to identify these variables uniquely? Consider first the identification of a nontabular variable such as Attribute 4. The cell under the leaf can be associated with a path label 0 and is thus uniquely identified by the path 1.1.3.1.1.0. To retrieve the data of the cell, a manager can issue a GET query to cell 1.1.3.1.1.0.

A more challenging task is to extend the identification scheme to tabular data such as the table of Subgroup 1. Each cell of the table can be identified by the respective column and row. Columns are identified by the path descriptors of their headers. How would one identify rows uniquely? Unlike the static column structure, rows of tables can be dynamically created and deleted and cannot, therefore, be assigned static identifiers. SNMP solves the problem simply by using the context of index columns to identify rows uniquely. For example, consider the table under Subgroup 1 and suppose the first column of the table, Attribute 1, includes data such as virtual-circuit numbers that uniquely identify each row. To identify the darker cell in the second column and third row, one can use the column identifier 1.1.1.2 and the row identifier 2731. SNMP simply concatenates these column and row identifiers to form a unique cell identifier 1.1.1.2.2731.

To retrieve the data of the dark cell, a manager can issue a GET query with identifier 1.1.1.2.2731. The agent parses this identifier to find the column and row identifiers and uses these to locate the cell. But, how would the manager find the cell identifier in the first place? While the column identifier is static and can be known a priori, the contents of Attribute 1 and the rows of the table can change dynamically and may not be known to the manager. A direct GET is insufficient to access tabular data.

To allow simple access to tabular data, SNMP introduced the GET-NEXT mechanisms to traverse the rows of tables without knowing a priori their identifiers. The manager can issue a GET-NEXT to the leaf cell 1.1.1.2. This will retrieve a variable binding pair ⟨1.1.1.2.379, 261⟩ for which 1.1.1.2.379 is the identifier of the cell next to the leaf 1.1.1.2, that is, the first cell of the second column, and 261 is the value stored in this cell. Thus, a single retrieval provides identification of both the cell and its contents. The manager can repeat the process, applying GET-NEXT to the cell 1.1.1.2.379. This will result in a variable binding pair identifying the second cell in Column 2, 1.1.1.2.431, and its value 53. Applying GET-NEXT to this cell will retrieve a variable binding pair ⟨1.1.1.2.2731, 256⟩, identifying the third cell in the column (dark cell) and its contents 256.

GET-NEXT thus provides a means to traverse tables and retrieve their contents a row at a time. While this is a satisfactory mechanism to resolve the problem of accessing small tables, it can be highly inefficient for retrieving large tables. The GET-NEXT response frames may contain a small amount of information and the manager must first retrieve and process a row before it can retrieve the next row. SNMPv2 has thus generalized this construct using a GET-BULK query to improve the efficiency of table traversal. A GET-BULK, similarly to GET-NEXT, retrieves successive fragments of a table, except GET-BULK request results in packing of data from as many rows as would fit in a response frame. Thus, a manager can retrieve large tables significantly faster.

CMIP, similar to SNMP, uses a directory tree to organize managed information. This directory tree is called a managed information tree (MIT) and is modeled after the X.500 directory structure. In contrast to SNMP MIBs, an MIT stores objects at intermediate nodes, not just at leaves. The MIT structure, unlike MIBs, can change dynamically as objects are created or deleted at nodes. The path labels cannot be statically assigned as in SNMP. Instead, each node of the MIT is labeled by distinguishing data attributes of the object attached to the node. These data attributes are called the relative distinguishing name (RDN) of the node. They must distinguish it among all other objects directly attached to its parent. By concatenating these RDNs along the path from the root to a node, one gets a unique distinguishing name (UDN) of the node.

A dynamic directory tree inspires completely different access techniques from those of a static one. A static tree permits a management application to request given data by specifying its static identifier. This identifier is known to the application designer or its user and can thus be directly provided. In contrast, when the directory tree is dynamic, an application designer or user typically cannot know the tree path to access specific data. To access MIT data, one typically needs to traverse large subsets of the tree to find the objects of interest. Therefore, whereas a GET request of SNMP results in predictable retrieval of specific data requested, a CMIP GET specifies a search of a subtree of the MIT and can result in an unpredictable amount of data retrieved. This leads to substantial differences in the semantics of GET requests of the two protocols. For example, SNMP uses direct GET and GET-NEXT requests, while a CMIP GET provides guidelines to search for the information to be retrieved. These and other differences in the query protocols are discussed below.

Event Notifications

CMIP GDMO ventures well beyond traditional OO software structures. Traditional object models view objects as a means to support synchronous manipulations of data and invocation of methods. External software (e.g., manager) can retrieve and change data attributes of an object or invoke any of its methods in synchrony with its computations. The GDMO introduces an asynchronous component to objects: event notifications. An event notification, unlike an attribute access or a method invocation, is assumed to be spontaneously generated by an object asynchronously with a manager's computations.

Asynchronous objects require mechanisms to monitor events, generate notifications, and send the notifications to managers. In addition, an agent must be configured to maintain event notification subscriptions by managers. CMIP managers create subscription objects in the agent's MIB to get notifications. Events are generated by managed objects and are then filtered and classified by the agent; the agent uses the subscription objects to generate notifications. This seemingly simple mechanism leads to the complex semantic questions and implementation issues considered below.

To begin, event monitoring requires an active computational thread at the agent that checks conditions and generates notifications. If monitoring were to be pursued on a per-object basis, each managed object would require a thread to detect and to communicate its events. This would lead to prohibitive implementation complexity. A more reasonable approach is to maintain a single thread per agent to monitor all local events. Then event information would be more effectively maintained per agent, as in SNMP, versus per object as in CMIP.

Event notification services require agents to maintain the subscription status of various managers. Synchronizing subscription databases on failures leads to complex issues. Consider first the failure of a manager. When the manager reboots, its reincarnation will post new subscriptions with various agents. The new reincarnation may have a different identity from the old one, and thus the agents will record these subscriptions as new. Should a manager crash and reboot often, it can exhaust agent resources and cause agents to crash, too. Consider the failure and rebooting of an agent. The MIT data is lost and must be reconstructed. Records of operational statistics can be reset on reboot entirely through local agent actions. Local configuration data also may be restored to default values from read-only memory (ROM) or a boot server. In contrast, configuration data created by different managers, such as event subscriptions, can be difficult to restore. One possibility is for each manager to monitor the status of the agent and restore its event subscriptions on failure and rebooting. Another possibility is for the agent to support persistent storage of its event subscription database. Both possibilities lead to significant complexity in organizing a management system.

These complexities led SNMPv1 to minimize support of traps and separate them entirely from its managed information model and MIB. Event notifications were left to agent implementations with the expectation that a trap-directed polling paradigm will minimize use of events primarily to agent events such as a cold or warm reboot. Events, contrary to these expectations, have remained very central components in management and were introduced by vendors in the form of a large number of private traps. SNMPv2 has thus introduced extensive facilities to support event notifications as integral components of its managed information services. The events monitored by an agent are defined in a special traps MIB. The parties that should be notified of an event are specified by a party MIB. An agent maintains an active computational thread to monitor event occurrence and notify parties according to the specifications recorded in these MIBs.

Query Protocols

The query protocol used by SNMP consists of the following commands:

- GET request: Specifies a list of variables to be retrieved
- GET-NEXT request: Specifies a list of variables in a table row to be retrieved
- GET-RESPONSE: Returns a list of variables and their values
- SET request: Changes the contents of a variable
- TRAP: Specifies the event, time of occurrence, and additional relevant information

SNMPv2 generalized the GET-NEXT request to a GET-BULK request, discussed in the previous section.

The query protocol used by CMIP consists of the following commands:

- GET request: Specifies a subset of the MIT and a filtering criterion to retrieve objects
- CANCEL-GET request: Specifies a GET request to be canceled
- SET request: Specifies objects to be modified
- CREATE request: Creates an MIT object
- DELETE request: Deletes an object
- ACTION request: Invoke an action (method) of an object
- EVENT-NOTIFICATION: Notifies a manager of an event

Each of these protocol commands has a respective response frame.

There are a number of significant distinctions between the query protocols. Notice first that CMIP offers a number of capabilities not directly supported by SNMP. These include creation/deletion of MIB objects and direct invocation of actions. SNMP supports remote invocation of agent actions only through implicit side effects of SET requests. Explicit invocations as supported by CMIP's action request are semantically more powerful and permit a manager to pass parameters to/from an action. The richer semantics of actions is difficult or impossible to accomplish through implicit actions of SNMP. It limits the power of SNMP in supporting actions such as configuration control or atomic changes of multiple MIB parameters.

SNMPv1 permitted MIB entries to be created and deleted only by an agent. With the introduction of RMON, it was necessary to support remote creation/ deletion of table entries by managers. For example, a manager wishing to activate an RMON monitoring function must create a table row that configures this activity. SNMPv2 thus included create/delete capabilities for table rows similar to CMIP. However, in contrast to CMIP, these functions are accomplished through MIB table structures rather than through additional protocol constructs. Furthermore, one can only create or delete table entries, but not modify the structure of the MIB tree. The MIB tree is entirely determined statically. This contrasts with CMIP's create/delete operations, which modify the very structure of the MIT. The create/delete commands of CMIP can thus change the access path to MIT objects dynamically. A management application may be unable to predict statically where MIT objects that it needs to access may reside. It must obtain this access path information dynamically and maintain it through possible dynamic changes. CMIP accesses to the MIT are thus typically organized to handle subtrees rather than individual managed objects as accessed by SNMP.

The distinctions between the GET operations of CMIP and SNMP are more subtle. An SNMP GET specifies directly the object to be retrieved. Traversal is restricted to a row (or a few rows) at a time using GET-NEXT. In all cases, the amount of data carried by the response frame is completely predictable statically and fits in a single frame. In contrast, CMIP GET requests are primarily oriented to traverse and search a subset of the MIT tree and can generate an unpredictable amount of response data.

This results in a number of distinctions of CMIP GET from that of SNMP. First, a GET query can result in a flood of data that may have to be aborted by the manager. The protocol thus includes a CANCEL-GET request to accomplish such an abortion. Second, since a query may result in responses that exceed a single datagram size, the protocol requires connection-oriented transport. Third, a broader range of conflicts can arise among concurrent accesses by multiple managers to the MIT, requiring a broader and more complex protection. For example, a GET query to a subtree may be in conflict with a concurrent command to delete the subtree. In contrast to SNMP, for which concurrency control can be limited to individual managed objects (i.e., represented by table rows), CMIP must support concurrency control over entire subtrees of the MIT. This results from the fundamental distinction in the units of access by the protocols — variables and table rows for SNMP versus MIT subtrees for CMIP. As a result, concurrency control of CMIP is substantially more complex.

Additional Considerations

The designs of SNMP and CMIP often make implicit assumptions of managed elements and the functions of the management protocol. It is useful to consider these assumptions to understand the possibilities and limitations of the designs.

A fundamental assumption of SNMP, for example, is that the size of MIB tables typically will be small. The rows of an MIB table represent resources of similar operational attributes, defined by columns. Typically, these are hardware (e.g., ports) or protocol (e.g., IP neighbors) resources for which only a small number of instances exist in a managed element. If tables are large, their retrieval a row at a time (or a bulk at a time) may be inadequate. For example, consider an ATM switch that supports several thousand virtual circuits (VCs) with operational data that are maintained in a table of AtomMIB. An application wishing to monitor VCs experiencing problems will have to retrieve the entire table and filter the entries. This may require too much time and too many resources. It would have been more usefully accomplished if the filters to detect a problem VC were to be executed at the agent, rather than at a remote manager, and were to generate appropriate traps. CMIP, in contrast, permits a single GET to apply a filter to all managed objects of a subtree of the MIT. Thus, all managed objects describing VCs can be scanned and filtered at the agent by a single GET, retrieving only problem VCs.

Of course, if all applications could be anticipated a priori, one could program their computations as agent instrumentation and retrieve the results. In the example above, the filters to determine VC problems could be instrumented as Boolean columns of the table and then a few GET-BULK requests to the correct column would retrieve the entire column and determine which VCs are experiencing problems. Obviously, this hard coding of applications into MIB tables is an inadequate solution to the problem of selective retrievals from large tables. Large tables in general emerge as a result of the growth in complexity of network elements and of the increasing virtualization of resources. Large tables also emerge when it is necessary to maintain relationships among resources. As network elements become more complex, the problems of organizing and manipulating large MIB tables are likely to become increasingly important.

A fundamental assumption of CMIP is that the agent can command sufficient computational resources to maintain the MIT and support the CMIP query protocol. The MIT is a fairly complex, object-oriented database, and the CMIP query protocol provides not only full-fledged query capabilities, but also includes a remote procedure call (RPC) mechanism of action requests. An agent must thus combine the functions of an object-oriented database server and those of an RPC server. The resources required to support these capabilities are substantial and of similar magnitude as those of a management platform. This gives rise to the following architectural paradox. If the agent commands substantial resources and can provide complex database and RPC server functions, why should management applications be executed at the platform rather than at the agent in the first place? Also, if the applications are to execute at the agent, why is an extensive query protocol even required (e.g., vs. remote display of application results at the platform using any of a number of window presentation protocols)?

In contrast to CMIP, SNMP maintained a sharp model that distinguished agent and manager roles. An agent is assumed to be a simple, (mostly) passive provider of access to management instrumentation. Active processing roles are given to managers. CMIP eliminates these boundaries, leaving an unclear, somewhat paradoxical, paradigm for architectural division of management functions.

The sharp division of manager/agent roles assumed by SNMP provides a clearer paradigm, but one that sometimes is unable to handle the increasingly complex manageability issues in emerging networks. In particular, a two-level hierarchy that centralizes all processing into managers and leaves agents in purely passive roles does not scale. SNMPv2, therefore, introduced support of hierarchical organization of managers and of middle managers to play dual roles of agents and managers. A discussion of these facilities, however, is beyond the scope of this article.

Current Directions

SNMPv1 has had very important successes. The protocol has been widely accepted and incorporated within thousands of different products. A number of SNMP MIBs have been very successful in establishing standards for management instrumentation of specific types of elements, such as Ethernet repeater hubs, remote LAN monitors, or ATM switches. This success has been recently extended from intermediate network systems to attached host computers. In particular, SNMP MIBs have been developed to instrument typical personal computer (PC) resources and general workstation resources to enable their management via SNMP.

This enormous success on the agent side contrasts with relatively minimal advances in developing and deploying management applications at the NMS side. Most SNMP agents installed in current systems are not managed at all;

those managed access a minimal fraction of their growing MIBs. A key reason for this gap is perhaps the absence of effective technologies and paradigms to leverage SNMP instrumentation to automate management functions. In the absence of such automation, the task of monitoring and interpreting MIB variables is left to the operations staff. With the growing pressure on operations staff and the growing complexity and number of MIBs, it is little wonder that the instrumentation layer offered by SNMP is hardly used. The problem of automating management functions is considered in a separate section below.

This growing applications gap could perhaps also explain the sluggish acceptance of SNMPv2 compared with the enthusiastic adoption of SNMPv1. The SNMPv2 effort resulted in very significant improvements of SNMPv1 in a few areas, while maintaining careful version compatibility and easing the process of transition. SNMPv2 incorporates effective mechanisms to support secure management, mechanisms for efficient access to tables, and mechanisms to support hierarchical organization of managers. It extended, streamlined, and resolved many limitations of SNMPv1. However, element vendors have not rushed to replace their SNMPv1 agents. A central reason for this is perhaps the lack of "applications pull" needed to justify the investment required for transition.

CMIP has enjoyed a more limited success. There have been a number of efforts to define GDMO models of certain types of resources, primarily used in WAN. However, the complexity of implementation of CMIP agents, coupled with the lack of platform software that can usefully leverage the extensive capabilities of these agents, led to very minimal deployment.

The emergence of broad distributed applications technologies and standards is likely to have a profound impact on the evolution of network management. For example, consider the recent Common Object Request Broker Architecture (CORBA) standard for remote object access by distributed applications. It may provide a more useful approach to accomplish the OO management paradigm envisioned by CMIP than was possible for CMIP.

The CORBA Interface Definition Language (IDL) supports a syntax for the access of remote objects. It permits an application to access data attributes and invoke methods of remote objects. The CORBA proposal already enjoys wide support because of its broad use in a large variety of distributed applications. A CMIP-like management agent could be implemented as a CORBA application requiring no protocol standards beyond those already provided by CORBA. The agent would maintain a repository of managed objects and use the IDL to support simple GET/SET/CREATE/DELETE access to objects. The management of naming by the MIT could be replaced by the object request broker of CORBA. Query interfaces such as filtered GET could be introduced as agent repository features, and event notifications could be handled via the CORBA publish/subscribe message mechanisms.

These possibilities raise questions about the need for specialized management protocols in the future. Management software could be potentially viewed in broader terms as distributed applications requiring efficient access to retrieve and manipulate remote data. Success of general mechanisms such as CORBA in providing broader capabilities for remote manipulation of objects could eliminate the need for application-specific standards. This does not imply that management protocol standards will be discarded. Perhaps a central accomplishment of SNMP has been the MIB frameworks and standards that it caused to instrument a broad range of network resources. These standards are likely to provide the basis for future network management regardless of the techniques by which such instrumentation is accessed and manipulated.

Management Applications

Consider again the overall organization of a network-management system as depicted in Fig. 3. This section describes typical management platform applications and the manner in which they utilize the management protocol. It is useful to note that, at this time, the main focus of management applications is to display network information and support configuration control of network elements by operations staff. There is a growing need to replace this labor-intensive operations paradigm with one that automates management functions through applications software.

Network-Monitoring Tools

A significant share of current management applications is devoted to displaying varied information about the network. A variety of tools is used to support such monitoring of network information: event monitoring, discovery, protocol analysis, and MIB browsing and monitoring tools.

Event Monitoring Tools

Vendors of various network elements typically provide application tools to monitor events associated with the elements. Older systems provided varied reports and text presentation of events. For example, the IBM NetView system developed in the 1970s to monitor teleprocessing networks presented text on operators' screens with messages describing events associated with various elements incorporated in the network. Network elements used a special protocol to report such events to the operators' platform. The protocol delivered vectors of pointers to a text database at the platform. These pointers identified the text pieces from which the screen message was to be composed.

With growing network complexity and event rates, current platforms typically use a GUI to present event information. Consider the scenario depicted in Fig. 3. A managed element generates traps sent to a management platform. These events are presented to operators through appropriate icons. For example, consider a router that experiences a port failure. The router will generate appropriate trap notification to a management platform. This trap will be captured in the management platform by a daemon that monitors and routes events to management applications. In this case, the daemon will forward the port-failure trap to an event-monitoring application provided by the router vendor. This event monitor will change the router icon color in the map database of the platform to "red." When an operator clicks on the icon to obtain greater details of the problem, the operator sees a detailed graphical image of the router with the respective port colored red. This permits effective monitoring of network events by operators.

It should be noted that, for this entire application model to work, the application must be intimately integrated with a variety of platform components: the event-monitoring services, the topology map database and display, and various other internal components. This intimate interaction between applications and platform services required development of standardized platform environments shared by multiple vendors. As a result, the growing proliferation of vendor-specific platforms of the 1980s contracted to a small number of platforms that became de facto market standards. The Open Systems Foundation (OSF) attempted, in the early 1990s, to establish a common standard for management platform services, the Distributed Management Environment (DME), as did other consortia such as X/Open. By and large, these efforts have not been successful.

Discovery Tools

Discovery tools aim to discover components incorporated in a network and their interconnection topology. Discovery tools are typically used to establish a topology map of the network at the routing/switching layer. For example, a discovery tool is used to establish a map of IP routers and hosts of an Internet domain and the links connecting them.

Discovery is often made complex because the configuration data that are required to identify resources and their interconnection are often buried in a large variety of configuration databases that are not readily or uniformly accessible. For example, to create a map of the physical layer of a LAN, one would need to identify various physical-layer interconnection resources (hubs, bridges, interfaces) and establish their connection via physical links. The information on various ports of a hub or a bridge is provided by an interface table of an SNMP MIB. The physical addresses of interfaces at a host and their mapping to higher layers (e.g., IP address) are maintained at various host operating system (OS) tables, which can vary greatly from system to system. The information on which host interface is attached to which hub physical port may be available through private vendor MIBs. A discovery application that would simplify identification of the physical structure of the network must be able to access this variety of information resources instrumented in a network and correlate it into a consistent picture.

With the rapid growth and change in a typical network, the significance of tools that can provide adequate representation of the network components is of increasing importance. Corporations invest a growing share of their capital expenses in building and maintaining networked system infrastructures. Merely accounting for these assets has thus become a problem of great importance that discovery and configuration management tools are struggling to solve.

Protocol Analysis Tools

Protocol analysis tools provide a means to capture and analyze network traffic. In a typical scenario, a protocol analysis tool is attached to a network link and captures the packet traffic transmitted over the link. This traffic is filtered through a variety of configurable filters to generate useful statistics.

Consider, for example, the scenario depicted in Fig. 1 and the failures discussed in the introduction through examples. Suppose Domain 1 attached a LAN protocol analyzer to its Ethernet. Packets broadcast over the Ethernet medium will be captured by the analyzer and classified to provide traffic statistics to operations staff. The analyzer may be configured, for example, to monitor TCP packets to or from the client workstation experiencing performance problems. To accomplish this, the analyzer uses filters that monitor IP packet headers to detect the address of the client workstation and determine whether the packet belongs to a TCP connection. Additional filters analyze various TCP header fields to determine connection statistics for various TCP connections maintained by the client. This permits one to monitor statistics of TCP segment loss, throughput, receiver window sizes, and other detailed data on TCP connection behaviors.

Protocol analyzers were initially developed as portable tools used for reactive problem diagnostics. When a segment experiences trouble, a protocol analyzer is attached to it to monitor traffic and diagnose the problem.

Recently, analyzers have shifted into new roles. With a growing number of segments incorporated in a typical LAN and with the increasing complexity of their traffic and interactions, reactive monitoring and searching through these segments is often inadequate. A transition began to a proactive embedded monitoring paradigm supported by remote analyzers, also known as remote monitors or probes. Probes are integrally embedded in the network and are attached permanently to segments. They provide continuous streams of traffic analysis used for proactive monitoring. Probes are remotely accessed by a management platform that can configure their traffic filters and retrieve the statistics that they collect. The Internet's remote monitoring (RMON) working group developed MIB standards to support configuration control and retrieval of traffic statistics.

The transition to switched networks requires yet another shift in protocol analysis. Current nonswitched LANs permit any interface to monitor all LAN traffic. Thus, a protocol analyzer can be attached to the LAN at any interface to monitor all of its traffic. In contrast, a switched LAN permits an interface to monitor only the traffic to or from that interface. To monitor LAN traffic, an analyzer must attach directly to the switching fabric. Switched LANs also make the task of diagnosing problems particularly complex since the relationships between physical connectivity and traffic flows are dynamically reconfigured. Thus, connectivity problems often result in dynamically changing manifestations. A number of recent switched Ethernet products have thus incorporated RMON monitors as integral components of the switch operations.

ATM LANs present particularly challenging network-monitoring problems. All current LANs encapsulate high-layer frames telescopically within headers of lower layers. This "onion skin" structure of frames permits an analyzer to monitor the physical-layer frame and analyze its headers produced by all protocol layers. In the example above, the analyzer could focus on the IP header and TCP header of a frame to collect statistics on the respective layer entities. An ATM LAN, in contrast, fragments frames to cells rather than encapsulate them. Protocol headers of higher layers are no longer directly observable at the physical layer. Monitoring of cells can only be used to analyze the cell layer. In order to monitor higher layers, it would be necessary to reassemble cells into frames and analyze their structures. This task is very difficult to accomplish at intermediate network nodes due to the high speed of traffic and the complexity of recognizing the global structure of virtual circuits. One possible solution is for protocol analyzers to be located at end-node interfaces where virtual circuits leading to or from the node can be effectively monitored. Indeed, analyzers of traditional (virtual) circuit-switched networks are typically attached at end nodes. The challenge in monitoring end-node traffic is, of course, how to correlate the massive traffic information collected to diagnose global network behaviors proactively.

Management Information Base Browsing and Monitoring Tools

There is an increasing range of SNMP MIBs of growing size instrumented in networked devices. Typical network elements contain MIBs with several hundred to several thousand variables. A management platform must enable the operations staff to search and monitor this vast source of operational data. Browsing tools permit an operator to traverse and search an MIB tree and activate periodic polling of MIB variables. Direct monitoring of an MIB variable is often insufficient. For example, an error counter may provide cumulative statistics of packet loss. One is typically interested in detecting changes in error rates. Such changes are reflected by the second derivative of the error counter. Platform-monitoring tools often permit the operations staff to monitor arithmetical-statistical functions of MIB variables graphically.

Configuration Management Tools

Configuration management is primarily concerned with establishing, maintaining, and controlling the configuration of network resources. Network operations involve, as illustrated in the section on configuration management problems, a significant range of configuration management problems. These problems center on maintaining consistent configuration databases associated with various related resources. These databases can range from internal tables maintained in memory at elements to large, persistent databases that maintain a vast amount of information about global resources and users.

Configuration management tools, at this time, are typically associated with specific elements and use element-specific structures and protocols. Consider again the example of the section on configuration management problems. Name and directory servers such as Domain Naming Service (DNS), Sun's NIS+, or Novell's Bindery, maintain configuration databases of resources at the routing and application layers of the network. These configuration databases are often organized to reflect domain resources and support mechanisms for federated multidomain organization of distributed networks. For example, DNS supports standardized mechanisms for distribution, caching, and replication of name databases. The maintenance of these databases is handled in part through manual mechanisms and in part through automated mechanisms. The division of these configuration management tasks and the mechanisms to support them are specific to DNS.

Configuration management often requires complex transactions with multiple configuration databases. For example, to provision service to a new telecommunications service customer, it may be necessary to add records to multiple global service and customer databases and to modify configuration of elements' resources associated with the service for multiple elements. Similarly, to provision a service to a new LAN user, one must configure host resources, the variety of servers to be used (e.g., file, mail, etc.), and name servers keeping track of domain addresses and other configuration databases of elements such as routers and hubs. Each of these configuration management transactions utilizes specialized tools associated with the specific resource. These transactions are typically handled through ad hoc processes that depend on great expertise and care. Even minimal inevitable errors can lead to inconsistent configurations and various failures.

Despite the advent of standardized management protocols and MIBs, configuration management remains, by and large, an ad hoc, nonstandardized, element-specific process supported by specialized tools. Streamlining, standardizing, and automating configuration management is a major challenge that network-management technologies have yet to accomplish. The need for such technologies will continue to increase as networks continue to incorporate more complex components and services, as well as permit a greater degree of dynamic changes.

At this time, it is unclear how a solution will be accomplished. It is useful to note, however, that certain configuration management areas are attracting interest, and technologies and standards are being developed to facilitate solutions. In particular, the Dynamic Host Configuration Protocol (DHCP) effort pursued by the Internet community illustrates an important attack on the problem. DHCP creates a standard that permits a host computer to dynamically attach to a domain and configure itself to integrate with and benefit from domain resources. It provides a useful example of how provisioning could be automated through a relatively simple protocol.

Problem Management Tools

Problem management tools fall under two categories: problem diagnosis and problem handling. Diagnosis tools aim to determine problems by either analyzing passively monitored symptom events caused by the problems and/or actively applying tests to isolate the problems. For example, to determine reachability problems in IP networks, one typically uses a PING tool. This tool generates active reachability tests to a target destination by sending echo request packets to it. The replies to these PING requests are analyzed to isolate reachability problems. Similarly, loopback test tools permit one to actively test reachability over a circuit.

Problem-handling tools presently focus on tracking the analysis and resolution of problems through a trouble-ticketing system. A trouble ticket is a database record that maintains the status of problem-handling activities. For example, when a reachability problem is detected by a PING tool, a trouble ticket is created with the data concerning the problem. This trouble ticket is routed to the operations staff responsible for resolving it. Often, a number of individuals will be responsible for different stages in the resolution, and the trouble ticket database schedules these stages and the flow of data among these activities.

With rapid growth in network complexity, effective analysis and handling of problems has become an area of significant concern. We discuss this below.

Current Directions and Challenges

This section provides a brief summary of network-management challenges arising in emerging networks and directions in which the field is moving.

Distributed Management

Current management systems, as depicted in Fig. 3, pursue a centralized paradigm. A centralized platform gathers and displays operations data from agents embedded in network elements. The scale and complexity of emerging networks, and their federated multidomain organization, render this centralized data collection and display inadequate. The rate at which a centralized platform can access and process data from agents sets strict limits on the scalability of management. A centralized management is sensitive to single-point failures at the platform or failures of its communications with managed elements. A multidomain network requires distribution of management functions and federated management models for their execution.

Distributed management architectures are of key significance in enabling economies of scale in operations management. A typical network is organized as multiple federated domains. Each domain requires management support typically provided by a dedicated operations staff. A large-enterprise network may involve scores or hundreds of domains in which operations tasks and expertise are currently replicated. A distributed management paradigm that automates these functions should enable a small group of expert staff to access and manage resources of an unlimited number of domains remotely. For example, a small team could manage all file servers or all physical-layer elements of a large-enterprise network. Creating such economies of scale in network management is of great importance.

A variety of studies have thus attacked the problem of distributing management functions. Management by delegation (MBD) technologies (5) provide an example of a successful approach to distribution. The idea of delegation is simple; instead of moving management data from a remote agent to centralized applications, the applications are dispatched to and executed at or near the agent. Management application programs are organized as mobile agents that can be dynamically dispatched by a network operations center (NOC) to remote locations. The delegation mechanisms support dispatching of such mobile agents, dynamic linking with the remote environment, and execution under local or remote control. For example, delegated applications can be used to monitor SNMP MIBs of a remote domain (or a complex element), detect and handle problems in the domain to provide self-healing, and report to the NOC. Delegated applications can be used to enforce operations policies, provide routine maintenance, or install and validate configuration changes. A small team of operations staff can use delegation to automate and manage remotely a large number of resources providing the economies of scale discussed above.

Automated Problem Management

Automation of management functions is perhaps the most central challenge facing the field. Automation could reduce dramatically the risks of failures and inefficiencies and the growing costs of operations. It is therefore useful to examine some of the central technical challenges involved in automating network management and the state of current research in addressing them.

Consider first the challenge of automating problem management. Problems typically consist of hardware or software failures, configuration inconsistencies, service failures, or other anomalous operations. When a problem occurs, it is typically manifested by a large number of symptom events such as traps generated by element agents or out-of-tolerance values for MIB variables. Automated problem management requires technologies to analyze and correlate these events to isolate the problem source. The example of the introductory section illustrates how events at various layers and systems must be correlated to a clock problem at a T3 link interface. Once the problems have been isolated, it is necessary to activate procedures that handle them. These handling procedures must be formulated and organized along operational policies of organizations. This requires technologies for management policy specifications and enforcement. Problem-handling procedures often require coordinated actions by multiple individuals. Trouble-ticketing technologies are used to track these activities. Next, we briefly consider these technologies for problem isolation and handling.

Automatic correlation of events to isolate problems presents difficult technical challenges. In a typical large network, the rate at which events are generated and the complexity of their causal relationships exceed the ability of operations staff to process them. A correlation system must be able to capture and maintain the knowledge about causal propagation necessary to analyze correlations. As the network changes, it must be easily adaptable to reflect new problems, symptoms, and causal propagation patterns. It must be able to monitor and analyze event streams of very high rates and accomplish robustness to false or lost alarms.

A number of studies have applied traditional knowledge-based techniques to address the correlation problem. Causal propagation knowledge is typically encoded in rules of the form "Event 1 causes Event 2," augmented with measures of uncertainty. Symptom events are correlated by searching this rule base to find problems that cause event chains leading to the symptoms. For network failure scenarios of limited complexity and scale, these techniques can be adequate. However, when the network becomes large and complex, the knowledge base of causal rules can become combinatorially explosive and the search involved in correlation can become intractable. Furthermore, false alarms or lost symptoms can misguide the search and present a very difficult problem even for small rule bases. Finally, changing the rules to adapt to network changes presents a difficult problem of maintaining knowledge bases.

Recent work has thus pursued alternative approaches to event correlation. A number of approaches target a more modest goal for correlation: merely classifying events that meet some similarity criterion. For example, events that occur in a similar time frame at a given resource are classified as correlated based on temporal and location similarity. In the example of the introductory section, events such as high bit error rates and corrupted packets at a router attached to the T3 links will be classified as correlated. However, they will not be classified as correlated with events at higher layers that they cause.

These simple classifiers can be somewhat extended by using finite-state machine (FSM) models to describe causal influence. States of the FSM reflect classifications of problem states and transitions reflect symptom events. Arriving symptoms are correlated by causing transitions to a given state. These FSM classification techniques are limited in correlating events at different resources. They are limited in handling false alarms or lost symptoms as these must be modeled by transitions of the FSM. The combinatorial complexity of reflecting all possible false alarms or lost symptoms in an FSM model precludes simple handling of these. An FSM model is also limited in handling dynamic changes in the network. One would require respective changes of the FSM model to reflect changes in the event propagation in the network. This is very difficult to accomplish.

Automated Configuration Management

Consider now the challenge of automating configuration management. Configuration management functions typically involve changes in the contents of various resource configuration databases. Configuration databases can vary from small, in-memory tables to large-scale persistent databases. Configuration management functions often require consistent modifications of multiple heterogeneous configuration databases.

Automation requires effective techniques to express configuration management functions in a machine-processable form. Computer system administrators have long used scripting languages such as PERL and tcl to specify configuration management scripts. Scripts are used to analyze and modify configuration files in order to install new end users or reconfigure system resources and services. Similarly, vendors of complex network elements typically use specialized configuration scripts to control the elements. A script describing a configuration is typically downloaded from a management platform to the element and executed to install the changes.

Direct modification of configuration data and use of scripts to handle low-level details provided an adequate paradigm for configuration management of systems with a limited range of resources and of limited complexity and configuration change frequency. Emerging networks often involve a very large number of heterogeneous resources of great complexity and with frequent configuration changes. Automation of functions thus requires techniques to abstract and coarse grain the data structures and methods used in configuration management. The coarse-grained structures and methods can be specified as policy-level abstractions that are compiled into lower-layer, fine-grained computations. Approaches such as these have been recently suggested by a number of researchers.

A configuration policy is typically formulated in terms of constraints on resource attributes and relationships. Such constraints can establish the range of permissible configurations (e.g., number of virtual circuits that may be allocated through a given port) or the relationships between configuration parameters. Technologies to formulate and enforce constraints on distributed heterogeneous configuration data are thus of central significance in support of configuration management.

Bibliography

This section provides selective references and a guide for additional reading in the various areas covered by this article. The goal of this section is to select a small number of references that provide useful, broad coverage and working knowledge of the field. The reader can find more extensive bibliographies in any of these references.

Introductory and General References

Aidarous, S., and Plevyak, T. (eds.), *Telecommunications Network Management into the 21st Century*, IEEE Press, New York, 1994.

> A useful book providing a broad coverage of various issues, technologies, standards, and approaches to network management.

Halsall, F., *Data Communications, Computer Networks and Open Systems*, 3d ed., Addison-Wesley, Reading, MA, 1992.

> Provides a broad introduction to networking technologies.

Sloman, M. (ed.), *Network and Distributed Systems Management*, Addison-Wesley, Reading, MA, 1994.

> A useful mix of introductory reviews and research articles.

Stevens, W. R., *TCP/IP Illustrated*, Vol. 1, Addison-Wesley, Reading, MA, 1994.

> Covers the Internet (TCP/IP) protocol stack broadly.

Tang, A., and Scoggins, S., *Open Networking with OSI*, Prentice-Hall, Englewood Cliffs, NJ, 1992.

> Covers the OSI protocol stack broadly.

Management Protocols

Rose, M., *The Simple Book: An Introduction to Management of TCP/IP Based Internets*, 2d ed., Prentice-Hall, Englewood Cliffs, NJ, 1994.

> SNMP and its applications to manage various networks are described extensively.

Stallings, W., *SNMP, SNMPv2, and CMIP. The Practical Guide to Network-Management Standards*, Addison-Wesley, Reading, MA, 1993.

> An extensive coverage of management protocols is provided.

Management Applications, Research, and Current Directions

The state-of-the-art research in network management is covered by several sources:

- The proceedings of the International Symposium on Integrated Network Management (ISINM) for 1989, 1991, and 1993, published by North Holland. The proceedings of ISINM 1995 are available from the IEEE Communications Society.
- The *Journal of Network and Systems Management* (JSNM), published by Plenum Press, New York.
- Various exchanges over network bulletin boards involving the network management research and development (R&D) community. A useful starting point is the mailing list snmp@psi.com. More specific activities pursued by various working groups of the IETF to standardize various MIBs can be pursued electronically through the IETF home page at http://www.ietf.net.

References

1. Stevens, W. R., *TCP/IP Illustrated*, Vol. 1, Addison-Wesley, Reading, MA, 1994.
2. Halsall, F., *Data Communications, Computer Networks and Open Systems*, 3d ed., Addison-Wesley, Reading, MA, 1992.
3. Rose, M., *The Simple Book: An Introduction to Management of TCP/IP Based Internets*, 2d ed., Prentice-Hall, Englewood Cliffs, NJ, 1994.
4. Stallings, W., *SNMP, SNMPv2, and CMIP. The Practical Guide to Network-Management Standards*, Addison-Wesley, Reading, MA, 1993.
5. Yemini, Y., Goldszmidt, G., and Yemini, S., Network Management by Delegation, In *Second International Symposium on Integrated Network Management* (K. Kappel and W. Zimmer, eds.), North Holland, Amsterdam, 1991.

YECHIAM YEMINI

Network-Management Technologies and Requirements in an Asynchronous Transfer Mode Environment

Network Model

In a high-speed digital switching network, the switching hardware is important, but network-management applications implemented in both hardware and software have a key role in successful networking solutions. If the solution implements asynchronous transfer mode (ATM) transport in a heterogeneous protocol environment, network management and systems management are important functional requirements. The promise of ATM is interactive multimedia applications; this promise will not be met without comprehensive network-management systems.

Network management in general is a set of interfaces to network devices, data to represent those devices, management applications, and an interface to a network operator. An overview of this is shown in Fig. 1.

Network Management

The whole task of network management becomes much more important and complex in an ATM network when compared to the management of a Transmission Control Protocol/Internet Protocol (TCP/IP) or System Network Architecture (SNA) network. ATM is a logical link layer protocol as defined in the Open System Interconnection (OSI) Layer 2 structure. This means that there are no network layer and address layer components for "networking." The significance of an ATM address is only between switches. An illustration of this is shown in Fig. 2, which is a schematic representation of the end-station-to-network interface. The MIBs (management information bases) contain the networking data that the management agent applications use to operate and control the interface. This type of interface is repeated in a similar form across the ATM network. This makes the network-management functions of problem determination and configuration management complex tasks.

What follows is a discussion of customers' functional requirements for a well-managed system and specifics for ATM management.

Topology management is the graphical view of the entire network, both logical and physical. This must be supported from a central point and console, but be capable of being distributed to different remote management workstations. In the case of ATM, the topology must include (local-area networks) LANs and other media networks that are integrated with the ATM network through ATM LAN emulation or other technology. The complexity of the heterogeneous environment is shown in Fig. 3.

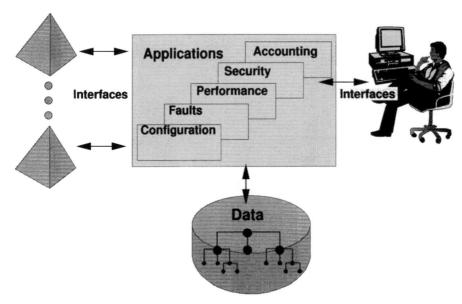

FIG. 1 Network management overview.

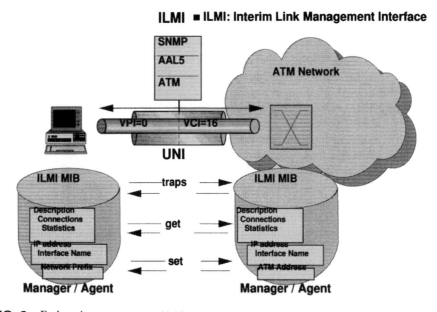

FIG. 2 End-station management (AAL = asynchronous transfer mode adaptation layer; ATM = asynchronous transfer mode; IP = Internet Protocol; MIB = management information base; SNMP = Simple Network Management Protocol; UNI = user–network interface; VCI = virtual channel identifier; VPI = virtual path identifier).

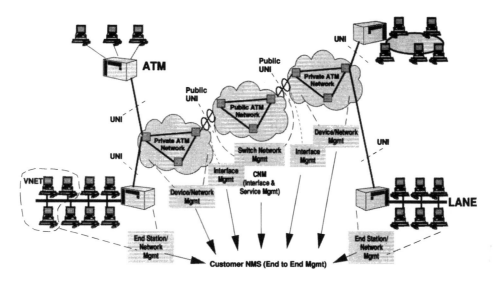

FIG. 3 Management perspectives showing the complexity of the heterogeneous environment (CNM = customer network management; NMS = network management system).

The end-station-to-end-station connectivity must pass through several networks. The management application must be able to present a topology representation of all the connections and protocols.

Fault management, linked with topology management, supports the primary network-management function of problem determination. ATM fault management needs to support, as close to the source as possible, the automated recovery from problems. Problems in a high-speed network need to be resolved, correlated, and have action taken as soon as possible to ensure network integrity.

Configuration and change management allows the network to operate and updates the topology with new resource data. This is the second basic function the network-management system must perform. This function includes the distribution of software and customer data.

Operations management includes the commands to a remote device and security for that type of operation.

Accounting management is the accounting for network usage.

Performance management is the collection of performance data and forwarding that data to a central point for management and analysis. Performance analysis has a major significance in an ATM network. With the ATM, preventive congestion control is required because of the amount and speed of cell transport. It is necessary to collect performance data to predict problems before they occur. In addition, tools need to be provided to model the network before it is configured and implemented. The objective is to provide an operational network when it first starts operations.

The management systems must use open and standard protocols. The use of proprietary systems to manage ATM resources will prevent effective interoperation between networks and other vendors' equipment.

The Asynchronous Transfer Mode Network Solutions Structure

The solution components for a well-managed ATM network consist of two core technologies and three supporting technologies. These technologies must work together as a complete system or the overall networking solution is not capable of supporting information flows critical to the mission or business. The scope of the problem is shown in Fig. 3; Fig. 4 represents the ATM Forum's reference configuration for managing an ATM network.

The first core technology is the set of applications for the development and maintenance of the network configuration. These applications include the configuration management of the switching hardware and operational software. The management systems in the ATM Forum reference configuration (Fig. 4) that control the M(2) and M(4) management flows would be the controlling systems for configuration management in the network. At the user–network

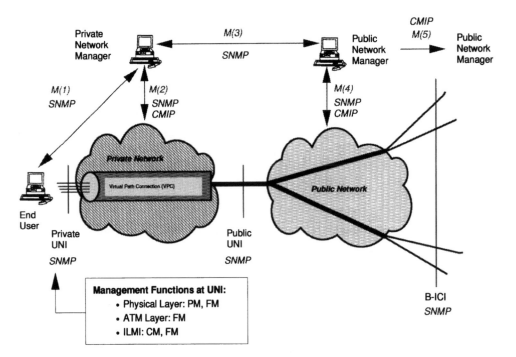

FIG. 4 Asynchronous Transfer Mode Forum network management (CM = configuration management; CMIP = Common Management Information Protocol; FM = fault management; PM = performance management; M(1) = private (or campus) network manager to end user management information flow; M(2) = private (or campus) network manager to private (or campus) network management information flow; M(3) = private (or customer) network manager to public (or service provider) network manager information flow; M(4) = public (or service provider) network manager to public network management information flow; M(5) = public (or service provider) network manager to public network manager information flow).

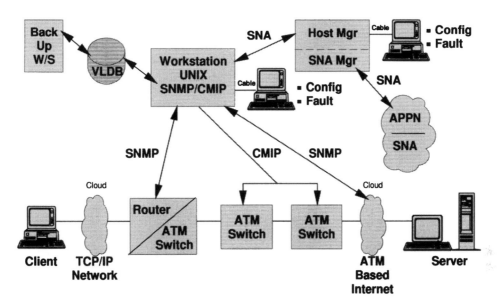

FIG. 5 Solution configuration.

interface (UNI) shown on the reference, configuration of the end station is done by the agent applications of the interim link management interface (ILMI). Configuration management also is accomplished across the public UNI and broadband inter-carrier interface (B-ICI) shown on the reference configuration by agent applications or, alternatively, through the management applications. Also required are the supporting applications to configure the databases of information that will not reside in the network itself. These supporting applications provide the capability to develop high-reliability network and operational management applications.

The second core technology is the set of network-management functions that support problem determination, performance monitoring, and accounting. The primary difference in ATM networks is that solutions must have good response times and a higher level of automation than current implementations. For wide-area ATM networks, the applications must support both Simple Network Management Protocol (SNMP) and Common Management Information Protocol (CMIP) management protocols. This is shown in the management systems of the reference configuration.

Operational ATM network management provides the control program for the switch, provides access, and manages the network links. These systems provide switching and the associated traffic management functions. Operational applications can also provide congestion control and rerouting in failure modes. They support network management by providing fault and performance information to the management applications.

Network performance management information is provided by the network and operational management systems. The information consists of transit timings, response times, and other related performance parameters specified by the traffic management configuration definition for the circuit. It also has a function in capacity planning. The performance management application collects and analyzes performance data to determine where additional bandwidth is required because of traffic growth or reported congestion. It also identifies problem areas where outages have occurred on a frequent basis and high-priority work needs to be initiated to maintain the availability specified in service agreements.

Support for the ATM standards framework is a necessary component. The key to interoperability is compliance with standards. Therefore, any ATM system must comply with the ATM Forum specifications and the International Telecommunication Union (ITU) standards when they apply to the systems and applications.

Heterogeneous Network Management

Network management for ATM cannot stand alone except in small networks. Figure 5 is more representative of an actual implementation of ATM. This solution configuration shows multiple management protocols and a mixture of ATM and other transport protocols. In addition, large host systems are needed to deal with remaining SNA and other protocols not based on ATM technology.

ROGER LEO KOSAK

Network Packet Delay Analysis Using Single-Server Queueing Models

Preface

One of the most important performance measures in the analysis of data networks is the delay required to deliver a packet from origin to destination. The packet delay will affect the choice and performance of several network algorithms such as routing and flow control. Queueing theory is the primary methodological framework for analyzing network delay. It provides a basis for adequate delay approximations, as well as valuable qualitative results and worthwhile insights.

The literature on queueing models for communications systems is abundant. In this article, we concentrate on single-server queueing models for communications systems. We describe some of the more recent models that were developed for modeling the behavior of nodes of modern communications networks. These models are fundamental and provide the foundation for more elaborate models with multiple servers.

We focus on the queueing and transmission delays of packets. First, we describe some classical queueing models, namely, the $M/M/1$, $M/G/1$, and $G/M/1$ queueing models. These models assume that all interarrival times and service times are independent. This assumption does not hold for communications systems; we describe extensions of the classical queueing models to allow for such dependency. It is demonstrated that with this dependency the delays in the buffers preceding communications links are smaller. Next, we describe completely different approaches for the characterization of arrival processes. These characterizations are motivated by the need to model real-time applications in high-speed communications networks in which it is important to capture the burstiness of the arrival processes. Finally, we describe another performance measure, namely, the message delay. The motivation for this measure arises from the process of segmentation and reassembly of messages into packets, which is natural for packet-switched networks.

Introduction

Queueing models and analytical techniques for evaluating the performance of communications systems evolved during the recent decades hand in hand with the progress of these systems. In its most general formulation, the problem addressed by a performance analyst is to characterize the service provided by a system when it is loaded with some given load. In this context, the system is known by a full description of the behavior of all its components (in a communications system, these are the nodes and the links of the system), the load is given as some known stochastic processes (in a communications system, the load corresponds to the traffic generated by the users of the system), and the service is to be characterized by the distribution of its parameters (such as the traffic in its links, the delays it causes, etc.).

The particular importance of the problem of performance analysis of communications systems stems from the variety of settings in which it is encountered. For instance, when a new communications network is designed that is supposed to guarantee some predetermined parameters of service, one wishes to calculate the amount and size of resources (e.g., buffer sizes at the nodes, capacity of the links, etc.) needed to fulfill the requirements. The difficulty of the problem arises from the competitive and heterogeneous demands for such resources from the users of the communications system. In fact, this problem is a special case of the resource allocation problem that is the root of most of the problems in the field of information processing and beyond this field.

The capacity assignment problem (see Ref. 1) is another variant of this problem in which one wishes to minimize the average delay of a packet in the network for a given topology, packet arrival processes, and the total capacity of the links of the network. Another example is when a new user wishes to join an already operational network. Since modern real-time applications require some minimal performance guarantees (such as very rare packet losses or very short delays), the question here is whether the new user can be admitted and receive the needed service, while the performance degradation other existing users sense will not violate their needs. One of the most difficult aspects of this problem is that of treating all users fairly when accepting a new user to the system. The problem of admission control in packet-switched networks was studied in the past; see, for instance, Ref. 2 for a discussion of this problem and a definition of fairness in such networks. Recently, many papers have studied this problem in the context of high-speed, packet-switched networks in which a fast-to-implement policy is needed (see, e.g., Refs. 3–5).

A certain special case of analysis of a communications system, which has been addressed and studied very intensively, is the case of single-node networks. Queueing models (see, e.g., Ref. 6) that consist of a single service station with a single server that is fed by a single stream of customers have been used. The input process of new customers is assumed to be some known stochastic point process, and the service is described by the distribution of the time to process each customer. The customers in these models correspond to packets in a communications system. Many other cases of single-node networks have also been studied; see Refs. 7–12 and many others. These works consider a variety of arrival processes, service distributions, and service disciplines, and address cases for which the system has several input links or output links. The fact of the matter is that it is impossible to discuss all the various models that appeared in the literature in a single article. Even a complete book would be too short to contain all of them. In this article, we try to describe some of the more recent models that were developed for modeling the behavior of nodes of modern communications networks.

The case in which the communications network consists of more than a single node has proved to be very complex, even in simple settings. This complexity is essentially due to the complicated way in which different traffic streams interact with each other within the network and the dependencies imposed by these interactions. One can distinguish between networks that are served by a single server and networks with a server in each of the nodes. The former usually corresponds to a communications system with users that share a common resource, such as a single-cell wireless system, an Ethernet, or a token-ring local-area network, and so on. Various queueing models are used to analyze the performance of such systems, such as polling models, priority models, and the like. Networks with servers in each node usually correspond to wide-area computer communications networks such as System Network Architecture (SNA), DECnet, ARPANET, and Tymnet; see Refs. 2 and 13 for a brief description of these networks. Closed-form solutions for the number of packets in the nodes of the network in this case are known only for a limited class of networks, known as product-form networks. A good example for this class is the Jackson network (see Refs. 6 and 14), in which the queue lengths of all the nodes behave as if they were independent.

It is apparent that queueing theory is the primary methodological framework for analyzing the performance of communications systems such as the delay of a packet in the system. Its use often requires substantial simplifying assumptions since, unfortunately, the theory is still limited and realistic assumptions make meaningful analysis extremely difficult. For this reason, it is sometimes impossible to obtain accurate quantitative performance measure predictions on the basis of queueing models. Nevertheless, these models often provide a basis for adequate delay approximations, as well as valuable qualitative results and worthwhile insights, as we attempt to describe in this article.

The literature on queueing models for communications systems is abundant. It is probably impossible to cover all subjects in a single article. Therefore, here we decided to concentrate on single-server queueing models for communications systems. These models are fundamental and provide the foundation for more elaborate models with multiple servers.

In the following sections, we describe several studies that analyze the packet delays in communications systems. In the section on classical queueing models, we describe the $M/M/1$, $M/G/1$ and $G/M/1$ systems. We describe several results that hold under very general assumptions and summarize the main results for each model. We proceed with more realistic models for communications systems in the section on correlated queue models, in which we describe models that relax the independence assumption of the interarrival and service times used in the classical models. In the section on burstiness characterization models, we describe completely different approaches for the characterization of arrival processes. These novel characterizations are motivated by the need to model real-time applications in high-speed communications networks in which it is important to capture the burstiness of the arrival processes. In the section on message delay processes, we describe another performance measure, namely, the message delay. We review some of the works that analyze the behavior of the message delay process in some classical queueing systems. Numerical examples are provided throughout the paper to clarify the differences among the various models.

General Model

We consider a communications link with a given transmission capacity of C bits per second. Several traffic streams (e.g., sessions or virtual connections) are multiplexed on the link. The manner in which the capacity is allocated among these traffic streams affects the performance characteristics (e.g., delay and loss) of each traffic stream. The most common scheme used in packet-switched networks is called *statistical multiplexing*. In this scheme, the packets of all traffic streams are merged into a single queue at the front of the communications link and transmitted on a first-come, first-served (FCFS) basis.

The variability of the arrival processes from the different traffic streams can cause the arrival rate of packets scheduled for transmission on the communications link to momentarily exceed the transmission rate, and hence a queue of packets can build up at the front of the link. Typical arrival rates range from one arrival per many minutes to one million arrivals per second. Simple models for the variability of the arrivals include Poisson arrivals, deterministic arrivals, and uniformly distributed arrivals. Typical packet lengths vary roughly from a few bits to 10^8 bits, with file-transfer applications at the high end and interactive sessions at the low end. Simple models for packet-length distribution include an exponentially decaying probability distribution, fixed-length packets, and uniform probability distribution.

One of the most important performance measures of a data network is the delay required to deliver a packet from origin to destination. The packet delay will affect the choice and performance of several network algorithms such as routing and flow control. In what follows, we focus on packet delay in a single-server queueing system representing the communications link. This delay consists of four components. First is the processing delay, which corresponds to the delay between the time the packet is received at the input link and the time it is assigned to an output link queue for transmission. This delay can be very large in today's communications networks, in which an average of a few thousand instructions are performed for each packet received in the node; the delay is negligible in high-speed networks such as asynchronous transfer mode (ATM) (15), for which the packet header is processed by very fast dedicated hardware. This delay is usually independent of the amount of traffic handled by the node. The second delay is the queueing delay, which corresponds to the delay between the time the packet is assigned to a queue for transmission and the time it starts being transmitted. During this time, the packet waits in the queue while other packets are being transmitted. Third is the transmission delay, which corresponds to the delay between the times that the first and the last bits of the packet are transmitted. The fourth delay is the propagation delay, which corresponds to the delay from the time the last bit of the packet is transmitted at one end of the link until the time it is received at the other end of the link. This delay depends on the physical characteristics of the link and is independent of the traffic carried on the link.

In what follows, we focus on the queueing and transmission delays of packets. First, we describe the classical queueing models $M/M/1$, $M/G/1$, and $G/M/1$. The first letter of these models indicates the nature of the arrival process: M stands for a Poisson process and G stands for general distribution of interarrival times. The second letter of the models indicates the nature of the probability distribution of the service (transmission) time: M and G stand for exponential and general distributions, respectively. The third letter indicates the number of servers; in our case, there is one server (transmission line). The servers do not idle when there are packets waiting in the queue (work-conserving server). Moreover, these models assume that all interarrival times and service times are independent. This assumption is the key difference between a communication system and its corresponding queueing system model, and it is related to the well-known independence assumption (see Kleinrock, Ref. 1). The independence assumption for Poisson arrival processes and exponentially distributed service times states that each time a packet is received at a node within the system, a new service time is chosen independently from an exponential distribution. This is clearly inadequate for a communications system since packets maintain their lengths as they pass through the network. However, it was suggested by Kleinrock that merging several packet streams on a transmission line has an effect akin to restoring the independence of interarrival times and packet service times.

To clarify this point further, we consider two transmission lines in tandem for which the interarrival times and service times of packets at the first transmission line are independent. The interarrival time between two successive packets on the second transmission line can certainly be no less than the service time for the second of these packets on the first transmission line. Since the service time for this packet on the second transmission line is directly related to its previous service time (and therefore highly correlated with the interarrival time between the two packets on the second transmission line), we see that the arrival process of packets to a node due to the internal traffic in the network is not independent of the service time these packets receive at that node. In what follows, we describe extensions of the classical queueing models to allow for this dependency between the interarrival and service times. We further describe additional motivations for such models and several results related to the packet delay distribution.

Another crucial point that we address in this article is related to the burstiness characterization of arrival processes. We describe several recent approaches for such a characterization that are motivated by the need to model real-time applications in high-speed communications networks. The common feature of these approaches is the bounding (either deterministic or stochastic) of the burstiness of the arrival processes that yields computable bounds on many performance measures in the network.

In many systems, the message delay (for which a message consists of a block of consecutive packets), and not the packet delay, is the measure of interest for the network designer. This is due to the fact that packets are data units that are only meaningful at lower layers and are created because of the network data unit size limitations. The ATM (15), Transmission Control Protocol/Internet Protocol (TCP/IP) (16), and systems based on time division multiple access (TDMA) (17) are examples of such systems in which the application message is segmented into bounded-size packets (cells) that are then transmitted through the network. At the receiving end, the transport protocol (or the adaptation layer) reassembles these packets into a message before the delivery to higher layers. In some applications, message delay is not the result of segmentation at the network layer, but of the nature of data partitioning in the storage. A file can be composed of multiple records stored at different locations in the disk. These records are read individually and may be transmitted as separate packets.

However, the entire file transfer delay is the measure of interest for the file transfer application. We describe some of the recent papers that analyze the behavior of the message delay for some classical queueing systems.

Classical Queueing Models

Here, we describe the $M/M/1$, $M/G/1$, and $G/M/1$ queueing models. The analysis of these models can be found in almost every book related to queueing theory. We refer to Kleinrock (6) for the details of the analysis of each of these classical models.

General Queueing Results

Before proceeding with the analysis of the packet delays in classical queueing models, we mention three general results that hold for a $G/G/1$ queueing system. The first general result is known as Little's theorem (18). It states that the average number of customers (packets) in a queueing system is equal to the average arrival rate of customers to that system times the average time spent in that system. The derivation of this theorem is simple, and it does not depend on any specific assumptions regarding the arrival distribution or the service time distribution, nor does it depend on the number of servers in the system or the particular queueing discipline within the system. Note that the system in this case can correspond to the queue and the server, the queue only, or the server only, in which case relations are obtained between the corresponding entities.

The second general result is related to the stability of the $G/G/1$ queueing system. Define ρ as the average arrival rate of customers times the average service time of a customer. Then, $\rho < 1$ is a sufficient condition for the stability of the $G/G/1$ queueing system, given that the arrival and the service processes are ergodic (19). Stability here refers to the fact that limiting distributions of all random variables of interest (such as the delay in the system) exist, and that all customers are eventually served. We assume that this condition holds for all queueing systems we consider unless otherwise specified.

Another result that holds under very general assumptions is that, in steady state, the system appears statistically identical to an arriving and departing customer. That is, in steady state, the number of customers in the system just before an arrival is equal in distribution to the number of customers just after a departure. The only requirement for this result to hold is that the system reaches a steady state with positive steady-state probabilities to have any n customers in the system, and that the number in the system changes by unit increments. These assumptions hold for all queueing systems we consider.

The $M/M/1$ System

The arrival process is Poisson with rate (inverse of average interarrival time) λ. The service time is exponentially distributed with rate (inverse of average service time) μ. The analysis of the $M/M/1$ system is based on the theory of Markov chains (20), for which the states of the Markov chain correspond to the number of customers in the system. In particular, it is based on a special case of a Markov process, named the birth-death process, in which two successive states can only differ by a unity. From the state transition rate diagram of the birth-death process, the so-called detailed balance equations are obtained from which the steady-state probability p_n of having n customers in the system is obtained. All quantities of interest can then be obtained from this probability distribution. Below, we summarize the results for the $M/M/1$ system.

- Utilization factor (proportion of time the server is busy) $\rho = \lambda/\mu$.
- Probability of n customers in the system $p_n = \rho^n(1 - \rho)$, $n = 0, 1, 2, \ldots$.
- Average number of customers in the system $N = \rho/(1 - \rho)$.
- Average customer time in the system $T = 1/(\mu - \lambda)$.
- Average number of customers in queue $N_Q = \rho^2/(1 - \rho)$.
- Average waiting time in queue of a customer $W = \rho/(\mu - \lambda)$.
- Probability density function (pdf) of the system time $t(y) = (\mu - \lambda) \exp^{-(\mu - \lambda)y}$, $y \geq 0$. That is, the system time is exponentially distributed with parameter $\mu - \lambda$.
- Probability density function of the waiting time $w(y) = (1 - \rho)u_0(y) + (\mu - \lambda) \exp^{-(\mu - \lambda)y}$, $y \geq 0$, where $u_0(y)$ is the unit impulse function.
- Laplace Stieltjes transform (LST) of the pdf of the system time $\Im(s) = (\mu - \lambda)/(s + \mu - \lambda)$.

From the expressions for N and T, note that increasing the arrival rate and the service rate by a factor of K, $K > 1$, does not change N but decreases T by a factor of K. In other words, a transmission line K times as fast will accommodate K times as many packets per second at K times smaller average delay per packet.

Finally, we describe two additional results related to this system. The first is known as the PASTA (Poisson arrivals see time averages) property (21). It states that when the arrival process is Poisson, an arriving customer finds the system in a "typical" state. That is, the probability distribution of the number of customers in the system just before an arrival equals the steady-state probability distribution. This holds for queueing systems with Poisson arrivals regardless of the distribution of the service time. The second is known as Burke's theorem (22). It states that the departure process of an $M/M/1$ system is itself Poisson with parameter λ and is independent of the other processes in the system.

The $M/G/1$ System

The arrival process is Poisson with rate λ. The service time has an arbitrary distribution with LST denoted by $\mathcal{B}(s)$. Let μ and x^2 be the average and the second moments of the service time. The analysis here is based on the method of the embedded Markov chain at the departure instants from service. Below, we summarize the results for the $M/G/1$ system.

- Utilization factor (proportion of time the server is busy) $\rho = \lambda/\mu$.
- The generating function $\mathcal{Q}(z) \triangleq \Sigma_{i=0}^{\infty} p_i z^i$ of the probability distribution p_i, $i \geq 0$ of the number of customers in the system in steady state (also at departure and arrival instants)

$$\mathcal{Q}(z) = \mathcal{B}(\lambda - \lambda z) \frac{(1 - \rho)(1 - z)}{\mathcal{B}(\lambda - \lambda z) - z}$$

- Average number of customers in the system

$$N = \rho + \frac{\lambda^2 x^2}{2(1 - \rho)}$$

- Average customer time in the system

$$T = \frac{1}{\mu} + \frac{\lambda x^2}{2(1 - \rho)}$$

- Average number of customers in queue

$$N_Q = \frac{\lambda^2 x^2}{2(1 - \rho)}$$

- Average waiting time in queue of a customer

$$W = \frac{\lambda x^2}{2(1 - \rho)}$$

- LST for the system time

$$\mathcal{T}(\delta) = \mathcal{B}(\delta) \frac{\delta(1 - \rho)}{\delta - \lambda + \lambda \mathcal{B}(\delta)}$$

- LST for the waiting time

$$\mathcal{W}(\delta) = \frac{\delta(1 - \rho)}{\delta - \lambda + \lambda \mathcal{B}(\delta)}$$

Since the $M/D/1$ system yields the minimum possible value for x^2 for a given μ, it follows that the values of W, T, N_Q, and N for the $M/D/1$ system are lower bounds to the corresponding quantities for an $M/G/1$ system of the same λ and μ.

It turns out that the average number of customers in the system is the same for any order of servicing customers (and not only for the FCFS discipline assumed throughout the article) as long as the order is determined independently of the required service times.

The *G/M/*1 System

The $G/M/1$ system is in fact the "dual" of the $M/G/1$ system. The interarrival times have a general LST $\alpha(\mathfrak{s})$ with a mean time between arrivals equal to $1/\lambda$. The service times of customers are exponentially distributed with mean $1/\mu$. The analysis is based on the method of the embedded Markov chain at the arrival instants to the system. All the results are expressed in terms of a root σ that is the unique root in the range $0 < \sigma < 1$ of the functional equation $\sigma = \alpha(\mu - \mu\sigma)$. Once σ is evaluated, the following results are immediately available:

- Utilization factor (proportion of time the server is busy) $\rho = \lambda/\mu$.
- Probability of n customers found in the system by an arrival $r_n = (1 - \sigma)\sigma^n$, $n = 0, 1, 2, \ldots$.
- Probability distribution function of the waiting time $W(y) = 1 - \sigma \exp^{-\mu(1-\sigma)y}$, $y \geq 0$.
- Average waiting time in queue of a customer $W = \sigma/[\mu(1 - \sigma)]$.

Note that the number found in the system by an arrival is geometrically distributed with parameter σ, independent of the form of the interarrival time distribution (except insofar as it affects the value for σ). We comment that the steady-state probabilities p_n of n customers in the system at an arbitrary instant differs from r_n in that $p_0 = 1 - \rho$ whereas $r_0 = 1 - \sigma$ and $p_n = \rho(1 - \sigma)\sigma^{n-1} = \rho r_{n-1}$ for $n = 1, 2, \ldots$ (see Ref. 23). In the $M/G/1$ system, we saw that $p_n = r_n$. In the $M/M/1$ system, $\sigma = \rho$ and the equations are the same as for the $M/M/1$ system. Note also that the waiting times are exponentially distributed (with an accumulation point at the origin), independent of the form of the interarrival time distribution.

Correlated Queue Models

The focus of correlated queue models is on a family of queues in which service and interarrival times exhibit some form of dependency. The initial motivation for such models was the modeling of a communications link in a packet-switched network carrying variable-size packets. The general issue of dependencies in queueing systems is clearly an important one and has been extensively studied in the literature. The reader is referred to Ref. 24 for a review of the various types of dependencies that exist in packet queues and a study of their impact on different system performance measures.

In what follows, we describe two types of dependencies that arise in communications systems. In the first type, the service time B_n of packet n depends on the interarrival time I_n between packets $n - 1$ and n. The discussion is based mainly on Ref. 25. In the second type, the interarrival time I_{n+1} between packets n and $n + 1$ depends on the service time B_n of packet n. The discussion is based mainly on Ref. 26.

Queues with Service Times Proportional to Interarrival Times

In this section, we focus on a particular type of dependency in which the service time associated with a packet (e.g., its transmission time on the link) is correlated with its interarrival time. Such correlations arise, for example, in the context of a packet-switched network in which variable-length packets are forwarded from one node to another. The finite speed of network links then results in large packets having correspondingly large interarrival times, that is, for a link of speed C, the amount of work received in a time interval τ cannot exceed $C \times \tau$. This strong positive correlation between interarrival and service times can greatly improve the delay characteristics in the buffers preceding communications links as demonstrated below. It is, therefore, important to provide models that account for this effect while remaining tractable.

One of the earliest works to investigate systematically the issue of correlation between service and interarrival times is Ref. 27. In Ref. 27, Kleinrock studied the impact of correlated packet lengths and interarrival times in the context of a queueing network model for communications networks. The intractability of the general problem led him to formulate the well-known and useful independence assumption (see discussion in the section on general models above), which amounts to ignoring correlations. This approach is reasonably accurate in the presence of sufficient traffic mixing in the network, but can significantly overestimate delays in systems in which there is a strong positive correlation between service and interarrival times, as in tandem queues (see Refs. 28–30) in which little or no traffic mixing is present.

The dependency between interarrival times and the amount of work that can be brought into a system has also been studied in the context of fluid-flow models (7,9–11,31–33), which assume that work arrives in and is removed from a system at continuous and possibly varying rates. A particularly popular and simple example is that of an ON-OFF source feeding a buffer, which is emptied at a constant rate. The finite input and output rates account for the dependency between the amount of data received and the elapsed time t, that is, the amount of data received is proportional to both the input rate and t.

While fluid-flow models capture some of the dependencies that exist between arrivals and service times in communications systems, they do not account for the granularity of arrivals and services. Rather, they assume that both arrivals and departures are progressive, with the work in the system being a continuous function of time. This may not always be an adequate assumption for communications systems (especially not in store-and-forward networks), that is, packets must typically be fully received before they can be forwarded. As illustrated in Fig. 1, where $W(t)$ stands for the amount of unfinished work in the system at time t, this can result in significant inaccuracies when estimating system performance (see also Ref. 34). In Ref. 25, the authors propose and analyze models that not only account for the type of dependencies captured by fluid-flow models, but also preserve the discrete nature of arrivals and services that is characteristic of many communications systems. Before proceeding with the description of these models, we complete our review of earlier works by discussing several papers that are directly relevant to these models (35–42).

One of the early works to consider a queueing system with explicit correla-

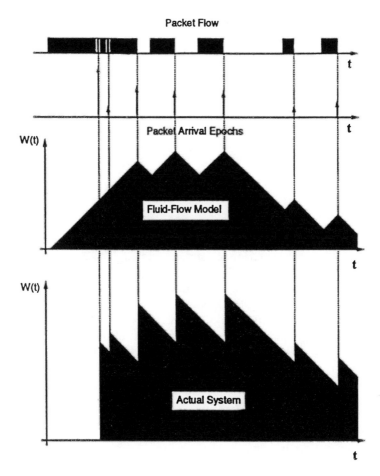

FIG. 1 Comparison of ON-OFF fluid-flow and discrete arrivals models.

tions between interarrival and service times is Ref. 35. It analyzes a system with Poisson arrivals at rate λ, where the service time B_n of the nth customer is proportional to the interarrival time I_n between the $(n - 1)$ and the nth customers. In other words, the service time is a deterministic function of the interarrival time, with $B_n = \alpha I_n$ ($\alpha < 1$ for stability). This system can be used to model a buffer connected to a unit speed communications link that receives an uninterrupted string of packets with exponentially distributed lengths from an upstream link of speed α. An explicit expression for the delay distribution is obtained in Ref. 35, while the initial busy period, the system state, and the output process are studied in Ref. 37. Below, we summarize the results for the correlated $M/M/1$ system.

- Utilization factor (proportion of time the server is busy) $\rho = \alpha$.
- The evolution equation of the system delay of a packet $t_{n+1} = (t_n - I_{n+1})^+ + \alpha I_{n+1}$, where $(x)^+ \triangleq \max(x,0)$.

- The LST of the system delay of a packet

$$\mathfrak{I}(\delta) = \prod_{i=1}^{\infty} \frac{1 - \alpha^i}{1 - (1 - \delta/c)\alpha^i} \qquad \text{where } c \triangleq \lambda/(1 - \alpha).$$

- The average of the system delay of a packet

$$T = \frac{1 - \alpha}{\lambda} \sum_{i=1}^{\infty} \frac{\alpha^i}{1 - \alpha^i}$$

Again, we see that a transmission line K times as fast will accommodate K times as many packets per second at K times smaller average delay per packet.

For a numerical example, Table 1 compares the average and the variance of the delay time of a packet in $M/M/1$ and correlated $M/M/1$ systems for average interarrival time $\lambda = 1$ and for different loads $\alpha = 0.2, 0.5, 0.9, 0.95$. Table 1 shows that the correlated system drastically reduces the average delay time compared with the $M/M/1$ system.

More general correlations were considered in Ref. 38 using a bivariate exponential distribution to characterize the correlation between interarrival and service times. This work was subsequently extended in several papers. The delay density was shown to have a hyperexponential distribution in Ref. 39, while Ref. 40 studied the sensitivity of this distribution to the value of the correlation coefficient. The busy period was investigated in Ref. 42, while a system with infinitely many servers was considered in Ref. 41. Combe, Borst, and Boxma analyzed a variant of the $M/G/1$ queue in which the service times of arriving customers depend on the length of the interval between their arrival and the previous arrival (43). They obtained the LST of the delay time of a customer and proved that the average delay time of a customer is smaller than or equal to the average delay time of a customer in the corresponding $M/G/1$ system without dependency.

Cidon et al. expanded the model of Ref. 35 in several new directions that make it more applicable to the modeling of actual communications systems (25). A system similar to that of Ref. 35 (see Fig. 2) was introduced and a simple derivation for the LST of the delay in the system was presented. This

TABLE 1 Averages and Variances of Delay Time of $M/M/1$ (t) and Correlated $M/M/1$ (t^c) Systems

α	$E[t]$	$E[t^c]$	$var[t]$	$var[t^c]$
0.2	0.250	0.241	0.063	0.041
0.5	1.000	0.803	1.000	0.248
0.9	9.000	2.709	81.00	1.164
0.95	19.00	3.470	361.0	1.365

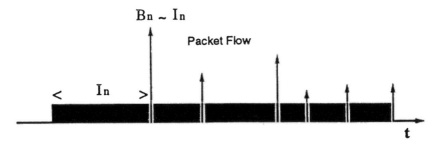

FIG. 2 System with proportional interarrival and service times.

simple derivation was obtained by directly focusing on the steady-state equations rather than on the transient evolution equations as was done in earlier works on similar correlated queues (35,37–39,41,42). The LST of the delay was then obtained by applying results from the theory of linear functional equations (44) and the analytic properties of the LST. This approach not only provides a formal framework for such problems, but it also results in a solution method that is applicable to a more general class of problems. In particular, it allows tackling more involved systems, as illustrated in Ref. 25 and below.

The first extension considered in Ref. 25 consists of the addition of an independent, generally distributed, nonnegative random variable to the service time. Using the notations introduced above, the service time of the nth customer is now of the form $B_n = \alpha I_n + J_n$, where J_n is an independent, nonnegative random variable with a general distribution. This extension is useful to model systems in which each packet needs additional service in excess of its raw transmission time. The additional service may be due to some overhead such as a header appended to the original data or correspond to some processing that needs to be performed for each packet. It was observed in Ref. 25 that the average delay time of a packet depends on the entire probability distribution of the service time and not only on its first and second moments as in the $M/G/1$ system.

The simple model of Ref. 35 is useful to capture the impact of dependencies between packet interarrival and service times. However, from a modeling point of view, it imposes a number of limiting constraints. In particular, it requires that the input correspond to a "saturated link" with a transmission rate lower than that of the output link ($\alpha < 1$). In order to overcome this limitation, the model is further extended in Ref. 25 to allow the input process to alternate between active and idle periods. This is achieved by allowing the proportionality constant α to be a random variable that takes the value $\alpha_1 > 0$ with probability g_1 and $\alpha_2 = 0$ with probability $g_2 = 1 - g_1$ (with $\alpha_1 g_1 < 1$ for a stable system). This results in an ON-OFF input process with exponentially distributed ON and OFF periods, a geometric number of packets in each ON period, and exponentially distributed packet sizes. Specifically, after an exponentially distributed time interval of duration I_n and mean $1/\lambda$, a packet of size $\alpha_i I_n$ is generated with probability g_i, $i = 1,2$. This creates exponentially distributed active and idle periods on the link with means $1/\lambda(1 - g_1)$ and $1/\lambda g_1$, respectively. The result-

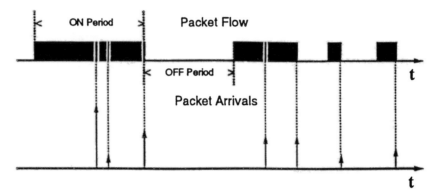

FIG. 3   System with ON-OFF source and multiple discrete arrivals.

ing arrival process is illustrated in Fig. 3. It is also possible to add an independent and generally distributed "overhead" to each packet. Explicit expressions for the LST and the average of the delay time of a packet were obtained in Ref. 25.

In a numerical example, Fig. 4 shows the average delay as a function of the proportionality parameter α for various values of g_1 assuming that $\lambda = 1$. As expected, the average delay grows monotonically with α and with g_1. Figure 4 also shows the average delay of a packet in an equivalent $M/G/1$ system in which the service time of a customer has the same distribution as in the corre-

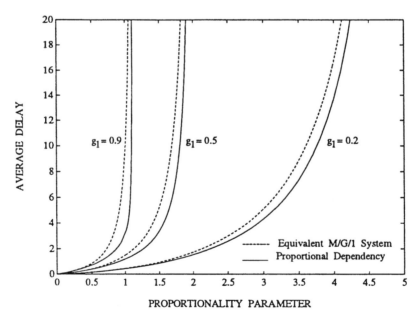

FIG. 4   Average delay versus proportionality parameter α.

lated system, but is sampled independently of any other event in the system. Figure 4 shows that the average delay of the equivalent $M/G/1$ system is always larger than the average delay of the system with random proportional dependency. When $g_1 = 0.9$, the difference becomes very large when the system is heavily loaded. For instance, for $\alpha = 1.08$ (1.1) the average delay in the correlated system is 6.55 (13.82), while in the equivalent $M/G/1$ system it is 38.46 (109.9).

This model, although reminiscent of a fluid-flow model for a two-state Markovian ON-OFF source, exhibits a number of key differences. First, data arrival does not take place gradually over the duration of an ON period. Rather, work accumulates for some interval of time, and it is only on its completion that a packet is generated to the system. This provides a more accurate representation of the discrete nature of packet arrivals. Second, the model allows the partition of a single ON period into multiple packets. This is in contrast to a fluid-flow model, in which data arrival is uninterrupted over the duration of the entire ON period and the transmission of bits rather than packets is considered. Despite its increased flexibility, this model still has a number of limitations. In particular, it requires that the average length of the active and the idle periods on the link be proportional, that is, within a factor $g_1/(1 - g_1)$. This implies that, for a given link utilization, the average duration of incoming bursts is fixed. Burst duration is, however, a key performance factor (4,34), and it is of interest to develop models that allow burst duration and utilization to vary independently.

In order to overcome this limitation, a model in which the arrival process corresponds to an extended ON-OFF process was presented and analyzed in Ref. 25. This model allows multiple packets with exponentially distributed lengths to be generated during a single ON period, but this is now achieved without imposing any constraint on the duration of the OFF periods and hence on the utilization. Specifically, the link is assumed to remain active for an exponentially distributed time I_n with mean $1/\lambda$, at the end of which a packet of size αI_n is generated. The link then starts a new ON period with probability $1 - p$ or enters an OFF period with probability p. The duration of an OFF period is exponentially distributed with mean $1/\mu$, and the link returns to the ON state at the end of an OFF period. The stability condition for this system is $\alpha < 1 + p\lambda/\mu$. This allows construction of ON periods in which the number (geometrically distributed) of consecutive packets that is generated is independent of the length of OFF periods.

Note that the arrival process of the first ON-OFF model can be viewed as a special case of this extended ON-OFF process with $p = g_2$ and $\mu = \lambda g_1$, with an analysis that is much simpler. Similarly, the more traditional ON-OFF process, in which each ON period corresponds to a single packet and is always followed by an OFF period, corresponds to the special case $p = 1$. The arrival process of the extended ON-OFF model is, therefore, quite general and provides the necessary flexibility to investigate the influence of different parameters on system performance. In addition, it is again possible to enhance the model further by allowing the addition of an independent and randomly distributed overhead to each packet.

An interesting numerical example that computes the average delay for the case $p = 1$ (i.e., a single packet is generated at the end of the ON state) and compares it to the values obtained assuming equivalent $G/M/1$ and fluid-flow models was provided in Ref. 25. The equivalent $G/M/1$ system has independent interarrival times with a probability distribution that has an LST that is $\alpha(\delta) = \mu\lambda/(\delta + \mu)(\delta + \lambda)$. The service times are independent of the interarrival times and exponentially distributed with parameter λ/α. The equivalent fluid-flow model is such that the output rate is 1 and the input rate in the ON state is $\alpha > 1$ (for $\alpha < 1$, the unfinished work in the system is always zero). For this model, the packet delay is defined from the time the last bit of the packet is received until the time it completely departs the system. Therefore, the average delay for the equivalent fluid-flow model is $(\alpha - 1)/[\lambda - \mu(\alpha - 1)]$ if $\alpha > 1$ and zero otherwise.

The average delays for all three models are plotted in Fig. 5 as a function of the proportionality parameter α. Note that $\alpha < 1 + \lambda/\mu$ is the stability condition for all three models. Two cases were considered: $\mu = 1$, $\lambda = 0.2$ and $\mu = 1$, $\lambda = 1.2$. The results for both cases illustrate the fact that the models developed in Ref. 25 in a sense "bridge the gaps" left by previous approaches. Specifically, while traditional point-process models such as the $G/M/1$ account for the granular nature of customer arrivals and departures, they typically ignore dependencies between interarrival and service times. As demonstrated in Fig. 5 and many previous studies, this often results in overly pessimistic estimates of system performance, especially at high loads. Conversely, fluid-flow models successfully capture the dependencies that exist between interarrival and service times, but they fail to preserve the discrete nature of these events. As alluded to

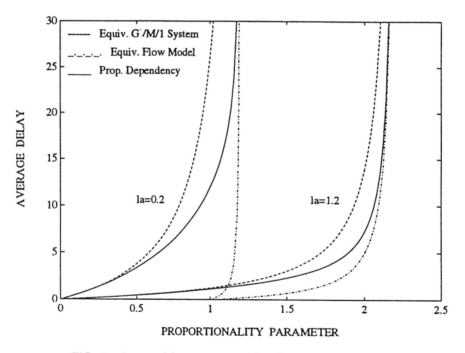

FIG. 5 Average delay versus proportionality parameter α; $\mu = 1$.

in Fig. 1 and illustrated in Fig. 5, this can in turn yield an overly optimistic view of system behavior, in particular at light and medium loads. The models developed in Ref. 25, because they are able to retain both aspects, provide more accurate estimates of actual system performance for all load values.

Queues with Interarrival Times Proportional to Service Times

In this section, we consider queueing systems in which the interarrival time between two packets depends on the service time of the first packet. We focus on proportional dependency, which is very natural in packet-switching networks. Here, the interarrival time between two consecutive packets arriving over a communications link is proportional to the size (in bits) of the first packet and consequently to the time it will take to forward this packet over the next link. The motivation for studying these queueing systems originated in the context of packet-switched networks, and especially in high-speed networks. The issue of interest was the effect of input policing functions that have been proposed to control the flow of packets in such networks. The basic goal of these policing functions is to ensure adequate network performance by regulating the amount of data that can arrive at a link within any given time interval. These controls result in significant dependencies between the amount of work brought in by packets and the time between the arrival of successive packets. Such dependencies have a significant impact on system performance, as described in Ref. 26.

While numerous previous papers have studied the effect of many different dependencies in queueing systems, very little work seems to have been done on the type of dependencies described in this section. In particular, the case in which the service time of a packet depends on the time since the previous arrival has been thoroughly studied (25,35–42). The first paper that addresses the dual problem in which the time to the next arrival depends on the service time of the arriving packet was Ref. 26. Previous work on the dual problem has been essentially limited to the study of general conditions for either stability (45) or finite moments of the busy period (46).

Cidon et al. consider systems in which the interarrival time I_{n+1} between packets n and $n + 1$ depends on the service time B_n of packet n (26). Specifically, they consider cases in which the dependency between I_{n+1} and B_n is a proportionality relation, and B_n is an exponentially distributed random variable. The proportional dependency was motivated by the modeling of some policing functions used in packet-switched networks.

For instance, let us illustrate how a simple spacer controller that is used to limit the peak rate at which a source can generate data into a network (see Refs. 3, 47, 48) introduces such dependencies. The enforcement of a maximum rate is achieved by requiring that, after sending a packet of size B, a space of duration B/R be inserted before the next packet can be sent. The rate R is then the maximum allowable rate for the source. (Note that the existence of a maximum network packet size is assumed here.) This rate R is typically equal to the source peak rate, but can be set to a lower value when low-speed links are present in the connection's path (47,48) or if the network traffic has to be smoothed (5).

Assuming that the above spacer is saturated by a source of rate R bits per second with traffic that is fed to a link of speed C bits per second, interarrival and service times (in seconds) at the link are then proportional with $I_{n+1} = \alpha B_n$ and $\alpha = C/R$. As shown in Fig. 6, which plots the evolution of the workload at the network link for the extreme case $\alpha = 1$, that is, equal source and link rates (stability requires $\alpha \geq 1$), the analysis of this simple case is of little interest. However, there exists a number of extensions to this basic model that make it nontrivial although still tractable and, more important, useful in modeling actual communications systems. These extensions were presented and analyzed in Ref. 26, in which the LST $\mathcal{W}(\delta)$ of the waiting time in steady state was obtained using the spectral analysis method typical of $G/G/1$ queues (1). In what follows, we describe some of these extensions and their importance in the modeling of communications systems.

The simplest extension consists of adding an independent random variable to the interarrival time. The presence of such a random component in the interarrival time allows more accurate modeling of how the traffic generated by a spacer controller arrives at an internal network link. First, such a model can capture the effect of interactions between packets from a given source and other traffic streams inside the network. In particular, the gaps that the spacer initially imposes between packets are modified according to the different delays that consecutive packets observe through the network. The arrival process at a link can then be modeled as consisting of a deterministic component (the spacing imposed by the spacer at the network access) to which a random network jitter has been added. Second, the addition of a random component also allows relaxing of the assumption of a saturated spacer queue since it can be used to model the time between packets that arrive at the spacer. Finally, another useful application is when the spacer itself randomizes the gaps between successive packets. This randomization in the spacer may be useful for avoiding correlation between traffic streams of distinct sources. In particular, it helps to prevent (malicious) sources from harming network performance by cooperating to generate a large burst of data into the network. Here, we summarize the results of this model for a system with positive jitter J_n exponentially distributed with parameter δ.

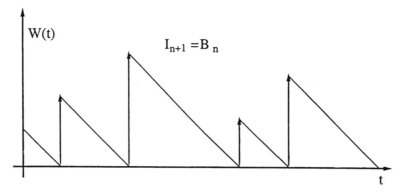

FIG. 6   Workload in queue with interarrival time proportional to service time.

- Utilization factor (proportion of time the server is busy) $\rho = \delta/(\alpha\delta + \mu)$, where B_n is exponentially distributed with parameter μ.
- The evolution equation of the waiting time of a packet $w_{n+1} = (w_n + (1 - \alpha)B_n - J_n)^+$.
- The LST of the waiting time of a packet

$$\mathcal{W}(\delta) = 1 - \gamma\delta + \gamma\delta \frac{(1 - \gamma\delta)/\gamma}{\delta + (1 - \gamma\delta)/\gamma}, \qquad \text{where } \gamma \triangleq (1 - \alpha)/\mu.$$

The waiting time in the system in steady state at the arrival instants thus behaves as the waiting time in the $M/M/1$ queue with arrival rate δ and service rate $1/\gamma = \mu/(1 - \alpha)$. However, the arrival rate to the system is actually $\lambda = \mu\delta/(\mu + \alpha\delta)$ and the service rate is clearly μ. Comparing this system with an $M/M/1$ system with parameters λ, μ, observe that both systems have the same stability condition. However, in the $M/M/1$ system, the term that governs the exponent (with a minus sign) of the pdf is

$$\mu\left(1 - \frac{\delta}{\mu + \alpha\delta}\right).$$

In this system, the term is

$$\mu\left(\frac{1}{1 - \alpha} - \frac{\delta}{\mu}\right)$$

which is larger when the stability condition ($\rho < 1$) holds. This implies that the tail probability decreases much faster in this system compared to the corresponding $M/M/1$ system. An extension to this system in which J_n can also take negative values (but the interarrival period is of course kept positive) can be found in Ref. 26.

Another extension of the basic model is to allow the proportionality constant to be a random variable. Specifically, cases in which the proportionality factor is randomly chosen from a finite set of values were considered in Ref. 26. This allows the modeling of a generalized spacer for which the factor used to compute the enforced spacing is allowed to vary. For example, the arrival of a high-priority packet that is sensitive to access delay could be handled by allowing earlier transmission. The interarrival times at the network link would then depend on both packet priorities and the size of the previous packet. The analysis of this system is more involved than that of the previous one (see Ref. 26).

The following numerical example was provided in Ref. 26. Consider a system with $I_{n+1} = \Omega_n B_n$, where Ω_n is a random variable with a finite support, independent of any other random variable in the system. Specifically, consider the case in which $\Omega_n = \alpha_i$ with probability a_i for $1 \le i \le N + M$ for some integers N, $M \ge 1$. Clearly, $\Sigma_{i=1}^{N+M} a_i = 1$. Consider the case $1 < \alpha_1 < \alpha_2 < \ldots < \alpha_N$ and $\alpha_{N+1} < \alpha_{N+2} < \ldots < \alpha_{N+M} \le 1$. The stability condition for the system is $\Sigma_{i=1}^{N+M} a_i\alpha_i > 1$.

Figures 7 and 8 show, respectively, the average and the variance of the waiting time of a system with random proportional dependency with $N = 3$, $M = 2$, and $a_i = 0.2$, $1 \leq i \leq 5$. They also show the same quantities in an equivalent $G/M/1$ system in which the service time is exponentially distributed with parameter μ, and the interarrival times are independent and sampled from a probability distribution with an LST that is

$$\alpha(\delta) = \sum_{i=1}^{N+M} a_i \frac{\mu/\alpha_i}{\mu/\alpha_i + \delta}$$

Namely, with probability a_i, the interarrival time is exponentially distributed with parameter μ/α_i. The sum $\Sigma_{i=1}^{N+M} a_i \alpha_i$ is kept constant. In particular, $\alpha_1 = 1.2$, $\alpha_2 = 1.3$, $\alpha_4 = 0.1$, and $\alpha_3 + \alpha_5$ are kept constant. The average and the variance of the waiting time are depicted as a function of the largest proportional parameter α_3.

Figures 7 and 8 show that the equivalent $G/M/1$ system exhibits much larger averages and variances of the waiting time. This implies that correlations between service times and interarrival times have a smoothing effect on the system. This has also been observed in Refs. 25 and 35–38 for different types of correlations. It is interesting to note that although $\alpha_3 + \alpha_5$ is kept constant, both the average and the variance increase with increasing α_3 (and decreasing α_5). This implies that decreasing α_5 has a more pronounced effect on the performance of the queueing system. The reason is that increasing α_3 and decreasing α_5 while keeping their sum constant increases the variability of the arrival pro-

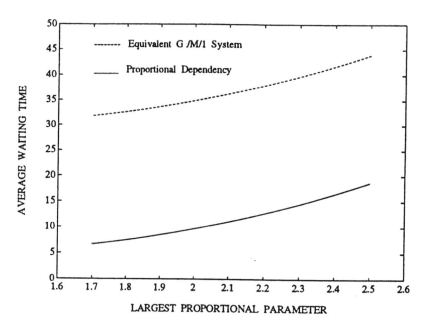

FIG. 7   Average waiting time versus largest proportional parameter: $a_i = 0.2$, $1 \leq i \leq 5$; $\alpha_1 = 1.2$, $\alpha_2 = 1.3$, $\alpha_4 = 0.1$, $\alpha_3 + \alpha_5 = 2.6$.

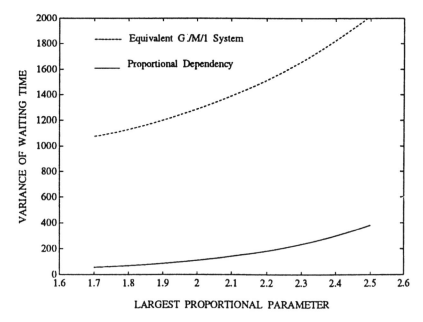

FIG. 8 Variance of waiting time versus largest proportional parameter: $a_i = 0.2$, $1 \leq i \leq 5$; $\alpha_1 = 1.2$, $\alpha_2 = 1.3$, $\alpha_4 = 0.1$, $\alpha_3 + \alpha_5 = 2.6$.

cess and hence increases the average and the variance of the waiting time as shown in Figs. 7 and 8.

Burstiness Characterization Models

In this section, we describe completely different approaches for characterization of arrival processes. These novel characterizations are motivated by the need to model real-time applications in high-speed communications networks in which it is important to capture the burstiness of the arrival processes.

The first approach we describe is deterministic and has been introduced by Cruz by characterization of the concept of a burstiness constraint (49–51). According to Cruz, a traffic stream with rate $R(t)$ at time t is said to satisfy such a constraint if there exist some constants ρ and σ such that

$$\int_{\delta}^{t} R(u)\,du \leq \rho(t - \delta) + \sigma$$

holds for all $0 \leq \delta < t$. In that case, $R(t)$ is said to be a (σ,ρ) process and the notation used is $R(t) \sim (\sigma,\rho)$. Note that this characterization ignores any stochastic nature of the traffic stream and must hold for any sample path of it.

The analysis Cruz presents achieves two major results regarding the performance of a single-server system. The first says that, if all input traffic to a single-server system satisfies burstiness constraints, then so does output traffic from that system (not necessarily with the same parameters, though). In other words, there exist constants ρ' and σ' such that the output process $R^{out}(t)$ satisfies $R^{out}(t) \sim (\sigma',\rho')$. This claim is proved by Cruz for a variety of systems with various service disciplines. For instance, for a work-conserving multiplexer with a general service discipline that is fed by two streams, $R_1(t) \sim (\sigma_1,\rho_1)$ and $R_2(t) \sim (\sigma_2,\rho_2)$, the output stream satisfies

$$R^{out}(t) \sim \left(\frac{C(\sigma_1 + \sigma_2)}{C - \rho_1 - \rho_2}, \rho_1 + \rho_2 \right)$$

where C $(C > \rho_1 + \rho_2)$ is the service rate of the multiplexer. The second result says that, in the above case, the delay suffered by each bit that enters the system is upper bounded by a constant D that depends, of course, on the parameters of the entering traffic streams and the nature of the examined system. For instance, for the above multiplexer, the constant D is given by

$$D = \frac{\sigma_1 + \sigma_2}{C - \rho_1 - \rho_2}$$

It is clear that the main advantages of the Cruz characterization are that the output processes of various systems have the same characterization as the input processes (with different parameters), and that it allows derivation of deterministic bounds on queue lengths and delays in the system. Further progress within this characterization has been presented by Parekh and Gallager for a special service discipline—the packet-based generalized processing sharing (PGPS) (52–54).

The second approach we describe is stochastic in nature and was introduced by Kurose (55). It attempts to overcome the deterministic nature of Cruz characterization. Suppose that for every $\tau > 0$ there exists a random variable $X(\tau)$ such that for all $0 \le s < t$ with $t - s = \tau$, the arrival stream $R(t)$ satisfies

$$\Pr\left\{ \int_s^t R(u)du \ge x \right\} \le \Pr\{X(\tau) \ge x\}$$

for all x. This means that $X(\tau)$ is a stochastic bound on $\int_s^t R(u)du$. We define $|X(\tau)| = sup\{x : \Pr\{X(\tau) \ge x\} > 0\}$. It is not difficult to see that $|X(\tau)|$ is a generalization of the term $\rho(t - s) + \sigma$ in Cruz characterization. Kurose introduced the characterization of a traffic stream in discrete time by stochastically bounding the amount of data it might carry in any fixed-length interval of time. In other words, a traffic stream with rate $R(t)$ $(t = 0, 1, 2, \ldots)$ is characterized by a series $\{(R_k,k); k \in \mathbb{N}\}$ of random variables if

$$\Pr\left\{ \int_s^t R(u)du \ge x \right\} \le \Pr\{R_{t-s} \ge x\}$$

holds for all $0 \leq \delta < t$ (δ and t are nonnegative integers) and all $x \geq 0$. This characterization is denoted by $R(t) \sim \{(R_k,k); k \in \mathbb{N}\}$. Kurose was also able to show that the output processes of various systems have the same characterization as the input processes (with different parameters). For instance, for a switch with two input streams $R_1(t) \sim \{(R_{n_1}^1, 1), (R_{n_1+1}^1, 2), \ldots \}$ and $R_2(t) \sim \{(R_{n_2}^2, 1), (R_{n_2+1}^2, 2), \ldots \}$, the output process of stream i ($i = 1,2$) satisfies $R_i^{out}(t) \sim \{(R_{n_i+l}^i, 1), (R_{n_i+l+1}^i, 2), \ldots \}$ where l is the least integer that satisfies $|R_{n_1+l}^1| + |R_{n_2+l}^2| \leq l + 1$. In addition, the delays suffered by each packet in the switch are upper bounded by some constant D. Note that, although this approach is stochastic in nature, the bound on l is a deterministic bound. In that sense, both approaches described so far require the intervals over which the sum of the input peak rates to a system exceeds its output capacity to be bounded. This is not the case most commonly used for arrival processes modeling input streams to a communications system, such as the simple Bernoulli process or the well-known Poisson process (which might have bursts of any length).

The third approach we describe is also stochastic and has been introduced by Yaron and Sidi (56). It attempts to overcome the limitations of the first two approaches. With this approach, rather than assuming that a traffic stream has a bounded burstiness, it is assumed that the distribution of its burst length has an exponential decay. Processes with such a characterization are said to have exponentially bounded burstiness (EBB). If $R(t)$ is the traffic rate of a stream, then the stream has EBB if there exist constants ρ, A, α such that

$$\Pr \left\{ \int_\delta^t R(u)\,du \geq \rho(t - \delta) + \sigma \right\} \leq A e^{-\alpha\sigma}$$

for all $\sigma \geq 0$ and all $0 \leq \delta < t$. In that case, we use the notation $R(t) \sim (\rho, A, \alpha)$. The basic advantage of this bound over the ones mentioned above is that it holds for natural processes such as Bernoulli, Poisson, and other processes. It is evident that it holds for most of the processes one might encounter in modeling communications networks. Two additional advantages make it applicable in many settings in which the aforementioned characterizations cannot be used. First, it imposes no deterministic requirements on the bounded processes and hence allows the analysis of systems in which the peak rates of the input streams might exceed the system's capacity over arbitrarily long intervals of time. Second, it allows the analysis of compound systems with correlated input processes—no independence assumption is needed when applying the analysis.

As with the two previous characterizations, Yaron and Sidi were able to show that, for various systems that are fed with EBB processes, the output streams are also bounded similarly. Furthermore, the delays these systems cause, and the length of the queues built up within them, all have exponentially decaying distributions. For instance, consider a work-conserving multiplexer that is fed by two streams $R_1(t) \sim (\rho_1, A_1, \alpha_1)$ and $R_2(t) \sim (\rho_2, A_2, \alpha_2)$. If the service rate of the multiplexer is C and $\rho_1 + \rho_2 < C$, Yaron and Sidi show that the output process $R(t)$ has exponentially bounded burstiness. In particular, $R(t) \sim (\rho', A', \alpha')$ with $\rho' = \rho_1 + \rho_2$, $A' = (A_1 + A_2)/(1 - e^{-\alpha'(C-\rho_1-\rho_2)})$

TABLE 2   Exponentially Bounded
Burstiness Characterizations
of a $P = 1/8$ Bernoulli Source

ρ	A	α
0.15	1	0.41
0.2	1	1.06
0.3	1	2.06

and $\alpha' = 1/(\alpha_1^{-1} + \alpha_2^{-1})$. Furthermore, the backlog $W(t)$ at the multiplexer satisfies $\Pr\{W(t) \geq \sigma\} \leq A'e^{-\alpha'\sigma}$ for all σ.

For a numerical example, consider a two-input multiplexer with output capacity $C = 1$ and with bounded input capacities $C_1 = C_2 = 1$ and assume it is fed with two Bernoulli input processes, each with parameter $P = \frac{1}{8}$. An appropriate EBB characterization of the input processes can be derived, which allows a tradeoff between the upper rate ρ and the decay factor α (see Table 2). Notice that the larger the gap one allows between the true mean rate P and the upper rate ρ of the characterization, the better the decay factor obtained.

Using these bounds for the input processes, one can compute exponential bounds for the queue length in the multiplexer. Figure 9 presents these bounds for the case of independent input processes, with some actual simulation result distributions. The dashed lines show the computed bounds for two different characterizations of the input processes (with $\rho = 0.2$ and with $\rho = 0.3$). The solid lines show the actual distributions at five different time stops ($t = 10, 50, 100, 500, 1000$) found by simulating the behavior of a multiplexer that has an empty queue at $t = 0$ for 100,000 times. Considering these actual distributions,

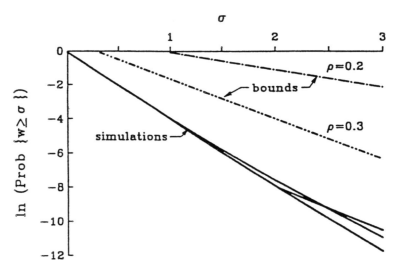

FIG. 9   Queue length distributions and its exponentially bounded burstiness bounds: independent sources.

it is apparent that the proposed analysis captures the transient, as well as the steady-state, behavior of the multiplexer queue. Similar analysis can be carried out for dependent sources (see Ref. 56).

Message Delay Processes

As mentioned in the general model section, in many systems the message delay, and not the packet delay, is the measure of interest for the network designer. Here, we describe some results for the message queueing delays in a node of a communications system. A *message* consists of a block of consecutive packets, and it corresponds to a higher layer protocol data unit. The *message delay* is defined as the time elapsing between the arrival epoch of the first packet of that message to the system until after the transmission of the last packet of that message is completed. We distinguish between two types of message-generation processes. The message can be generated as a batch, that is, all the packets that compose the message arrive at the system at a single instant of time (which corresponds to the well-known batch arrival model), or the generation can be dispersed, that is, the packets that compose the message arrive at the system at different times. Due to the finite speed of the communications links and the multiplexing of packets from different sessions, the dispersed generation model is more adequate for communications networks.

Understanding the message delay behavior is important for the proper design of time-out mechanisms for data applications, such as end-to-end protocols for reliable transmission in ATM networks in which the retransmitted quantity is the message and not the individual packets. Another design example is the time-out for message retransmissions in data-link protocols such as the go-back-N protocol, in which the whole window (or message) is retransmitted (see, e.g., Ref. 2). Individual packet delay distributions are usually not sufficient for proper understanding of system behavior. In general, the delays of two consecutive packets are strongly correlated, that is, the delay of the second packet is conditioned on the event that the first packet delay is larger (smaller) than the delay of an arbitrary packet.

Traditionally, the analysis of the message delay was associated with batch arrival processes (see, e.g., Refs. 6, 17, 57); that is, each batch corresponds to a message. In this case, the message delay coincides with the delay of the last packet of the message (batch). This fact facilitates the analysis of the message delay distribution. Kleinrock (6) analyzed a bulk (batch) arrival queueing system in which a single-server queueing system with bulk Poisson arrivals (i.e., the arrival instants are Poisson and at each arrival instant a bulk of arbitrary size sampled from an independent integer random variable is brought to the system and corresponds to a message) and exponentially distributed service times was considered. Halfin analyzed a discrete-time queueing system with batch arrivals, in which batch sizes have geometric distribution and the queue discipline is indifferent to batch sizes and service times (57). He proved that the packet delay distribution is the same as the batch (message) delay distribution, with delay defined as the delay of the last-served packet in the batch. The proof was based on a discrete-time analog of the PASTA theorem. The message delay distribution for TDMA systems with a generalized arrival process was presented in Ref. 17. The analysis was based on a generating functions approach.

However, in packet-switched networks, packets that belong to the same message may arrive at different instants of time (be dispersed) and may be interleaved (due to statistical multiplexing) by packets that belong to other messages. The difficulty that arises in the analysis of the message delay distribution for the dispersed generation model is that there is a correlation between the system states seen by different packets of the same message. The effect of the correlation between successive arrivals to the system on the average packet delays was studied in Ref. 58 for Poisson cluster processes (PCPs). Here, messages arrive at the system according to Poisson process, but unlike the batch Poisson arrival process, for which all the packets of the batch (message) arrive at the same time, the members of a cluster are separated by a random variable. In Ref. 58, the average delay of packets was approximated for a *PCP/D/*1 system.

In Ref. 59, a new technique to analyze the message delay in systems with dispersed arrivals was introduced. It has been shown that the correlation between the delays of packets that belong to the same message has a strong effect on the performance of the system. For example, it was shown that evaluating the time-out for message retransmissions under the assumption that the packet delays are independent is quite pessimistic. The model used in Ref. 59 for ascertaining the correlation in the packet delay process consists of a source that generates packets and sends them through a single server with an infinite number of buffers, which represents the communications system. For this, *n* consecutive packets are grouped into a message. An exact analysis of the message delay was presented. In particular, an efficient recursive procedure to obtain the LST of the message delay for different arrival models and different numbers of sessions was presented. It was shown that the assumption that the delays of packets are independent from packet to packet can lead to wrong conclusions. This fact was demonstrated by comparing the exact variance of the message delay with the variance of the message delay as obtained from the above independence assumption. Numerical examples were provided to show that the variance of the message delay may be overestimated by the above independence assumption for a wide range of message sizes.

It was observed in Ref. 59 that the message delay is composed of two components. The first is the time elapsing between the arrival epoch of the first packet of the message to the system and the arrival epoch of the last packet of that message to the system. The second is the time delay of an arbitrary packet (the last packet of that message) in the system. These two components are, of course, dependent random variables. However, the average message delay can be obtained directly from the sum of the averages of these two components. Cidon, Khamisy, and Sidi suggest a simple way to approximate the message delay by assuming that these two components are independent random variables (59). The numerical results in Ref. 59 demonstrated the relative error of such an approximation and the so-called negative feedback effect that governs the message delay process. If the message's packet arrival happens to concentrate over a short time interval, then the message arrival time becomes short. On the other hand, this causes a larger queue to build up, resulting in a larger queueing delay for the last packet of the message. Similarly, if the message's packet arrivals happen to be more dispersed, then the queueing delay of the last packet tends to become shorter. Thus, the message delay distribution in the dispersed generation model tends to concentrate around the average much more than can be expected using the above independence assumption of the message arrival time and the last packet delay time.

The following numerical example was provided in Ref. 59. Consider an $M/M/1$ system with arrival rate λ, service rate μ, and messages of fixed size n. The average message delay equals $[(n-1)/\lambda] + [1/(\mu - \lambda)]$. The variance of the message delay using the above approximation equals $[(n-1)/\lambda^2] + [1/(\mu - \lambda)^2]$. The relative variance error of the message delay, defined as $100 \cdot [(approximated\ variance)/(exact\ variance) - 1]$ is plotted in Fig. 10 versus the message size n for $\mu = 1$ and for different values of λ ($\lambda = 0.5, 0.8, 0.9$). For all cases, observe that the approximated variance of the message delay is much larger than the exact one. Observe also that the approximation becomes worse for heavy loads in a wide range of message size.

Two other important quantities, namely, the maximum delay of a packet in a message and the number of packets in a message with delays that exceed a prespecified time threshold, were analyzed in Ref. 60 for the dispersed generation model. These quantities are important for the proper design of playback algorithms (61) and time-out mechanisms for retransmissions. In Ref. 60, a new

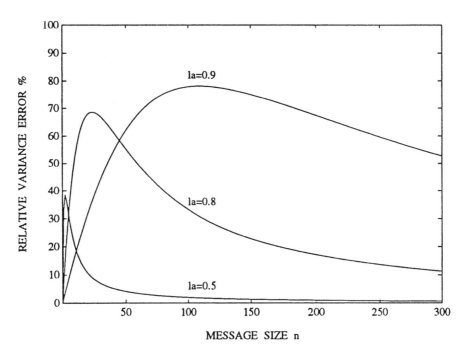

FIG. 10   The relative variance error of the message delay versus the message size n.

analytical approach that yields efficient recursions for the computation of the probability distribution of each quantity was presented. It has been shown that the correlation between packet delays of the same message has a strong effect on each of these quantities.

References

1. Kleinrock, L., *Queuing Systems, Volume 2: Computer Applications*, John Wiley and Sons, New York, 1976.
2. Bertsekas, D., and Gallager, R., *Data Networks*, Prentice-Hall International Editions, Prentice-Hall, Englewood Cliffs, NJ, 1987.
3. Elwalid, A. I., and Mitra, D., Analysis and Design of Rate-Based Congestion Control of High-Speed Networks, I: Stochastic Fluid Models, Access Regulation, *Queuing Systems Theory Appl.*, 9:29–64 (September 1991).
4. Guérin, R., Ahmadi, H., and Naghshineh, M., Equivalent Capacity and Its Application to Bandwidth Allocation in High-Speed Networks, *IEEE J. Sel. Areas Commun.*, SAC-9(7):968–981 (September 1991).
5. Guérin, R., and Gün, L., A Unified Approach to Bandwidth Allocation and Access Control in Fast Packet-Switched Networks, *Proc. Infocom '92*, 1–12 (1992).
6. Kleinrock, L., *Queuing Systems, Volume 1: Theory*, John Wiley and Sons, New York, 1975.
7. Anick, D., Mitra, D., and Sondhi, M. M., Stochastic Theory of a Data-Handling System with Multiple Sources, *Bell Sys. Tech. J.*, 61(8):1871–1894 (October 1982).
8. Gaver, D. P., Jr., and Miller, R. G., Jr., Limiting Distributions for Some Storage Problems. In *Studies in Applied Probability and Management Science* (K. J. Arrow, S. Karlin, and H. Scarf, eds.), Stanford University Press, Berkeley, CA, 1962, pp. 110–126.
9. Kaspi, H., and Rubinovitch, M., The Stochastic Behavior of a Buffer with Non-Identical Input Lines, *Stochastic Processes and Their Applications*, 3:73–88 (1975).
10. Kosten, L., Liquid Models for a Type of Information Storage Problems, *Delft Progress Report: Mathematical Engineering, Mathematics and Information Engineering*, 11:71–86 (1986).
11. Mitra, D., Stochastic Theory of a Fluid Model of Producers and Consumers Coupled by a Buffer, *Adv. Appl. Prob.*, 20:646–676 (September 1988).
12. Rubinovitch, M., The Output of a Buffered Data Communication System, *Stochastic Processes and Their Applications*, 1:375–382 (1973).
13. Tanenbaum, A. S., *Computer Networks*, Prentice-Hall International Editions, Prentice-Hall, Englewood Cliffs, NJ, 1989.
14. Kelly, F. P., *Reversibility and Stochastic Networks*, John Wiley and Sons, New York, 1979.
15. ATM Forum: Technical Committee, *UNI Specification Draft 2.4, Technical Report*, August 1993.
16. Comer, D., *Internetworking with TCP/IP, Principles, Protocols, and Architectures*, Prentice-Hall, Englewood Cliffs, NJ, 1988.
17. Rom, R., and Sidi, M., *Multiple Access Protocols; Performance and Analysis*, Springer-Verlag, New York, 1990.
18. Little, J., A Proof of the Queuing Formula $L = \lambda W$, *J. Operations Research*, 18:172–174 (1961).

19. Baccelli, P., and Bremaud, P., *Elements of Queuing Theory*, Springer-Verlag, New York, 1994.
20. Ross, S. M., *Introduction to Probability Models*, New York, 1980.
21. Wolff, R. W., Poisson Arrivals See Time Averages, *Operations Research*, 30(2) (March–April 1982).
22. Burk, P. J., The Output of a Queuing System, *J. Operations Research*, 4:699–704 (1966).
23. Cohen, J. W., *The Single Server Queue*, North-Holland, 1969.
24. Fendick, K. W., Saksena, V. R., and Whitt, W., Dependence in Packet Queues, *IEEE Trans. Commun.*, COM-37(11):1173–1183 (November 1989).
25. Cidon, I., Guérin, R., Khamisy, A., and Sidi, M., Analysis of a Correlated Queue in Communication Systems, *IEEE Trans. Information Theory*, 39(2):156–465 (March 1993).
26. Cidon, I., Guérin, R., Khamisy, A., and Sidi, M., On Queues with Inter-Arrival Times Proportional to Service Times, *Infocom '93*, 237–245 (April 1993).
27. Kleinrock, L., *Communication Nets*, McGraw-Hill, New York, 1964; reprint, Dover, New York, 1972.
28. Boxma, O. J., On a Tandem Queuing Model with Identical Service Times at Both Counters, Parts I and II, *Adv. Appl. Prob.*, 11:616–659 (1979).
29. Calo, S. B., Delay Properties of Message Channels, *ICC*, 43.5.1–43.5.4 (1979).
30. Calo, S. B., Message Delays in Repeated-Service Tandem Connections, *IEEE Trans. Commun.*, COM-29(5):670–678 (May 1981).
31. Elwalid, A. I., Mitra, D., and Stern, T. E., Statistical Multiplexing of Markov Modulated Sources: Theory and Computational Algorithms, *Proc. ITC-13*, 495–500 (1991).
32. Hashida, O., and Fujiki, M., Queuing Models for Buffer Memory in Store-and-Forward Systems, *Proc. ITC-7*, 323/1–323/7 (1973).
33. Stern, T. E., and Elwalid, A. I., Analysis of Separable Markov-Modulated Models for Information-Handling Systems, *Adv. Appl. Prob.*, 105–139 (March 1991).
34. Roberts, J. W., Variable-Bit-Rate Traffic Control in B-ISDN, *IEEE Commun. Mag.*, 29(9):50–56 (September 1991).
35. Conolly, B. W., The Waiting Time Process for a Certain Correlated Queue, *Operations Research*, 16:1006–1015 (1968).
36. Conolly, B. W., and Hadidi, N., A Comparison of the Operational Features of Conventional Queues with a Self-Regulating System, *Appl. Stat.*, 18:41–53 (1969).
37. Conolly, B. W., and Hadidi, N., A Correlated Queue, *J. Appl. Prob.*, 6:122–136 (1969).
38. Conolly, B. W., and Choo, Q. H., The Waiting Time Process for a Generalized Correlated Queue with Exponential Demand and Service, *SIAM J. Appl. Math.*, 37(2):263–275 (October 1979).
39. Hadidi, N., Queues with Partial Correlation, *SIAM J. Appl. Math.*, 40(3):467–475 (June 1981).
40. Hadidi, N., Further Results on Queues with Partial Correlation, *Operations Research*, 33:203–209 (1985).
41. Langaris, C., A Correlated Queue with Infinitely Many Servers, *J. Appl. Prob.*, 23:155–165 (1986).
42. Langaris, C., Busy-Period Analysis of a Correlated Queue with Exponential Demand and Service, *J. Appl. Prob.*, 24:476–485 (1987).
43. Combe, M. B., Borst, S. C., and Boxma, O. J., Collection of Customers: A Correlated M/G/1 Queue, *Performance Evaluation Rev.*, 20:47–59 (1991).
44. Kuczma, M., Choczewski, B., and Ger, R., Iterative Functional Equations. *Ency-*

clopedia of Mathematics and Its Applications, Cambridge University Press, Cambridge, England, 1990.

45. Loynes, R. M., The Stability of a Queue with Non-Independent Inter-Arrival and Service Times, *Proc. Cambridge Philos. Soc.*, 58(3):497–520 (1962).

46. Ghahramani, S., and Wolff, R. W., A New Proof of Finite Moment Conditions for GI/G/1 Busy Periods, *Queuing Systems Theory Appl.*, 4(2):171–178 (June 1989).

47. Bala, K., Cidon, I., and Sohraby, K., Congestion Control for High-Speed Packet Switch, *Proc. Infocom '90*, (1990).

48. Cidon, I., Gopal, I., and Guérin, R., Bandwidth Management and Congestion Control in PlaNET, *IEEE Commun. Mag.*, 29(10):54–63 (October 1991).

49. Cruz, R. L., A Calculus for Network Delay and a Note on Topologies of Interconnection Networks, Ph.D. thesis, University of Illinois, Urbana-Champaign, 1987.

50. Cruz, R. L., A Calculus for Network Delay, Part I: Network Elements in Isolation, *IEEE Trans. Information Theory*, 37(1):114–131 (January 1991).

51. Cruz, R. L., A Calculus for Network Delay, Part II: Network Analysis, *IEEE Trans. Information Theory*, 37(1):132–141 (January 1991).

52. Parekh, A. K., A Generalized Processor Sharing Approach to Flow Control in Integrated Services Networks, Ph.D. thesis, Massachusetts Institute of Technology, Cambridge, 1992.

53. Parekh, A. K., and Gallager, R. G., A Generalized Processor Sharing Approach to Flow Control – The Single Node Case, *IEEE/ACM Trans. Networking*, 1:344–357 (June 1993).

54. Parekh, A. K., and Gallager, R. G., A Generalized Processor Sharing Approach to Flow Control – The Multiple Node Case, *MIT Laboratory for Information and Decision Systems Technical Report 2076*, 1991 (also in *Proc. Infocom '93*, 521–530, 1993).

55. Kurose, J., On Computing Per-Session Performance Bounds in High-Speed Multi-Hop Computer Networks, Performance '92, Newport, June 1992.

56. Yaron, O., and Sidi, M., Calculating Performance Bounds in Communication Networks, *IEEE/ACM Trans. Networking*, 1:372–385 (June 1993).

57. Halfin, S., Batch Delays Versus Customer Delays, *Bell Sys. Tech. J.*, 62(7) (September 1983).

58. Sohraby, K., Delay Analysis of a Single Server Queue with Poisson Cluster Arrival Process Arising in ATM Networks, *Globecom '89*, 611–616 (1989).

59. Cidon, I., Khamisy, A., and Sidi, M., On Queuing Delays of Dispersed Messages, *Queuing Systems – Theory and Applications* (1994) (also in *Proc. Infocom '93*, 843–849, April 1993).

60. Cidon, I., Khamisy, A., and Sidi, M., Dispersed Messages in Discrete-time Queues: Delay, Jitter and Threshold Crossing, *Proc. Infocom '94* (June 1994) (also in *Technion EE PUB No. 894*, August 1993).

61. Cochennec, J.-Y., et al., Asynchronous Time-division Networks: Terminal Synchronization for Video and Sound Signals, *Proc. Globecom '85*, 791–794 (December 1985).

MOSHE SIDI
ASAD KHAMISY

Network Protection

Introduction and Overview

As a principle, the goal of information protection is to assure that the right information gets to the right place at the right time. The term *assure* in this context is key to understanding that we actively work to achieve the unattainable goal of perfection, while trading off assurance with other issues.

To understand the difficulty in getting the right information to the right place at the right time, we have to consider that, in today's environment, there are many potential impediments to this goal. The impediments are commonly classified in two dimensions. One dimension is the range from accidental to malicious, and relates to the cause of the impediment. The other dimension typically includes corruption of information, denial of services, and leakage of information, and relates to the effect of the impediment.

In order to be effective, protection must provide some degree of assurance that these causes do not result in these effects. This is normally done through a combination of efforts in the areas of

Protection policy	Protection management
Standards and procedures	Documentation
Protection audit	Technical safeguards
Incident response	Testing
Physical protection	Personnel issues
Legal considerations	Protection awareness
Training and education	Organizational structure

These efforts combine and intertwine to provide an environment in which the likelihood of getting the right information to the right place at the right time is enhanced. The extent to which this likelihood is enhanced is generally related to the quality and quantity of the joint effort put forth in these areas, but there is not a simple relationship between these efforts and the impact on assurance. One of the reasons this relationship is complex is that there are synergistic effects among causes, effects, and protective measures (1,2).

Many attempts have been made to quantify protection, but these attempts have generally failed to a greater or lesser extent because of their dependence on statistics. A fundamental limitation on the use of statistics in this field is that, under malicious attack, statistics may be meaningless. The reason for this is that statistics are based on a set of assumptions that are usually easy to invalidate by the use of coherent attack techniques (3,4). For example, once a successful attack is found, it can be repeated until it is countered. A statistical defense would assume that the likelihood of the same attack being repeated would be astronomically small, while in the coherent environment of adaptive learning systems, the likelihood is almost unity.

One alternative approach that has been successful in mitigating the flaws of purely statistical protection is the perspective of strategy and tactics. In this approach, we view protection as a game played between causes and defenders, with the goal being control over effects. There is no perfect offense or defense, and thus winners and losers are determined by qualitative differences in their strategies and tactics.

The Goal of Protection

The goal of information protection is generally to provide assurance that the right information gets to the right place at the right time. Many other views on this goal have been advanced, but they seem to end up collapsing into this goal in one form or another. To give a sense of this, it is helpful to provide an example.

A common goal in the age of electronic signatures is to provide nonrepudiation services. This sort of service is designed to provide a receipt that can be used as proof that particular information was sent by a particular sender. If that information was a legally binding agreement, then the nonrepudiation function serves as proof that the sender agreed to the terms of the contract in the case that the sender should attempt at a later time to claim that they never agreed to those terms. But, with a system with high enough assurance that the right information gets to the right place at the right time, a judge could simply ask the system what was sent from whom to whom when and get the right answer. No other proof would be needed because the fact that a system with such high assurance provided the information would be proof enough. From this example, you might want to add the clause "and nothing else" to our protection goal and that would be fine.

The term *assure* means to give confidence in or to make certain. Except for the fatalists among us, we all believe that few things are truly certain. In order to attain increasingly more stringent levels of certainty, we generally have to be increasingly rigorous. The study of information protection is therefore a study in tradeoffs. In order to attain a particular level of assurance, we have to pay a particular price, not only financially, but in terms of time and effort.

A good example for which people have historically been unwilling to pay the price is in the use of good passwords (5). In this case, it has been widely known for many years that the vast majority of user-selected passwords are easily guessed by a relatively unsophisticated attacker. There are alternative technologies that are more convenient than typing long passwords; they cost on the order of a few dollars per user, yet as a society, we have not chosen to use them.

Information protection is closely related to a number of other fields, including but not limited to, computer security, communications security, and fault-tolerant computing. In this use, information protection is intended to be encompassing rather than limiting, but there are a few distinctions worth noting here. The first one is that information indicates symbolic representations in the most general sense. It is not necessarily limited to computer systems or information stored in electronic form. *Protection* means keeping from harm. Since it is impossible, by definition, to harm a computer or information, we are considering the impact of information systems on people and are trying to protect people from any resulting harm.

Information protection in a networking environment is complicated because of the physical distribution of information and computational capabilities and the intermixing of protective techniques. The physical distribution dramatically changes the cost tradeoffs associated with different protective techniques and, in many cases, makes techniques that would be very effective in a nonnetworked environment essentially useless. The mixing of techniques is quite complex because of synergistic effects. For example, two independently sound protective schemes may yield unsound protection when networked together (2).

Dimensions of Protection

Information protection is a multidimensional issue that spans all aspects of systems design and implementation. This is in part because protection, like a chain, tends to fail at the weakest link. Here are the nine dimensions used in analyzing protection in a recent study of a cable television network:

Causes	Effects	Services
Value	Protection	Cost
Policy	Exposure	Assurance

To give a flavor for these dimensions, some sample details are given here, but it is important to remember as you read them that this scheme represents only one view of the protection issue and that the details provided are by no means comprehensive.

Causes

There are many causes for protection failures, and there is no hope for enumerating them all here. Having said that, we now attempt to enumerate the most common causes. We begin with some of the accidental ways information systems can fail:

Errors and omissions	Power failure
Fire	Flood
Earth movement	Solar flares
Volcanos	Severe weather
Static electricity	Air conditioning loss
Moving of computers	System maintenance
Testing	Cleaning chemicals
Inadequate maintenance	Humidity
Smoke	Gases and fumes
Dust	Heat
Temperature cycling	Electromagnetic interference

As a specific example, increasing the temperature of semiconductor junctions by 10°C typically reduces the mean time to failure by a factor of two. So a component that would have lasted for an average of 10 years will only last for an average of 5 years at the higher temperature.

Most accidental causes of failure can be mitigated by using statistical analysis of the likelihood of events and providing coverage of failures (6,7). This is done by a combination of design standards, operational controls, training of workers, and combinations of a host of other techniques.

Next, we list some of the different sorts of people that commonly launch intentional attacks. It is important to understand the sorts of attackers in order to get a grasp on the length to which they may go to attack systems and the techniques they are likely to use and to balance the cost of defense against the perceived threat.

Programmers	Private detectives
Reporters	Executives
Consultants	Whistle-blowers
Hackers	Club initiates
Crackers	Tiger teams
Competitors	Maintenance people
Professional thieves	Hoods
Vandals	Crackers for hire
Organized criminals	Drug cartels
Terrorists	Spies
Police	Government agencies
Infrastructure warriors	Nation states and economic rivals
Military organizations	Information warriors

An example of reporters launching attacks occurred during the 1994 Winter Olympics, when reporters broke into computer mail accounts in the Olympic athlete computer system to read Tonya Harding's mail to see if they could dig up a story about her (8).

Next, we look at some of the motives for an attack. Motives are particularly important because we can act to eliminate many motives by management policy and practices, because understanding motives leads to methods of prevention, detection, and correction, and because motives are related to the people that may attack and the methods they may be willing to use (9).

Money or profit	Fun, challenge, and acceptance
Vengeance or justice	Sadism or other mental illness
Religious or political beliefs	Self-defense
Testing	Coercion
Military or economic advantage	Gathering or destruction of evidence

An example of a common motive for the so-called hacker is acceptance. In order to gain acceptance in a group of peers, the young hacker may launch an attack to demonstrate computer prowess (10). Once we understand the motive, it may lead to prevention (e.g., create a high profile "honey pot" for hackers to go after instead of the real systems you are trying to protect), detection (e.g., look for particular sorts of activities typically used by hackers), and eradication (e.g., put in a set of controls that will be effective against people with perseverance but little or no funding).

Finally, we look at some of the methods used in attacks to get a sense of the breadth of technical issues that may be involved in defenses:

Trojan horses	Time bombs
Condition bombs	Dumpster diving
Human engineering	Bribery
Get hired and spy	Password guessing
Invalid values on calls	Computer viruses
Data diddling	Packet insertion
Packet watching	Van Eck bugging
Electronic interference	Private branch exchange (PBX) bugging
Open microphone listening	Video viewing
Wire closet attacks	False update disks
Shoulder surfing	Toll fraud groups
Process bypassing	Backup theft
Log-in spoofing	Hangup hooking
Call forwarding fakery	Misfiling
Exploiting design flaws	Input overflow
Illegal value insertion	Induced stress failures

Open microphone listening may offer some insight into the technical challenges in the telecommunications area. It turns out that many modern PBX systems have a design flaw (feature?) that allows the microphone on speakerphone inputs and paging devices to be controlled from the PBX. In other words, hanging up the handset does not physically disable the microphone. This is very handy for operating a speakerphone, but it also leads to an attack in which the PBX is programmed to leave the microphone on even though the handset is on hook. The voice line is then connected to another telephone, perhaps via a conference call setup, and the telephone acts as a bugging device for listening to conversations in the room with the hung-up telephone. How does the attacker program the PBX for this attack? It turns out that most substantial PBX systems have dial-in maintenance lines from the supplier, and those lines are rarely well protected. Attackers get the telephone number of the maintenance line, read the maintenance manuals, and learn how to reprogram the switch.

Effects

The effects of accidental or malicious incidents are limited only by the number of possible states and state transitions available in the system being considered. In other words, any information can potentially be modified, stolen, or destroyed. The implication of these effects can be quite startling.

For example, in systems that manage electronic funds transfers (EFTs), changes in a few bits have caused hundreds of millions of dollars to be irretrievably lost or stolen. In systems that deliver controlled doses of radiation, patients have been killed because of seemingly minor errors (11). In aerospace applications, single-bit errors have caused multibillion dollar systems to be destroyed (12). Single-digit errors have caused innocent people to be arrested and held in jail for days. Data entry errors have caused prisoners to be released from jail early. A two-bit change in a telephone switching system release caused telephone service in several cities to be lost for several days in 1991 (13). The nationwide AT&T network failure in 1990 was caused by a few wrong bits in a communications protocol (1).

The list of effects is seemingly endless and, in the information age into which we are now moving, the impacts will become greater because our dependency on these systems will become greater. Consider that today essentially all finance, communications, transportation, sales, delivery, manufacturing, design, development, and every other component of your life is intimately tied into information technology. The potential harmful effects are directly related to this dependency.

Services

Information systems provide a wide range of services, from entertainment to real-time control of critical functions. Each service has different requirements for performance, reliability, availability, accuracy, timeliness, space consumption, and any number of other parameters. For example, looking up government documents from a national database for a high school project requires dramatically different protection than carrying out a remote surgical operation on a patient with a brain injury on a battlefield halfway around the world.

The range of services provided in today's information systems are far too broad to describe here, but for a starter, here is a list extracted from one on-line information service:

Business and finance	News, weather, and sports
Computing groups	Reference and education
Conferences	Shopping
Entertainment and games	Travel and leisure
Groups and clubs	Using Delphi
Internet services	Workspace
Mail	Member directory

Each of these categories includes numerous subselections. For example, the business and finance selection leads to this menu:

UPI business news	PR newswire — press releases
Business wire — press releases	Rategram CD reports
Commodity quotes	Register of public corporations
Dow-Jones averages	SOS — stock + options advisors
Financial and commodity news	Stock quotes
Forum	Translation services
Futures focus	Trendvest ratings + custom analysis
Marketpulse	Money fund report from Donaghue

If such a menu system goes 3 levels deep with an average of 15 items per menu, the bottom level ends up with something like 15^3, or 3375, different services. Each service may in itself be relatively complex as well. Meanwhile, the cable television system offers on the order of 50 channels, each with a different program every half hour (or 2500 different selections per day) and will soon expand to 500 channels in many cable systems.

To get an idea of the implications of a protection failure, consider the impact on your organization if you were a local commodities trader and the commodities information on your purchases were electronically corrupted in transit. In the commodities market, you can easily lose everything you have invested in a matter of seconds. Even loss of access for a matter of minutes might cost you a great deal, but at the same time, all of the trades that have taken place are public knowledge. So, in this environment, secrecy about trades has no value except in the planning stages, while integrity and availability are of paramount importance.

Value

In designing protection, we must consider the value of the information being protected. For example, the commodities trading described above totals billions of dollars a day in value, but any one transaction by the average person in the markets is only on the order of thousands of dollars. The value of information is a very tricky issue to consider.

There have been times when only a few bits of information were worth millions of dollars and thousands of lives. A good example was the invasion of Normandy beach, which would have been a disaster had the Germans known only a few bits of information (e.g., the date, time, and location). On the other hand, that same information would have had no value whatsoever as little as one day after the invasion took place. So, the value of information may change dramatically with time.

There are a number of different accounting principles used to value information. *Sunk cost* relates value to the time and effort taken to gather, enter, store, and retrieve it. *Replacement cost* relates value to the cost of regeneration. *Fair market value* relates value to what people are willing to pay for it. The first of these measures is time independent, while the other two are highly dependent on time and market conditions.

Protection technology has not yet reached the point at which the value of each bit of information is constantly rated and protection is then metered out in proportion to the rated value, but you can imagine that the day may soon come when this is common. The main problem in valuation today is the rating process, particularly as it relates to time-dependent changes. There is currently no clear and automatable basis for such ratings (14), and thus protection experts normally call on domain experts to rate values for different classes of information in order to do analysis.

Protection

There are a lot of different protection techniques in use today, and there are a potentially infinite number of different techniques and ways to combine them. Furthermore, many of these techniques can be used with varying degrees of assurance. Several examples of protection techniques are described below.

Cost

Protection is not free. There are design and development costs, implementation costs, operation and maintenance costs, and conversion and decommissioning costs. Each of these may involve capital investment, time, and materials. It is not very unusual for people to pay more for protection than it can possibly save. This is almost universally evident in the area of computer virus defense, for which many defenses cost only a few dollars to purchase, but cost several hundred dollars per computer per year to operate and maintain (2). The average damage from viruses turns out to be far less than the cost of using most defenses, but life-cycle costs are not analyzed by most purchasers.

Policy

Policy is a driving force behind protection; there are a lot of different policies used for different purposes. Here are some examples of different policy components used in different situations:

- *Fairness*: It seems reasonable that the set of policies applied to any one information transporter, provider, or consumer of a particular type should be applied to all equivalent transporters, providers, and consumers. In other words, the principle of fairness should be applied within these groups to equivalent members. Any other policy will tend to lead to disagreements, conflicts of interest, and legal problems.
- *Legality*: As a matter of course, all policies should be checked for legal implications. It turns out that, in today's dramatically changing legal environment, the information protection field is rapidly becoming a morass, decisions are often contradictory, and state and local laws may be different from federal and international laws. This is particularly true in the areas of cryptography, public access, interconnections with common carriers, financial transactions, attachments to Federal Interest Computers, and intellectual property laws.
- *Information ownership*: Each individual owns and controls all information generated by them. At their discretion, they may release the information to other parties. Parties may require information from other parties in order to do business with them, in which case it is the responsibility of the parties to come to an agreement about the exchange of information.
- *Informed consent*: In the informed consent policy, all providers and customers are constantly rated by each other and the results of the ratings are provided to everyone on the network as part of their purchasing process. Hence, the parties are mutually informed and, knowing this, they may consent to transactions.
- *Authorized access*: In the authorized access policy, there are legally authorized agents who have the right to observe and/or modify information in the network. For example, law enforcement or particular government agencies might be granted access under subpoena.

- *Voting schemes*: In voting schemes, there are a multitude of authorities available for various transactions, and the value and type of the transaction dictates which authority sets are required for the transaction to take place.
- *Waste not, want not*: In the waste not, want not policy, information systems that are not otherwise in use may be used by people throughout the network to perform useful work. This could be used to implement a giant, parallel processing, supercomputing facility accessible throughout the network for performing highly complex computations.
- *Detect and respond*: Rather than trying to prevent all of the protection problems that may occur, we take the tactic of implementing as many detection capabilities as are technically feasible. In particular, we wish to detect known anomalous behavior and changes in behavior that might indicate anomalous situations. After detection, we should respond to incidents rapidly and vigorously and publicly pursue those who violate others' rights and should enhance the quality of the detection system in terms of reducing false positives and false negatives.
- *Junk information screening*: People do not like junk mail, and they downright hate junk facsimiles and solicitation calls. Junk information screening reflects that possibility by requiring a set of markers to be placed on each bundle of information (not packets) by the provider. Users can then automatically screen in or out information based on classification. If substantial misclassification comes from any provider, they can be removed from the network.

Clearly, there are an unlimited number of different policies. In addition, policies may intermingle so as to form a system that makes them inconsistent or unusable.

Exposure

Whenever there is a potential for loss, that potential is exposed through potential protection failures. These exposures are the things we seek to cover with protection. For example, a weak password scheme that allows any sequence of characters and digits up to 10 symbols long to be used to authenticate an identified user to a system represents a substantial exposure because, on the average, users will select easily guessed passwords. We can cover this exposure by requiring special features of these passwords, by assigning users pseudorandom passwords, by using a different authentication mechanism, by restricting the physical location from which access may be attained, and so on.

The existence of an exposure alone does not justify coverage. For example, this password scheme might be perfectly acceptable for controlling access to a nonnetworked computer that contains only publicly accessible, noncopyrighted information. In order to justify coverage, the exposure must create a potential for loss sufficient to justify protection. We often discuss exposures in terms of the potential for loss rather than as existential entities.

Assurance

The extent to which we wish to assure that protection goals are met is a function of the sort of events we wish to provide protection from and the lengths to which we are willing to go to provide that protection.

Against accidental events, probability and statistics are commonly used with great effect. We can generally categorize accidental events, determine probabilities of those events based on historical data, associate loss figures with those events, and generate expected loss figures for each event. We can then evaluate techniques for reducing the probability of loss when these events occur, associate costs with those techniques, and determine which techniques are cost effective. This is commonly called *risk analysis* and leads directly to a cost-effective implementation of protection. Unfortunately, this whole scheme falls apart when faced with malicious attacks.

Under malicious attack, most of the assumptions about the probability of events are invalid because attackers are adaptive. When they find a point of attack, they exploit it until it is repaired. Thus, the probability of attacking in a particular way is nearly 0 until an attack works, at which time the probability goes to very nearly 1.

In this view, any inadequately covered exposure will eventually be found and exploited to the maximum potential for loss. The question of assurance then goes to the question of what is adequate coverage for an exposure. One way to consider this is to estimate how much an opponent is willing to spend to attack, determine the worst case attacks for that dollar amount, and provide coverage such that the required resources for a successful attack exceed the estimated resources available for that attack.

An Organizational Perspective

Organizations must deal with the information protection issue from many perspectives. Since there is no universally accepted set of protection perspectives, many organizations develop their strategies in a piecemeal fashion. One perspective that has proven useful over a long time is described here.

Protection Management

Protection management is the management part of the process by which protection is implemented and maintained. In order for protection to work, adequate resources must be applied in an appropriate fashion. Protection management works toward optimizing the use of resources to make protection as effective as possible at or below budgeted cost levels. In most large commercial organizations today, the highest level information protection management personnel report directly to corporate officers. Protection management typically takes place at all levels of the organization. It is fairly common for lower level technical information protection management to be carried out by relatively untrained systems and network administrators who also have nonprotection roles in maintaining the operation of information systems. Top-level information protection managers are usually dedicated to the information protection task.

Protection Policy

Protection policy forms the basis on which protection decisions are made. Typical policy areas include, but are not limited to,

- guiding principles and motivations behind corporate protection policy
- specific requirements for integrity, availability, and confidentiality of various sorts of information and the basis for differentiating them
- policies regarding hiring, performance, and firing of information workers
- statements of the responsibility of all parties

Protection policy is normally the responsibility of the board of directors and operating officers of a company and has the same stature as any other official statement of corporate policy.

Protection policy is the basis for considering integrity, availability, and privacy. The reason policy is so important is that it is impossible to achieve a goal unless you know what the goal is.

The first and most vital step in achieving protection in any environment is to define the protection policy; this is something that should be done very carefully. For example, suppose we just use "I want to get the right information to the right place at the right time" as the policy. If this is the only guiding force behind decisions, we may be put out of business very quickly and we will never achieve our goal, which should be a balance between cost and protection.

Standards and Procedures

Standards and procedures are used to implement policy at an operational level. It is not unusual to have employees who are unaware of or ignore standards, and procedural compliance is rarely adequate to implement the desired protection effectively. In assessing standards and procedures, it is important not only to determine that they are adequate to meet operational requirements, but that they are adequate in light of the level of compliance, training, and expertise of the people implementing them. Standards and procedures apply at all levels of the personnel structure, from the janitor to the board members.

A typical example of an information protection standard is the International Organization for Standardization's (ISO's) Open System Interconnection (OSI) standard for security, which is an addendum to the normal OSI standard document. The fact that protection is considered an addendum is a bad sign since it is commonly recognized that protection cannot be added cost effectively once a system is designed. An even larger concern is the lack of enough detail in the standard to make systems designed to meet the standard compatible. On the opposite end of the spectrum, the Data Encryption Standard (DES) is so well specified that any two properly implemented versions are guaranteed to work together. This standard even includes a test suite for verifying proper operation.

Procedures, except in military and paramilitary organizations, are devised locally and implemented unevenly. A typical procedural safeguard is always to keep the doors to telephone rooms and wire closets locked. It is fairly common to leave wire closets unlocked; this can easily be exploited to tap into communications, destroy physical connections, and corrupt information flowing through the wire closet.

Documentation

Documentation is the expression of policy, standards, and procedures in a usable form. The utility of documentation is a key component of the success of protection. Protection documentation is intended to be used by specific people performing specific job functions and as such should be appropriately tailored. Documentation should be periodically updated to reflect the changing information environment; these changes should be reviewed, as they occur, by all involved parties to assure relevance. Documentation is normally in place at every level of a corporation, from the document that expresses the corporate protection policy to the help cards that tell information workers how to respond to protection-related situations.

A typical document that every user should have at hand whenever using an information system tells how to respond to different emergency conditions. Such a document should be very simple (sixth-grade level or lower) to read and understand, should give step-by-step instructions on what to do in each of the situations addressed, should have easy-to-use tabs to get the user to the right procedure fast, and should be kept up to date at all times. In its simplest form, this consists of a telephone number to call for assistance.

Protection Audit

An audit is vital to assuring that protection is properly in place and detecting incidents not detected by other operating protection techniques. An audit is also important to fulfilling the fiduciary duty of corporate officers for due diligence, detecting unauthorized behavior by authorized individuals, and assuring that other protective measures are properly operating. An audit is normally carried out by internal auditors and verified by independent, outside personnel with special knowledge in the fields of inquiry, who work for and report to corporate officers and have unlimited access to examine, but not modify, information.

A typical audit function is verifying that all user identities authorized for use of an information system are valid and that changes in authorizations are up to date. It is common for internal auditors to verify this sort of information for all systems on at least an annual basis, and for external auditors to select systems at random to verify that the internal auditor is doing a proper job.

Technical Safeguards

Technical safeguards are commonly used to provide a high degree of compliance in addressing specific, known classes of vulnerabilities. Technical safeguards must not only meet the requirements of addressing the vulnerabilities they are intended to mitigate, but must also be properly implemented, installed, operated, and maintained. They are also subject to abuse in cases for which they are inadequate to the actual threats or create undue burdens on the users. Technical safeguards are typically implemented by systems administrators based on guidance from protection managers.

There are literally tens of thousands of information protection products on the market today that implement different sorts of technical safeguards. A typical example is a hardware device placed on a personal computer's printer port to verify that a software program is properly authorized before allowing it to be used. Other examples include the use of a Faraday box to reduce or eliminate certain electromagnetic emanations, audit trails and automated analysis to detect intrusion, operating system features to prevent unauthorized modification, hardware devices to prevent use of the keyboard when no attendant is present, and cryptographic mechanisms to conceal the meaning of transmitted information.

Incident Response

Incident response is required whenever a protection-related incident is detected. The process of response is predicated on detection of a protection-related event, and thus detection is a key element in any response plan. The response plan should include all necessary people, procedures, and tools required in order to limit the incident effectively and mitigate any harm done to as large an extent as possible and appropriate to the incident. In many situations, time is of the essence in incidence response; therefore all of the elements of the response should be in place and operating properly before an incident occurs. This makes planning and testing very important. Incident response is normally managed by specially trained central response teams with a local presence and cooperation of all users on which there has been an impact.

Incident response is commonly handled by a special incident response team. These teams typically consist of a local systems administrator and several organizationwide experts. These people are typically involved in organizational protection decisions and have training and education responsibilities as well as those for incident response.

Testing

Any system that is expected to work properly must be adequately tested. The testing requirement applies to human as well as automated systems and to the protection plan itself. It is common to implement new or modified computer hardware or software without adequately testing the interaction of new systems with existing systems; this often leads to downtime and corruptions. Similarly, disaster recovery plans are often untested until an actual disaster, at which point it is too late to improve the plan to compensate for inadequacies. Testing is normally carried out by those who have operational responsibility for functional areas.

A regular program of testing is normally required just to keep systems operating and to assure that new software does not have an adverse impact on operations. It is very common to have a laboratory in which all configurations are tested before purchases are completed or products are put into use.

Physical Protection

There is no effective information protection without physical protection of information assets and the systems that store and handle those assets. Physical protection is expensive and thus must be applied judiciously to remain cost effective. Physical protection concentrates on finding and applying the most cost-effective physical protection given the assets to be protected and the systems used to store and handle those assets. There is a strong interaction between the design of systems and the physical protection requirements. Physical protection is typically implemented and maintained in conjunction with operations personnel who are responsible for other aspects of the physical plant and sometimes involves an internal security force of one form or another.

A typical physical security situation involves alarms, a central alarm response facility, and a set of roving guards. The guards check in at specified times and cover the entire area under guard on a regular basis. Alarms typically result in live responses that take guards off their normal search patterns. A well-designed system will not allow an enemy to predict behavior in a consistent manner because, once behavior can be predicted, the reflexive response can be exploited in attack.

Personnel Issues

Information systems are tools used by people. At the heart of effective protection is a team of trustworthy, diligent people. Although there are no sure indicators of what people will do, individuals who have responsibilities involving unusually high exposures are commonly checked for known criminal behavior, financial stability, susceptibility to bribery, and other factors that may tend to lead to inappropriate behavior. Although most people provide references when submitting a resumé, many companies do not thoroughly check references or consider the impacts of job changes over time. Personnel security considers these issues and tries to address them to reduce the likelihood of incidents involving intentional abuse. In most corporations, there is a personnel department that is supposed to handle these issues, but it is common to have communications breakdowns between personnel and technical protection management that result in poor procedural safeguards and unnecessary exposures.

Legal Considerations

Legal requirements for information protection vary from state to state and country to country. For example, British law is very explicit in stating the responsibility to report computer crimes, while U.S. laws do not punish executives who fail to report incidents. Similarly, software piracy laws, privacy laws, Federal Trade Commission regulations, recent federal statutes, contracts with other businesses, health and safety regulations, worker monitoring laws, intellectual property laws, and many other factors have an impact on the proper implementation of information protection. Legal matters are normally handled in conjunction with corporate legal staff and involve all levels of the organization.

A typical legal problem is the lack of adequate warning on entry (3). In many states, there are specific statutes that prohibit civil penalties for illegal entry into information systems in which adequate warning was not provided. Similarly, worker monitoring without adequate notice commonly leads to law suits. These issues are often complex and poorly defined, and there is not yet enough case history to establish clearly what is legal and illegal in the information industry.

Protection Awareness

Protection awareness is often cited as the best indicator of how effective an information protection program is. Despite the many technological breakthroughs in information protection over the last several years, it is still alert and aware employees that first detect most protection problems. This is especially important for systems administrators and people in key positions.

A typical example of awareness causing detection is the case documented by Cliff Stoll in his book, *The Cuckoo's Egg* (15). In this case, an international spy had been entering computer installations throughout the United States and was only detected when Stoll noticed a 75-cent accounting difference between his accounting program and the system's accounting program. Upon investigation, he found that someone had been using his computer to launch attacks against other computers, called in the Federal Bureau of Investigation (FBI) (who laughed at him because of the small financial damage to his system), and eventually helped track the attacker.

Training and Education

Training has consistently been demonstrated to have a positive impact on performance, especially under emergency conditions. For that reason, training programs are a vital component in the overall protection posture of a company. It is common for people who once helped the systems administrator by doing backups to become the new systems administrator based on attrition, and they end up in charge of information protection by accident. The net effect is an environment that relies on undertrained and ill-prepared people who are often unable to cope adequately with situations as they arise. The need for training increases with the amount of responsibility for protection. Every employee should have an exposure level of at least a few hours per quarter, while those who are responsible for protection should attend regular and rigorous professional educational programs.

Training programs should include a few hours per quarter for each employee to keep awareness high, to point out recent incidents and how employee responses have helped or hindered in their resolution, and to provide a venue for discussing policy issues. Recent events are commonly used, and many companies use these opportunities to bring in outside speakers who can give a fresh perspective to an interesting topic receiving coverage in the media.

Educational programs are undertaken either in-house or through a short course firm. The decision to use in-house education is generally dictated by cost factors, and as a rule, it is less expensive to educate 3 or more employees in-house than to send them to a short course in another city. This figure is heavily affected by air fares and hotel bills. As an alternative to short courses, conferences offer a cost effective way to get broad exposure for key employees.

Organizational Suitability

Information protection spans the corporate community. It crosses departmental and hierarchical levels and has an impact on every aspect of corporate operations. This implies that the corporate information protection manager should be able to communicate and work well with people throughout the corporation. To be effective, the mandate for information protection must come from the board of directors and operating officers.

Most organizations with substantial information systems groups are reasonably well suited for implementing protection, while most organizations with small and overworked information systems groups are poorly suited for implementing protection. The reason is that information protection involves a broad range of organizational activities that are partially in place in most large information systems groups and rarely in place otherwise. The expertise required for an electronic data processing (EDP) audit, for example, is rarely available in organizations without large professional information systems staffs.

Another issue in organizational suitability that often comes up is the general attitude of the people in the organization. In a carefree sort of organization with little emphasis on intellectual property, very casual work rules, and relatively unsophisticated systems, it may be a big step to provide substantial information protection. On the other hand, a highly disciplined and structured organization with stringent work rules and highly sophisticated systems tends to lend itself to highly coordinated protection efforts. A major key to attaining proper information protection is devising a plan that is well suited to the organization.

Other Perspectives

There are clearly other ways of organizing protection issues. The one above is not particularly better than any others except that it seems to provide broad coverage from a perspective that has proven useful to many organizations.

Analytical Techniques

There are many analytical techniques available for trying to cost effectively match protection with desired assurance levels in a particular environment. The most popular techniques are outlined here with some simple examples.

Risk Analysis

Risk analysis is commonly used to find cost-effective measures to mitigate probabilistic risks. The basic technique assumes that we can associate probabilities with attacks per unit time $P(a,t)$, expected losses per attack $L(a)$, reductions in probabilities through the use of defenses $dP(a,t)(d)$, and costs per unit time with defenses $C(d,t)$. We then compute an overall expected loss per unit time as

$$\sum_{\forall a} P(a,t) \times L(a)$$

For each defense, we assess a cost and reduction in successful attack probability and then compute a cost effectiveness of loss reduction for each defense as

$$CE(d) = frac(P(a,t) \times dP(a,t)(d) \times L(a)) - C(d,t)C(d,t)$$

We start implementing protection with the most cost-effective defensive technique — the one with the highest value for $CE(d)$ — and implement each protective technique with a positive $CE(d)$ in order.

There are some rather obvious problems with this technique. The first one is that assessing $P(a,t)$, $L(a)$, $dP(a,t)(d)$, and $C(d,t)$ is not trivial. For example, $P(a,t)$ assumes that attacks can be categorized with probability distributions as if they were random events. $L(a)$ assumes that we can assess the value of different sorts of losses that can result from different sorts of attacks and meaningfully quantify them with a single number. Also, $dP(a,t)(d)$ assumes we can determine the reduction in attack probability given a particular sort of defense, which is highly speculative. $C(d,t)$ is not easy to compute even in limited circumstances because it involves a thorough understanding of life-cycle costs in the particular environment.

The second major problem with this technique is that it inherently assumes that attacks and defenses are independent statistical phenomena when they are in fact highly correlated. When a particular sort of attack starts to take place, the likelihood of further attacks greatly increases in today's environment. There are synergistic effects of attack and defense that make the independence assumption inaccurate. Furthermore, after we implement one defense, $CE(d)$ changes for all of the defenses because the reduction in expected loss from the first technique may have an impact on the other techniques' $dP(a,t)(d)$.

The third major problem with this technique is that it ignores the threshold issue of acceptable losses. At some point, regardless of our ability to cost effectively reduce losses, they reach a low enough level to simply become unimportant. On the opposite end of the spectrum, a very low probability event may cause catastrophic losses. If we do not cover it and we bet wrong, the loss may mean the end of the organization, while the cost of defense may be in a range that warrants its use as a defense against this loss regardless of the value of $CE(d)$.

Simple Covering Analysis

An analytical technique has recently been developed based on the covering table methods commonly used in computer engineering. The basic idea is to find a minimum cost set of techniques to cover a set of identified exposures. In its simplest form, the covering analysis starts with a two-dimensional table in which one dimension lists exposures and the other dimension lists defenses. The table is filled with marks to indicate which defense covers which set of exposures.

For example, the use of cryptographic checksums in an integrity shell to detect change might cover all computer viruses, outsider modifications of programs, and the introduction of new Trojan horse programs into the execution path. This technique does not prevent the leakage of secrets. We would then have one column (integrity shell) with check marks in each of three rows (viruses, outsider modifications, and new Trojan horses). If access controls could be used to limit all but viruses, we would put in similar check marks, leaving the following table:

	Integrity Shell	Access Control	Other	Other 2	Other 3
Viruses	✔				
Outsider modifications	✔	✔			
New Trojan	✔	✔	✔		
Leak secrets		✔			
Other attack				✔	✔

In this case, in order to cover all of the exposures, we need both access controls and integrity shells, but in a more complex table, there are systematic rules for finding the lowest cost cover. The rules are

1. If only one check "covers" any given row, the technique in that column is required in order to cover the corresponding vulnerability. Select the column with the check as "necessary." Since all rows covered by the necessary column are now covered, remove those rows from further consideration. In the example above, integrity shell is necessary because it is the only defense that covers viruses and access control is necessary because it is the only defense that covers leak secrets. Since this technique also covers other exposures, all of these rows are removed.
2. If there are columns left without a check, remove them from the matrix. In this example, the "Other" column is removed since it provides no required coverage. We are now left only with tradeoffs between competing techniques.
3. For each remaining technique, if any one technique both costs less than or the same as another technique and covers everything the other technique

covers, remove the more expensive technique. This selects in favor of techniques with equal or superior coverage and lower or equal costs.

4. For each remaining technique, select that technique, act as if it were necessary under the criterion above, and proceed through these steps until there is nothing left to cover. This process will generate a set of possible minimum cost covers.

5. Add the costs of the list of items selected for each possible minimal cost cover and select the cover with the lowest total cost. This will produce the lowest cost cover.

Full-Blown Covering Analysis

A very complex extension of this simple covering analysis is a multidimensional covering analysis in which we take all of the dimensions discussed above into account and try to reach an optimal solution for a particular situation. To get a sense of the complexity of this problem, there are at least 12 dimensions we have identified here, and each dimension can take on numerous values. Even a system in which we have only 10 different possible values for each of 12 dimensions leads us to 10^{12} different table locations to fill before the analysis can begin. In general, the number of covers we might have to analyze is 2^n, where n is the number of defensive techniques under consideration.

In one recent study in which a simple covering analysis was done for a small local-area network, over 30 defensive techniques were analyzed, for a total of about 10^9 possible covers. Fortunately, the covering algorithm dramatically reduced the complexity of the problem, and it was solved by hand in under an hour.

Although some algorithms may reduce the problem of computing an optimal cover in most cases, nobody has yet done a full-blown covering analysis, and no thorough analysis of the implications of such an effort have been made.

Even in such a thorough analysis as this, other issues have to be considered. A sensitivity analysis is usually appropriate to determine the sensitivity of the solution to minor changes in parameters. This is important because, in many cases, even a slight change in a few table values may result in a completely different optimal cover with dramatically different cost implications. Protection is normally implemented over a life-cycle, and the changes in modern information systems over their life-cycles tend to be quite dramatic. Meanwhile, defenses and costs of information technology change very quickly.

Experience

The best technique available today for overall protection analysis seems to be a joint effort by a human expert and the people in the organization requiring protection (16). Although the best people use a variety of tools to aid them in their decision process, the fitting of protection to an environment is essentially a negotiation process between the protection expert and the people in charge of protection at the organization being protected.

Strategy and Tactics

There are two critical planning perspectives for information protection: strategic planning and tactical planning. The difference between strategy and tactics is commonly described in terms of time frames. Strategic planning is planning for the long run, while tactical planning is planning for the short run. Strategic planning concentrates on determining what resources to have available and what goals we should try to achieve under different circumstances, while tactical planning concentrates on how to apply the available resources to achieve those goals in a particular circumstance.

In the planning process, so many things should be considered that we cannot even list them all here. They tend to vary from organization to organization and person to person, and they involve too many variables to draw general conclusions without sufficient facts. Some of the major issues in planning are listed here, but this cannot possibly substitute for analysis by experts with both technical and organizational knowledge of the actual requirements.

General Principles

First and foremost, planning a defense is a study in tradeoffs. No single defense is safest for all situations, and no combination of defenses is cost effective in all environments. This underscores a basic protection principle: Protection is something you do, not something you buy.

As an example, suppose we want to protect our house from water damage. It does not matter how good a roof we buy for our house, it is not going to protect our house forever. We have to maintain the roof to keep the water out. It is the same with protecting information systems: you cannot just buy it, you have to do it.

General Strategic Needs

At the core of any strategy is the team of people that develop it and carry it out, both in the long run and in tactical situations. The first and most important thing you should do is gather a good team of people to help develop and implement strategy.

Any strategy that is going to work requires resources, both over the long run and during tactical situations. We commonly see strategies that fail because of insufficient resources, resources poorly applied, a lack of consideration of the difference between strategic and tactical needs, and insufficient attention to detail.

In order to assure that strategies will work in times of need, it is necessary to test tactical components ahead of time. The most common problem in tactical situations is events that were not anticipated during strategic planning and not discovered during testing. The effect is almost always a very substantial added expense.

General Tactical Needs

Efficient response in tactical situations normally calls for the complete attention of some number of trained experts carrying out well-defined roles. In addition, it commonly requires that a substantial number of nonexperts act in concert at the direction of those trained experts.

Tactical situations require sufficient resources on hand to deal with the situation. If sufficient resources are not present, the cost of obtaining those resources on an emergency basis tends to be very high, the delays caused by the acquisition process may be even more costly, and the inexperience with the newly obtained resources may cause further problems.

A small amount of resources properly applied almost always beats a large amount of resources poorly applied. You do not necessarily need to spend a lot to have a strong defense, but you have to spend wisely and react quickly.

Some Widely Applicable Results

Even though there is a vast array of different environments, there are some strategies and tactics that seem to be almost universally beneficial and widely applicable.

- *Consider highest exposures first.* In planning a strategy for defense, it is important to consider the potential harm when assessing what to do. Specifically, the highest exposures should receive the most attention and should thus be addressed with the highest priority.
- *Use the strongest defenses feasible.* All other things being equal, it is better to use a stronger defense than a weaker one. As a rule, sound prevention is better than detection and cure, and general-purpose detection and cure is better than special-purpose detection and cure, but a false sense of security is worse than any of them.
- *Proactive defense is critical.* "Be prepared" is the Boy Scout motto, and it applies in information protection. People who take proactive action get away with lower cost, smaller incidents, and faster recovery than those who simply wait for a problem and then react.
- *Rapid central reporting and response works.* Epidemiological results are clear in showing that rapid centralized reporting and response reduces incident size, incident duration, and organizational impact (17).
- *Incident response teams work.* You need a team that receives incident reports and responds to them on an emergency basis. This team is virtually always the same team that helps determine proper proactive strategy, implements tactical response, and performs analysis after incidents to improve future performance.
- *Keep good records and analyze them.* Record keeping is one of the most valuable aids to improving tactical and strategic response. When properly analyzed, records of incidents allow planners to devise more cost-effective tactical responses, which in turn provide more accurate information for strategic planning to reduce costs while improving response.

- *Do not punish victims.* One of the most common strategic mistakes is punishing those who report incidents or accidentally become a vector for their spread. Punishment may be direct or indirect and may include such subtleties as requiring the victim to do extra paperwork or such abusive treatment as getting a lecture on following written policies. When you punish victims, you increase response time because others become hesitant to report incidents. Statistics support the improved response of organizations with policies not to punish.
- *Procedural policies fail in prevention.* Many organizations make the mistake of relying on procedure as a preventive defense. The fact is, people are not perfect. When they make mistakes, assessing blame is not an effective response. You should plan on procedural failures and plan a response that considers people as they really are.
- *Training and education works.* You have to train your users to use your tactical capabilities or they will be ineffective in live situations, and you have to educate your experts in order to develop an effective plan and carry out the plan effectively in emergencies.
- *All your eggs in one basket.* Diversity works. Given a choice of buying 100 copies of either of two nearly equivalent products, buy 50 of each and haggle to keep the prices low. When one fails, the other may succeed, and epidemiology shows that covering more attacks is far more important than covering every machine identically.
- *Defense in depth works.* The synergistic effects of layering multiple defenses make defense in depth the most effective solution to date. The synergistic effect is analogous to the advantage of the moated wall. A wall can be battered with a ram and a moat can be crossed by boats or swimmers, but trying to use a battering ram against a stone wall from a boat or while swimming is far more difficult.

Summary and Conclusions

Information protection is a very complex and vital problem. The degree of dependency on information technology is high, while protection in today's environment is generally poor. Understanding protection takes a lot of time and effort, and there are many pitfalls for the inexperienced practitioner. The large number of different causes, effects, and tradeoffs, in conjunction with the nonstatistical nature of malicious attackers, makes most analytical techniques fail. Protection has an impact on and involves all aspects of an organization and, in order to be effective, the entire organization has to work in concert against attacks. Optimizing cost and effectiveness is a very complex problem and, despite the availability of analytical techniques, human experts are still the best way to find appropriate defenses. Strategy and tactics are vital to the success of a protection program, and the people who are involved in this planning and action are critical to the success of the protection program.

Bibliography

Schneier, B., *Applied Cryptography: Protocols, Algorithms, and Source Code in C*, John Wiley and Sons, New York, 1994.

> A tutorial overview of modern cryptography and its solutions for digital age problems. The author points out pitfalls not obvious to the inexperienced and provides a very extensive bibliography.

Cohen, F., *A Short Course on Computer Viruses — Second Edition*, John Wiley and Sons, New York, 1994.

> A broad and authoritative analysis of important implications of computer viruses and defenses for them, including networks. Along with a formal definition and mathematically supported conclusions, there are numerous explanations of experimental and practical results.

Denning, D. E., *Cryptography and Data Security*, Addison-Wesley, Reading, MA, 1982.

> A graduate text covering many areas of computer security, including information flow and database inference. The author's concentration is on confidentiality of data rather than integrity.

National Research Council, *Computers at Risk: Safe Computing in the Information Age*, National Academy Press, Washington, DC, 1991.

> This study on the state of current computer security and expectations for the future contains a number of examples and recommendations at a broad level for information system security.

Longley, D., and Shain, M., *Data and Computer Security*, Macmillan, New York, 1987.

> This dictionary of standards, concepts, and terms is extensively indexed and cross-referenced. It is useful when acquiring background in this area and includes many terms from U.S. government and military sources.

Information Systems Security Association, *Guideline for Information Valuation*, 1993.

> This is a committee consensus on procedures for valuation of data. It is intended to result in defensible values that include consideration of the costs of replacement, revenue stream generated, denial of use, and other costs for which monetary values can be assigned.

Rivest, R., Shamir, A., and Adleman, L., *On Digital Signatures and Public-Key Cryptosystems*, MIT Laboratory for Computer Science, Technical Report, MIT/LCS/TR-212, Boston, MA, January, 1979.

> This is the paper describing the most widely used public key cryptosystem, and one considered to be extremely strong with appropriate key lengths. This system has been endorsed by most major computer companies and is available in their commercial products.

Landwehr, C. E., The Best Available Technologies for Computer Security, *IEEE Computer* (July 1983).

A summary of computer security techniques, including those used in military computer information systems. Since systematic research on computer viruses was not published until later, it does not include defenses and implications.

Landreth, B., *Out of the Inner Circle*, 2d Ed., Tempus Books of Microsoft Press, Redmond, WA, 1989.

This book clearly describes techniques used to break into computers and illustrates the gap between what system defenders understand and the level of sophistication of a dedicated amateur attacker. It describes "hacker" motivations and social engineering.

Denning, P. J., (ed.), *Computers Under Attack: Intruders, Worms, and Viruses*, ACM Press, New York, 1990.

This is a collection of almost 40 papers covering computer virus and other computer security issues and attacks. Methods, defenses, hacker debates, and social and legal issues are subjects of examination.

References

1. Neumann, P. G., The Computer-Related Risk of the Year: Weak Links and Correlated Events, COMPASS '91, IEEE 6th Annual Conference on Computer Assurance, 1993.
2. Cohen, F., *A Short Course on Computer Viruses*, 2d ed., John Wiley and Sons, New York, 1994.
3. National Research Council, *Computers at Risk: Safe Computing in the Information Age*, National Academy Press, Washington, DC, 1991.
4. Minoli, D., Cost Implications of Survivability of Terrestrial Networks Under Malicious Failure, *IEEE Trans. Commun.*, V28#9: 1668–1674 (September 1980).
5. Jobusch, D. L., and Oldehoeft, A. E., A Survey of Password Mechanisms: Weaknesses and Potential Improvements, *Computers and Security*, 8(8) (1989).
6. Gray, J., and Siewiorek, D. P., High-Availability Computer Systems, *IEEE Computer* (September 1991).
7. Siewiorek, D. P., Architecture of Fault-Tolerant Computers: An Historical Perspective, *Proc. IEEE* (December, 1991).
8. Sullivan, T., Curious Reporters Tap into Harding's E-Mail in Olympic Computer, *USA Today*, February 25, 1994, final edition.
9. Campbell, D. E., The Intelligent Threat, *Security Management* Special Section, 19A–22A (March 1989).
10. Sterling, B., *The Hacker Crackdown*, Bantam, 1992.
11. Fortier, S. C., and Michael, J. B., A Risk-Based Approach to Cost-Benefit Analysis of Software Safety Activities, COMPASS '93, IEEE 8th Annual Conference on Computer Assurance, 58 (1993).
12. Neumann, P., Mariner I—No Holds Barred, *RISKS-LIST: RISKS-FORUM Digest*, May 30, 1989, 8(Issue 75).
13. Taff, A., BOCs Divulge Genesis of CCS7 Crashes, *Network World*, July 15, 1991.
14. Information Systems Security Association, *Guideline for Information Valuation*, 1993.
15. Stoll, C., *The Cuckoo's Egg*, Doubleday, New York, 1989.

16. Filsinger, J., and Heaney, J. E., Identifying Generic Safety Requirements, COM-PASS '93, IEEE 8th Annual Conference on Computer Assurance, 41 (1993).
17. Kephart, J., White, S., and Chess, D., Computers and Epidemiology, *IEEE Spectrum*, 20–26 (May 1993).

FREDERICK B. COHEN
CHARLES M. PRESTON

Network Protocol and Performance Measurements for Packet Switching

Introduction

The main idea behind packet switching is to create a network of links and interconnecting switching nodes with the function of providing communications services. At the source, the stream of information is divided into small digital units, called *packets*, each consisting of several fields. The main field contains the user's actual information. The other parts are the header field, which contains such information as a packet's source and destination addresses, its priority and its error codes, and flags for synchronization. In general, the key functions of a packet-switched network are (1):

- *Routing*: Since the source and destination station are usually not directly connected, the network must route each packet, from node to node, through the network.
- *Flow or traffic control*: The amount of traffic entered and transmitted through the network must be regulated for efficient, stable operation and acceptable performance.
- *Error control*: Inevitably, packets may be lost in the network. Some networks ignore this contingency; most take measures to at least partially alleviate the problem at the attached stations.

Some of these aspects are not usually mentioned in connection with a circuit-switched network. It is perhaps an advantage that packet switching combines these features into the network services. At least, it makes the communications network responsible for error control, though it does not take the responsibility entirely away from the network's end user.

The packet-switching technology uses transmission media to carry packets from multiple users, in a statistically multiplexed fashion, through the network via a series of switching nodes. Advances in technology have lowered the unit cost of transmitting packets. The great advantage of statistical multiplexing is to combine bursty data traffic from many sources into aggregate flows that can be accommodated economically over the transmission facilities.

For many possible reasons (e.g., errors, buffer overflow at switching nodes, time-outs, etc.), the packets may arrive at the destination out of order, or duplicated, or lost. So, the network must be designed such that it can handle these eventualities in an acceptable way for the end users or applications. The general characteristics of packet-switching networks as initially viewed for a new technology can be summarized as follows (2):

- Random delay
- Random throughput

- Out-of-order packet arrivals at the destination
- Lost and duplicated packets
- Nodal (switch) storage
- Speed matching between network and attached systems

In order to respond to these characteristics, the packet-switching network must provide many of the following functions (3):

- Packetizing
- Buffering
- Pipelining
- Routing procedures
- Sequencing and numbering
- Error control (noise, duplicate, and lost packet detection)
- Storage (resource) allocation
- Flow control

Packet-switching technology was first introduced around 1970 in such networks as the Advanced Research Projects Agency Network (ARPANET) of the U.S. Department of Defense (4). Once the technology advanced, it was widely implemented in telephone networks for the common channel signaling (CCS) system and currently is implemented as part of Integrated Services Digital Networks (ISDNs).

Today's larger networks have been designed to optimally handle a vast amount of digital 64-kilobits-per-second (kb/s) voice calls. As such, these networks use circuit switching in which a dedicated path is usually allocated between source and destination for the duration of call connection. Such dedicated voice networks have rather homogeneous traffic with fairly known characteristics. Contrast this with the traffic mix expected for the future communications networks: an unpredictable mixture of voice, data, image, and video communications. In the near term, networks are well served with the fundamental building blocks of 64-kb/s digital bearer channels, whereas a more flexible paradigm of communication will be needed to handle the future network traffic.

Many studies have focused on the characteristics of the future integrated networks (5). A major conclusion, so far, has been that most integrated services may best be served with some type of packet switching that utilizes "fixed packets" called *cells*. Cell switching, which naturally led to a large study of the asynchronous transfer mode (ATM), in contrast to the circuit switching, which may be considered a type of the synchronous transfer mode (STM), is an outgrowth of experience with packet switching and variations of it such as frame relay and the like.

In order to understand and appreciate the tools used for network modeling and performance analysis, it is helpful to understand, in general terms, the principals of data communications using packet switching. We summarize these principals by describing the International Organization for Standardization/Open System Interconnection (ISO/OSI) protocols, some examples of their implementations, and networks actually utilizing packet switching. Then, we embark on the notion of modeling and performance analysis of packet switching.

Packet-Switched Networks

Communications networks fall into three very general categories:

1. Circuit (or line) switching
2. Message switching
3. Packet switching

A circuit-switching network provides a dedicated channel to be set up from the source to destination of the call, namely, dedicating an actual end-to-end physical path to the users at either end. In message switching, only one channel is used at a time for a given message transmission. The message first travels from its source node to the next node on its path and when the entire message is received at this node, then the next step (which is an outgoing channel for transmission to the next node) is selected; if this selected channel is busy, the message waits in a queue and, finally, when the channel becomes available, transmission begins. Thus, the message "hops" from node to node through the network using only one channel at a time (3). Packet switching is basically the same as message switching except that messages are decomposed into smaller units called *packets*, each of which may have a maximum length. These packets are usually numbered and addressed and make their way through the network in a packet-switched mechanism of store-and-forward fashion.

Packet switching can be more economical than circuit switching in its use of transmission lines because it interleaves packets from different streams according to their demand for channel capacity. This economy is offset by the overhead of packet headers and control packets of various kinds. It produces an environment in which the use of circuit switching would be intermittent or sparse — the case with much computer communications and most kinds of terminal interactions.

Generally speaking, digital data transmission lends itself very nicely to message and packet switching. Given the diversity of user applications and the myriad approaches to efficient resource sharing by application of distributed telecommunications networks, it is important to understand the concept of packet-switched networks. Packet switching has emerged as a telecommunications technique with unlimited potential for many applications. Public packet-switching networks have been built in most countries around the world, and numerous private and experimental networks are currently planning to migrate toward the high-speed, ATM-based, cell-switching technology. As we discuss below, the ATM transport mechanism has matured to become one of the most promising approaches for implementing Integrated Services Digital Networks.

There are numerous advantages with regard to packet-switching technology. For example, since packets are stored as they pass through switching nodes, it is possible to conduct speed, format, and code conversion during the switching process. This, however, is not possible with circuit switching, which requires complete end-to-end compatibility and no storage at the intermediate nodes in this regard. The use of a circuit-switched connection only makes good sense if there is a need for transmitting a long, continuous stream of data with relatively constant bandwidth. On the other hand, if the data flow is bursty in nature, then some form of resource sharing can be used to great advantage; packet switching is an effective choice here. Another key feature of packet switching is its ability to adaptively select good paths for packet transmission as a function of the network congestion. This is referred to as *bandwidth allocation* or *bandwidth on demand*. Clearly, major disadvantages of circuit switching are the difficulties in tailoring the bandwidth allocated to the instantaneous requirements of a call and the ensuing poor utilization of bandwidth for some types of applications.

Generally, packet switching can easily adapt to a wide range of user services and user demands because it permits communications resources to be used very efficiently. As data-processing and networking technologies advance the computer processors that form the heart of packet switches, packet switching will find an even wider range of applications.

Performance Measures

The emphasis of performance parameters of packet switching networks is mainly defined under the following network measures: delay, throughput, error, and blocking, discussed below. Cost and reliability are other important network performance measures and are not discussed here. The reliability issue is discussed only in passing. Delay and throughput measures are closely related, but usually are applied as performance criteria to different kinds of traffic. In particular, interactive traffic must be delivered quickly and is usually rather short, in which case throughput is not a central issue. On the other hand, a long file transfer is not so much concerned with the initial delay in getting the first few bits across the network, but rather with how many bits per second can be pumped through the network simultaneously.

Delay

One of the most important measures of performance in a packet-switched network is the average delay (and, in some cases, delay distribution) required to deliver a packet from a source to a destination. Furthermore, delay considerations strongly influence the choice of performance of network algorithms such as routing, flow control, and congestion control. For these reasons, it is important to understand the nature and mechanism of delay and the manner in which it depends on the characteristics of the packet-switched network.

The overall end-to-end delay within packet-switched communications media consists of four components (6):

1. *Processing delay*: The delay between the time the packet is received at a switch and the time the packet is assigned to an outgoing link queue for transmission.
2. *Queueing delay*: The delay between the time the packet is assigned to a queue for transmission and the time it starts being transmitted.
3. *Transmission delay*: The delay between the transmission of the first and last bits of a packet.
4. *Propagation delay*: The delay from the time the last bit is transmitted at the source switch until the time it is received at the destination switch. Generally, this is proportional to the physical distance between transmitter and receiver and is ordinarily small except in the case of a satellite link. In the case of fast packet switches, this component is much larger than the other components mentioned above.

Due to the possibility that a packet may require retransmission on a link because of transmission errors or various other causes, one needs to incorporate them in a typical delay model. The propagation delay depends on the physical characteristics of the link and is independent of the traffic carried by the link. The processing delay, however, is dependent on the amount of traffic handled by the corresponding node. Generally, in lower speed packet-switching networks, most delay occurs due to the nature of queueing and transmission delays. Furthermore, these delays depend on the overall congestion in the network, which in turn relate to both network access and backbone traffic characteristics.

In a system handling packets of variable length, one finds that it needs only a small proportion of long packets to influence the average delay considerably. Because the messages in queues are handled in sequence, one long packet may impose extra delay on all those, including short ones, that are behind it. It is for this reason that a strict maximum size is generally imposed on packets. A rough idea of the effect of packet length on delay can be obtained by an application of queueing theory (3,7).

As mentioned above, we note that, in the case of a very high speed packet network such as Broadband ISDN (B-ISDN), propagation delay is comparatively larger than the cell/packet transmission delay. In this case, the issue of control is much different from those of low-speed packet-switched networks. We discuss some of these issues below.

Throughput

The *end-to-end throughput* is defined as the total rate of data being transmitted between the source and the destination stations. The throughput analysis is commonly done in terms of the total number of bits transferred, including overhead (header and trailer) bits; the calculations are a bit easier, and this approach isolates performance effects due to the network alone. So, one must work backward from this to determine effective throughput.

The end-to-end throughput behavior of a data communications link is more closely akin to the behavior of the packet-switched network's background traffic. The relationship between offered load and network throughput is well known (7). As the offered load increases to the point of onset of congestion, the throughput increases uniformly with the offered load. As congestion becomes noticeable, the rate of increase of the throughput falls and eventually a maximum is reached. In a poorly controlled network, further increase in applied load produces a reduction in throughput.

For a certain class of traffic (e.g., voice and interactive data with short packets), the delay should be as small as possible consistent with performance objectives, whereas for certain other types of traffic (e.g., file transfer), larger delays may be tolerated. In the first case, throughput is not a central issue, whereas the second case is concerned with how many bits per second can be pumped through the network simultaneously. A part of the total end-to-end delay (e.g., packet processing delay) is generally fixed, while the rest (e.g., queueing delay) may be variable.

Error

When transmitting information over long distances using transmission media (typically wire, microwave, satellite, and optical fiber), the probability of error being introduced during the transmission depends on the medium and can be relatively large in some cases. For instance, errors can be introduced on a transmission medium by electromagnetic radiation, static noise, lightning strikes, physical damage, or human interface. The severity of error could result in losing individual bits, bytes, or blocks of data such as whole packets. Depending on where in a packet, header, or information field an error has occurred, that packet could be misdelivered, dropped, or duplicated, resulting in retransmissions. Thus, mechanisms are necessary to detect and, it is hoped, recover from transmission errors. The same concern applies when transmission is affected through more complex systems, such as packet-switched networks, which may carry out their own internal error control. End-to-end service quality may be considerably improved over that of a non-error-controlled system, but residual errors are often encountered that can be traced to the interface problems, installation changes, operation, management, hardware and software failures, and so on.

Packets transmitted between two network nodes appear as bit strings, sent and received in byte-size blocks. Generally, an error means that a bit that was transmitted as a 0 is received as a 1 or a 1 is received as a 0. Bit integrity may be assumed with a probability of one bit in error in 10^{10} to 10^{11} bits transmitted, which is customarily accepted to be satisfactory. Note that such a bit error rate is achieved using error-detection and -correction procedures in data communications networks (8,9). Most error-detection techniques operate on the following principle: For a given frame of bits, additional bits that constitute an error-detection code are added by the transmitter. This code is calculated as a function of the other transmitted bits. The receiver performs the same calculation and compares the two results. A detected error occurs if, and only if, there is a mismatch.

Generally, these error rates could be made as low as required using any of several error-detection and -correction codes (1). Note that it is possible that packets could be lost or duplicated without being trapped by error-detection codes.

Among various error-detection techniques, a very powerful, but easily implemented, approach known as the cyclic redundancy check (CRC) is usually used in today's digital communications networks. The error-correction codes are rarely used in data transmission since retransmission of lost or dropped packets is generally more efficient. Error-correction codes are used in situations for which retransmission is impractical.

Blocking

In a packet-switched network, the basic transmission resources are link bandwidth and node buffer storage, both fundamental to the operation of a store-and-forward system. Generally, the network facilities can carry offered traffic up to their maximum allowable capacity. However, there are circumstances for which the externally offered load becomes larger than the maximum allowable capacity. Then, if no actions are taken to restrict the entry of traffic into the network, node buffer sizes at bottleneck links may grow indefinitely and eventually exceed the buffer space at the corresponding nodes. When this happens, packets arriving at nodes with no available buffer space will have to be discarded and later retransmitted. Such actions at the nodes are usually referred to as *buffer blocking events*. As a result, a *congestion* phenomenon similar to a highway traffic jam occurs by which, as the offered load increases, the actual network throughput decreases while packet delay becomes excessive. The more severe the degree of congestion is, the greater is the proportion of network resources devoted to retransmission; therefore, the delivered traffic will be reduced. This, further, causes the network protocols to take extra actions in preventing any additional calls to be accepted at the source nodes. Such a task in network design is referred to as a *call-blocking measure.*

So, buffer and call blockings and, more specifically, congestion in packet-switched networks are more a matter of exhaustion of the switch resources. Free flow of traffic is dependent on the existence of a pool of available buffers in each node to receive the forwarded traffic. In the case of a full buffer, packets arrive in a node only to find no buffer available to store them. They are, therefore, depending on the network protocol, either subject to retransmission from the previous node or from the originating node.

Standards of Packet-Network Performance

Extensive work has been carried out by researchers, developers, and standards bodies to develop performance criteria for the existing packet-based and the future cell-based communications networks. In choosing appropriate performance criteria and parameters, consideration has been given to those network connection elements that provide services to a user. Such network performance parameters are defined and measured in terms of parameters that are meaningful to the network providers and are used for the purpose of system design, configuration, operation, and maintenance. Additional attention is given to those performance parameters that are directly experienced by, perceived by, or relevant to the user, irrespective of the networks involved. These attributes of performance criteria should cover both subjective and objective effects on services provided to the users.

Historically, the performance of telecommunications and data communications services has been largely determined by the supplier of those services. This is, however, no longer true in today's communications industry, which has become a much more competitive market; surely, this trend is accelerating. The result of such growth has given customers a variety of alternatives—among services, products, and suppliers—for fulfilling their communications needs. Thus, customers now have the opportunity to determine and purchase the quality of communications services that they need. Note that the quality of services closely reflects the broad array of performance parameters experienced by users.

User-oriented performance considerations for packet networks have been based on two different but related approaches: (1) the grade-of-service (GoS) concept for voice and (2) end-to-end service requirements for data, still image, and video applications. The GoS concept accounts for the subjective effects of network impairments on voice-related services. The service requirement approach accounts for proliferation of different types of packet-switched networks and attempts to derive a set of performance criteria such as (10)

- User oriented: Refers to the user's experience with the network and not to the technical performance of the network
- Complete: Encompasses all of the user's significant performance factors
- Universally applicable: Not restricted to any type of network or any particular network
- Assessable by an experienced user
- Sufficiently sensitive to deterioration in performance to aid the user in applying the corrective measures

Note that both the GoS and service requirement approaches are end to end, top down, and must eventually be mapped to the network-oriented performance parameters that are more applicable to the design, maintenance, and operation of the network components.

The standards bodies such as the American National Standards Institute (ANSI) (11–13) and International Telecommunication Union (ITU) study groups (14,15) have recommended using a performance matrix in assessing the quality of data communications products and services. This matrix was later expanded to broaden the quality of the communications services framework so that it would capture any criteria that customers use in judging the quality of the service they receive (10,16). For meaningful definitions to both the customer and the service provider, this matrix (see Table 1) is designed by having its rows list all communications functions experienced or performed by customers when using a communications service. Also, the columns of the matrix list all quality

TABLE 1 A Quality of Service Defining Matrix

Communications Function	Quality Criteria						
	Speed	Accuracy	Availability	Reliability	Security	Simplicity	Flexibility
Technical sales & planning							
Provisioning							
Technical quality							
*Connection establishment							
*User information transfer							
*Connection release							
Billing							
Network/service management by customer							
Repair							
Technical support							

Source: From Ref. 16.

criteria used by customers to judge the quality with which each row of defined communications functions is executed.

The communications functions include categories ranging from technical sales and planning, provisioning, billing, and technical support to technical quality, which consists of connection establishment, user information transfer, and connection release. The quality criteria cover speed, accuracy, availability, reliability, security, simplicity, and flexibility parameters. So, the main idea here is to identify any and all relevant performance parameters associated with a new or requested service and assign them to appropriate cells of such a matrix. Such a defined matrix for quality could sometimes represent only a subset of overall quality depending on a specific requested service. Furthermore, one should keep in mind that most service attributes visible to customers involve a combination of network and customer equipment. At any event, the use of such a quality framework to a particular service and customer application might require a good understanding of these concepts and terminology.

In Ref. 17, *network performance* (NP) is defined as the ability of a network or network portion to provide the functions related to communications between users. Network performance parameters derived from the above quality of service (QoS) table are divided into primary and derived NP parameters. A *primary* performance parameter refers to a parameter determined on the basis of direct observations of events at service access points or connection element boundaries. A *derived* performance parameter is measured on the basis of observed values of one or more relevant primary performance parameters and decision thresholds for each relevant primary performance parameter.

Table 2 illustrates the relationship between primary and derived performance

TABLE 2   Relationship Between Primary and Derived Network Performance
Parameters

Function	Criteria		
	Speed	Accuracy	Dependability
Access			
User information transfer			
Disengagement			

Availability Function

Source: From Ref. 17.

parameters in a 3 × 3 matrix approach format (17). This matrix provides a
systematic method of identifying and organizing candidate NP parameters with
the objective of defining a concise set of parameters and their QoS counterparts.
Specifically, the matrix parameters are defined on the basis of events at connection element boundaries representing the following nine generic primary performance parameters:

1.   Access speed
2.   Access accuracy
3.   Access dependability
4.   Information transfer speed
5.   Information transfer accuracy
6.   Information transfer dependability
7.   Disengagement speed
8.   Disengagement accuracy
9.   Disengagement dependability

The derived performance parameters are defined on the basis of a functional
relationship of primary performance parameters, outage thresholds, and an
observation interval. The generic derived performance parameter associated
with such a function is availability. The term *availability* is further identified as
transitions between the available and the unavailable states.

The scope of such a framework has been further expanded to assess the
quality of B-ISDN service and its network performance parameters. The major
performance-related issues are presented in terms of QoS and NP parameters
(17,18). Reference 19 provides even more focus on the assessment of the techni-

cal quality of B-ISDNs at points at which ATM cells are present. This is done based on technical quality areas of connection establishment instead of access, user information transfer, and connection release instead of disengagement. Each of these categories is assessed for speed, accuracy, and availability (17).

Broadband ISDN/ATM Networks

Recent advances in computing and communications have produced an increased demand for new and fast communications services. This is because success and growth of any industry heavily depend on how efficient its communications infrastructure is. Broadband ISDN is viewed as the next generation of telecommunications and data communications networks, which will support diverse applications ranging from low bit rate communications between terminals and host computers to the broadcast of high-resolution videos. This means not only that it needs to offer an end-to-end digital, efficient, and cost-effective means of communication for the existing range of communications services, but it also should easily evolve toward a fully integrated solution for a wide variety of communications applications in the future.

In order to provide such a flexible operating environment in a B-ISDN, the asynchronous transfer mode has been proposed by the ITU (20,21). The ATM is a fast packet-switching technique that can be used for a broad spectrum of services, ranging from low-speed to high-speed, from constant bit rate to bursty traffic, and from connection-oriented to connectionless applications. In ATM networks, information (such as voice, data, and video) is divided into fixed-length data blocks, called *cells*; these cells are asynchronously transmitted through the network.

The inherent characteristics of the ATM technique have been recognized as the uniqueness of the transporting cell units and dynamic sharing of network resources. Taking advantage of these characteristics, much effort has been focused on investigating the required basic techniques, such as switch architecture, bandwidth allocation algorithm, access network architecture, and congestion control.

Packet-Switched Network Protocols

International Organization for Standardization Model

The International Organization for Standardization's Open System Interconnection (OSI) Model consists of seven layers: physical, link, network, transport, session, presentation, and application. Each layer is responsible for performing a portion of the functions that ultimately result in transportation of data without error and congestion. Although the standard was originally considered for data communications in the context of packet switching, there is no reason to assume that the fundamental concepts may not apply equally to the future integrated services and/or other specialized switching techniques.

Because voice communication is a familiar technology with easily recognizable services, we use it here in our discussion of packet switching for analogy and ease of understanding. In packet switching, small units of data called *packets* are transported through the network. Packet switching involves sharing of resources such as buffers, transmission facilities, and switching resources. Generally, there are two modes of transmission: connection oriented (virtual circuit), with guaranteed sequential delivery of packets, and the connectionless (datagram) mode. In the latter, packets do not follow the same paths as they do in the virtual circuit mode and as such sequential delivery is not guaranteed. In contrast, in circuit switching a dedicated path should first be established. Resource sharing is limited to switch call processing and facilities among users. Generally, no error correction, detection, sequential delivery of information units, or flow/congestion control is involved in circuit switching.

The function of the packet-switching protocols is to ensure timely delivery of data that are correct and recognizable. It is desirable that this is done in a cost-effective manner. For data applications, adherence to the functions of the protocol is more critical than for voice applications. The packet-switching protocol is layered. The lower three layers are responsible for network services, delivery functions that are implemented in the network nodes and possibly end-user devices. The upper layers perform end-user functions and are implemented in end-user devices. The purpose of the layered architecture is openness. If a change occurs in a function of a layer while interfaces are held constant, the changes will be applied only to the layers affected and not to the interfaces or other layers.

Open System Interconnection Standards Architecture and Protocols

The objectives of the seven-layer OSI protocol architecture are discussed next (7). The first is simplicity; each layer should not be too complex. Second, there should not be such a large number of layers that they would have difficulty interfacing with each other across the network. Third, there should be natural boundaries such that similar functions are grouped together into one layer. Within the seven layers, each layer provides services to the layer immediately above it (and each layer receives services from the layer immediately below it). The OSI Model can be thought of as an abstract model or construct. In Ref. 22, details of these layers are provided.

Unified View of Open System Interconnection Protocols

In the hierarchical OSI Model, the layer above is called the *service user*, the one below the *service provider* (7). Interface between these two is accomplished through the *service access points*. High-level functions of three phases of data transmission that facilitate the understanding of the service preemptives are

- Call establishment (setup): Negotiation of a set of parameters for use during data transfer

- Data transfer: Error detection, sequence numbering, flow, and priority assignment of the actual transmitted data
- Call disconnect: Agreement on how to disconnect a call

At each layer, the Protocol Data Unit (PDU) is defined as the sum of the Protocol Control Information (PCI) and the actual data fields. PCI is also referred to as the "overhead" field because it generally does not contain the actual data.

Four "service primitives" for interaction between the service user and service provider (layers $N + 1$ and N, respectively) are defined:

1. Request
2. Indication
3. Response
4. Confirm

These primitives are not necessarily implemented among all layers or all phases of communication. Each primitive may result in the exchange of one or multiple PDUs. These primitives are exchanged in PDUs. However, they may not be implemented in all cases.

Priority and Scheduling

At certain layers of the protocol hierarchy and/or any implementation of protocols in the network or inside a switch, there are functions to assure priority handling of data among end points. For example, in the ATM networks, a bit in the control portion of the cell header is used to implement cell loss priority (CLP) (see the section, "Asynchronous Transfer Mode") which has the function of setting the proper priority level for the cells belonging to a particular connection. Implementation of the priority concept for different services and products may have different implications. For example, in ATM networks a low-priority cell may be subject to no delivery when congestion or other network problems occur. In other implementations, a low-priority packet (or data entity) will be delivered, however late, perhaps out of sequence from other data messages. For example, in message switching systems, a low-priority message may ultimately be delivered later than a similar message at a higher priority. Generally, as these examples indicate, priority handling of the information exchange affects the GoS of the services provided. These GoS measures might be delay, blocking (call or cell/packet), and/or other measures that may be found appropriate for a given network and/or service.

Packet-Network Traffic Characteristics

The detailed characterization of packet-switching network traffic needs to be described for a range of telecommunications and data communications services.

Description of services and associated bandwidth requirements and switching implications for the broadband services are discussed in Refs. 23 and 24. The extent of the diversity in telecommunications services has resulted in categorizing them according to the ITU classification (25). The first category of services is divided into

- Interactive services, with transmission in both directions
- Distribution services, with one-way transmission with no backward channel

The interactive services are further subdivided into the following services:

- Conversational: Sender and receiver are simultaneously present
- Messaging: Sending information to a database for subsequent retrieval
- Retrieval: Extraction of information from a database

The distributed services are broken into

- Distribution services with user presentation control
- Distribution services without user presentation control

This classification further results in a large number of services in each of these groups as highlighted in Ref. 26. The characteristics of traffic types and their transport requirements depend on the specific applications and/or services. For the purpose of packet-network traffic characterization in this summary, all services are classified here under the three classes of voice, data, and image/video applications.

Future communications networks must carry a wide variety of services with widely differing characteristics. In a communications network, all packets may not be of equal "value." A packet that represents part of a data file transfer may be less sensitive to delay and time jitter, but of far greater value than a packet from a digitized voice connection. In this case, by *value* we mean a threshold for requesting retransmission in case the packet is lost or its criticality if it is delayed. Some packets may have far more stringent requirements than those packets that carry color-related scales of a pixel of an image. Further, data communications services require greater accuracy than voice or image/video services. Voice calls are delay intolerant, whereas many data applications are delay tolerant. This statement, however, is reversed when we are considering error instead of delay. As discussed above, performance measurements of delay, throughput, and packet loss are usually applied as performance criteria to different types of traffic.

Voice

Voice transmission is still the most common type of communication worldwide. In the early 1960s, AT&T pioneered the first commercial introduction of a digital carrier system. This has led to deployment of the T1 digital carrier systems, transmitting at 1.544 megabits per second (Mb/s), designed to handle 24 voice channels at 64 kb/s each. The range of voice applications has been expanded to voiceband data, voice with silence suppression, videotelephony, voice mail, and spoken information services during the past few decades.

Delay is a critical performance parameter of voice traffic. Because of the real-time nature, voice services need bandwidth on demand with low delay. The characteristic of voice traffic requires voice packets to arrive within a fixed interarrival time at the receiving end.

In a packet-switching network, voice packets may be class marked, or routed in a manner so that they can be handled as expeditiously as possible. However, voice packets need not be error checked since there is no time for retransmission of errored packets. At the destination, packets received with errors will be processed and then passed through the speech synthesizer. The worst result will be a short noise burst if the errors are severe.

The performance objectives for voice traffic should address problems of bit dropping (which in some systems is used as a means of reducing bandwidth demand) and packet dropping in packetized voice transmission. In addition, the quality of voice calls in terms of subjective measures, such as mean opinion scores (MOSs), as well as objective measures such as mean bits per sample, may be introduced.

Data

The major part of data applications include LAN interconnection, personal computer (PC) and intermainframe file transfers, mainframe load sharing, electronic mail, remote database access, distributed processing, and more. In general, most of such applications belong to interactive services of the conversational, messaging, and retrieval categories discussed above. The characteristics of such data traffic services further differ on the basis of their bit rates, burst sizes, performance requirements, retransmission criteria, and other parameters that are media related. For example, performance requirements such as delay, delay variation, throughput, bit error rate, and packet loss contribute to traffic characteristics of the network.

The principal characteristic of most data traffic types is that they are generated unpredictably and, in many situations, lead to a bursty type of traffic in the network. Interactive data traffic is mostly transmitted in short bursts of a few bytes to as many as several hundred bytes between terminals or between terminals and computers. The distribution of packet sizes on a LAN indicates that most packets are short (32–64 bytes) and most data are in the longer packets (512–1024) bytes (27). Empirical measurements show that 65% of the data have a packet size of 512–1024 bytes and 35% have 32–64-byte packets. Note that a LAN interconnect is an important example of a service that requires both a low bit error rate and a low end-to-end delay.

The data transmission between computers or between mass storage systems could involve the transfer of large files of up to several million bytes of data. A comparison of intermainframe file transfer with PC-to-PC file transfer indicates that, although the file size may be larger for the mainframe, the end-to-end delay requirement may be more stringent in the PC case, thus affecting the bandwidth requirement (28).

Data traffic can be categorized as interactive or batch from the end-user's orientation of the service description. End-to-end delay is an important performance criterion for interactive data communication such as LAN interconnection or PC-to-PC interaction similar to voice conversation. In particular, interactive traffic must be delivered quickly and is usually rather short. On the other hand, a long file transfer is not so much concerned with the initial delay in getting the first few bits across the network, but rather with how many bits per second can be transported through the network.

Image/Video

By image communication, we mean the transmission of still pictures or motionless objects. The most common form of this communication is presently known as *facsimile transmission*, in which an entire page of information is transmitted as is, rather than as the digital representation of the letters or characters that were on the page. The facsimile transmission can be interactive or batch. *Interactive facsimile* refers to usage like a telephone conversation, whereas *batch facsimile* refers to the presently used mode of operation in which documents are accumulated and then sent as a batch. Depending on how an interactive facsimile service is defined, one may require use of a rather intelligent facsimile machine on each end, perhaps in conjunction with a personal computer, in order to handle such service properly. Such a facsimile machine may be used in both cases; however, in the interactive case, end-to-end delay is an important performance criterion, whereas in the batch case it is not. Most interactive image services require relatively low throughput and delay, versus the batch mode image services which require relatively high throughput and allow higher delays. The information integrity requirements in both cases are generally more critical than for voice services.

Generally, image services produce very large bursts of data in the network. For instance, an image transfer service used for a medical application may create a large-size burst (120 Mb/s of packets for a 14 × 17 inch x-ray film (29), compressible to about 8 Mb/s) compared to the same service used for sending draft fashion designs (8 Mb/s compressible to about 1 Mb/s) (30). The size of burst traffic increases even further when color-catalog-type production images are transmitted using today's high-resolution screens of 1 Megapixel with either 256 colors (8 Mb/s) or 256 levels of three colors (24 Mb/s). It should be noted that this may be compressed by a factor of 3 to 15, depending on the type of image. Such an application clearly creates a total burst size that is larger than the previous example, but the end-to-end delay requirement is less important. Instead, the packet loss rate is a more important criterion.

Packet-Switched Network Modeling

There are many phases in the modeling and performance analysis of communications networks. Some important network design issues need to be considered during the planning and the implementation phases of a new network that is expected to carry computer system traffic among sites and locations at which jobs are executed. Switch architecture and cost/performance tradeoff are among such important issues that require a great deal of attention during this period. For an existing network in which network components are already in place and the network is operational, such modeling and performance analysis is performed in order to predict the network growth (31,32), reallocation of resources due to the change in traffic patterns, and/or characteristics and network applications, among other time-varying parameters. Depending on whether planning is for a new network or an existing one, the task of modeling may vary in terms of the sequence of the process and which step should be performed first. However, the steps described below are common among both. A variety of methodologies and techniques are available and are described here.

Modeling and performance analysis of packet-switched networks start with the design, in which network components including packet switches, trunks (links), concentrators, multiplexers, and other peripherals may be sized to meet the network performance objectives. These objectives are typically called GoS or QoS and they may encompass all performance measures from quality of service and transmission to call admission procedures (which may result in call blocking), delay, throughput, and methods of control that may be incorporated in the network (e.g., congestion and flow controls). Based on a set of objectives, one determines the design requirements that would ordinarily meet such objectives. Many tools and techniques (including simulation and analytical tools such as queueing theory) are available for the performance evaluation.

Given that a set of tools are used and a particular network design is completed to meet a given set of objectives, the question is whether there would be a need to perform modeling and performance analysis after completion of the design. Generally, the answer is yes. Such procedures are used to check the network performance under overloads and other exceptional conditions for which the network may not have been designed. For example, a network may meet its GoS under engineered traffic for which it is designed. What if there are operational circumstances for which the network was not designed? Examples of these circumstances are trunk failures and overloads that are caused by special events that may activate such occurrences in the network. Generally, the techniques offered for performance analysis and modeling of packet- and circuit-switched networks which are used, respectively, for data and voice communications are similar, if not the same.

Modeling Methodologies

During the performance modeling of communications networks in general and packet-switched networks in particular, several steps need to be taken and several possibilities exist for methodologies for performance and modeling.

Values of the parameters that should be input into a model should first be determined. Where do these values come from? In "real" systems, these values come from measurements and observations during network operation. A network that is not operational yet is hardly a system for measurement and observations. However, components of the network, including packet switches and links, are usually separable entities that may be subject to measurements. A packet switch consists of a control/call processing unit (CPU), which schedules tasks and performs algorithmic functions of the switch. A CPU in a packet switch is similar to a central processing unit in any computer system. Measurements of its processing speed and time to perform tasks (such as to access a routing table in the case of static routing or calculating the route in the case of dynamic routing, etc.) are quite straightforward and known techniques are available to perform them (33). Several statistical methods are available in order to structurally organize the measurements and possibly fit some type of known distribution to them so that the task of analytical or simulation modeling can be performed more easily.

With the network parameters known, the next step is to create analytical and/or simulation models to accomplish the overall performance analysis.

Analytical Methods

A historically efficient method of predicting the outcome of interaction among system components and elements is through modeling using mathematical interrelationships that describe the dynamics of the system. Queueing models are among the forerunners in this category. Usually, a wide range of decomposition and/or simplification of the original system is performed before it can be handled using simple queueing systems (34). In recent years, other methods that also lend themselves to queueing principles have been adopted and shown to be efficient. Among them are fluid flow techniques, which have been shown to track very effectively the simulation, and/or more complicated analytical models representing the network under study. Indeed, as our understanding of networks and systems expands, it is the case that other, perhaps new, tools and techniques will be introduced in this area.

Simulation Methods

When the complexity of network control techniques is beyond the tractability of mathematical methods, then usually simulation is used. In the past, simulation models were so specialized to the particular system under study that their cost made them a last option for studying networks and systems. With the introduction of object-oriented programming techniques and the development of modular software methodologies to build models out of existing modules, it is becoming more prevalent that such tools are more cost effective than before and are used in a wider scope of problems than in the past (35). The primary task in the simulation method is development of confidence intervals in studying models of networks and systems. It is now possible to develop estimates of the simulation run time and the number of events that may be needed to simulate in order to develop confidence in the results as produced by the simulation models (36,37).

Network Design

In the network design phase, methodologies such as heuristics and algorithms may be used to size the network components. As mentioned above, such components may include switching devices (actual nodal switches, concentrators, multiplexers, etc.) and network trunks. Algorithms and heuristics are targeted at meeting the network performance objectives and, in particular, GoS. Unconstraint problems with no performance or resource limitations are considered unrealistic. Thus a sensible approach is to apply certain constraints (derived from practical limitations and specifications) to the problem. In these cases, while the constraints are met, a cost objective function is minimized or revenue collected from the network is maximized.

Usually, many parameters and factors may enter the design phase; they may include routing, network topology and capacity constraints, reliability and availability requirements, and control mechanisms that may have an impact on network component sizing. For example, if the network switching functions allow dynamic routing by which packets may use selective network paths in real time, rather than on a fixed path basis, this may have an impact on the heuristic that is used to size packet-switched network trunks (in this particular example, one would think that the required capacity of a dynamic-based-routing network for the same GoS and/or performance requirements should be smaller than that required in a fixed-routing procedure). Such differences usually imply different sizing methodologies and algorithms for the switches and other network components.

Noncoincidence of Network Traffic

In typically large networks, there may be noncoincidence of traffic in different parts of the network due to the varying time and activity zones. This fact can usually be used to reduce the required network capacity/sizing requirements for a given set of performance requirements. For example, when there is noncoincidence of traffic between the East and West Coasts because of timing differences, in order to complete a voice or data call, one may route the East Coast traffic to the West Coast and take advantage of the idle resources and vice versa. Such resource sharing concepts are usually implemented using dynamic rather than static control mechanisms. This does not mean that static or "semistatic" controls are excluded from being implemented in such a network. A good example of a semistatic control mechanism in a circuit-switched network is the dynamic nonhierarchical network routing (DNHR) (7) (this is in contrast with the dynamic real-time routing methodologies). In this routing method, network routing changes according to the time of day; however, while implementing a particular routing method in a given time of day, the routing is rather static and not totally dynamic, as is the case in real-time methods. (In this particular example, DNHR allows use of a predetermined list of alternatives, while real-time routing methods may not have such limitations and any available alternative may be used (38).)

In most cases, one may not be able to completely isolate the inherent interrelationship among different factors in separate steps of the design process as a lot of research effort has gone into finding the best way to handle some of the difficulties resulting from complexities of interactions among factors and parameters that may affect the design.

Topology

Network topology refers to the connectivity of switching devices and other network peripherals. This has traditionally played an important role in the cost of network implementations. A major portion of the cost of a network would correspond to the cost of links (trunks) that are incorporated in the network. Although with the advent of fiber optics and ample bandwidth that such media offers, the cost of network links may not be as critical as it used to be, but because it ultimately corresponds to the difficulties of getting right-of-way to lay down cables and other legal matters, it is an important factor in the design step.

Topological optimization would correspond to describing a network connectivity (based on constraints and other objectives) such that the cost of the network is minimized. In most cases, the layout of the cables may already be in place (such as leased lines, etc.) and the important exercise is to choose properly from among many options so that performance objectives are met. Most carriers need to choose from many options of cable layouts already in place and may be disconnected/reconnected from or to the Digital Access and Cross-Connect Systems (DACSs) that provide facility connectivity in their networks. In a typical tradeoff study for choosing among the options, one may also consider use of "other providers," such as using the facilities of other carriers rather than building one's own facilities.

Multidrop Networks

A class of packet-switched networks utilizes multidrop links for communications between one or multiple host computers and the corresponding terminals. This class of networks was created during the early phase of the implementation of packet-switched networks in consideration of the "high" cost of bandwidth and the advantage of bandwidth sharing among multiple terminals. Here, several terminals share the same communications line, not generally in a contention mode, but usually in a polling fashion. In a group of such networks, multidrop lines may terminate on a concentrator (in a point-to-point or multidrop fashion) and in turn several concentrators communicate with the host computer in a point-to-point or multidrop (multipoint) fashion (see Fig. 1).

Several schemes of polling are possible. In one method, the host computer may have a polling list for polling all terminals individually, regardless of their connectivity to the concentrators. In this case, concentrators only have a role in reducing the cost of communications lines that connect terminals to the host computer. In another case (a more pragmatic one), concentrators poll their individual terminals (which may in turn connect in a multidrop fashion) and,

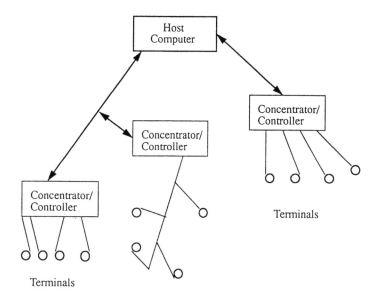

FIG. 1   A multidrop network with concentrators/controllers.

when polled by the host computer, respond to it with all the requests/responses from all their terminals. Responses from terminals may only include those with messages for the host or they may include negative responses to the polls by the host computers.

The first and the most important problem to solve in multidrop network design is that of terminal/concentrator location. In this problem, one should decide, given the cost of links connecting different terminals/concentrators directly to a host and/or to one another, what is the "optimum" network topology. In this regard, many algorithms are available, mostly heuristics that come within close approximation of the optimum designs. They may be classified into two groups: unconstrained and constrained. Among the unconstrained algorithms, the minimum spanning tree is at the top of the list. As the name implies, this class of algorithms does not consider the network design/performance constraints and thus is best for benchmarking the heuristics. The constrained class, on the other hand, may take into account such limitations as the response time, capacity of lines, and so on. It represents a more realistic class of algorithms for the actual design and implementation of multidrop communications networks. Below, we describe one such algorithm in an example from Ref. 40.

In this example, the Esau Williams heuristic algorithm, which incorporates the flow constraint, is utilized to design a multidrop network. It is assumed that the cost of links for connecting terminals/concentrators and/or the host computer is known. Also, the total flows on the multidrop links that may have resulted from a number of design constraints such as delay or capacity are imposed. The objective is to design a minimum-cost multidrop network that interconnects all terminals to the host computer. This example is from Ref. 40 without modifications.

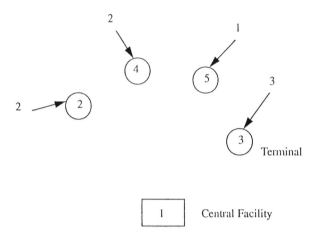

FIG. 2   A multidrop network with the corresponding node flow demands.

See Fig. 2 for layout of the network. The cost matrix below (Table 3) shows the corresponding communications links costs.

For the demand flows from each of the nodes (it is assumed here that all flows are unidirectional from terminals/concentrators to the host) shown in Fig. 2 (which assumes a maximum flow of six units on the multidrop lines), we describe the Esau Williams algorithm as follows (40).

Step 1.   Calculate the tradeoffs (between connecting terminals/concentrator, hereafter referred to as *node*, to one another and connecting to the central host, number 1 in Fig. 2). These are the differences between C_{ij}, which represents the cost of connecting Nodes i and j, and C_{i1}, which corresponds to the cost of connecting Node i to the host 1. The tradeoffs are shown as t_{ij} in Ref. 40.

$$t_{ij} = C_{ij} - C_{i1} \qquad \text{for all } i, j = 2, 3, \dots$$

We note that when $t_{ij} > 0$, it implies that it is most cost effective to connect Node i directly to the host rather than connecting it to Node j as part of a multidrop link. When $t_{ij} < 0$, on the other hand, then it implies that we are better off connecting Nodes i and j directly together for a multidrop connectiv-

TABLE 3   Multidrop Network Example —
Cost Matrix

Nodes	1	2	3	4	5
1	—	3	3	5	10
2	3	—	6	4	8
3	3	6	—	3	5
4	5	4	3	—	7
5	10	8	5	7	—

ity, given that the flow constraints are met, that is, if $f_i + f_j < F$. In this expression, f_i represents the traffic flow generated at Node i and F represents the constraint on the traffic flow. Table 4 contains the summary of the tradeoffs. In this table, Row i refers to the first index of t_{ij} and Column j refers to the second index.

Note that from this table, we conclude that Node 2 should be directly connected to the central facility (i.e., Box 1 in Fig. 2), while Node 5 should connect to Node 3. This conclusion is because the tradeoff between connecting Node 5 is minimum when it connects to Node 3.

Step 2. Using the observation above, choose to connect those i and j nodes together that result in the minimum t_{ij}. In this example, Nodes 5 and 3 should be connected.

Step 3. For the selected ij in Step 2, check the constraint and, if satisfied, go to Step 4; otherwise, set $t_{ij} = \infty$, and go to Step 2. In this example, total flow from Node 5 to the host is 1; thus, the total flow is below the constraint of $F = 5$.

Step 4. Interconnect i and j and relabel Node i as j. Calculate the total flow on this new link, and verify the constraint. Create a new tradeoff table and go to Step 2.

In the example just described, the new flow for Node 3 is now $3 + 1 = 4$. Thus, when going to Step 2, we will use this new flow for Node 3. The next node pair to be considered now is the one with the smallest t_{ij}. In this case, we note $t_{54} = -3$ is the smallest. However, because Node 5 already is connected to Node 3, we eliminate it from being connected to any other node and choose the next smallest tradeoff, which is $t_{43} = -2$. This implies checking whether it is more cost effective to connect 4 to the center directly or to Node 3. The conclusion is that it may be more cost effective to connect it to Node 3. Because the flow from Node 4 is 2 and currently total flow from Node 3 is $3 + 1 = 4$, we observe that, with this node added, the total flow will be $4 + 2 = 6$, and the flow constraint of $F = 5$ will be violated. Therefore, we set $t_{43} = \infty$.

From Table 4, the next candidate is the node pair 4–2 ($t_{42} = -2$ and is the smallest). Thus, we connect Nodes 2 and 4 (note that flow constraint is not violated). These setups are repeated until all nodes are connected to each other or to the host. Fig. 3 shows the final connectivity.

TABLE 4   Cost Tradeoffs for the Multidrop Network Example

	1	2	3	4	5
1	—	3	3	5	10
2	3	—	3	1	5
3	3	3	—	0	2
4	5	−1	−2	—	2
5	10	−2	−5	−3	—

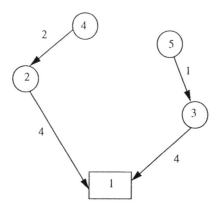

FIG. 3   Constrained solution for the multidrop network example.

In the Esau Williams algorithm, as well as many others for the multidrop networks, the implicit assumption is that we know what the constraints for the connectivity of nodes are. We note that such constraints may appear in the form of flow or delay and they can be checked using available methods such as queueing analysis in determining the delay. Some of the analysis may lend itself to the polling case, which is considered next.

A Simple Analysis of a Polling System.   Many variations of polling systems have been designed, implemented, and analyzed. In this section, we present a simple analysis of such systems that may be tailored for simple response time calculations of such systems.

As shown in Figure 4, suppose there are several terminals polled by a center. Suppose the rate of arrival of messages at each terminal destined for the center is λ_{out} and that the total arrival rate of messages at the center destined for the

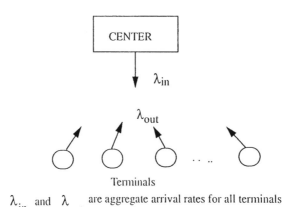

FIG. 4   A simple polling system.

terminals is λ_{in}. We assume that these rates correspond to the rate of arrivals following a Poisson process with appropriate interarrival times of $1/\lambda_{out}$ and $1/\lambda_{in}$, respectively. Suppose that the expected (average) cycle time in this system is $E(C)$. This time corresponds to the average delay between two consecutive cycles to poll a particular station. We denote the average "no-load" cycle time as $E(h)$. No-load cycle time refers to the time that is required to poll all terminals in one polling cycle, with the assumption that neither the terminals nor the central facility have data to send. Similarly, in this case, the center facility has nothing to send to the terminals. If $E(W)$ denotes the average duration of a cycle when the terminals and the center have messages to send, excluding the "no-load" cycle time, then

$$E(C) = E(h) + E(W)$$

We note that $E(W) = [\lambda_{in}/\mu_{in} + \lambda_{out}/\mu_{out}] \cdot E(C)$.

In this expression, μ_{in} and μ_{out} refer to the average service time of messages originated at the center or at the terminals, respectively. In this simple analysis, we assume that messages are exponentially distributed in size. We refer to $\rho = \lambda_{in}/\mu_{in} + \lambda_{out}/\mu_{out}$ as the system utilization. Thus,

$$E(C) = E(h) + E(C) \cdot \rho$$

and

$$E(C) = E(h)/(1 - \rho)$$

This simple expression is sufficient to help in understanding many of the inherent characteristics in some polling systems and is of value when difficulties of more complicated models can be bypassed.

Capacity

Traditional ways of estimating link and processor capacities in a packet-switched network were based on models that would predict network performance such as total, average, and percentile delays. It is expected that such methods and similar ones using, say, queueing theory and fluid flow approximations, as well as simulations, will continue to be used in the context of high-speed packet and the B-ISDN networks.

In order to give a frame of reference for how these methodologies may be used, in the following we give examples of $M/M/1$ queueing models that may be used for the performance analysis of packet-switched networks. The development that follows is from Ref. 7, and we use same notations when possible.

First, we introduce some notations. C_i is the capacity of Link i in bits per second (b/s). Note that, based on this convention, each network link might have a different capacity and it is not necessary for all links to be of equal capacity. Also, b is the packet length in bits and $1/\mu_i$ is the packet transmission time on Link i in seconds. Note that $1/\mu_i = b/C_i$, that is, packet transmission time on the ith link is equal to the ratio of packet size over the link capacity.

Finally, λ_i is the rate of packet arrival on the ith link (say, packets per second) and $E(T)$ is the total expected delay of a packet-switched network consisting of M links.

Then, for the simple case of Poisson arrival of packets on the ith link and the exponential service time of packets (which means packets are also exponentially distributed in size) and an infinite buffer size at the ith link, one can show the following expression for the average delay in a packet-switched network with M links:

$$E(T) = \sum_{i=1}^{M} 1/(\mu_i - \lambda_i) \qquad (1)$$

This is an immediate result of what is known as an $M/M/1$ queueing system with infinite storage. This expression indicates that the total average delay of a packet-switched network is equal to the sum of the average delay on individual links. The question of determining the λ_i's is generally not a trivial one and depends on the network topology, routing, and also flow distribution and possibly optimization process that may have been adopted in the network design phase.

For example, suppose γ_{ij} represents the rate of packet arrival to the network at a source node i and destined for the Node j. The unit of this quantity is the same as λ_i, namely, packets per second (or whatever the time units may be). Then, it can be shown (7)

$$E(T) = 1/\gamma \left[\sum_{i=1}^{M} \lambda_i \times T_i \right]$$

where T_i is the average delay on the ith link and is equal to

$$T_i = 1/(\mu_i - \lambda_i)$$

Also, γ represents the total traffic entering the network, that is,

$$\gamma = \sum_{i=1}^{M} \sum_{j=1}^{M} \gamma_{ij}$$

$E(T)$ can now be written as follows:

$$E(T) = 1/\gamma \left[\sum_{i=1}^{M} \lambda_i/(\mu_i - \lambda_i) \right]$$

The task of determining λ_i from γ_{ij}, as mentioned above, is based on the routing method and general theory of flow optimization and is briefly covered below.

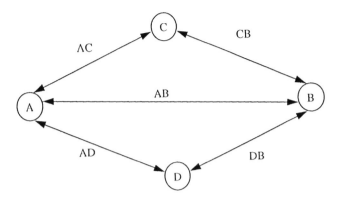

FIG. 5 Network topology without the actual facility layout consideration (links are labeled).

Reliability

As mentioned above, a network designer tries to choose from many options of facility layouts such that the network performance objectives are met and the minimum cost is incurred. One such objective is a measure of network reliability. For example, as shown in Fig. 5, while from the network topology point of view it seems that three diverse paths between Nodes A and B are available (a direct path AB, one through Node C, ACB, and the other through Node D, ADB), in reality (Fig. 6), choices of diversity are much more limited because

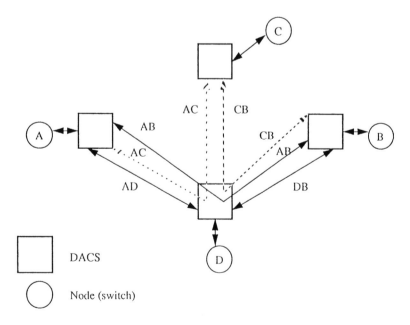

FIG. 6 Actual facility layout for the network topology shown in Fig. 5.

of the way facilities are connected through DACSs (41,42). Incorporating the reliability objective or constraints becomes an increasingly important factor in such circumstances.

Routing

Unlike circuit switching, which in the traditional networks used hierarchical structures, packet networks are usually "flat," at least as far as routing is concerned. (There are many exceptions, and it is not quite true that in terms of routing all packet networks are flat. In terms of practical implementations, one can find packet networks that are also implemented in a hierarchical fashion.) A good source for the methods of circuit-switched routing is Ref. 43. Some packet-switched networks may employ the same type of hierarchy that circuit-switched networks follow. Also, many of today's circuit-switched networks are not hierarchical anymore and use real-time routing techniques that lend themselves to flat routing structures. For an example, see Ref. 38.

Packet routing may have structural constraints implied by the maximum number of switch hops that a packet may take between any given source and destination, areas of the network that, for whatever reason (e.g., security), a given source/destination may not be allowed to use. There may be a number of concepts in packet-routing structure used in some of the actual implementations.

Among the more familiar routing methods one may mention the following: fixed, flooding, random, and adaptive (1). As the name implies, in the fixed routing method all packets between a source/destination node pair use a fixed path all the time. In the flooding method, as packets enter a source, they are copied according to the number of outgoing trunks at a given node and transmitted on all outgoing links. Each receiving node will do the same, copy the packet and transmit on all outgoing links at that node. It is the responsibility of the receiving nodes to determine if the packet (or its copy) has visited the node before and whether to destroy the newly arriving copy. Timestamps and hop count methods may be used to prevent the network from getting congested with many copies of packets en route. In the random method, at each node, the outgoing link is determined with some probability and the packet is only transmitted on that link. These probabilities may depend on the existing link loads and/or delay, or they may be predetermined (on some basis). In the adaptive (or dynamic) routing method, at each node, up-to-date information received from a central location or from other network nodes (in the case of distributed algorithms) is used in order to determine the outgoing route or the path for each packet.

Most packet-switched networks implement some type of least-cost routing algorithm. Among the more interesting ones are Dijkstra and its variations, such as the Bellman–Ford (7). In the Dijkstra algorithm for the least-cost routing, for every source node, the least-cost paths to all other nodes are determined in several iterations as follows.

In the kth iteration, the shortest paths for the k nodes closest to the (selected) source node will be determined. At the $(k + 1)$ iteration, a node that is not already selected that has the shortest path from the source node is added to the set of selected nodes. These iterations are repeated until (for a given source node) all network nodes are included in the set of already-selected nodes. When repeated for all network nodes as sources, a complete set of paths for all source/ destination pairs will be developed.

Whether the routing is predetermined or "on-the-fly," most network-routing methods lend themselves to one of the two following methods (or their variations): distributed and centralized. In the distributed routing method, network nodes participate in determining the routes; for this purpose, they may exchange information among themselves. The information that is exchanged depends on the objectives that are being set for the network performance. For example, in the ARPANET routing (7), nodes exchange their nodal delays (average over some period of time and for each link) and may use the Dijkstra algorithm to calculate the shortest paths. Such information as that for ARPANET is updated periodically and thus the routes might change depending on the network load fluctuations (adaptive). In the centralized routing, either a fixed or dynamic set of routes are downloaded to all network nodes (usually periodically) and are used for packet routing before they are updated again by a control center.

One approach to the determination of packet-network routing is to determine the link flows that are used in Eq. (1) in order to predict network performance. In the following, we provide a method for determining link flows and network routing.

As mentioned above in the average delay calculations, given the link capacities C_i and the traffic entering network at Node i destined for Node j, γ_{ij}, one should determine the link flows λ_i for all network links $i = 1, 2, 3, \ldots, M$ such that the total network delay expressed in Eq. (1) is minimized. In flow optimization terms, the flow between source/destination node pair i,j is referred to as a *commodity*. Thus, in a network with N nodes where there are a total of $H = N \times (N - 1)$ node pairs, we say that there are H commodities. The task of flow distribution in order to optimize network performance to minimize total average delay is then to distribute the H commodities based on a routing method and to achieve the performance objectives. As such, the assumption of the forthcoming development is that the packet-switched network routing can be developed after we have determined the contribution of node pair traffic to each link's traffic. The total flow on link k, λ_k, is

$$\lambda_k = \sum_{i=1}^{N} \sum_{j=1}^{N} \gamma_{ij}^{k}$$

Here, γ_{ij}^{k} refers to the component of the γ_{ij} that, due to an assumed network routing, flows on Link $k = 1, 2, \ldots, M$. Note that there is the possibility that, for certain node pairs ij, there will be no traffic flow on a given link k and thus the contribution in the equation above from that link will be zero. This simply implies that for the node pair ij, the routing path does not include the link k. We also note that

$$\gamma_{ij}^{k} = function\ (\gamma_{ij})$$

where the *function* depends on the assumed routing method to be developed. In order to solve the optimization problem (say, minimization of total average network delay), one simple observation is referred to as the "flow conservation" law. According to this law, the total flow that enters a node should equal the total flow that exits the same node. Sometimes, an additional constraint of $\lambda_k \leq C_k$ for all network links $k = 1, 2, \ldots, M$ is also imposed. This constraint implies that the total flow on Link k should be less than (and, at worst, equal to) the capacity of the link. To solve this problem, many methods (all interactive) are available; the most notable one is the flow-deviation method described in Ref. 44.

Flow and Congestion Control

In packet-switched networks in general and in high-speed networks (such as ATM/B-ISDN) in particular, flow control and congestion avoidance is critical to proper operation. If congestion develops and is not properly handled, data units could be lost and may result in inefficiencies in the network. To prevent congestion from developing, one needs to implement proper flow control methods in the network. A variety of alternatives is available; by the protocol hierarchy that ISO/OSI recommends, many layers do have mechanisms to impose flow control. For example, in High-Level Data Link Control (HDLC), a window-based flow control is available that, in addition to implementing an acknowledgment procedure on each network link between network nodes, also provides mechanisms for flow control on those links. Similar methods may also be implemented at the network layer between end points and also at other layers, such as the transport or even session and application layers.

We note that these control mechanisms are not duplicate functions and each perform their own responsibilities. For example, while a network link that is en route for a session may not be congested, the session or application layers may still apply such controls at their layers because the sessions/applications themselves (such as their respective processors) could be congested and vice versa.

General analytical techniques used for delay analysis such as queueing theory and the like may be used here also for evaluation of one method or another. For example, suppose at a node to prevent congestion, one implements a threshold mechanism by which, in an input queue in a packet-switched node (see Fig. 7), may discriminate against the low-priority packets and preferentially treat the higher priority ones. If the two streams of packets arrive at the queue and queue occupancy is below a given threshold T, then both the high- and low-priority packets will be accepted.

However, when the queue contents exceed T, low-priority packets are discarded and only high-priority packets are allowed to enter the queue at the packet node. By adjusting this threshold, one may impose the desired GoS (the delay or packet loss rate) for the packet node. This method of control is referred to as *space priority* (45); other versions may be found in Ref. 46. A version of this problem with modulated Markov Poisson process (MMPP) input traffic is

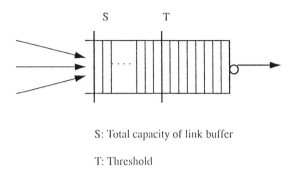

S: Total capacity of link buffer

T: Threshold

FIG. 7   A threshold-based congestion control method.

analyzed in Ref. 47. A similar method of congestion control that discriminates between the transit and input traffic at a node is also described in Ref. 7 and is called *input-buffer limiting for congestion control.*

The importance of congestion control methods focused on B-ISDN and ATM networks is due to their higher speed and the fact that propagation delay (which dominates the delay of packets) in these networks is much larger than the packet (or cell) transmission times. In this case, most flow control mechanisms devised for ATM/B-ISDN networks are not based on the feedback control mechanisms. In high-speed networks, a feedback control packet carrying information about the state of the network (or window size, as is the case in low-speed packet networks) takes a longer time in propagation delay to travel than the packet transmission time. Meanwhile, the congestion state in the network may have changed before such information is received at the source. Thus, the control is obsolete before it can affect the proper state of congestion. For these reasons, in high-speed networks source shaping and source-based control strategies are devised so that before a call enters the network, the condition of the network is evaluated and a call is allowed (or accordingly shaped) to prevent congestion. Among the techniques under consideration is the leaky bucket method (48). In this method, in order to make sure that different sources of broadband network services do not exceed their contracted rates, they are subjected to input rate control at that entry point to the network. Traffic units (e.g., cells) that exceed the contracted rate are discarded (and/or delayed) before they enter the network.

Performance Issues of Multiplexers/Concentrators in Packet Switching

Multiplexers and concentrators are among the devices used for carrying traffic from customer devices into the packet-switched networks. A multiplexer provides several input links and one output trunk (link) in the time division or frequency division multiplexing mode. Generally, the number of time slots and/or frequencies available at the output of the multiplexer is equal to the number of input links; as such, no call blocking is experienced. However, assuming that calls may be carrying randomly interleaved packets on each input link into the multiplexer, there is the possibility of queueing inside the multiplexer during traffic bursts and in periods when the output capacity of the time/frequency slots may not be sufficient to handle the call burst. In such circumstances, it is then critical to size the buffer inside the multiplexer so that packets are not lost (or lost with an acceptably small probability) (49). In the case of concentrators, usually fewer output lines than input are provided, and the task is to accommodate the incoming calls with an acceptable probability of blocking. The general method for analysis and evaluation of concentrators and multiplexers is the use of the Markov processes (50).

In the following, we give a hypothetical example of a concentrator in a packet-switched environment. Suppose such a concentrator has the capacity for M incoming links and N outgoing links ($N < M$). Thus, at any time, a maximum of N calls may be in progress in this system. Also, suppose that sessions/calls arrive at this system at the rate of λ_c calls per unit time. Within each call, suppose packets arrive as a Poisson process with the rate λ per unit time. The objective here (in contrast with a similar analysis performed in Ref. 7 in which the concern is calculating the probability of call blocking) is to calculate the content distribution of all buffers associated with each of the concentrator output links.

If we assume that there are N calls in progress in this system, then the packet arrival rate to each of the output links is also λ. Assuming that packets are exponentially distributed with a mean $1/\mu$ and that each output link has K buffers allocated to receive packets from a session/call assigned to that output link, we find that the buffer content distribution follows that of an $M/M/1/K$ queueing system. Many references (e.g., Refs. 7 and 50) give explicit expressions for this simple queueing model. The probability of n packets in each of the output link buffers is

$$p_n = (1 - \rho) \times \rho^n/(1 - \rho^{K+1})$$

and the probability that the buffer is full and thus upon arrival a packet may be lost is

$$p_K = (1 - \rho) \times \rho^K/(1 - \rho^{K+1})$$

where $\rho = \lambda/\mu$. In Ref. 7, for a circuit-switched concentrator such as the above, the probability of call blocking, among other distributions, is calculated.

We notice that, in this simple example, we assumed complete partitioning of output buffers among the N output links. Other schemes of buffer sharing among the output links, such as partial sharing/partial partitioning and total sharing, are also available in the literature (51).

Integrated Network Model

Service integration in packet switching means that possibly the switching fabric and most likely the transmission facilities are shared among all services without distinction. There are many possibilities of network integration in which the modeling tools for performance analysis would be specific to the type of integration. A hybrid method of integration has also been proposed and analyzed in which packet- and circuit-switched services may coexist on the switching and transmission media. Examples are given in Refs. 52–54 for the transmission cases and in Ref. 55 for the cases of circuit-switched services.

The alternatives of integration are many, and the asynchronous transfer mode (ATM) is the latest switching and transmission technique that has been considered for implementation. For example, in this technique, all entities, data, voice, video, and the like are transmitted as fixed units called *cells*, and the switching of these entities may be handled uniformly among all services. There are many ways to differentiate among the services that may use ATM in terms of their performance requirements and offered GoS. However, the general notion is that all entities will be treated in a similar fashion as far as switching and transmission are concerned. In other integration methods, a controller may decide (perhaps dynamically) on the bandwidth that needs to be allocated to the different services that share the network facilities and switching (56).

Among the methods recently adopted for the analysis of ATM-type networks is the fluid flow approximation technique. The general theory of these methods is similar to those described in Ref. 57. A number of recent publications, most notably by D. Anick et al. (49), analyze these systems using fluid flow models. Such models are shown to be accurate and effective in terms of processing/computation time and accuracy of results (58). In examples of such models, it is assumed that sources are in one of several states and in each state may emit fluid at a given rate. While in a given state, fluid flow remains constant at that rate. The duration of the fluid flow at the given rate is generally assumed exponentially distributed. After a source departs this state, it enters another state with another fluid flow rate. The process follows a Markov process with multiple states and is analyzed with buffer content as a continuous variable, which results in a tremendous reduction in the state space of the system (in contrast to the discrete state space representation). Such fluid flow approximation models have been successfully applied in the analysis of ATM/B-ISDN networks.

Future Networks and Related Performance Issues

In the last decades, separate communications networks have been deployed to support specific sets of services such as voice communications and low-to-medium-speed data services. As discussed above, during the 1980s an enormous worldwide research effort was made to demonstrate the feasibility of the packet-switched network in supporting narrowband services as well as broadband services. The challenge of the forthcoming B-ISDN as envisioned by researchers and the ITU is to deploy a unique transport network that provides a B-ISDN interface flexible enough to support all of today's services, as well as future narrowband and broadband services (59,60). In addition, the envisioned B-ISDN is expected to offer high throughput, reasonable delay, and sufficient flexibility to support the wide spectrum of services efficiently. This would require high-speed transmission and switching and efficient network-management procedures.

Broadband Integrated Services Digital Network Principles

Broadband ISDN (B-ISDN) is a high-bandwidth digital communications network that is capable of supporting bit rates in the order of several megabits per second and reaching to the range of a few gigabits per second. According to the ITU Recommendation I.121 such networks need to provide the following capabilities:

> B-ISDN supports switch, semi-permanent and permanent point-to-point and point-to-multipoint connections and provides on demand, reserved and permanent services. Connections in B-ISDN support both circuit mode and packet mode services of a mono- and/or multi-media type and of a connectionless or connection-oriented nature and in a bidirectional or unidirectional configuration.
>
> A B-ISDN will contain intelligent capabilities for the purpose of providing advanced service characteristics, supporting powerful operation and maintenance tools, network control and management. (61)

To provide such a flexible operating environment in a broadband ISDN, the ATM transport technique has been recommended by the ITU. The ATM is a fast packet-switching technique that can be used for a broad spectrum of services ranging from low speed to high speed, from constant bit rate traffic to bursty traffic, and from connection-oriented to connectionless applications. Even though the development of the ATM has been toward a broadband technology for the public network (62), it is more likely that the first real application of B-ISDN will be in a private-network domain.

Asynchronous Transfer Mode

The ATM transport technique is recommended by the ITU as a target solution for implementing the B-ISDN network (61). The term *transfer* comprises both transmission and switching aspects, so a transfer mode is a specific way of transmitting and switching information in a network. This technique will influence digital hierarchies, multiplexing structure, and switching interfaces for broadband networks. It multiplexes the wide range of traffic by dividing it into fixed-size blocks/units, called *cells*, and sharing the transmission capacity among many connections.

In ATM-based networks, multiplexing and switching of cells is independent of the actual application. So, an ATM equipment piece can handle low- and high-bit-rate connections with different burst natures. Such capability requires dynamic bandwidth allocation on demand with a fine degree of granularity. The ATM concept requires addressing new performance issues such as the impact of possible cell loss, cell transmission delay, and cell delay variation requirements by different services.

In the late 1980s, there was considerable debate and discussion in ITU Study Group XVIII (SG XVIII) and within research communities over the selection of appropriate sizes for the header and the information fields of the ATM cell (20,21,63). Specifically, the major disagreement was over the choice of either the ATM fixed-length cell (FLC) or the ATM variable-length cell (VLC) format. References 64 and 65 analyze the performance of a broadband network at the ATM adaptation layer and the ATM layer and provide some performance tradeoffs associated with the ATM FLC versus VLC formats. Note that, in the case of VLC format, the adaptation layer maps arriving messages directly into the information field of the ATM cell. In the case of FLCs, regardless of their sizes, the adaptation layer breaks all arriving messages into equally sized ATM cells. The main reasons for choosing FLC are based on the fact that it provides low delay and high throughput. Furthermore, this approach appears to provide a simple method of managing the bandwidth on the broadband channel (66). The VLC cell format, on the other hand, seems to minimize access and network utilization. This is due to the fact that a better match between cell size and message size results in less cell overhead. The final recommendation was to deploy a small FLC cell format considering various performance-related issues such as the impact of isolated and burst error events, lost/dropped and misrouted cells, clock recovery and synchronization, bandwidth allocation and congestion control, and implementation complexity.

The selected ATM cell has a fixed length of 53 octets, consisting of a 5-octet header field and a 48-octet information field. In an ATM network, the user information is divided into cells and transmitted continuously between two ends. When no information is to be transmitted, idle cells need to be sent instead to have the connection active. The idle cells are identified by a special preassigned header of nearly all zeros.

There are other cells with special headers that are specifically used for the purposes of ATM network operations and maintenance. ITU Recommendations I.150 (67) and I.136 (68) include the functional characteristics of the ATM layer and the specifications of the ATM cell structure. The cell header used between the user and the ATM network differs from that used within the ATM network. These cell headers are specified as the B-ISDN user–network interface (UNI) and the B-ISDN network-node interface (NNI) (69). Their cell header structures (shown in Figs. 8 and 9) consist of the following fields:

- generic flow control (GFC)
- virtual path identifier (VPI)
- virtual channel identifier (VCI)
- payload type (PLT)
- cell loss priority (CLP)
- header error control (HEC)
- reserved (RES)

As shown in Figs. 8 and 9, the main difference between the BISDN UNI and the B-ISDN NNI cell headers is with Bits 5–8 of Octet 1 assigned to the GFC field.

Bit

8	7	6	5	4	3	2	1

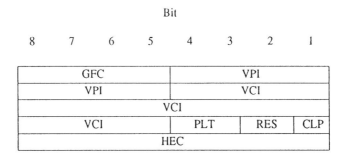

FIG. 8   Cell header at the Broadband Integrated Services Digital Network user–network interface (GFC = generic flow control; VPI = virtual path identifier; VCI = virtual channel identifier; PLT = payload type; RES = reserved; CLP = cell loss priority; HEC
= header error control).

The GFC mechanism assists in control of the traffic flow from ATM connections at the B-ISDN user–network interface for both point-to-point and point-
to-multipoint configurations. The VPI and the VCI fields provide enhanced
capabilities at the network interface. In particular, the virtual path permits
global routing of all its virtual circuits, and the virtual circuit within a virtual
path identifies the distinct connection. The PLT field carries user information
and network information. The payload of user information cells contains user
information and specific service adaptation functions. The network information
of the PLT is used to carry the network operation and maintenance information. The RES field, Bit 2 of Octet 4, is reserved for future use. It may be used
to enhance the currently defined cell header functions. The CLP is explicitly
used to indicate the cell loss priority. When low priority is selected, the cell can
be discarded, depending on the network conditions. In the case of high priority,
the network will allocate sufficient resources to handle such agreed quality
of services (QoSs). ATM (HEC) protects only the header field (not the user
information field). The HEC error-detection field is capable of detecting and
correcting single-bit errors and detecting certain multiple-bit errors in the ATM

Bit

8	7	6	5	4	3	2	1

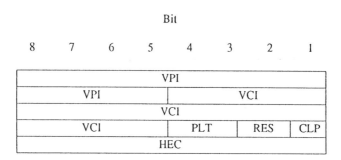

FIG. 9   Cell header at the Broadband Integrated Services Digital Network network-node interface.

cell header. When a single-bit error is detected, it will be corrected. However, when a multi-bit error is detected, the cell will be discarded.

Broadband Integrated Services Digital Network Architecture

The general architecture of the B-ISDN is described in the ITU Recommendation 1.327 (70). This recommendation proposes that the information transfer and signaling capabilities of the B-ISDN be composed of basic ISDN and B-ISDN capabilities and network-specific signaling capabilities. The layered protocol of B-ISDN is shown in Fig. 10.

The physical layer is provided by the Synchronous Optical Network (SONET) structure (71), which is an international standard for the transport of a family of signals on a fiber-based network. Its functionalities include some basic tasks (e.g., timing, cell delineation, cell header verification, etc.). The ATM-service-dependent layers of the B-ISDN protocol consist of the ATM layer and the ATM adaptation layer. These two layers are located between the physical layer and the link layer. As discussed above, the ATM cell is the basic information-transfer unit at these layers. The information field of each cell carries the user's data and is transparent to the ATM layer, and the header field, which is specific to the ATM layer, carries the necessary information associated with a specific virtual channel. The ATM layer functionalities include the VCI/VPI translation, the header generation and extraction, and the multiplexing/demultiplexing of cells with different VCI/VPI onto the same physical layer connection. The ATM adaptation layer supports those functionalities that are needed to interface the ATM layer to the non-ATM link layer. They include functions such as clock transparency, segmentation, and reassembly.

At the transmitting end, for example, the arriving messages from the link

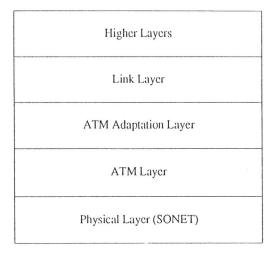

FIG. 10   Layered protocol of the Broadband Integrated Services Digital Network (SONET = Synchronous Optical Network).

layer are broken into ATM cells by the ATM adaptation layer. The receiving end reassembles the adaptation layer messages from the information field of ATM cells before delivery to the higher layer. The ATM layer performs the multiplexing and demultiplexing of several traffic streams into the payload of the interface. The ATM layers guarantee cell sequence integrity under normal and fault-free conditions. The multiplexing and switching of cells is independent of the actual application. This indicates that ATM-based equipment can handle low-bit-rate connections as well as high-bit-rate connections. Such capabilities further provide dynamic bandwidth allocation on demand with a fine degree of granularity.

Switch Architecture

A key element in the migration toward BISDN is the creation of a high-speed switch that provides a high-performance packet interconnection from its input ports to its output ports. Such an ATM switch may need to support total throughput ranging from hundreds of Mb/s to several Gb/s, with a latency measured in the tens of microseconds and cell loss probability of 10^{-9} or smaller. Such a switch architecture further should be tailored to grow gracefully to achieve the important performance characteristics of high throughput, low delay, and small cell loss probability as dictated by ATM networks.

Note that to match the transmission speed of the network links and to minimize the overhead due to processing of network protocols, the switching of the cell is done in the hardware switching fabrics in ATM networks, unlike the traditional packet-switching approach, in which switching is done by computers running communications processes. For this, very-large-scale-integrated (VLSI) technology is used to implement the switching protocols in hardware rather than software.

The ATM switching system is responsible for two main tasks: the routing function (i.e., VPI/VCI translation) and cell transport from input links to appropriate output links. Such a switch may be constructed in a three-stage configuration of input, center stage, and output modules. The first two stages provide a self-routing interconnected fabric. A simple, distributed path-assignment control is executed at the input modules to avoid internal switch fabric congestion and to minimize the switch fabric complexity (5,72). Also, the simple routing of cells avoids the complexity of optimal scheduling (73). The output stage, composed of modest-size packet-switch modules, can be implemented with a shared-memory switch, the Starlite switch (74), or the knockout switch (75). ATM cells are queued only in the output stage in order to achieve the possible throughput and delay performances (76). The ATM switch fabric operates synchronously, with fixed-length cells arriving at its input ports in a time-slotted fashion. Each cell needs to be routed to any one of the input lines of the center stage switch that contains its destined outport port. The center stage output module will then send the cell to its appropriate output stage port using the address in the cell header. Output cell switch modules also buffer cells if there are several waiting for the same output line.

Signaling Network

The ATM signaling protocol is used to establish, maintain, and release ATM VC and VP connections on which user information is transported across a B-ISDN network. ITU has specified standards for signaling protocols across both the UNI and the NNI. The ITU recommended signaling network for B-ISDN is based on the current principle of out-of-band signaling used for the 64-kb/s ISDN under the physical D-channel signaling. Across the UNI, the existing standard for ISDN signaling employs Digital Subscriber Signaling System No. 1 (DSS1) (77). The upper layer protocols of DSS1 include the Q.931 and the Q.932 signaling protocols (78,79). These protocols specify the information elements and procedures needed to control connections across the UNI. The ITU *Specification of Signaling System No. 7*, known as the SS7 Protocol, may be used to support the NNI signaling needs (80).

Since a B-ISDN network needs to support the basic ISDN applications and new high-speed services, several new functionalities have been identified for the ATM signaling protocols (81). One newly considered UNI signaling function is based on extending the current Q.931 signaling protocol. This has resulted in extending the corresponding SS7 for the NNI Protocol. Note that the main focus of the newly considered protocols is to provide the functionality required to support broadband services.

Services

The main benefit of ATM use in B-ISDN technology is its highly flexible mechanism for supporting multimedia services. Even though such technology is aimed for public network services, it is likely that the first real application of ATM-based service will be offered in the private customer network domain. In addition to the basic voice traffic, the network will need to efficiently support a variety of traffic types from LAN and multimedia traffic to perhaps multigigabit per second broadband services.

Current ITU recommendations on B-ISDN identify the following four service classes to support connection-oriented and connectionless services (59):

1. Conversational services
2. Messaging services
3. Data-retrieval services
4. Distribution services

Even if the QoS parameters to be guaranteed for each class have yet to be defined, the B-ISDN network should be able to provide certain performance characteristics to each service class in terms of quality of service defined above. Three key issues must be fulfilled to prove that the B-ISDN is about to arrive in the near future (82): (1) showing that ATM switching nodes for such a B-ISDN network are within the reach of current VLSI technology, (2) developing a suitable protocol architecture for the communications network capable of fulfilling the required targets for the different service classes, (3) envisioning a control strategy from the overall B-ISDN that enables the network providers to provide each of the supported services with its own QoS parameters independent of the network traffic conditions.

Performance Issues

The success of the proposed B-ISDN network depends on meeting the existing user and network performance needs, as well as the ATM-related performance parameters. The guarantee of performance-related issues is crucial in promoting various telecommunications services and their acceptance by both potential end users and major network providers. So, an ATM-based broadband network must meet certain performance requirements in order to be accepted by both the service users and the network providers. Specific performance parameters need to be defined for the ATM layers in addition to the performance parameters already introduced for existing high-speed packet-switching networks. Further, the relationship between ATM-layer-specific performance and the performance of the neighboring layers in B-ISDN needs to be specified.

Extensive effort has been made in recent years to develop performance criteria for ATM-based B-ISDN networks (17,18). The ATM-related performance issues are defined in terms of QoS and network performance parameters. The basis of deriving a performance measure framework for B-ISDN networks is from expanding the quality of service matrix described in the section, "Standards of Packet-Network Performance." Under such criteria, Refs. 17 and 19 provide an assessment of the technical quality of B-ISDNs at points at which ATM cells are observable. Note that the technical quality measure consists of connection establishment, user information transfer, and connection release (see Table 1). In the following section, we briefly describe the quality of B-ISDN service related to each of these categories.

Connection Establishment and Connection Release

Connection establishment and connection release are two major performance measures that are crucial to the QoS in a B-ISDN network. Signaling protocols across the UNI and the NNI are used to establish, maintain, and release connections (77). Connection establishment further depends on network resource availaability (such as bandwidth and GoS traffic engineering parameters of blocking, delay, etc.) (19,83). Speed, reliability, and availability of connection establishment and connection release are other important performance measures that are more relevant to the user of a B-ISDN service.

User Information Transfer

Note that the QoS may be influenced not only by the ATM transport network performance, but also by higher layer mechanisms. The ITU Draft Recommendation I.35B looks at the overall performance of the ATM layer and the accuracy of ATM cell transfer between B-ISDN network nodes (84). The quality of the ATM cell transfer is defined under the following categories:

- Successfully delivered cells
- Errored cells and severely errored cells
- Lost cells
- Inserted cells

Efforts are underway to define the corresponding performance parameters of the ATM layer, as well as its relationship with the performance of the neighboring layers for both the service users and the network providers (85,86).

From the QoS point of view (discussed in the section, "Standards of Packet-Network Performance"), here we briefly assess the performance of B-ISDN user information transfer considering the quality criteria of accuracy, reliability, and speed in particular.

Accuracy of User Information Transfer. The accuracy of user information transfer is defined as the number of errored cells that occurred within the total number of delivered cells between two ends. The corresponding ATM performance parameters are cell error ratio and multiply errored cell block ratio. The ratio errored cells to the number of successfully delivered cells is defined as the *cell error ratio* (84). The *multiply errored cell block ratio* is defined to address those cases when the errored cells are clustered together (87).

Reliability of User Information Transfer. The network performance corresponding to reliability of user information transfer is defined on the basis of the cell loss ratio and the cell insertion rate. Undetected errors in the destination address field of an ATM cell header cause either loss or misrouting of the cell, creating potential performance effects in both cases. When such errors are detected, the cell is discarded to minimize the risk of misrouting. The effect of discarded cells is likely to degrade the performance of the service being supported. The *cell loss ratio* is defined as the ratio of the number of lost cells to the sum of the number of lost and successfully delivered cells. The number of misrouted cells within a specified time interval is called the *cell insertion rate* (84).

Speed of User Information Transfer. The ITU standards bodies have defined the following delay performance measures to study the speed of user information transfer and the delay sensitivities of the services expected to be supported by an ATM-based network:

- Cell transfer delay or absolute delay
- Mean cell transfer delay
- Cell delay variation

- Differential delay
- Cell transfer capacity

The cell transfer delay is directly related to the overall end-to-end delay presented in the section, "Delay," for ATM cell transfer instead of packet transfer between two reference points. The *mean cell transfer delay* is defined as the average of a specified number of cell transfer delays. On the same service connection, the difference between a single observation of cell transfer delay and the mean cell transfer delay is called *cell delay variation*. The difference between the delay experienced on two different connections between the same end points is measured by the differential delay.

The cell transfer capacity is defined to ensure the desired bandwidth is provided for a requested service, as well as to help network providers to allocate such needed capacity with an efficient method (19,84). This parameter is described as the maximum possible number of successfully delivered cell outcomes occurring over a specified ATM connection during a unit of time.

Summary

In this article, we examined the fundamental protocols and performance aspects of packet-switched communications networks. In the first section, we discussed the basic idea behind packet-switched networks. Then, various performance-related issues were defined and studied in terms of measures that are meaningful to the network providers and the users alike. The standards of packet-network performance issues were presented on the basis of the GoS concept and those end-to-end performance requirements that are perceived by or are relevant to the user. Under the scope of standards bodies, a framework for defining the quality of communications services was discussed. We also presented some of the classical and new methods of performance analysis in packet-switched networks.

In the last section, we discussed various performance issues related to an ATM-based B-ISDN network. We observed that the ATM-cell-based technique, which is of the family of traditional packet-switched networks, has reached such a level of maturity in its research, development, and standardization that it is now a strong contender for application in the local-area network environment.

References

1. Stallings, W., *Data and Computer Communications*, Macmillan, New York, 1991.
2. Crowther, W. R., et al., Issues in Packet Switching Network Design, *AFIPS Conf. Proc., 1975 National Computer Conf.*, 44:161–175 (1975).
3. Kleinrock, L., *Queueing Systems, Volume II: Computer Applications*, John Wiley and Sons, New York, 1976.

4. Roberts, L. G., and Wessler, B. D., Computer Network Development to Achieve Resource Sharing, *Proc. AFIPS Spring Joint Comput. Conf*, 36:543–549 (1970).

5. Ahmadi, H., and Denzel, W. E., Survey of Modern High-Performance Switching Techniques, *IEEE J. Sel. Areas Commun.*, 7 (7):1091–1103 (September 1989).

6. Bertsekas, D., and Gallager, R., *Data Networks*, Prentice-Hall, Englewood Cliffs, NJ, 1987.

7. Schwartz, M., *Telecommunication Networks: Protocols, Modeling and Analysis*, Addison-Wesley, Reading, MA, 1987.

8. Viterbi, A. J., Error Control for Data Communication, *ACM Sigcom Comput. Comm. Rev.*, 6(1):22–37 (January 1976).

9. Rosner, R. D., *Satellites, Packets, and Distributed Telecommunications*, Lifetime Learning Publications, Belmont, CA, 1984.

10. Gruber, J. G., Performance Requirements for Integrated Voice/Data Networks. *IEEE J. Sel. Areas Commun.*, SAC-1 (6):981–1003 (December 1983).

11. American National Standards Institute, *American National Standard for Information Systems—Data Communications Systems and Services—User-Oriented Performance Parameters*, ANSI X3.102-1983, ANSI, 1983.

12. American National Standards Institute, Standard T1.503-1989, *Network Performance Parameters for Dedicated Digital Services—Definitions and Measurement Methods*, ANSI, 1989.

13. American National Standards Institute, Standard T1.504-1989, *Packet-Switched Data Communications Services—Performance Parameters*, ANSI, February 1989.

14. International Telecommunication Union, Study Groups VII and XVIII Have Proposed Using a Performance Matrix in Draft Recommendations X.134–137 and I.QNP, Respectively, ITU, Geneva, 1987.

15. International Telecommunication Union, *Measurement of Performance Values for Public Data Network When Providing International Packet-Switched Services*, ITU Recommendation X.138, ITU, Geneva, April 1992.

16. Richters, J. S., and Dvorak, C. A., A Framework for Defining the Quality of Communications Services, *IEEE Commun. Mag.* (October 1988).

17. International Telecommunication Union, General Aspects of Quality of Service and Network Performance in Digital Network, Including ISDN, ITU Recommendation I.350, Melbourne, 1988.

18. International Telecommunication Union, ITU Recommendation G.106 (Red Book), ITU, Geneva, 1987.

19. Noorchashm, M. R., and Glossbrenner, K. C., Major Performance Issues in Broadband ISDN, Broadband Communications, *Proc. of International Federation of Information Processing* (*IFIP*), 1992.

20. International Telecommunication Union, *Broadband Aspect of ISDN*, XG XVIII, ITU Draft Recommendation I.121, Seoul, 1988.

21. Sinha, R., T1S1 Technical Sub-Committee, Broadband Aspect of ISDN, Baseline Document, May 1988.

22. Znati, T. F., Communication Protocols for Computer Networks: Fundamentals. In *The Froehlich/Kent Encyclopedia of Telecommunications*, Vol. 3 (F. E. Froehlich and A. Kent, eds.), Marcel Dekker, New York, 1992, pp. 323–393.

23. Armbruster, H., Application of Future Broadband Services in the Office and Home, *IEEE J. Sel. Areas Commun.*, 4(4):429–437 (1986).

24. Spears, D. R., Broadband ISDN Switching Capabilities from a Service Perspective, *IEEE J. Sel. Areas Commun.*, 5(8):1222–1230 (1987).

25. International Telecommunication Union, *Part C of the Report of the Seoul Meeting* (*Jan. 25–Feb. 5, 1988*), Study Group XVIII—Report R 55 (C), ITU, Geneva, 1988.

26. Wright, D. J., and To, M., Telecommunication Applications of the 1990s and Their Transport Requirements, *IEEE Network Mag.*, 34–40 (March 1990).

27. Seguin, H., and Flichy, P., Interactive Services in French Videocommunication Networks, *IEEE Globecom '87*, 50.5.1–50.5.5 (1987).

28. Lidinsky, W. P., Data Communications Needs, *IEEE Network Mag.* (March 1990).

29. Gray, J. E., Karsell, P. R., Becker, G. P., and Gehring, D. G., Total Digital Radiology: Is It Feasible? *Amer. J. Radiol.*, 143:1345–1349 (1984).

30. Wright, D., *Business Effectiveness of Communication Systems in the 1990s*, Admin. Sci. Assoc. of Canada Conf., Montreal, 1989.

31. Mowafi, O. A., and Sohraby, K. A., MULTINET — A Computer Aided Network Design Tool, IEEE Conf., Gaithersburg, MD, December 8, 1981.

32. Chung, K., Mowafi, O. A., and Sohraby, K. A., PERFORM — WWMCCS Intercomputer Network (WIN) Performance Optimization Research Model, IEEE Comcon Conf., 351–359, September 1982.

33. Kobayashi, H., *Modeling and Analysis: An Introduction to System Performance Evaluation Methodology*, Addison-Wesley, Reading, MA, 1978.

34. Kleinrock, L., *Communications Nets: Stochastic Message Flow and Delay*, McGraw-Hill, New York, 1964.

35. Funka-Lea, C. A., Kontogiorgos, T. D., Morris, R. J. T., and Rubin, L. D., Interactive Visual Modeling for Performance, *IEEE Software*, 58–68 (1991).

36. Law, A. M., and Kelton, W. D., *Simulation Modeling and Analysis*, McGraw-Hill, New York, 1982.

37. Ferrari, D., *Computer Systems Performance Evaluation*, Prentice Hall, Englewood Cliffs, NJ, 1978.

38. Ash, G. R., et al., Real-Time Network Routing in the AT&T Network — Improved Service Quality at Lower Cost, *IEEE Globecom '92* (1992).

39. Schwartz, M., *Computer-Communication Network Design and Analysis*, Prentice-Hall, Englewood Cliffs, NJ, 1977.

40. Esau, L. R., and Williams, K. C., A Method for Approximating the Optimal Network, *IBM Sys. J.*, 5(3) (1966).

41. Monterio, J. A. S., and Gerla, M., Topological Reconfiguration of ATM Networks, *IEEE Infocom. '90* (1990).

42. Anderson, J. O., Digital Cross Connect Systems — A System Family for the Transport Family, *Ericson Review*, No. 2 (1990).

43. Girard, A., *Routing and Dimensioning in Circuit-Switched Networks*, Addison-Wesley, Reading, MA, 1990.

44. Gerla, M., The Design of Store-and-Forward Networks for Computer Communications, Ph. D thesis, Dept. of Computer Science, UCLA, 1973.

45. Kroner, H., Comparative Performance Study of Space Priority Mechanisms for ATM Networks, *IEEE Infocom '90* (1990).

46. Yegani, P., Performance Models for ATM Switching of Continuous Bit Rate (CBR) and Bursty Traffic with Threshold-Based Discarding, *Proc. IEEE ICC '92* (1992).

47. Sohraby, K., Selective Cell Discard with MMPP Traffic Model for B-ISDN Congestion Control, *IEEE Globecom '93* (1993).

48. Sidi, M., et al., Congestion Control Through Input Rate Regulation, *IEEE Globecom '89* (1989).

49. Anick, D., Mitra, D., and Sondhi, M. M., Stochastic Theory of a Data Handling System with Multiple Sources, *Bell Sys. Tech. J.* (1982).

50. Kleinrock, L., *Queueing Systems, Volume I: Theory*, John Wiley and Sons, New York, 1975.

51. Rathgeb, E. P., Theimer, T. H., and Huber, M. N., Buffering Concepts for ATM Switching Networks, *IEEE Globecom '88* (1988).

52. Sriram, K., Varshney, P. K., and Shanthikumar, J. G., Discrete Time Analysis of Integrated Voice/Data Multiplexers with and without Speech Activity Detectors, *IEEE J. Sel. Areas Commun.* (December 1983).

53. Williams, G. F., and Leon-Garcia, A., Performance Analysis of Integrated Voice and Data Hybrid-Switched Links, *IEEE Trans. Commun.* (June 1984).

54. Yamaguchi, T., et al., An Integrated Hybrid Traffic Switching System Mixing Preemptive Wideband and Waitable Narrowband Calls, *Electronics and Communications in Japan*, 43–52 (1970).

55. Gimpelson, L. A., Analysis of Mixtures of Wideband and Narrowband Traffic, *IEEE Trans. Commun. Tech.*, 258–266 (September 1965).

56. Schwartz, M., and Kraimeche, B., An Analytic Control Model for an Integrated Node, *IEEE Infocom '83* (1983).

57. Cox, D. R., and Miller, H. D. *The Theory of Stochastic Processes*, Methuen, London, 1965.

58. Elwalid, A. I., and Mitra, D., Analysis and Design of Rate-Based Congestion Control of High Speed Networks, I: Stochastic Fluid Models, Access Regulation, *Queueing Systems*, 9:29–64 (1991).

59. International Telecommunication Union, *I-Series Recommendation on B-ISDN*, ITU, Geneva, 1991.

60. Byrne, W. R., Kafka, H. J., Luderer, G. W. R., Nelson, B. L., and Clapp, G. H., Evolution of Metropolitan Area Networks to Broadband ISDN, *Proc. XIII International Switching Symposium*, 2:15–22 (1990).

61. International Telecommunication Union, *Broadband Aspects of ISDN*, ITU Recommendation I.121, ITU, Geneva, 1991.

62. International Telecommunication Union, *B-ISDN Asynchronous Transfer Mode Functional Characteristics*, ITU SG XVIII, TD 65, ITU Draft Recommendation I.150, ITU, Geneva, June 1992.

63. International Telecommunication Union, *Asynchronous Transfer Mode (ATM): Considerations About the Cell Size*, RTT Belgium ITU SG XVIII, June 1987.

64. Salahi, J., Modeling the Performance of a Broadband Network Using Various ATM Cell Formats, *IEEE ICC '89*, 1208–1213 (June 1989).

65. Parekh, S. P., and Sohraby, K., Some Performance Trade-offs Associated with ATM Fixed-Length vs. Variable-Length Cell Formats, *IEEE Globecom '88* (1988).

66. Coudreuse, J. P., and Boye, P., Asynchronous Time-Division Techniques for Real Time ISDN's, Seminaire International sur Les reseaux temps reel, Bondol, April 16–18, 1986.

67. International Telecommunication Union, *B-ISDN ATM Functional Characteristics*, ITU Recommendation I.150, ITU, Geneva, 1991.

68. International Telecommunication Union, *B-ISDN ATM Layer Specification*, ITU Recommendation I.361, ITU, Geneva, 1991.

69. Handel, R., and Huber, M. N., *Integrated Broadband Networks*, Addison-Wesley, Reading, MA, 1991.

70. International Telecommunication Union, *B-ISDN ATM Functional Architecture*, ITU Recommendation I.327, ITU, Geneva, 1991.

71. Bell Communications Research, Synchronous Optical Network (SONET), T1X1.4-006, February 27, 1985.

72. Hluchyj, M. G., and Karol, M. J., Queueing in Space-Division Packet-Switching, *Infocom '88*, 334–343 (March 1988).

73. Tobagi, F. A., Fast Packet Switch Architectures for Broadband Integrated Services Digital Networks, *Proc. IEEE*, 78(1):133–167 (January 1990).

74. Hung, A., and Knauer, S., STARLITE: A Wideband Digital Switch, *Proc. Globecom '84*, 121–124 (November 1984).

75. Yeh, Y. S., Hluchyj, M. G., and Acompora, A. S., The Knockout Switch: A Simple Modular Architecture for High Performance Packet Switching, *IEEE J. Sel. Areas Commun.*, SAC-5:1274–83, October 1987.

76. Suzuki, H., et al., Output Buffer Switch Architecture for Asynchronous Transfer Mode, *Proc. International Switching Symp.*, 99–103 (March 1989).

77. International Telecommunication Union, Recommendation Digital Subscriber Signaling System No. 1. In *ITU Blue Book*, Vol. 6, Fascicles VI.10, ITU, Geneva, 1989.

78. International Telecommunication Union, ITU Recommendation Q.931. In *ITU Blue Book*, V1.11, ITU, Geneva, 1989.

79. International Telecommunication Union, ITU Recommendation Q.932. In *ITU Blue Book*, V1.11, ITU, Geneva, 1989.

80. International Telecommunication Union Study Group XI, Specification of Signaling System No. 7. In *ITU Blue Book*, ITU, Geneva, 1989.

81. International Telecommunication Union Study Group XI, *Baseline Text for the Harmonized Signaling Requirements*, Temporary Document X1/4-37, ITU, Geneva, March 1992.

82. Pattavina, A., Nonblocking Architectures for ATM Switching, *IEEE Commun. Mag.*, 38–48 (February 1993).

83. International Telecommunication Union, *Recommendation E.721*, ITU, Melbourne, 1988.

84. International Telecommunication Union, *Broadband ISDN Performance*, COM XVIII-TD 31 (XVIII), ITU Draft Recommendation I.35B, Matsuyama, December 1990.

85. International Telecommunication Union, *Draft Recommendation G.82x*, ITU, Geneva, 1991.

86. International Telecommunication Union, T1Q1.3 Contributions 077R2, ITU, Geneva, October, 1991.

87. International Telecommunication Union, *New Draft Recommendation I.35B*, ITU, Geneva, 1991.

JAVAD SALAHI
KAZEM SOHRABY

Network Protocols for Packet Switching (X.25) and Service Characteristics

Introduction

The late 1960s and early 1970s were marked by the beginning of the development of data communications networks either in the form of experimental networks, sponsored by government agencies, or in the form of commercial networks created by other organizations. The data traffic services provided by these networks would parallel the voice traffic services offered by the telephone system.

The basic tenet of these networks is that packet switching, also referred to as *store and forward*, is an appropriate technology for public data networks. Contrary to circuit switching, the packet-switching method does not require the establishment of a dedicated physical path between the sender and the receiver prior to traffic exchange. Instead, packet switching views data to be exchanged between a sender and a receiver as a sequence of small segments of limited length, referred to as *packets*. Packets are then exchanged without the need to establish a dedicated physical path between the communicating stations. A header associated with each packet contains all the information necessary to route the packet from source to destination. The packet is stored in a buffer at each switching node of the network, and the header is processed to determine the next hop in the route to the destination. When the appropriate outgoing link becomes available, the packet is forwarded to the next switching node along the path to the destination.

The major benefit of packet switching is the pipelining effect resulting from the simultaneous handling of several packets by the intermediate nodes. Furthermore, packet switching does not require preallocation of the required bandwidth to handle the traffic exchanged between the source and destination. Bandwidth is allocated dynamically and traffic is handled on a delay basis. New traffic may be accepted, even when the current traffic load is high. Channel capacity is used only when traffic sources are active, thereby resulting in a more efficient use of the network bandwidth. This characteristic makes the switching technique very suitable for bursty traffic inherent in data traffic.

The flexibility of packet switching to allocate bandwidth to different applications, coupled with its efficiency for handling burstiness, prompted the design and development of several data packet networks tailored to meet specific needs and requirements of the supporting organization. In the early stages of development, however, no effort was made by the network designers to adhere to a common set of communications protocol conventions. As a result, the design of each network incorporated substantially different terminal access procedures for both host computers and slow-speed character terminals that made open communications among different organizations virtually impossible.

As the need for data communications and information exchange increased, network designers recognized that the commercial viability of data packet networks hinged largely on the development and adoption of a set of standards to

145

ease the task of interfacing dissimilar vendor products with each other and facilitating international internetworking.

The goal of defining standards for a device-independent interface between packet data networks and user devices was undertaken by the CCITT (International Telegraph and Telephone Consultative Committee). CCITT is an organ of the International Telecommunication Union (ITU), which is a treaty organization of the United Nations. In March 1993, the name of the group changed to the International Telecommunication Union – Telecommunications Standardization Sector, whereupon CCITT was officially superseded by the acronym ITU-TS. The X.25 Recommendations were issued under the auspices of CCITT. This acronym is used in relation to the X.25 standards specifications throughout this article.

The CCITT is primarily concerned with telephone and data communications systems. The members of the CCITT are the national post, telegraph, and telecommunications (PTT) administrations. In countries where the PTT administrative functions are not provided by a single entity, an agency of the national government leads the delegation to the CCITT. In the United States, for example, this representation is housed in the Department of State.

The charter of the CCITT is to study and issue recommendations on technical, operating, and tariff issues relating to telephony and telegraphy. Its primary objective is to achieve compatibility of international communications connections, regardless of the countries of origins and destinations. This objective is achieved through a cooperative standardization process that aims at producing technical recommendations about telephone, telegraph, and data communications interfaces. CCITT recommends the implementation of the protocols and interfaces specified by these documentations to its members. The global coverage that CCITT members have over practical implementation of communications protocols usually results in these recommendations becoming internationally recognized standards.

The original document of the CCITT Recommendation X.25 was based on proposals from Tymnet, Telenet, and Datapac. Tymnet and Telenet are two major public packet-switching data networks developed in the United States. Datapac is a packet-switching network developed by Telecom Canada. The first draft of the X.25 specification document, commonly referred to as the *Gray Book*, was issued in 1974. The document was officially approved in 1976 as a standard for an interface specification between the user equipment and the packet data network. A significant revision of the protocol was approved in 1982. The revised X.25 specification was more complete than its predecessor and eliminated a number of protocol specification ambiguities that led to implementation inconsistencies among different vendors. In accordance with its study periods of four years, CCITT produced subsequent revisions of the X.25 Recommendations in 1980, 1984 (*Red Book*), 1988 (*Blue Book*), and 1992. These versions aimed at further enhancing the capabilities of the X.25 interface and adding new end-to-end services.

The X.25 interface specification has reached a high level of maturity and become one of the best known and widely used protocol standards in packet-switching networks. The protocol specification has become synonymous with public packet-switched networks, and most vendors support an X.25 interface.

The purpose of this article is to describe the basic functionalities of the CCITT Recommendation X.25 Protocol architecture and present a consolidated view of the characteristics of the end-to-end services accessible through the X.25 interface. The future of X.25 in data communications is also discussed.

The X.25 Network

A public data network (PDN) is a large data network offering a range of services to any users willing to pay. In the United States, access to PDN capabilities are essentially provided by "common carriers." The services offered by these networks and their prices are regulated by the Federal Communications Commission (FCC). In Europe, PDNs are part of the PTT. A user in France, for example, subscribes to the country's PDN (Transpac).

The needs and requirements of PDN users are heterogeneous in nature and vary from one country to another. In order to prevent different countries from developing incompatible network interfaces, CCITT developed Recommendation X.25, which defines a standard interface for connection of data terminals, computers, and other systems to PDNs.

The term *X.25 network* refers to a packet-switching network that is accessed using an X.25-compliant interface. The X.25 interface specification document stipulates that local and remote systems can exchange information as long as the interface they use to connect to the PDN adheres to Recommendation X.25. The X.25 interface does not describe a specific type of a PDN and does not address the internal mechanisms of a network. The X.25 Recommendation, for example, does not contain any specification of routing algorithms. Routing is considered a network internal issue that must be addressed by the network designer. Furthermore, support of the X.25 interface does not necessarily imply that the protocol between internal network nodes adheres to the X.25 Protocol specifications. In fact, most PDN service providers implement their internal proprietary protocols, which are usually designed to optimize network performance.

X.25 Network Configuration

A typical configuration of an X.25 PDN is depicted in Fig. 1. The basic components of the network configuration include data terminal equipments (DTEs), data circuit terminating equipments (DCEs), and packet-switching exchanges (PSEs).

DTEs refer to the devices at the user's side of the network. They usually include such devices as visual display units, computer systems, and office workstations. DTEs are classified as either packet-mode DTEs or character-mode terminal DTEs (DTE-Cs). Packet-Mode DTEs have the capability of executing the X.25 Protocol to interface to a synchronous packet network. DTE-Cs are asynchronous terminals or asynchronous personal computers (PCs). These de-

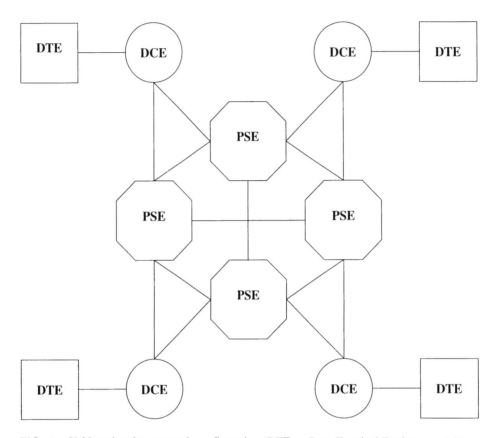

FIG. 1   X.25 packet data network configuration (DTE = Data Terminal Equipment; DCE = Data Circuit Terminating Equipment; PSE = packet-switching exchange).

vices are not equipped with the sophisticated X.25 logic to handle packets and properly interface to a synchronous packet network. Support for DTE-Cs is provided by a packet assembler/disassembler (PAD), which performs the functions of protocol conversion and packet assembly and disassembly at each end of the network.

   Data circuit terminating equipments (DCEs) refer to the equipment that provide all functions required to establish, maintain, and terminate a connection to the network, including the signal conversion and coding between the DTE and the common carrier's line. The DCE may be equipment or an integral part of either the DTE or the intermediate equipment. DCEs usually include devices such as data sets and modems.

   PSEs are switching devices that forward packets from the source DTE toward the destination DTE. The connections between the PSEs themselves and between PSEs and DCEs can use any established transmission facility, such as high-speed trunk circuits, microwave channels, or satellite links. The PSEs form the communications network backbone. As discussed above, Recommendation

X.25 does not define the internal operations within the network. Consequently, the designer of the PDN must address issues such as congestion control, routing, and failure management.

X.25 Protocol Architecture

Recommendation X.25 specifies the interaction between a DTE and a DCE of a PDN. The interaction occurs at the DTE–DCE interface and is divided into three distinct levels: physical, link, and packet.

This architecture is depicted in Fig. 2. The X.25 Protocol architecture, which was designed prior to the publication of the Open System Interconnection (OSI) Reference Model, is based on the concept of layering. The X.25 architecture corresponds to the lower three layers of the OSI Reference Model. Each layer of the architecture performs a set of specific functions required for communication and adds value to the services provided by the layer directly below it.

The Physical-Level Protocol specifies the physical interface between the DTE and the DCE. Procedures at the physical level specify the electrical, mechanical, and low-level protocols for the establishment, maintenance, and release of communications circuits to transmit binary digits of information.

The Link-Level Protocol specifies the mechanism to access the link between the DTE and DCE. Procedures at the link level specify the protocol used by the DTE and DCE to exchange data messages, referred to as *frames*, across a data link. The basic functions of a data-link layer protocol include the control of data flow over the communications link, the detection of errors created by the physical layer, the correction and retransmission of packets in error, and the suppression of duplicate messages.

The Packet-Layer Protocol allows the establishment and clearing of a networkwide connection for data transfer between DTEs. The main functions of

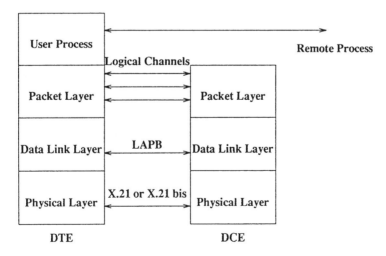

FIG. 2 X.25 Protocol architecture (LAPB = Link Access Protocol–Balanced).

the Packet-Layer Protocol include the management of virtual circuits between the user and the DCE, the specification of the manner in which control information and user data are structured into packets, and the provision of the network services between the communicating DTEs and the intermediate DCEs.

The service provided by the X.25 Protocol architecture enable higher layer protocols to operate independently from the operational characteristics and topology of the transmission facilities. The data units exchanged between packet layer peer entities in different end systems, implementing the X.25 Protocol specifications, are called Protocol Data Units (PDUs). PDUs are commonly referred to as *packets*. The PDU includes a Service Data Unit (SDU) and a header containing the Protocol Control Information (PCI). The SDU represents the user data to be transmitted by the packet layer entity to the remote service user. The PCI contains control information that identifies the data to be transmitted, the address of the destination, and the type of service required.

In order to transmit the data packet, the packet layer entity invokes the services of the Link-Level Protocol. The Link-Level Protocol encapsulates the packet into a frame. The *frame* constitutes the basic unit of transfer between two adjacent nodes. The basic structure of the frame includes a frame delimiter field, usually referred to as a *flag*, a frame control and address field, an information field, and a frame-check sequence. At the Physical-Level Protocol, the frame is viewed as a bit stream. The Physical-Level Protocol uses the specification of the transmission medium interface to transmit the bits of the frame. The packet, frame, and bit relationship in X.25 is depicted in Fig. 3. The characteristics of the three levels of protocols of the X.25 architecture are further described in the following sections.

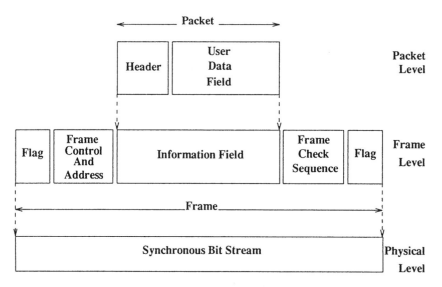

FIG. 3   Packet, frame, and bit relationship.

X.25 Physical Layer

The X.25 Physical-Level Protocol in the X.25 network architecture plays a minor role in the network architecture. It specifies an electrical or optical path between DTE and DCE. The interaction between the DTE and DCE at the physical level is described by a set of standards that define the mechanical, electrical, functional, and procedural attributes of the physical interface.

Several standards have been developed to describe different types of physical interfaces. The X.25 Recommendations state that X.21, X.21 bis, or the V series recommendations are to be used at the physical level. In practice, however, other types of physical-layer interfaces are frequently used in X.25 networks. These include RS-232-C, EIA-232-D, and RS-422 interfaces, modem and digital interfaces, proprietary satellite interfaces, and ISDN (Integrated Services Digital Network) channels.

X.21

The X.21 interface, depicted in Fig. 4, specifies a 15-pin connector that defines both balanced and unbalanced transmission modes. Among the specified pins, only 8 are currently in use.

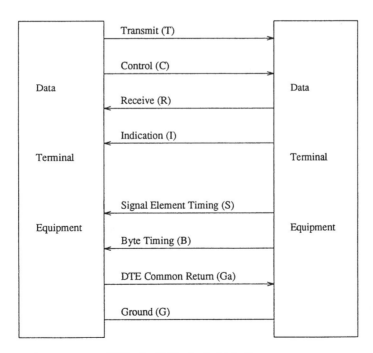

FIG. 4   X.21 physical interface.

The transmit (T) circuit of the interface carries the data from the DTE to the DCE during the data-transfer phase. However, the T circuit may also carry control information from the DTE to the DCE during the call setup or the call clear. Similarly, the receive (R) circuit carries the data and control information from the DCE to the DTE. During the data-transfer phase, the control (C) circuit remains on. The C circuit is controlled by the DTE and is used to indicate to the DCE the meaning of the data being transmitted. The DCE uses the indication (I) circuit to indicate to the DTE the type of data being transmitted on the receive circuit. Notice that during the control phase, both C and I signals may be either on or off, depending on the current state of the protocol.

Two states are allowed for the DCE and three for the DTE. When the DCE is in the *ready state*, the signal R is 1 and I is off. In this state, the DCE is ready to operate. When the DCE is in the *not ready state*, the T signal is 0, I is off, and the service is not available. The DTE may be in the *ready state*, the *uncontrolled not ready*, and *controlled not ready*. In the first state, the readiness of the DTE to operate is indicated by a steady I on the T circuit and a control OFF on C. The second state indicates that the DTE is unable to enter operational phases owing to an abnormal condition such as the fault condition. In this state, the control circuit T is 0 and C is off. Finally, in the third state, the DTE is operational, but is not ready to accept calls at the current time. This is indicated by a pattern of alternating binary 0 and binary 1 on the T circuit, while maintaining the C circuit on the off state.

The signal element timing provides the DTE with timing information. This information is used to receive the incoming data properly over the receive circuit. Byte timing provides the DTE with eight-bit byte element timing. The signal is normally on, but changes to off at the same time the circuit S signifies the last bit of an eight-bit byte. The service is optional and may be agreed on during the data-transfer phase by the DTE and the DCE.

V.24 Interface

The V.24 Recommendation is a physical standard mainly used in modems, line drivers, multiplexers, and digital service units. The V.24 Recommendation provides the functional definition of the interchange circuits between the DCE and the DTE, but does not provide the specification of the mechanical and electrical attributes of the interface. The recommendation refers to V.28 and International Organization for Standardization (ISO) 2110 for the specification of the mechanical and electrical characteristics of the physical interface.

The V.24 Recommendation defines two sets of interchange circuits. The first set, referred to as the 100 series, consists of 43 circuits that provide the ability to transfer data and a range of control information. The second set, referred to as the 200 series, consists of 12 circuits that allow the DTE to control the equipment connected to the system.

The functionalities provided by the V.24 interface circuits are defined to address the needs of any data communications environment. Most applications, however, do not require all the functionalities provided by the V.24 interface. Consequently, a subset of the V.24 circuits is normally used to address the needs of these applications.

X.21 Bis

Despite being referred to in the X.25 Recommendation document, the use of X.21 is very limited in actual installations. Only a small number of modem types and communications equipment actually implement the full functionality of the X.21 interface specification. This lack of widespread usage is due to the fact that most of the states defined by X.21 for connection establishment, data transfer, and connection termination are not all required by a network implementing the X.25 Recommendation at the packet layer level. Consequently, many installations only implement a subset of these states. The recommendation X.21 bis was designed to define an X.21 type of interface that could be used by existing modems and wiring equipment and still provide the same benefits as the X.21 interface. The X.21 bis interface uses the electrical specification of the V.28 interface and the V.24 interchange circuits' definitions.

Modems and Digital Interfaces

As stated above, the role played by the physical layer in the X.25 Protocol architecture is minimal. This made it possible to use physical media interfaces other than those recommended in the X.25 document. For example, X.25 allows the use of CCITT V series recommendations at the physical layer. These series are defined by CCITT for use with public telephone networks. They are concerned with the connection of DTEs to a modem connected to a public switched telephone network (PSTN). These standards include a precise definition of both the type of modulation scheme that must be used and the number and use of the additional interface control lines. Several of these standards can be used for modem-based access to X.25 networks.

The use of modems to connect to X.25 networks has also been achieved based on the Bell modem standards. These standards usually require some reconfigurations in order to communicate with modems that adhere to the V series recommendations. The enhanced functionalities provided by these modems, however, make them an attractive alternative to achieve X.25 physical-layer connectivity. The same observation applies to Hayes modems, which are considered a de facto standard for PC-based environments. The Hayes modems and their compatibles adhere to the CCITT V series recommendations. Equipped with a processor, these modems are designed to respond to a set of instructions referred to as *AT commands*. These commands direct the modem to carry out certain actions, such as dialing or answering the phone. Hayes "intelligent"

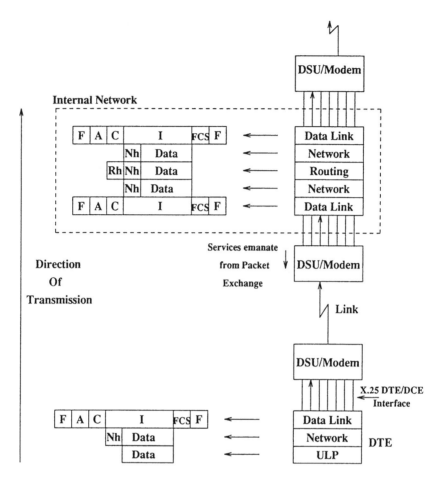

FIG. 5   X.25 digital interface (A = address field; C = control field; DSU data service unit; F = flag; FCS = frame-check sequence; I = information field; Nh = network header; Rh = routing header; ULP = upper layer protocol).

modems provide an attractive approach to X.25 connectivity in a PC-based environment.

The access to X.25 networks has not been limited to analog interfaces. The use of digital interfaces is proliferating as these interfaces become more available and affordable. A typical configuration is depicted in Fig. 5. Access is achieved based on channel service units (CSUs) and data service units (DSUs). These units achieve DTE signal conversion into bipolar digital signals, perform clocking and signal generation, and achieve conditioning, signal reshaping, and loop-back testing. Access to an X.25 network can also be achieved using the T1 digital interface technology. Furthermore, provisions have been made for an ISDN interface to X.25 networks.

X.25 Data-Link Layer Protocols

The data-link layer in X.25 networks aims at achieving reliable data transmission over a potentially unreliable transmission link. The basic functions of the data-link layer include

- establishing communication between two ends of the data link
- providing a reliable frame delivery service between the ends of the data link
- initiating the termination process to relinquish the control of the link
- coping with abnormal conditions such as loss of response or link failure

In order to support these functionalities efficiently, the data-link layer uses mechanisms such as frame segmentation, frame synchronization, error control, and flow control. *Frame segmentation* consists of breaking long streams of data into more manageable, less-error-prone data frames. *Frame synchronization* allows the destination to identify correctly the beginning and end of each received frame. *Error control* refers to the procedures used to detect and correct errors that occur during the transmission of frames. *Flow control* refers to the procedures used to regulate the traffic on an individual connection between a source and a destination and prevent the sender from overflowing the buffers of the receiver.

At the link level, the X.25 Recommendation allows either the Link Access Protocol (LAP) or Link Access Protocol–Balanced (LAP-B). Both of these protocols are derived from the High-Level Data-Link Control (HDLC), a bit-oriented protocol developed by the International Organization for Standardization (ISO 3309) for communication in an open system environment. HDLC is designed to satisfy a wider variety of data-link requirements, including point-to-point and multipoint links, half-duplex and full-duplex operations, primary-secondary and peer-to-peer interactions, and links with different propagation and transmission delay characteristics.

LAP and LAP-B are subsets of HDLC. These recommended link-layer protocols are tailored to address the needs of specific network environments and achieve greater reliability and throughput. LAP was designed for environments in which the interaction between the end nodes is based on a master-slave relationship. The decrease of this type of environment, along with the development of LAP-B (an extended version of LAP), reduced the use of LAP in X.25 network environments. Most vendors support the offering of an X.25 network interface based on a LAP-B Data-Link Layer Protocol. Integrated circuits are now available that implement LAP-B in firmware preprogrammed memory. The availability of these circuits has significantly increased the use of LAP-B.

Recently, however, access to X.25 networks based on digital interfaces as specified by the Link Access Protocol, D Channel (LAPD) is increasingly gaining widespread support of different vendors. LAPD is a subset of HDLC for use with Integrated Services Digital Networks.

Similarly, the proliferation of local-area networks (LANs) in corporate environments prompted the development of several standards that specify the use of the Logical Link Control (LLC) Protocol to transport X.25 packets. The LLC layer, developed by the Institute of Electrical and Electronics Engineers (IEEE) 802 Standards Committee, is a derivative of HDLC. The layer provides a service for moving data frames between stations attached to a LAN. The following sections provide a description of the basic functionalities of LAP-B and a discussion of the main features of LAPD and LLC as they apply to X.25 network environments.

Link Access Protocol–Balanced

The Link Access Protocol–Balanced (LAP-B) is a bit-oriented protocol designed to achieve code independence, adaptability, high reliability, and high efficiency. These requirements are satisfied based on two concepts: positional significance and coded control fields. Positional significance allows the structuring of a frame into a set of fields, each of which has a fixed position relative to a frame delimiter and has a particular meaning. The concept of coded control fields allows the use of different combinations of bits in positionally significant subfields to represent specific meanings of control functions.

The LAP-B is used to control the transfer of information frames across a balanced, point-to-point, full-duplex link configuration. The DTE and DCE at each end of a LAP-B link are classified as combined stations. The interaction between these combined stations is based on an *asynchronous response mode*. Based on this approach, either the DTE or the DCE may issue both command and response frames without soliciting permission from the other combined station. Such an approach is well suited for a balanced, point-to-point, full-duplex link configuration as it eliminates unnecessary polling delay and overhead.

LAP-B uses a single frame format for all types of data and control exchanges. The structure of a LAP-B frame is depicted in Fig. 6. Each frame is delimited by two flag fields and contains an address field, a control field, an information field, and a frame-check sequence field. The frames carry commands and responses issued by stations at each end of the link.

Flag Field

The flag field, also referred to as the *interframe signal*, delimits the frame at both ends with a 01111110 bit pattern. A single flag, however, may be used as

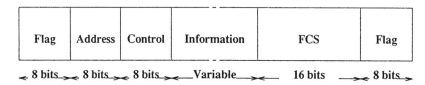

FIG. 6   Link Access Protocol–Balanced packet format.

the end of the current frame and the beginning of the next one. Stations at each end of the link are required to monitor continuously the link and hunt for the flag sequence in order to synchronize on the start of the frame.

The success of the synchronization scheme depends on the unique occurrence of the flag field within an LAP-B frame. The LAP-B, however, is a bit-oriented, code-transparent protocol that allows arbitrary bit patterns to be part of the user's data stream. The occurrence of the flag bit pattern within a frame may destroy the frame-level synchronization.

In order to prevent this pattern from occurring in the middle of a frame, a procedure known as *bit stuffing* is used. Based on this procedure, the sender inserts a zero after the occurrence of five consecutive ones within a frame. Bit stuffing applies for any field within the frame, including the address, control, information, and frame-check sequence fields. At the receiving end, the destination station detects the starting flag and monitors the rest of the bit stream. When it encounters a pattern of five consecutive ones within a frame, the receiver examines the sixth bit. If this bit is set to zero, the receiver considers it to be a stuffed bit and deletes it. A value of one causes the receiver to examine the seventh bit and accepts the pattern as a delimiter if this bit is a zero. If the sixth and the seventh bits are both set to one, the receiver continues monitoring the bit stream. At the occurrence of at least 7, but no more than 15 consecutive ones, the receiver interprets the bit sequence as an *abort* signal, which indicates a link error or failure. A pattern of 15 or more consecutive ones is interpreted by the receiver as an *idle* pattern.

Address Field

The address field is used to identify a station uniquely on a data link. The field does not contain any reference to higher layer protocols. The X.25 Recommendations specify that, in an LAP-B configuration, the address field identifies the DTE as A and the DCE as B, where A = 00000001 and B = 00000011. Furthermore, the protocol requires that a station sets the address field of the frame to the address of the receiving stations when it transmits a command and sets to its own address when it transmits a response. The LAP-B addressing rules are summarized in Table 1.

TABLE 1 Link Access Protocol–Balanced Addressing Convention

Data Flow		Frame Type	
From	To	Command	Response
DTE → DCE		Address = 01 hex (B)	Address = 03 hex (A)
DCE → DTE		Address = 03 hex (A)	Address = 01 hex (B)

	1	2	3	4	5	6	7	8
Information	0		N(S)		P/F		N(R)	
Supervisory	1	0	S		P/F		N(R)	
Unnumbered	1	1	M		P/F		M	

Control Field

1	2	3	4	5	6	7	8	9	10	11	12	13	14	15	16
0		N(S)						P/F			N(R)				
1	0	S	0	0	0	0	P/F				N(R)				

Extended Control Field

FIG. 7   Control field format (M = unnumbered function bit; N(R) = receive sequence number; N(S) = send sequence number; P/F = poll/final bit; S = supervisory function bit).

Control Field

The control field specifies the type of the frame. LAP-B defines three types of frames: information, supervisory, and unnumbered. Each frame type has a different control field. The format of the control field associated with each type of frame is depicted in Fig. 7. A summary of the frames used in LAP-B is provided in Table 2.

The information frame (I frame) carries the data to be transmitted to the end user. Each I frame contains a send sequence number, N(S), a receive sequence number, N(R), and a poll/final (P/F) bit.

TABLE 2   Link Access Protocol–Balanced Frame Summary

Type	Commands	Responses
Information	I	
Supervisory	RR	RR
	RNR	RNR
	REJ	REJ
Unnumbered	SABM	UA
	DISC	DA
		FRMR

The N(S) indicates the sequence number associated with the currently transmitted frame. The N(R) represents the sequence number of the next frame expected from the receiver. The N(R) is also interpreted at the receiver side as a positive acknowledgment for the receipt of all transmitted frames up to frame N(R) − 1.

The basic control field uses three bits to store the sequence numbers, allowing maximum send and receive windows of seven frames. This field, however, can be extended to include seven bits. The send and receive windows are thus extended to 128 frames. The extended sequence number has been added to LAP-B in order to improve the throughput of satellite and high-bit-rate links by reducing the amount of time stations attached to these types of links spend waiting for acknowledgments when the seven sequence numbers allowed by the reduced control field are exhausted.

In order to detect lost and out-of-sequence frames, each station maintains both a send and a receive sequence variable, V(S) and V(R), respectively. V(S) represents the send sequence number associated with the next I frame to be transmitted by the station. V(R) represents the send sequence number of the next in-sequence I frame expected by the station. A frame carrying a value of N(S) that does not match the current value of V(R) is considered to be out of sequence. The receiver discards the out-of-sequence frame, but does not increment the value of V(R). The discrepancy between the value of V(R) returned by the receiver in the N(R) field of the next transmitted frame and the value of V(S) alerts the sender to "go back" and retransmit all frames starting from N(R) and forward.

The P/F bit is known as the poll/final bit. The P/F bit is called a poll bit P when used in a command frame and a final bit F when used in a response frame. The sender sets the P bit whenever it requires the receiver to acknowledge the command frame. The receiver acknowledges the frame by returning an appropriate response frame with the F bit set. LAP-B permits only one P bit to be outstanding at any time. Furthermore, the protocol does not permit an information frame to be returned in the response. Response frames can only be sent in supervisory or unnumbered frames. A summary of the use of the P/F with different types of frames is provided in Table 3.

The supervisory frames (S frames) are used for flow and error control. The S frames supported by LAP-B include the receive ready (RR) frame, the receive not ready (RNR) frame, and the reject (REJ) frame.

As described above, acknowledgments for transmitted frames are usually

TABLE 3 Link Access Protocol–Balanced Poll/Final Bit Usage

Data Flow		Frame Type	
From	To	Command	Response
DTE → DCE		Address = 01 hex (B)	Address = 03 hex (A)
DCE → DTE		Address = 03 hex (A)	Address = 01 hex (B)

piggybacked in I frames by setting properly the value of the N(S) field. However, in order to account for the case for which no I frame is available for transmission and prevent the sender from timing out, S frames are used.

The receive ready (RR) frame is a positive acknowledgment. The frame acknowledges the outstanding unacknowledged frames. A receive not ready (RNR) frame is also considered a positive acknowledgment for outstanding unacknowledged frames. The frame also indicates a receiver busy condition, which forces the sender to stop sending frames until a subsequent RR frame is issued. With the P bit, the RR and RNR frames are considered to be requests for the current status of the receiving station. The REJ frame is a negative acknowledgment. The frame forces the station receiving it to retransmit all frames starting from N(R) and forward.

The unnumbered frames (U frames) are used for a variety of control functions, including link setting and disconnection and error recovery. The set asynchronous balanced mode (SABM) frame allows the link to be set up. The extended mode of operation can be selected by issuing the set asynchronous balanced/extended mode (SABME) frame. Either the DTE or the DCE can set up the link. Link setup is usually preceded by the transmission of a disconnect (DISC) frame. The DISC frame terminates the previous mode, potentially clearing any pending traffic. If the station is not logically operational or the link cannot be set up, the receiving station replies with a disconnected mode (DM) frame. The DM frame indicates a refusal to set the requested mode. If the receiving station is capable of establishing the link, it replies with an unnumbered acknowledgment (UA) frame. The UA frame acknowledges the setting of the link in the requested mode. The frame reject (FRMR) frame is used to report errors in a received frame, including invalid control field, data field too long, and invalid count and error conditions that are not recoverable by the retransmission of the frame.

The rules governing the exchange of commands and responses specify the protocol of the interaction between the DTE and DCE. The LAP-B requires exact actions to be taken with respect to these commands and responses.

Information Field

The information field contains user data. This field is present only in I-frames and some unnumbered frames. The length of the field is not defined in the standard. Frequently, however, the length is limited to a specified maximum number of bytes, which depends on the implementation.

Frame-Check Sequence Field

Transmission impairments usually result in data errors. The procedure used by LAP-B to control errors is for the receiver to detect the error and request the retransmission of the corrupted frame. Error detection is achieved by appending to the original frame an additional set of bits that constitutes an *error-detecting code*. In LAP-B, this code is created by a *cyclic redundancy check* (CRC) of the original frame. The FCS procedure is implemented in hardware using shift registers. The transmitter divides the generator polynomial on the bit representation of the address, control, and information fields. The generator polynomial used is $x^{16} + x^{12} + x^5 + 1$. The ones complement of the remainder is transmitted as the FCS.

Link Access Protocol, D Channel

The Link Access Protocol, D Channel (LAPD), is a subset of HDLC for use with ISDN. The purpose of ISDN is to offer, over the same network, a wide range of services to support different types of applications, including voice, digital data, and video. The economy of scale achieved by integration allows the offering of these services at a lower cost compared to the cost required to provide them separately.

The transmission structure of ISDN is based on a digital pipe between the central office and the customer's ISDN interface. The capacity of the digital pipe determines the number of channels carried between users. Any access link is constructed from three types of channels: B, D, and H. The B channel is the basic user channel. Three types of connection services (circuit switched, packet switched, and semipermanent) can be set up over the B channel to carry digital data and voice. A circuit-switched digital service uses the D channel for connection establishment and the B channel to carry digital data. The packet-switched service allows users to exchange data packets over an X.25 network. A semipermanent switch is similar to a leased-line service. No connection setup is required.

The D channel is used to carry signaling information to control circuit-switched calls on the associated B channel. At times when no control signal is waiting, the D channel can be used for packet-switching traffic or low-speed telemetry.

The H channels are provided to carry information at high bit rates. The user may use these channels as a high-speed trunk to carry applications with high bandwidth requirements (e.g., video, high-quality audio, video, large volumes of data streams, and fast facsimile).

The LAPD has been defined to control the flow of information frames associated with the signaling and call setup over the D channel. The LAPD is based on the HDLC Protocol. The protocol provides for unnumbered supervisory and information-transfer frames. The procedures used by LAPD for acknowledged data-transfer services are similar to the HDLC equivalent service. The unacknowledged data service, however, uses the information field of an unnumbered information (UI) frame to carry the data. Notice that the UI frame is not supported by LAP-B.

The LAPD frame structure is similar to the LAP-B frame format. The only exception is the structure of the address field. The LAPD has to deal with two levels of multiplexing. The first level occurs at the subscriber site, at which multiple users may be sharing the same physical interface. The second level of multiplexing occurs with each user device, in which multiple types of traffic, such as data and control signaling, may coexist. To accommodate these two levels of multiplexing, the LAPD address includes two parts. The first part, the

Terminal End Point Identifier (TEI), identifies a specific terminal. It is possible, however, for a single device, such as a terminal concentrator, to be assigned multiple TEIs. The second part, Service Access Point Identifier (SAPI), identifies the point at which the data-link layer services are provided to the layer above, namely, a Layer 3 entity within the device.

Despite the few minor differences between LAP-B and LAPD, it is possible to use a LAPD channel instead of a LAP-B link to transport X.25 packets between the DTE and the packet handler of the ISDN node. A most commonly used approach, however, uses LAPD to establish an ISDN B channel between the DTE and the ISDN node. The B channel is then used to exchange X.25 packets, encapsulated into LAP-B frames, between the DTE and the packet handler of the ISDN node.

Logical Link Control

The main purpose of the LLC sublayer is to shield the physical characteristics of the underlying network and provide a uniform network access interface to the higher layer protocols. The functionalities provided by the LLC sublayer are equivalent to the upper sublayer of the OSI data-link layer.

The LLC sublayer provides three types of services. (A fourth type of LLC, LLC Type 4, is under development. It aims at offering a stream-oriented, high-speed data transfer for point-to-point and multicast-based environments. The LLC allows a sequence of bytes, rather than packets, to be exchanged between multiple concurrent connections between the same set of source and destination addresses.) The first type, *unacknowledged connectionless service* (LLC Type 1), allows the user to initiate the transfer of service data with a minimum of protocol overhead. Typically, this service is used when functions such as error recovery and sequencing are provided by higher layer protocols. The second type of service, *connection-oriented service* (LLC Type 2), provides the user with the means to set up a link-level logical connection before initiating the transfer of any service data. This service guarantees reliable sequential delivery of service data units. The service also performs flow control to regulate traffic between the sender and the receiver. The third type of service, *acknowledged connectionless service* (LLC Type 3), provides a mechanism for the user to acknowledge transmission without having to establish a connection. This service has been created to cater to the requirements of a class of real-time applications that cannot tolerate the time overhead of setting up a logical connection prior to sending service data units.

Based on the above service types, the IEEE 802.2 standards define four classes of LLC sublayers. The first class, Class I, supports only LLC Type 1 services. The second class, Class II, supports LLC Type 1 and LLC Type 2 services. The third class, Class III, supports LLC Type 1 and LLC Type 3 services, while the fourth class, Class IV, supports all LLC service types. In practice, however, only Class I is widely implemented, especially in the technical and office environment.

The LLC Protocol defines a data unit carried in the information field of the media access control (MAC) frame. The format of the LLC Protocol Data Unit is shown in Fig. 8. The SSAP and DSAP octets contain the source service access point and destination service access point, respectively. Service access points can be viewed as communications ports provided by the LLC Protocol to the higher layer protocols. Service primitives can be used by protocols of the higher layers to require services provided by the LLC sublayer. The I/G bit of the DSAP indicates whether the address is an individual SAP or a group SAP. Individual addresses refer to a specific SAP, while group addresses usually refer to several SAPs. The C/R bit of the SSAP indicates whether the protocol data unit carries a command or a response.

The control field of the Protocol Data Unit indicates the type of frame currently being sent. The control fields bear strong resemblance to the control fields of the HDLC Protocol. As with HDLC, three frame formats are defined: information frames, supervisory frames, and unnumbered frames. The use of these frames is defined by the type of operations, namely, Type 1 (connectionless service), Type 2 (connection-oriented service), and Type 3 (acknowledged connectionless service).

The information field contains a packet from a higher level layer. The IEEE 802.2 Standard does not specify a limit on the length of the information field.

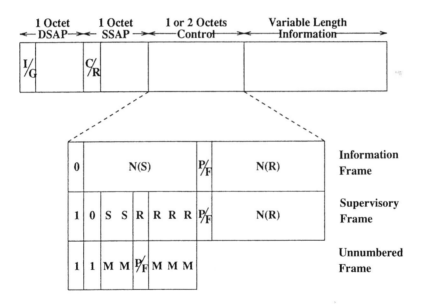

N(S) : Transmitter Send Sequence Number

N(R) : Transmitter Receive Sequence Number

P/F : Poll / Final bit

S : Supervisory Function Bit

M : Modifier Function Bit

R : Reserved and Set to Zero

FIG. 8 Logical Link Control Protocol Data Unit (DSAP = destination service access point; G = group; I = individual; SSAP = source service access point).

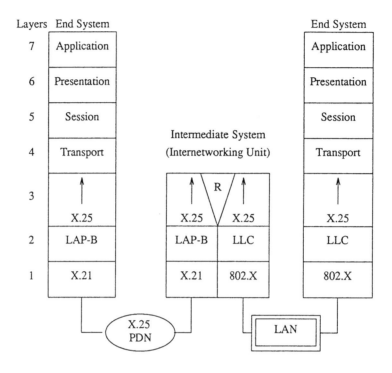

R : Router

FIG. 9   X.25 and local-area network (LAN) internetworking (PDN = public data network).

The specifications of the MAC layer, however, may impose a maximum size on the MAC information field.

The services provided by the LLC layer allow for the establishment of a symmetrical logical connection between two DTEs over a broadcast medium. The X.25 specification, however, allows for an asymmetrical, point-to-point connection between the DTE and DCE at each side of the network. Despite this major difference, it is possible to achieve internetworking between LANs and X.25 networks.

Several standards describe how the LLC Protocol can be used to transport X.25 packets across LANs. A typical configuration is depicted in Fig. 9. In this configuration, the internetworking unit performs the functionalities of the X.25 network. These functionalities include processing of call setup and clear packets and mapping the MAC addresses into the corresponding X.25 network addresses.

X.25 Packet Layer Protocol

The Packet Layer Protocol (PLP), defined by the X.25 Recommendations, specifies a virtual-circuit service to the end user. The primary functions required to provide this service include

- the establishment, supervision, and clearing of virtual circuits between the local and remote DTEs
- the formation of control and data packets
- the exchange of control and data packets between the local DTE and DCE and between the remote DTE and DCE

The concept of virtual circuits and logical channels are fundamental to all PLP procedures. The description of the basic characteristics of an X.25 virtual circuit and a discussion of the mapping between virtual circuits and logical channels are provided next.

X.25 Virtual Circuits and Logical Channels

The X.25 PLP operates on the premise of virtual circuits and logical channels. A *virtual circuit* is a logical association between a local DTE and a remote DTE. The association has an end-to-end significance and defines a logical path between the calling DTE and the called DTE. Users of the X.25 network perceive a virtual circuit as a dedicated physical circuit. In reality, however, the underlying physical circuit may be multiplexed among several users. Furthermore, packets between the calling and called DTEs may follow different physical paths. It is the responsibility of the network internal routing protocol to select the most expeditious route through the intermediate nodes for each packet to take. As stated in "The X.25 Network," the specification of the routing protocol is outside the scope of the X.25 Recommendations.

A *logical channel* is the local connection relationship between the user DTE and the network DCE. A logical channel has significance only at the DTE–DCE interface on either side of the network. The concept of a logical channel allows a DTE to establish multiple virtual circuits with other DTEs simultaneously. Packets generated by these virtual circuits are carried over a single physical DTE–DCE link.

The process of a local DTE establishing a virtual circuit with a remote DTE is referred to as *making a call*. An instance of a virtual circuit between a *calling* DTE A and a *called* DTE B and its association to a logical channel at each end is depicted in Fig. 10. Notice that DTE A can simultaneously communicate with multiple remote DTEs on the network as long as virtual circuits exist and can be set up to connect DTE A to the remote DTEs. Each virtual circuit is associated with a logical channel uniquely identified by its Logical Channel Identifier (LCI). The LCI is composed of a four-bit Logical Channel Group Number (LCGN) and an eight-bit Logical Channel Number (LCN). For each established connection, the network administrator is responsible for mapping a logical channel into a virtual circuit. Packets from various sources, on their way to a destination, are identified based on the virtual circuit with which they are associated, checked for potential errors, and delivered in order by the X.25 PLP controlling the DTE–DCE interface.

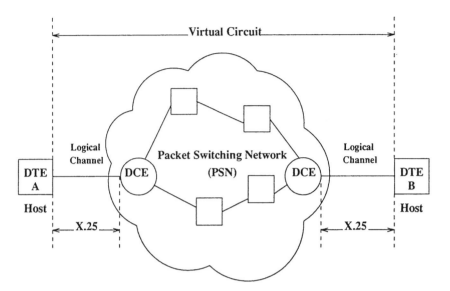

FIG. 10 Logical channels and virtual circuits.

X.25 Service Options

The X.25 service interface allows for two types of virtual circuits between DTEs: a permanent virtual circuit (PVC) and a switched virtual circuit (SVC). Furthermore, in response to demand for connectionless network service, the standard includes two additional interface options, datagram service and the fast-select facility. The datagram service, however, was subsequently dropped in the 1984 edition of the X.25 Standard specification because of lack of support by the community. A description of the basic characteristics of these services is provided next.

Permanent Virtual Circuit

The permanent virtual circuit (PVC) service is a permanent association between two DTEs that is established by a DTE user when subscribing to a PDN. The service is similar to a point-to-point private or leased telephone line. Consequently, the service does not require connection establishment prior to data transfer and connection termination on data transmission completion. Furthermore, the end-to-end connection between the local and remote DTEs remains present as long as both DTEs remain active.

In order to obtain a PVC, the user negotiates a set of particular requirements with the network service provider. These requirements are usually specified in terms of the speed of the access line between the DTE and the network interface, the volume of data transferred, and the duration of the connection. These parameters are also used by the PDN service provider to determine the cost of the connection. The tariff structures usually vary from one PDN service provider to another. Consequently, PDN service subscribers must carefully choose the parameter values so that the subsequent PDN connection is cost effective for the user's application.

In response to a PVC service request, the PDN administration allocates the resources necessary to meet the user-specified requirements. The resources remain allocated for the duration of the subscription period. The local and remote DTEs can use these resources to exchange data without the need to establish a virtual circuit.

The PVC service is usually used by applications that require that a connection is always available between the network and the DTE and may not tolerate a long response time. Since resources are reserved in advance and no call establishment is required for a PVC, the response time is usually minimal. Permanent virtual circuits are also used to carry network-management traffic.

Switched Virtual Circuit

The second type of virtual circuit, the switched virtual circuit (SVC), is frequently referred to as a *virtual call*. The service is comparable to dial-up telephone lines and represents a temporary association between two DTEs. Prior to data exchange, a connection must be established between the DTEs. The connection is dynamically set up on one of the available channels and will exist until one of the DTEs clears it or stops running its communications link.

In order to set up a switched virtual circuit, the source DTE sends to the remote DTE, via the local DCE, a call request packet. Notice that the DTE must choose an idle virtual circuit number that is either confirmed by the DCE or modified if the virtual circuit chosen is in use at the destination DTE.

On response to the SVC establishment request, the local DCE forwards the packet to the remote DCE, which in turn passes it to the destination DTE. The destination DTE may accept the connection and respond by a call-accepted packet or may reject the connection. When the source DTE receives the accepted call packet, the virtual circuit is established. Both DTEs may then use the full-duplex connection to exchange data. Either DTE may terminate the connection by sending a clear request packet, which is answered by a clear confirm packet from the other DTE.

Virtual calls at a particular location are classified as one-way incoming calls, one-way outgoing calls, and two-way calls. The division of virtual calls into categories only applies during call establishment. On call establishment, communication is full duplex between the associated DTEs.

In a one-way incoming call, circuits may only be used to receive calls from other DTEs. No calls can be initiated from the location. These virtual calls are usually used to provide autoanswer types of services. One-way outgoing calls allow circuits to be used to initiate calls to other DTEs, but do not accept incoming calls. These virtual calls usually correspond to autodial types of services. In a two-way virtual call, circuits have the full capability to both receive and initiate calls.

The switched virtual circuit service is useful for applications that do not require permanent allocation of network resources to guarantee the availability of a virtual circuit when connection establishment is attempted. Furthermore, this class of applications can tolerate small delays in data transfer. Therefore, the delay resulting from a connection setup has no significant impact on the overall performance of the application.

Fast Select

In addition to permanent and switched virtual circuits, the standard provides another form of establishing a connection, namely, fast select. The main purpose of the fast-select call is to support a transaction-oriented type of traffic and short network sessions. These types of traffic cannot effectively use either a switched virtual call, because of the overhead involved in the connection establishment and termination, or a permanent virtual circuit, because their occasional use does not warrant the permanent use of the resource.

Fast select provides for two options: fast-select call and fast-select call with intermediate clear. Both modes of transmission allow a DTE to transmit up to 128 octets of data by means of the appropriate request packet. The fast-select facility allows the remote DTE to respond with a call-accept packet that may also contain data. If desired, the data-exchange session may continue between the two DTEs until completion of the transfer. Fast select with immediate clear, however, requires the remote DTE to transmit a clear request packet immediately on receiving a call request packet.

Virtual Circuits' Configuration

As stated in the description of service options, X.25 allows a DTE to multiplex several virtual circuits over the same physical link. Each virtual circuit is identified by a 12-bit LCI. (In this context, it is normal to refer to the LCI as the LCN, even though the LCN properly identifies part of the virtual circuit identifier.) Consequently, up to 4095 virtual circuits can be established over 1 physical link connecting the local DTE and DCE. In practice, however, the number of virtual circuits available to a DTE is determined by an agreement between the user and the X.25 network service provider at subscription time.

Typically, a user subscribes for a predetermined number of PVCs and a predetermined number of SVCs. In response, the network administrator configures the DCE and informs the user of the LCIs to be used. The user uses the assigned numbers to configure the DTE if this has not been supplied by the network service provider. In private networks in which the user owns both the DTE and DCE, the allocation of LCIs is entirely the responsibility of the user.

The X.25 Protocol specification defines the rules for the assignment of LCIs to PVCs and to different classes of SVCs. These classes include one-way incoming calls, one-way outgoing calls, and two-way virtual calls.

To accommodate all categories, the rules state that the LCI numbers are divided into four groups corresponding to the PVCs and the three SVC classes. The size of any group is at the discretion of the user. The X.25 specification, however, recommends that the assignment of LCIs within a group be contiguous so that no unassigned LCIs can exist within a group's boundary. Furthermore, the X.25 Standard recommends that a range of LCIs be left unassigned between groups to allow a network administration to assign new circuits to a subscribed user without changing the existing LCI assignment and configuration.

Based on the LCI assignment convention, LCI 0 is always reserved for control and diagnostic packets common to all virtual circuits. The next range of contiguous numbers, starting at 1, is assigned to the PVC category. The next range of contiguous numbers is allocated to one-way incoming calls. If no PVCs are required, however, the lowest LCI of the one-way incoming calls starts at 1. The next range of contiguous numbers is allocated to two-way calls. A gap usually exists between the ranges of numbers allocated to one-way incoming and two-way calls. The last category of SVCs is assigned a range of contiguous numbers, the smallest of which is larger than the largest number of the two-way calls' range of numbers. Similarly, a gap exists between the two-way and one-way outgoing calls.

The assignment of LCIs to virtual circuits may be achieved by either the DTE or the DCE, depending on the direction of the call. LCIs for incoming calls are assigned by the DCE, while LCIs for outgoing calls are assigned by the DTE. As a result, a collision may occur when an outgoing DTE call establishment request is assigned the same LCI as an incoming DCE call establishment request. To minimize the likelihood of collision, a specific search method for the next available LCI is used by the DTE and DCE. The method, intended to avoid the simultaneous selection of the same LCI for two different virtual circuits by the DTE and DCE, is illustrated in Fig. 11.

Based on the search method, the DCE assigns the lowest available LCI, if any, for the incoming calls. If no LCI is currently available for the one-way incoming category, the DCE searches, from lowest to highest, the numbers allocated to two-way calls. For outgoing calls, the DTE assigns the highest available logical channel, if any. If none, the DTE searches, from highest to lowest, the numbers allocated to two-way calls. If all numbers are busy, the call request is cleared and no virtual circuit is established. Furthermore, if a collision, resulting in a call request for an outgoing call from a DTE being assigned the same LCI as an incoming call from the DCE, cannot be avoided, the X.25 Protocol specifications stipulate that the incoming call be cleared and the call request processed.

The assignment of LCI values has only local significance and is only pertinent to the local DTE and its packet-exchange DCE. At the other end of the network, the virtual circuit may be assigned a different LCI as agreed on by the remote DTE and its associated packet-exchange DCE. Consequently, a virtual circuit constitutes a logical association between the local and remote DTEs, which is viewed by the network as an association between two LCIs. The first LCI identifies a logical channel at the source DTE–DCE interface, while the second identifies a logical channel at the destination DTE–DCE interface.

Packet Layer Protocol Procedure Overview

The basic procedure for exchanging packets between the local DTE and the remote DTE can be described in terms of three phases: call setup, data-transfer,

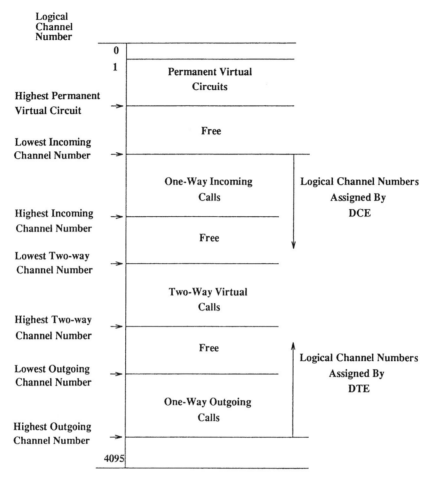

FIG. 11   Virtual channel identifier assignment.

and call clearing. A simplified illustration of the basic steps involved in each phase is depicted in Fig. 12.

The execution of any of the three phases depends on the type of virtual circuit supporting the end-to-end communication between the local and remote DTEs. In the case of an SVC service, all three phases are executed. A PVC service, however, does not require either a call-setup phase or a call-clearing phase. A PVC and its required resources are allocated automatically when the network components start up. Furthermore, the DTEs at each end of the network remain permanently associated. Consequently, a PVC supporting end-to-end communication between two DTEs is always in the data-transfer phase, eliminating the need for call establishment and clearing.

Call-Setup Phase

The call-setup phase concerns the establishment of a virtual circuit and the allocation of the communications carrier facilities to support communication between the calling DTE and the called DTE.

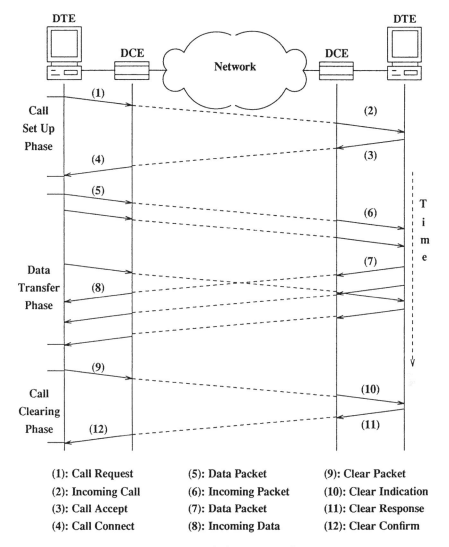

(1): Call Request   (5): Data Packet   (9): Clear Packet

(2): Incoming Call   (6): Incoming Packet (10): Clear Indication

(3): Call Accept   (7): Data Packet   (11): Clear Response

(4): Call Connect   (8): Incoming Data   (12): Clear Confirm

FIG. 12 X.25 timing sequence diagram.

The calling DTE indicates that it wishes to establish a virtual circuit with a remote DTE by sending a call request packet across the local DTE–DCE interface. The header of the packet includes the address of the remote DTE and a logical channel number selected by the calling DTE.

On receiving the call request packet, the remote DCE selects a logical channel for the incoming call and indicates to the called DTE the presence of the call establishment request by sending an incoming call packet. The called DTE indicates its acceptance of the call establishment by sending a call accept packet that contains the same logical channel number as the one contained in the incoming call packet. The called DTE then enters the data-transfer phase.

On receiving the call accept packet, the DCE issues a call connected packet to the calling DTE. The receipt of this packet completes the call-setup phase and causes the calling DTE to enter the data-transfer phase. The remote DCE, however, may advise the calling DTE that the call establishment request cannot be completed. Refusal to accept a call is usually due to lack of resources and results in a clear indication packet being issued to the calling DTE. The packet indicates the cause of the call establishment failure. The most common causes of call-setup failure include lack of free logical channels, busy number, access barred, network congestion, service out of order, invalid call, call not obtainable, reverse charging not allowed, local procedure error, and remote procedure error.

In order to keep track of the state of an ongoing virtual call, the local and remote DTEs store in their mapping tables the amount of information required to identify the call and its associated physical resources. The stored information may include the calling and called DTE addresses, the LCN associated with the virtual call, the physical link identifier that carries the virtual call traffic, and timers to keep track of the local and network connections' status. The timers are turned on upon transmission of a call-setup packet to the local remote DTE. The DTE must include a recovery action if the timer expires and no call connected packet is received.

The X.25 Recommendations do not specifically define the type of information stored in the mapping table. Consequently, the details of the mapping procedure and the type of information stored are vendor specific. A generic example, illustrating the method and information used to maintain the logical mapping of a virtual call, is depicted in Fig. 13. In this figure, the local DTE is referred to as DTE L, while the remote DTE is referred to as DTE R. DTE L is connected to the local DCE through Physical Link 1. Physical Link 10 connects the remote DCE to DTE R. At the DTE L side, the outgoing call is assigned LCN 10, while at the remote side LCN 20 is selected to identify the incoming call. The mapping information is stored at both ends of the virtual circuit and is used to identify and route the packets associated with the call.

Data-Transfer Phase

In the data-transfer phase, traffic flows in either direction, at any time, between the calling and the called DTEs. The type of traffic includes user data and control data. User data carry the user's information, while control data allow the DTE and DCE to regulate the flow of traffic between each other.

User messages that are longer than the maximum packet size are segmented into several packets. Each packet, except the last one usually, carries the maximum amount of information allowed. Each packet is identified by a sequence number. These numbers allow the DCE and DTE to detect the loss or duplication of packets and to control the flow of data across the network.

Call-Clearing Phase

A DTE or a DCE may request the clearing of a switched virtual circuit by issuing a clear request packet. Notice that the request can be issued by either

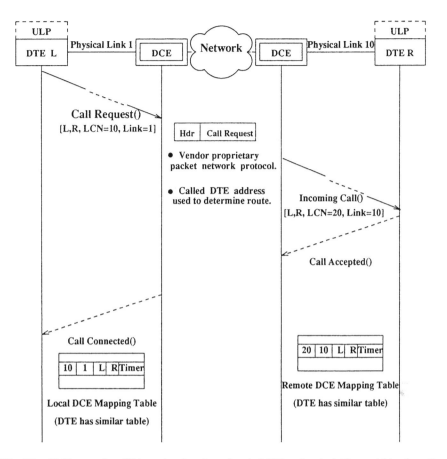

FIG. 13 X.25 mapping (Hdr = header; L = local; LCN = Logical Channel Number; R = remote).

end of the virtual circuit. In response to a clear request packet, the remote DCE issues a clear indication to the attached DTE, which replies with a clear confirmation packet. When received by the call-clearing initiator, the clear confirmation packet signals the completion of the virtual call and causes the LCN to become available for another call establishment.

X.25 Packet Types

In order to provide the functionalities required to initiate, maintain, and clear a DTE-to-DTE call, the X.25 Protocol defines a number of packet types. These packets, which can be grouped based on the functions they perform, allow

- establishment and clearing of the switched virtual circuit
- data transfer and interrupt

- flow control
- reset and restart of virtual circuits
- error indication and management

The main types of packets and a brief description of their basic functionalities are listed in Table 4. The first group of packets handles call setup and clearing. Packets in the second group are used to carry data and interrupts. Data packets carry user information, while interrupt packets allow the transmission of unsequenced data packets to handle unusual conditions. The third group

TABLE 4   X.25 Packet Types

Call-Management Packet Types	
DTE ⇒ DCE	DCE ⇒ DTE
Call request	Incoming call
Call accepted	Call connected
Clear request	Clear indication
DTE clear confirmation	DCE clear confirmation

Data-Transfer and Interrupt Packet Types	
DTE ⇒ DCE	DCE ⇒ DTE
DTE data	DCE data
DTE interrupt	DCE interrupt
DTE interrupt confirmation	DCE interrupt confirmation

Flow Control Packet Types	
DTE ⇒ DCE	
DTE RR	DCE RR
DTE RNR	DCE RNR
DTE REJ	

Virtual Circuit Management Packet Types	
DTE ⇒ DCE	DCE ⇒ DTE
Reset request	Reset Indication
DTE reset confirmation	DCE reset confirmation
Restart request	Restart indication
DTE restart confirmation	DCE restart confirmation

Error Indication Packet Types	
DTE ⇒ DCE	DCE ⇒ DTE
	Diagnostic

of packets is dedicated to handle control flow. Flow control packets allow the DTE and DCE to regulate the flow of traffic between them so that one side does not transmit packets faster than the other side can receive them. The fourth group of packets handles the reinitialization of switched or permanent virtual circuits and resets packet sequence numbers to zero. Packets in the fifth group are issued by the DCE to handle fault diagnosis. The following sections describe in more detail the encoding of these packets and examine the main functionalities of these packets.

X.25 General Packet Format

X.25 packets are carried in the information field of an LAP-B frame. The general format of the X.25 packet is depicted in Fig. 14. Each X.25 packet is identified by its *header*. The information contained in the header is used by the network to identify the packet type and carry the functionalities required to achieve control, routing, and reliable delivery of packets to their destinations.

The user data field consists of user and control information created at the upper layers. This information is transferred transparently by the X.25 Packet Layer Protocol from the local DTE to the remote DTE through the network. The Packet Layer Protocol guarantees that the content of the user data field is preserved from one end of the virtual circuit to the other, thereby ensuring that the peer upper layer at the remote end receives the content of the user data field in an unaltered form.

FIG. 14   X.25 packet format.

Packet Header Format

The basic packet header is made of three octets (see Fig. 15). This header is used in all packet types. However, certain types of packets, notably those used during call setup, use an extended header to specify additional information, such as DTE addresses, user facilities, and possibly call setup data.

General Format Identifier

The General Format Identifier (GFI) field of the packet header is a four-bit, binary-encoded field. The GFI carries vital information that indicates the layout of the remainder of the packet. The format of the GFI is depicted in Fig. 16. Bit 8 of the GFI represents the qualifier bit (Q bit) data packets, the address bit (A bit) in call setup and clearing packets, and is set to 0 in all other packets. Bit 7 of the GFI represents the delivery bit (D bit) for delivery confirmation procedures, while Bits 6 and 5 determine the modulo of the packet sequence numbering scheme used.

Qualifier Bit.   The Q bit is only relevant to data packets. The only CCITT standardized use of this bit is to distinguish between packets containing user

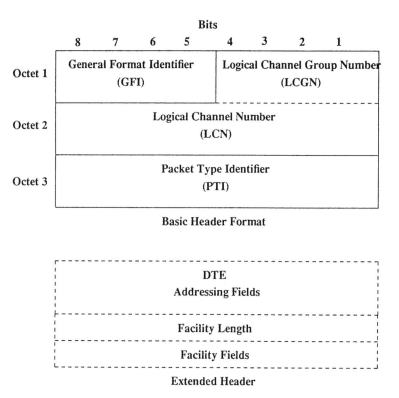

FIG. 15   X.25 packet header encoding.

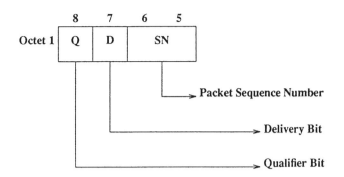

FIG. 16 General Format Identifier (D = delivery; Q = qualifier; SN = sequence numbering).

information and packets containing information dedicated to control devices. The value of the Q bit remains the same for all data packets in a complete sequence.

The functionality provided by the Q bit is particularly useful to handle the interaction between an asynchronous terminal, connected to the network via a control device such as a packet assembler/disassembler (PAD), and a host system. This interaction is governed by X.29, a CCITT protocol standard concerned with the exchange of control information across the PDN between the PAD and the host DTE. The host DTE exercises control over the PAD using control packets. The host may also alter the PAD parameters to accommodate the requirements of a particular application. The Q bit can be used to distinguish X.29 control packets from packets containing user data. When set by the local DTE, the Q bit indicates that the data being sent do not contain user information and are only significant for the control device connected to the remote DTE.

It is important to note that X.25 is totally unaware of the Q bit-related activities. This has prompted some vendors to use the Q bit to create their own "qualified" proprietary protocol within the packet.

Address Bit. The A bit is a new feature that was added to the CCITT 1988 X.25 Protocol recommendations document. The A bit is the same as the Q bit. Contrary to the Q bit, which is relevant only for data packets, the A bit is meaningful only for call setup and clearing packets.

The A bit identifies two possible formattings of the address fields. When set to 0, the A bit identifies a non-Type of Address/Network Plan Identification (non-TOA/NPI). This format, which must be supported by all PDNs, conforms to the CCITT X.121 Recommendations. When set to 1, however, the A bit indicates that a TOA/NPI address format is being used. The format of TOA/NPI addresses conforms to a structure with components that specify a Type of Address (TOA) subfield, a Numbering Plan Identification (NPI) subfield, and an address value subfield.

The TOA/NPI format is optional and usually is supported by those networks wishing to communicate with ISDNs for which the non-TOA/NPI address format does not provide sufficient address capacity. A discussion of the overall international addressing and numbering plan, devised and administered by the CCITT, is provided in sections below.

Delivery Confirmation Bit.   The delivery confirmation bit (D bit) allows the local DTE to specify end-to-end acknowledgment of data packets. The D bit is normally set to 0 and indicates that data packet acknowledgments have only local significance. Local acknowledgments indicate that the local DCE has received the packet. Delivery of the packet to the remote DTE is "guaranteed" by the network. Some networks, however, allow the local DTE to set the D bit to 1 and obtain a remote acknowledgment. Remote acknowledgments have end-to-end significance and represent delivery confirmations rather than delivery guarantees. The manipulation of the D bit to request either a local or a remote acknowledgment is illustrated in Fig. 17.

The D bit is only significant in call setup and data packets. In all other types of packets, the bit is set to 0. The D bit may be turned on and off selectively on a per packet basis.

Use of the D bit is not universally supported on all networks. Some believe that the inclusion of this function in a network layer protocol violates the rules of Open System Interconnection architecture and protocols. They argue that end-to-end acknowledgments are best handled by the transport layer. Furthermore, its use may restrict throughput by imposing a round-trip acknowledgment delay for every packet transmitted.

Sequence Numbering Information Bits.   Two of the codes of Bits 5 and 6 (sequence numbering [SN] bits) identify the format of the packet sequence numbering scheme. The third code of these bits is used to indicate an extension to an expanded format for a family of general format identifier codes that is a subject of further study. The fourth code is reserved for other applications.

When used in the context of packet sequence numbering, the setting of the SN bits to 01 indicates that the sequence numbers are modulo 8. Consequently, the packet sequence number varies from 0 to 7. When set to 10, however, the SN bits indicate that the packet sequence numbers are modulo 128. Packet sequence numbers modulo 128 vary from 0 to 127. This is sometimes described as the *extended packet sequence numbering* facility. The majority of PDNs, however, support only a modulo 8 packet sequence numbering; the lack of support of the extended packet sequence numbering is due to the considerable increase in buffer capacity needed to store potentially up to 128 outstanding packets on every active virtual circuit in the network.

The packet sequence numbering, either modulo 8 or 128, must be the same at both sides of the DTE–DCE interface. Furthermore, the selected scheme is used for all logical channels. The Packet Layer Protocol maintains its own set of send and receive sequence counters. These counters operate independently from the data-link layer counters.

Logical Channel Identifiers

The Logical Channel Identifier (LCI) identifies the virtual circuit being used between the local and remote DTEs. The LCI is defined over 12 bits, thereby

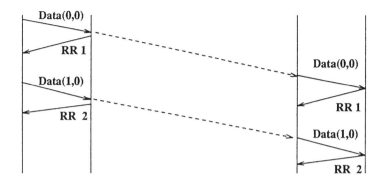

Local Acknowledgment

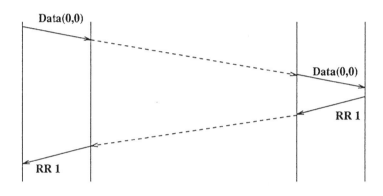

End To End Acknowledgment

Data(P(s), P(r)): Data packet containing a send and receive sequence numbers, P(s) and P(r), respectively.

RR(P(r)): Receive Ready packet containing send sequence number of the next expected packet to be received.

FIG. 17   Delivery confirmation bit semantics.

allowing a total of $2^{12} = 4906$ virtual circuits over the same physical link connecting a DTE to a DCE. The LCI is a combination of two fields: the Logical Channel Group Number (LCGN) and the Logical Channel Number (LCN). The LCGN, a four-bit, binary-encoded field, identifies 1 among 16 possible groups. The LCN, an eight-bit, binary-encoded field, identifies 1 among 256 possible channels within a group.

Packet Type Identifier

The third octet of the packet header contains the Packet Type Identifier (PTI). This field is used to differentiate among X.25 packet types. The standard specifies up to 28 packet types. In practice, however, only 15 packet types are need-ed since half of the original set of the specified packet types are duplicates. The encoding of the PTI field for different types of packets is summarized in Table 5.

The PTI contains a unique eight-bit binary identifier. For most packet types, the encoding of the field represents a unique identifier of a particular packet type. However, for those packets that require sequence numbering, such as data packets, the field is also used to carry information related to flow control, fragmentation, and reassembly. A data packet, for example, is simply identified by the setting of the low-order bit of the PTI field to 0. The remaining bits of the field, shown as x in Table 5, represent other information, including the current values of the send and receive packet sequence numbers and the indica-tion of the subsequent transmission of more packets related to the same mes-sage.

Packet Type Descriptions

The sections above provide a description of the packet's general format and a detailed description of the basic field of the packet header. The following sec-tions describe in more detail each packet type and examine the specific differ-ence among actual packet types.

Packet types can be classified into different groups based on the similarities

TABLE 5 Packet Type Identifier Field Encoding

	PTI Field	Packet Types	
Bit	8 7 6 5 4 3 2 1	From DTE to DCE	From DCE to DTE
Call setup	0 0 0 0 1 0 1 1	Call request	Incoming call
clearing	0 0 0 0 1 1 1 1	Call accepted	Call connected
	0 0 0 1 0 0 1 1	Clear request	Clear indication
	0 0 0 1 0 1 1 1	DTE clear confirmation	DCE clear confirmation
Data &	x x x x x x x 0	DTE data	DCE data
interrupts	0 0 1 0 0 0 1 1	DTE interrupt	DCE interrupt
	0 0 1 0 0 1 1 1	DTE interrupt confirmation	DCE interrupt confirmation
Flow control	x x x 0 0 0 0 1	DTE RR	DCE RR
	x x x 0 0 1 0 1	DTE RNR	DCE RNR
	x x x 0 1 0 0 1	DTE REJ	Not meaningful
Reset	0 0 0 1 1 0 1 1	Reset request	Reset indication
	0 0 0 1 1 1 1 1	DTE reset confirmation	DCE reset confirmation
Restart	1 1 1 1 1 0 1 1	Restart request	Restart indication
	1 1 1 1 1 1 1 1	DTE restart confirmation	DCE restart confirmation
Diagnosis	1 1 1 1 0 0 0 1		Diagnostic
Registration	1 1 1 1 0 0 1 1	Registration request	
	1 1 1 1 0 1 1 1		Registration confirmation

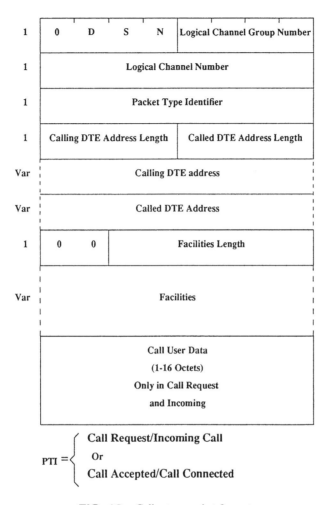

FIG. 18 Call setup packet format.

of their formats. This classification conveniently separates the packet types into identifiable groups and facilitates the functional description of the packet type within each group. The identifiable groups include call setup, call clearing, supervisory, data, flow control, interrupt, confirmation, diagnostic, and registration.

Call Setup Packets

The call setup packets consist of call request, incoming call, call accepted, and call connected. These packets only appear during the setup phase of a switched virtual circuit. Permanent virtual circuits do not use call setup packets as these circuits do not require channel establishment.

The format of call setup packets is shown in Fig. 18. In addition to the standard GFI, LCI, and PTI fields, a call setup packet contains a number of other significant fields.

The octets following the PTI field contain information describing the length, expressed in semioctets, and the actual network addresses of the calling and called DTEs. These addresses allow the identification of the DTEs involved in the switched virtual call and the establishment of the initial contact. Once the virtual circuit is established, however, the LCI in the data packets will sufficiently identify the two associated DTEs.

The addressing scheme used by X.25 networks is described in a related standard known as the CCITT X.121 Recommendation. The standard aims to provide worldwide addressing compatibility among manufacturers and network service providers. It describes an international addressing scheme that specifies that any DTE connected to a public packet data network will have an International Data Number (IDN) that may contain up to 14 digits. Private data packet networks, however, do not typically require 14 digits and generally use much shorter addresses. The details of the CCITT X.121 Recommendation are provided in later sections.

Immediately following the address fields, the format of a call setup packet allows for a facility length field and a user facility field. These fields allow users to request "optional user facilities" and obtain special network services that are not part of the normal default offering.

The first octet of the facility fields defines the length of the user facility field. The following octets contain the facility code of the selected facilities and, if necessary, a facility parameter field that contains a value for the requested facility. The actual facilities available to the users depend on their subscription profile.

The most commonly offered user facilities include reverse charging, closed user group, fast select, window size negotiation, packet size negotiation, and throughput class negotiation. Many network service providers require that users request these facilities at subscription times. Based on the user subscription profile, the calling DTE can specify any combination of facilities each time a call is set up. The possible permutations of user facilities and associated values are extensive and vary considerably from one network service provider to another. The scopes of most of the widely available user facilities are described in sections below.

The final field of a call setup packet is the optional call user data field. This field only appears in call request and incoming call packets, and its presence is optional. The main purpose of this field is to enable users to transmit a small amount of actual data to the called DTE at the call setup time. The field may contain information such as the user's identification, password, or even additional routing information for destination subnetworks that do not adhere to the X.25 network specifications. The information inserted by the calling DTE in the call user data field is transmitted transparently by the network to the called DTE.

In conventional call request packets, the call user data field may contain a maximum of 16 octets. When present, the purpose of the information contained in this field is identified by Bits 8 and 7 of its first octet. If Bits 8 and 7 are set to 00, a portion of the remaining field may contain the Protocol Identifier (PID) and supply additional protocol information. This is particularly useful when the virtual circuit involves a connection between a host DTE and a packet assembler/disassembler (PAD). The PAD uses the call user data field in the call request to identify itself and state that it requires the X.29 Protocol.

When the user facility fast select has a subscription, the maximum length of the call user data field can be extended. This special type of call request packet may contain up to 128 octets. The called DTE is allowed to respond with a call accepted packet that can also contain user data.

Call-Clearing Packets

Call-clearing packets are used to disconnect the switched virtual circuit identified in the LCI contained in these packets. A DTE, attempting to clear a switched virtual call, issues a clear request packet containing the LCI of the circuit to be cleared. The DTE receiving the clear indication replies by issuing a clear confirmation.

Call-clearing packets can only affect switched virtual circuits. They do not apply to permanent virtual circuits, which require that the logical association between the DTEs is permanently held by the network. Following a call-clearing phase, the LCI associated with the cleared switched virtual circuit becomes available for the establishment of a new virtual call.

A clearing request affects only a single virtual circuit. Following the request, all packets in transit are lost. Furthermore, resuming communication between the DTEs requires the reestablishment of the virtual circuit.

The general format of a call-clearing packet is depicted in Fig. 19. This format applies to clear request and clear indication packets. The clear confirmation packet, however, does not contain the cause field, the diagnostic field, and the user fields.

The first three fields of the packet format represent the encoding of the standard GFI, LCI, and PTI fields. The clearing cause field contains the reasons for the rejection of the call. The diagnostic code field is optional. When present, it provides more specific information about the cause of rejection.

The reasons for call clearing cover a wide variety of circumstances. Calls may be rejected due to improperly formed call request packets, incorrect DTE addresses, an invalid facility request, network congestion, or security authorization failure. The CCITT has developed clearing and diagnostic codes to cover most typical situations. Listings of the CCITT clearing and diagnostic codes are provided in Tables 6 and 7, respectively. Network equipment manufacturers have also developed their own extensions to the basic set of CCITT codes. These extensions are usually network specific and may not be correctly interpreted across PDN boundaries.

Supervisory Packets

Supervisory packets include restart request, restart indication, reset request, and reset indication. The general format of the supervisory packets is depicted in

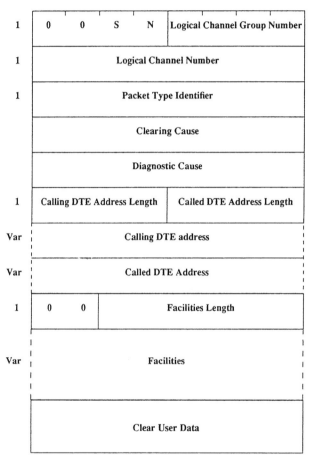

1	0	0	S	N	Logical Channel Group Number
1	Logical Channel Number					
1	Packet Type Identifier					
	Clearing Cause					
	Diagnostic Cause					
1	Calling DTE Address Length			Called DTE Address Length		
Var	Calling DTE address					
Var	Called DTE Address					
1	0	0	Facilities Length		
Var	Facilities					
	Clear User Data					

PTI : Clear Request, Clear Indication or Clear Confirmation

FIG. 19   Call-clearing packets' format.

Fig. 20. In addition to standard GFI, LCI, and PTI fields, the packet contains cause and diagnostic fields. The cause code field provides additional information about the operation undertaken, while the diagnostic code field, when present, provides the DTE with further information about the problem.

The restart packets are used to initialize the DTE–DCE interface. Initialization of the interface usually takes place either at start-up time or following a severe hardware or software failure, such as a network crash or control center failure.

The restart procedure provides the only means of placing all the virtual circuits of a DTE–DCE interface, irrespective of their previous logical state, into a known state. When issued by a DTE, a restart request packet causes all the switched virtual circuits to be cleared and all the permanent virtual circuits, currently held by the DTE, to be reset. Consequently, the LCI field of a restart request packet is always set to 0 as the actions of the request are enforced

TABLE 6 Call-Clearing Indication Codes

Clear Indication	Mnemonic	Description
DTE originated	DTE	Remote DTE cleared the call.
Number busy	OCC	Called number is busy.
Out of order	DER	Called number is out of order.
Remote procedure error	RPE	Procedure error caused by remote DTE.
No reverse charging	NRC	Collect call refused.
Incompatible destination	ICD	Incompatible called and calling DTEs.
Fast select not subscribed	NFS	Fast select not supported by called DTE.
Invalid facility request	INV	Invalid call.
Access barred	NA	Calling DTE cannot call remote DTE.
Local procedure error	ERR	Procedure error caused by local DTE.
Network congestion	NC	Congestion prevents call establishment.
Not obtainable	NP	Called number is unassigned or unknown.
RPOA out of order	RP	RPOA unable to forward call.

RPOA = recognized private operating agency

indiscriminately on all the subscribed virtual circuits. Furthermore, when the restart procedure is undertaken, all outstanding packets may be lost and can only be recovered by higher level protocols. Because of the severity of the wide-scale effect a restart procedure may have on the network operations at a given user site, the restart request packet must be used carefully.

In response to the restart request packet, the network sends a clear request to every DTE involved in a virtual call with the DTE that invoked the restart request packet and sends a reset command to all affected permanent virtual connections. The packet exchange of a restart procedure is depicted in Fig. 21. As shown in the figure, in response to a restart request, the remote DCEs that are involved in an SVC react by issuing a clear indication to their attached DTEs, while the remote DCEs that are involved in a PVC issue a reset indication to their associated remote DTEs. Meanwhile, the local DTE that issues the restart request packet receives a restart confirmation as soon as the local DCE accepts the request. The local DCE does not wait until all remote DCE–DTE interfaces complete their clearing or resetting procedures before it issues the restart confirmation packet.

The reset request packet is used to reinitialize a switched or permanent virtual circuit. The identification of the virtual circuit to be reset is carried in the LCI field of the packet. A reset request is always issued during the data-transfer phase to reset the send and receive packet sequence numbers of a particular virtual circuit to 0. As a result, all data and interrupt packets submitted to the network by a DTE at either end of the virtual circuit are removed. The LCNs at each end, however, remain active, and data transmission can be resumed immediately following the completion of the resetting procedure without the need to reestablish the virtual circuit.

The reset procedure is usually invoked when the virtual circuit experiences packet loss or duplication or when the transmitted packets cannot be re-

TABLE 7 Error and Diagnostic Codes

Code	Description	Decimal Code Range
No additional information	Invalid sender sequence number, invalid receiver sequence number	0, 1–2, 15
Packet type invalid	Invalid packets for different states	16, 17–29, 31
Packet not allowed	Unidentifiable packet, call on one-way logical channel, invalid packet type on a PVC, packet on unassigned logical channel, reject not subscribed to, packet too short, packet too long, invalid general format identifier, restart with nonzero bits 1–4 and 9–16, packet type not compatible with facility, unauthorized interrupt confirmation, unauthorized interrupt, unauthorized reject	32, 33–45, 47
Timer expired	For incoming call, for clear indication, for reset indication, for restart indication	48, 49–52, 63
Call setup problem	Facility code not allowed, facility parameter not allowed, invalid called address, invalid calling address, invalid facility/registration length, incoming calls barred, no logical channel variable, call collision, duplicate facility requested, nonzero address length, nonzero facility length, facility not provided when expected, invalid CCITT-specified DTE facility, maximum number of call redirections or call deflections exceeded	64, 65–78, 79
Miscellaneous	Improper cause code from DTE, not aligned octet, inconsistent Q bit setting, network user identification (NUI) problem	80–84, 95
Not assigned		96, 111
International problem	Remote network problem, international protocol problem, international link busy, international link out of order, transit network facility problem, remote network facility problem, international routing problem, temporary routing problem, unknown called data network identification code (DNIC)	112, 113–121

sequenced properly. Furthermore, the procedure can be initiated by the DTE, which issues a reset request packet, or by the DCE, which issues a reset indication packet.

Data Packets

Data packets carry user information during the data-transfer phase of a virtual call. The format of the data packet is depicted in Fig. 22. Two packet formats

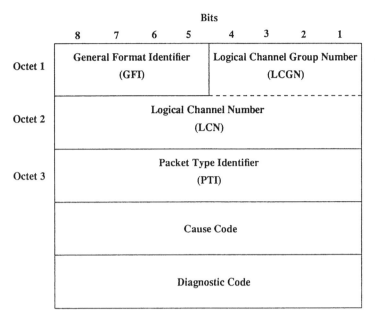

FIG. 20 Supervisory packets' format.

are allowed for data packets. The first format, identified by the setting of the SN bits to 01, is used when the packet sequence numbers are computed modulo 8. The second format, identified by the setting of the SN bits to 10, allows for an extended packet sequence numbering scheme that numbers packets modulo 128.

In addition to the SN bits, the data packet header GFI field contains a Q bit and a D bit. When required, these bits can be used to obtain special network services. The Q bit is used to qualify the purpose of the user data carried by the data packet, and the D bit is used for specifying either a local or an end-to-end acknowledgment.

The LCI field is used by the network to associate the data to a subscribed virtual circuit. Notice, however, that data packets need not carry address fields. The network relies on the association between the virtual circuit and the physical routing path, established during call setup, to route data packets between the local and remote DTEs.

The PTI for a data packet is simply the setting of Bit 1 to 0. The rest of the bits of the PTI data packet represent the packet sequence numbering information and the fragmentation bit information.

The user data field carries the information submitted to the network for transmission. The CCITT specifies the standard maximum length of the user data field to be 128 octets. The specification, however, allows the network service provider to offer other optional lengths, including 16, 32, 64, 256, 512, 1024, 2408, and 4096 octets. The user can select the desired user data field length at subscription time or by negotiation on a per-call basis. The selection of a proper packet size is crucial to achieving high throughput. Selecting a small

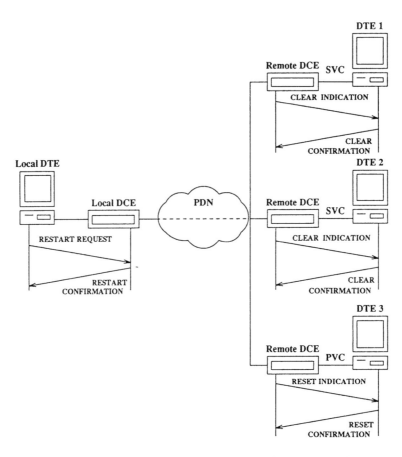

FIG. 21   Restart procedure packet exchange; local Data Terminal Equipment invokes the restart procedure (PDN = public data network; PVC = permanent virtual circuit; SVC = switched virtual circuit).

packet size may cause excessive overhead when transmitting large messages. From the buffering perspective, however, the transmission of a unit composed of a large amount of data may take as long as transmitting several shorter units. Furthermore, the transmission of a large size data unit is more likely to result in error, necessitating retransmission of the entire data unit. With smaller size data units, bit errors are less likely to occur, thereby reducing the need for data to be retransmitted.

Data Packet Sequence Numbers.   Sequence numbering of packets is used to detect potential data packet loss, enforce ordering, and achieve flow control. Most often, data packets are numbered modulo 8. In this case, each data packet includes a three-bit packet send sequence number P(S) and a three-bit packet receive sequence number P(R). Optionally, however, a DTE may request, via the user facility mechanism, the use of an extended seven-bit packet sequence number.

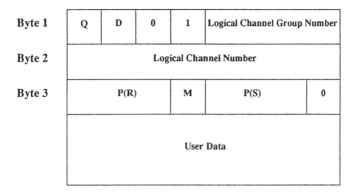

Data Packet Formats (Modulo 8)

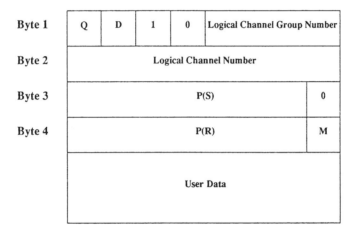

Data Packet Format (Modulo 128)

FIG. 22 Data packets' format.

The P(S) and P(R) values, modulo 8 and modulo 128, are used to number outgoing and incoming data packets consecutively. The value of P(S), assigned by the DTE on a virtual circuit basis, represents the sequence number of the outgoing data packet. The value of P(R) carried by an outgoing packet, however, represents the sequence number of the next packet expected to be received by the sending DTE on the virtual circuit. This is interpreted by the receiving DTE as a piggybacked positive acknowledgment of packet P(R) − 1 that it sent to the DTE at the other end of the virtual circuit.

The P(S) and P(R) values, which apply to packets of a particular virtual circuit, are used in addition to the similar N(S) and N(R) values managed by the by LAP-B Data-Link Layer Protocol. The difference, however, between the link and packet layer sequence numbering is significant. The link-level sequence numbers account for the traffic generated by all logical channels on the physical link and are used to control the traffic flow over the link. The packet layer

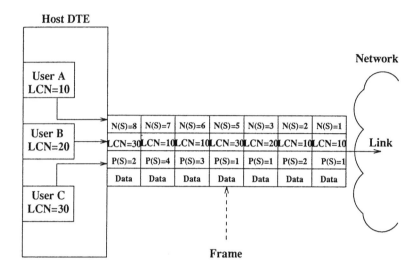

N(S): Sequence Number of Link Layer Protocol

P(S): Sequence Number of Packet Layer Protocol

FIG. 23   Link- and packet-layer sequence numbering.

sequence numbers are used to manage the traffic flow of a particular logical channel. The distinction between the link and packet layer packet sequence numbering is depicted in Fig. 23.

Fragmentation and Reassembly.   Fragmentation allows the reliable transmission of large messages by breaking these messages into multiple data packets of the maximum network allowable length. The M bit, or *more bit*, provides a mechanism to identify a contiguous sequence of packets as a complete packet sequence. This capability allows the identification of the original data blocks to be preserved, even if fragmented into smaller packets. This capability is useful as a packet must travel across a set of interconnected networks, each supporting a different packet size. The capability can also be used to provide a service to higher level applications, such as a database application, which require that a large segment of data be presented to the receiving DTE as one logical unit. If the size of the original data segment is larger than the permissible packet size on the X.25 network, the segment is fragmented into smaller data blocks carried separately in data packets. The M bit can be used to control fragmentation and reassembly.

To specify the mechanism of fragmentation and reassembly, the X.25 Protocol specifications use the M bit, in conjunction with the D bit, to define two types of packets. The first type identifies packets as A packets. An A packet is a full-length packet in which the M bit is set to 1 and the D is set to 0. The second type identifies packets as B packets. A B packet is any packet that is not an A packet. Only a B packet can have a D bit set to 1 for an end-to-end acknowledgment.

Based on the packet type categories, a complete packet sequence consists of zero or more A packets followed by a B packet. The network can combine the Category A packets and the immediately following B packets of a complete packet sequence into a larger packet. The B packets themselves must maintain separate entities as separate packets. The way in which the B packet is handled depends on the setting of the M and D bits. An example of packet sequences with intermediate end-to-end acknowledgments is depicted in Fig. 24. The figure also illustrates how packets from the original packet sequence are combined into a new sequence. Figure 25 shows how segmentation and reassembly are handled to overcome the discrepancy in packet sizes at each end of the virtual circuit.

Flow Control Packets

The flow control packets include receive ready (RR), receive not ready (RNR), and reject (REJ). These packets are only used during the data-transfer phase. The general format of these packets is depicted in Fig. 26. Depending on the modulo used for packet sequence numbers, two packet formats are possible. The normal sequencing packet format uses modulo 8 to number packets, while the extended sequencing packet format uses modulo 128. The modulo 8 version, however, is more commonly used.

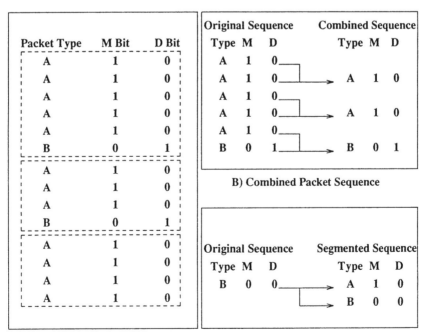

FIG. 24 Packet sequence manipulation.

In addition to the GFI and LCI fields, each type contains a P(R) field that carries the sequence number, augmented by 1, of the most recently received data packet. Consequently, the P(R) field conveys a received data packet acknowledgment. Usually, P(R) are piggybacked in the header of a data packet. Flow control packets are used to update P(R) when no data are available.

The RR packet can be used by either a DTE or a DCE. When issued, the packet indicates the readiness of its sender to receive data packets, starting with the sequence number encoded in the P(R) field. The packet can also be used to acknowledge any packets previously received. The RNR can be issued by either a DTE or a DCE. When used, the packet indicates that its sender is temporarily unable to accept data packets. The P(R) value of the packet is also used to acknowledge any packets previously received. Submission of a RNR packet by a specific DTE usually causes the network to issue RNR packets to all associated DTEs in an effort to "choke" the packet flow into the network and prevent

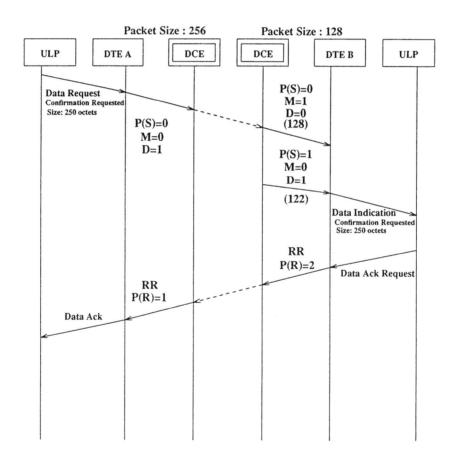

ULP : Upper Level Protocol

FIG. 25   Packet segmentation and reassembly.

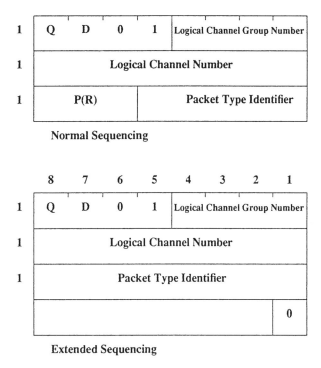

FIG. 26 Flow control packets' format.

buffers from overflowing. The RNR condition is usually cleared by the transmission of a RR packet when the conditions of the network improve.

The REJ packet can only be issued by a DTE to a DCE. The packet indicates the rejection of the received packet. Packets are usually rejected if they carry the incorrect sequence number. This usually occurs when the packet with the expected sequence number has been corrupted in transit and therefore discarded. The REJ packet is considered as a request for the retransmission of all missing packets, starting with the sequence number contained in the P(R) field.

The REJ packet is an optional user facility selected at subscription time. Normally, such a feature is not required as the data-link layer is expected to deliver the packet reliably and in sequence. Furthermore, more severe problems are expected to be resolved by resetting the virtual circuit. Nevertheless, this facility may be useful in a setting in which the DTEs are not equipped with a large enough memory to cope with a full window of packets delivered by the data-link layer. This feature, however, is not supported by all network implementations.

The packet layer flow control scheme supplements the flow control scheme supported by the LAP-B at the data-link control level. The two schemes are virtually similar in format and procedures. The main purposes of the packet layer flow control scheme are to control the traffic flow between the sender and the receiver and prevent one side from transmitting data faster than the other side can handle. The transmission of data packets is controlled separately for each direction. New data packets can only be transmitted on authorization from the receiver. The authorization is contained in the packet sequence number field of a data packet, an RR packet, a RNR packet, or a REJ packet.

The data flow control uses a sliding window protocol. Based on this protocol, packets are numbered cyclically, ranging from 0 to the value of the packet sequence numbering modulo minus 1. The extended modulo 128 packet sequence numbering scheme, for example, allows the packet sequence numbering fields to contain the maximum value of 127.

The window size, however, determines the maximum number of data packets that can be outstanding across the DTE–DCE interface. From the sender's perspective, the window size specifies the number of data packets a DTE can transmit before it must wait for an acknowledgment. Similarly, from the receiver's perspective, the window size specifies the number of data packets a DTE can receive before it must send an acknowledgment. A window size of 3, for example, allows a sender to send at most three packets before waiting for an acknowledgment. Consequently, if the value contained in the P(R) field of the most recently received packet is 6, the sender can send a first packet with a P(S) value of 6, a second packet with a P(S) value of 7, and a third packet with a P(S) value of 0, assuming a modulo 8 packet sequence numbering scheme. On transmitting the third packet, the sender must wait for the receiver's authorization before transmitting new packets. The main features of the window-based flow control protocol, using different window sizes at each end of the virtual circuit, are depicted in Fig. 27.

The value of the window size may be allocated on a per virtual circuit basis. The standard window size value for each direction is 2. Other values, however, can be negotiated at the subscription time.

Interrupt Packets

The normal exchange of data packets between two DTEs may not be adequate to cope with urgent situations that may arise during the data-transfer phase. The need to transmit urgent data may occur, for example, when a terminal user decides to "abort" the execution of a faulty program at the remote host. Using the regular data packet channel to send the abort request may produce an undesirable delayed effect owing to the time it may take to deliver the amount of data currently queued and waiting transmission and delivery.

To deal more efficiently with the delivery of urgent data, X.25 offers an *interrupt* procedure. The procedure allows a DTE to send a limited amount of "out-of-band" data irrespective of the current packet sequence numbers or flow control constraints. The operations of an interrupt procedure are depicted in Fig. 28. Only one interrupt procedure can be in progress for a particular virtual call and DTE at any time. Urgent data are carried in control packets, referred to as *interrupt* packets. These packets are used during the data-transfer phase and are delivered to the destination DTE at a higher priority than normal data packets. The general format of these packets is depicted in Fig. 29.

The interrupt packet contains the usual packet header, except for the packet sequence number fields. Interrupt packets are unsequenced and do not contain

Window Size modulo 8

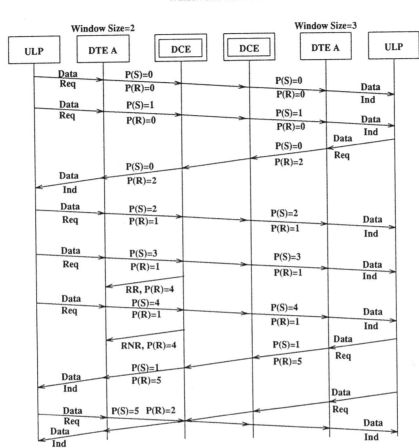

FIG. 27 Packet layer flow control.

either a P(S) field or a P(R) field. They may be thought of as privileged data packets as they can overtake the sequenced data packets waiting transmission in the network. Furthermore, interrupt packets cannot be blocked by a RNR or constrained by flow control. The packet, however, does not affect the state of the DTE-DCE interface and it does not alter the packet sequence numbering count.

The interrupt user data field contains interrupt data. Up to 32 octets of user data are permitted in an interrupt packet. Acknowledgment of such a packet is always end to end. A DTE, however, may not send a second interrupt packet on a virtual circuit until an interrupt confirmation packet of the previous interrupt packet is received. This restriction simplifies the interrupt procedure and reduces the amount of overhead required to exchange the information.

Confirmation Packets

The confirmation packets are used to acknowledge that a previously requested action has been undertaken. There are four types of confirmation packets:

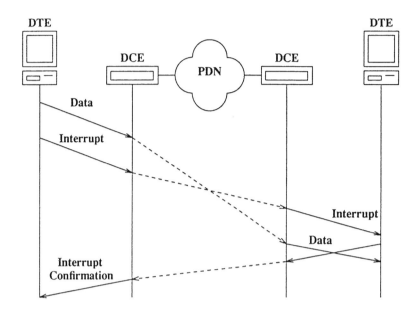

PDN: Packet Data Network

FIG. 28   Interrupt procedure.

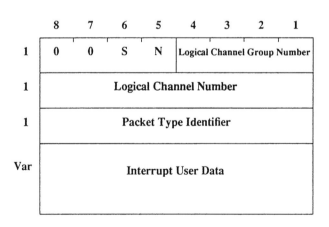

Packet Type Identifier=INTERRUPT

FIG. 29   Interrupt packets' format.

restart confirmation, reset confirmation, interrupt confirmation, and clear confirmation. The general format of the first three packets is depicted in Fig. 30.

The clear confirmation packet uses a different format, described in the section "Call-Clearing Packets." The confirmation packets play a crucial role in bringing the DTEs involved in the virtual call to a known state and detecting the occurrence of severe problems on the virtual circuit. An interrupt procedure, for example, is only completed when an interrupt confirmation packet is received. If the interrupt packet remains unacknowledged by the remote DTE, the local DTE becomes aware that a severe failure, which is not simply due to network congestion, may have occurred on the virtual circuit. This may prompt the local DTE to start an error-recovery procedure by resetting or clearing the logical channels or by issuing a restart packet in the case of a severe problem.

Diagnostic Packets

The diagnostic packet is a special packet issued by the DCE's fault diagnosis. The packet indicates certain error conditions that are usually not covered by other procedures, such as reset or restart. The diagnostic packet is issued only by the network to report a particular problem and does not require a confirmation packet.

The general format of the diagnostic packet is depicted in Fig. 31. The LCI field of the packet is set to 0. The diagnostic code field identifies the diagnostic reported in the packet. The X.25 Recommendations specify a large number of diagnostic codes to cover a wide range of problem situations. A brief description of these codes is provided in Table 7. The diagnostic description field provides further information about the result of the diagnosis.

Registration Packets

The registration packets are used to carry user facility registration operations. Using these packets, a user can alter a DTE subscription profile in an on-line

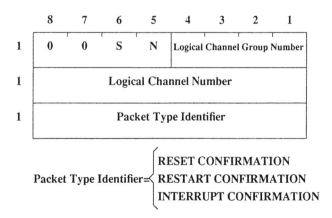

FIG. 30 Confirmation packets' format.

	8	7	6	5	4	3	2	1
1	0	0	S	N	0	0	0	0
1	0	0	0	0	0	0	0	0
1	Packet Type Identifier							
	Diagnostic Code							
	Diagnostic Description							

Packet Type Identifier=DIAGNOSTIC

FIG. 31   Diagnostic packets' format.

mode without the manual intervention of the network service provider. The user can also use these packets to obtain details about the current subscription profile of a particular DTE. The details may include the names of the selected use facilities, their associated parameters, and the logical channel allocation taken by the DTE.

The general format of a registration request packet is depicted in Fig. 32. In addition to the standard header fields, the packet contains the address information fields and the registration information fields. Note that a registration request is not channel selective as its LCI is set to 0. Consequently, no data transfer can occur on any virtual circuit when the registration request is issued.

In response to a registration request, the DCE issues a registration confirmation. In this packet, the DCE reports on the status of the request and indicates the action taken. A list of the user facilities still enabled may also be included.

X.25 Addressing Recommendations

As stated previously, DTEs connected to a PDN are uniquely identified by their addresses. In early releases, the address format of the calling and called DTEs were not specified in the X.25 Recommendations. The lack of specifications prompted users and network service providers to develop their private addressing conventions. With the global proliferation of DTEs and the increasing importance of internetworking, CCITT recognized the need for the development of a uniform addressing scheme. The specifications related to the format and encoding of the X.25 addressing fields are provided in the 1984 and 1988 CCITT Recommendations.

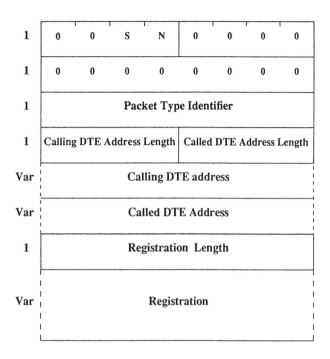

FIG. 32   Registration packets' format.

X.25 Data Terminating Equipment Address Specification

The address of a DTE in an X.25 network may include a main address and a complementary address. The format of the main address depends on the encoding of Bit 8, the A bit, of the general format identifier.

When the A bit is set to 0, the address format of the DTE conforms to the specification of the X.121 and X.301 Recommendations. When the A bit is set to 1, the format of the main address includes three fields: the type of address (TOA) field, the numbering plan identification (NPI) field, and the address field. This format is shown in Fig. 33. The possible values and the semantics of the TOA field and NPI field are listed in Table 8 and Table 9, respectively.

The complementary address represents address information provided in ad-

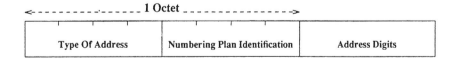

The A bit, the general format identifier, is set to 1.i

FIG. 33   Main address format.

TABLE 8 Type of Address Subfield Encoding

Type of Address Encoding				Type of Address
0	0	0	0	Network-dependent number
0	0	0	1	International number
0	0	1	0	National number
To be defined				Complementary address
Other values				Reserved

The TOA subfield is encoded in Bits 8, 7, 6, and 5 in the called DTE address field and Bits 4, 3, 2, and 1 in the calling DTE address field.

dition to the addressing information specified in the X.121 Recommendations. When present in a DTE address field, the complementary address is passed transparently by the network. Furthermore, when the DTE address field does not contain a main address (thereby carrying only a complementary address), if the A bit is set, the addressing information must be preceded by the TOA and NPI subfields.

X.121 Recommendations

In order to address the needs for a uniform addressing scheme, the CCITT developed a set of recommendations describing related numbering plans for the public data network (PDN) based on Recommendations X.25 and other types of public networks. These networks include the public switched telephone network (PSTN), the international Telex, and, most recently, the international Integrated Services Digital Network (ISDN).

The address scheme for PDNs is specified in the CCITT Recommendations X.121. The document describes an international addressing convention for public data networks. Based on a telephone number assignment mechanism, the

TABLE 9 Numbering Plan Identification Subfield Encoding

Numbering Plan Identification Encoding				Numbering Plan
0	0	1	1	X.121
To be defined				Network dependent
Other values				Reserved

The NPI subfield is encoded in Bits 8, 7, 6, and 5 in the calling DTE address field and Bits 4, 3, 2, and 1 in the called DTE address field.

addressing scheme divides the entire global set of network addresses into hierarchically grouped domains. A certain number of these domains are managed directly by the CCITT, while others are administered by different government agencies, such as the country's local PTT agency.

The X.121 address consists of a sequence of up to 14 digits. The address format is depicted in Fig. 34. The first four digits represent the data network identification code (DNIC). The leading digit of DNIC, which must not be of value 1, has a special meaning. Its value specifies either a world geographic zone or a network type. A value of 8, for example, specifies that the address is based on the F.69 Telex numbering plan; a value of 9 specifies that the address conforms to the E.163 public switched telephone network's numbering plan; and a value of 2 identifies Europe as the world geographic zone.

The first three digits of the DNIC represent the data country code (DCC). A DCC value of 208, for example, identifies France as the country code, while a value of 302 represents Canada's country code. Canada is part of the zone of North America, as identified by the leading digit 3 of the DNIC.

The last bit of the DNIC identifies a specific public data network within the country. In some countries, the number of public data networks exceeds the limit of 10. In the United States, for example, many of the Regional Bell Operating Companies have created their own public networks, thereby causing the number of networks to exceed 10. To accommodate these situations, the CCITT grants multiple DCCs to the same country.

Following the DNIC, the 10 remaining digits of the X.121 address represent the national terminal number (NTN). The NTN identifies a specific DTE within the packet data network. The use of these digits is at the discretion of the network service provider and is not standardized. An example of an X.25 address is depicted in Fig. 35.

Addressing in an Open System Interconnection

The numbering plan recommendations produced by CCITT for public networks (namely, X.121, F.69, E.163, and E.164) are different for each network and

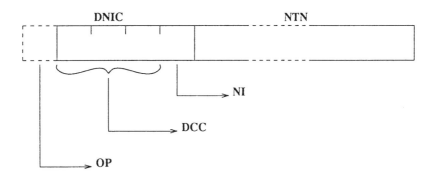

FIG. 34   The X.121 numbering plan (DCC = data country code; DNIC = data network identification code; NI = network identifier; NTN = national terminal number; OP = optional prefix).

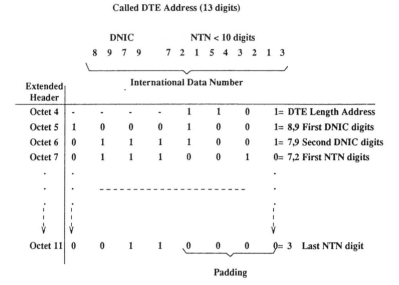

FIG. 35   Address encoding example.

generally identify a single point of attachment to the public network. Furthermore, the address length specified in Recommendations X.121 is inadequate to meet the requirements of densely populated and interconnected networks. These shortcomings made the interconnection of multiple private networks, although belonging to the same organization, difficult. In addition, the interconnection of private networks to public networks could only be achieved if the size of the private network was small enough for the private network to be considered as a simple extension to the public network.

To address these shortcomings, the CCITT defined an optional facility termed *calling and called address extension*. The main purpose of the facility is to allow more addresses and accommodate the interconnection of densely populated networks. The address extension facility, however, does not address other considerations related to ease of configuration, changes in management, routing efficiency, and large interconnected networks in which the number of addresses to be supplied can increase beyond reasonable bounds. What is needed is a global addressing scheme that allows the use of a single address as the only necessary requirement to reach any destination, regardless of the location of the caller and the number of subnetworks between the calling and the called DTEs. Furthermore, the scheme must not be rigid, must support the national requirements of different countries, and must provide for different kinds of address schemes, including geographic and nongeographic schemes and existing worldwide schemes. These considerations were the focus of the Open System Interconnection addressing scheme termed the *network service access point* (NSAP). The scheme is intended to be a global mechanism under which geographic and nongeographic address schemes, as well as worldwide schemes such as X.121, can be accommodated.

The ISO standard addressing scheme allows the abstract syntax of a network address value to be encoded in either decimal or binary form. Furthermore, the standard does not restrict the external representation of the address to conform either to the abstract syntax or the to transfer syntax.

The format of the network address, defined by the ISO standard, divides the address into two major parts: the initial domain part (IDP) and the domain specific part (DSP). This format is depicted in Fig. 36.

The IDP is further subdivided into two parts, the authority and format identifier (AFI) and the initial domain identifier (IDI). The AFI is a two-decimal-digit encoded field. These digits identify the format of the IDI, the authority responsible for allocating the IDI values, and the abstract syntax of the DSP. More specifically, the AFI value determines the length of the IDI and the significance of any leading zeros in the IDI value. Furthermore, the AFI value indicates whether the DSP is formatted using values expressed as decimal numbers or values expressed as binary numbers.

The relevant addressing and numbering plan standards currently associated with the two-digit code values of the AFI are defined by either the ISO-administered addressing plans or the CCITT-administered addressing plans. The IDP values associated with the ISO address administration identify a particular country or an international organization. The IDP values associated with the CCITT address administration identify individual subscribers in a manner similar to telephone numbers.

The IDI identifies the entity allowed to assign values to the DSP. More specifically, in the case of the ISO AFI assignments, the value of the IDI specifies, using a variable number of digits, the network addressing domain from which values of the DSP are allocated and the network addressing authority responsible for allocating the DSP values. In the case of the CCITT AFI assignments, the IDI field encodes a specific addressing value identified in the AFI. Note, however, that the IDI field can be null, in which case the entire address is contained in the DSP.

The DSP has no predefined semantics or length. It is the responsibility of the network addressing authority identified in the IDI to define the semantics of the DSP values and the length of the allocated values. The syntax of the DSP

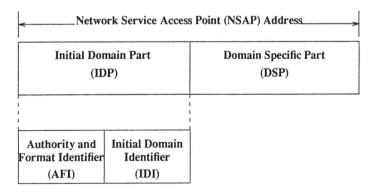

FIG. 36 International Organization for Standardization network address format.

can be binary octet, decimal digits, characters, or national characters. The length of the DSP may be null, as in the case of the X.121 addressing standard. In the general case of a DSP with an arbitrary length, the last octet of the least significant octet usually represents the selector field. The value of this field defines the particular type of the transport layer entity currently using the network services.

The encoding of the NSAP can be binary, decimal, or a combination of both. The preferred encoding for the ISO implementation is binary. The binary scheme uses a maximum number of 20 octets to encode the network address. In the decimal representation, mostly used by CCITT, the network address uses a maximum of 40 digits to encode the network address of the end system. The CCITT is considering the change from decimal to binary as the preferred address encoding method.

X.25 User Facilities

Appreciable differences in terms of performance objectives, availability, security, and accounting are likely to exist among different packet data networks. Furthermore, the user's profile varies considerably based on the user's expertise and both need and entitlement of access to various kinds of information. Consequently, a key element in the success of the X.25 Recommendations is that the access interface it specifies is readily adaptable to accommodate varied types of networks and technologies. Furthermore, the interface must be flexible enough to be tailored to meet a wide spectrum of user requirements.

In order to address the above needs, the X.25 Recommendations define a set of user services and facilities for packet-switched data-transmission service. These facilities not only allow users to customize their PDN connections, but also provide both users and network service providers with very powerful services. Selected at subscription time, these facilities can be applied selectively and incrementally. A list of all the possible facilities is provided in the CCITT X.2 Recommendations. The availability of user facilities varies considerably from one network service provider to another.

All user facilities are optional. Some of the facilities are considered "essential," while others are considered "additional." Essential user facilities are to be made available internationally in all PDNs supporting the X.25 Recommendations. Additional user facilities, however, are left to the discretion of the network service provider. Furthermore, the subscription to a certain group of user facilities can be achieved for a predetermined fixed period of time, while user facilities can be negotiated on a per-call basis. A list of the packet-switched data-transmission service user facilities is provided in Tables 10 and 11. Table 10 lists the user facilities for which subscription can be obtained for an agreed-on contractual period of time, while Table 11 lists the user facilities that can be negotiated on a per-call basis. The following sections discuss the most significant and widely supported user facilities.

TABLE 10   Subscription-Based User Facilities

Optional User Facility Name	Optional User Facility Name
Extended frame sequence numbering	Closed user group
Multilink procedure	Closed user group with outgoing access
On-line facility registration	Closed user group with incoming access
Extended packet sequence numbering*	Incoming call barred within a closed user group
D-bit modification	Outgoing call barred within a closed user group
Packet transmission	Bilateral closed user group
Incoming calls barred	Bilateral user group with outgoing access
Outgoing calls barred	Local charging prevention
One-way logical channel incoming	NUI subscription
One-way logical channel outgoing	NUI override
Nonstandard default packet sizes†	Charging information
Nonstandard default window sizes	RPOA subscription
Flow control parameter negotiation	Hunt group
Throughput class negotiation	Call redirection
Default throughput class assignments	Call deflection subscription
Fast-select acceptance	TOA/NPI address subscription
Reverse charging acceptance	Direct call

*Modulo 128
†16, 32, 64, 256, 512, 1024, 2048, and 4096

Essential User Facilities

As stated above, essential user facilities are internationally provided on all PDNs that follow the X.25 Recommendations. These facilities include call barring, closed user group, fast select, flow control parameter negotiation, one-way logical channels, throughput class negotiation, and transit delay selection and indication.

TABLE 11   Call-Based User Facilities

Flow control parameter negotiation
Throughput class negotiation
Closed user group selection
Closed user group with outgoing access selection
Bilateral closed user group selection
Reverse charging
Fast select
NUI selection
Charging information
RPOA selection
Call deflection selection
Call redirection or call deflection notification
Called line address modified notification
Transit delay selection and indication
Abbreviated address calling

Call Barring

The call barring facility affects the DTE–DCE interface and allows control of the network access. When subscribed to this facility, it applies to all logical channels used for virtual calls. Permanent virtual circuits, however, are not affected by this facility.

Two options are available on a per-call basis. The first option, incoming calls barred, prevents incoming calls from being presented to the DTE. The second option, outgoing calls barred, prevents the DCE from accepting outgoing virtual calls from the DTE. Consequently, a DTE subscribing to incoming calls barred can initiate calls, but cannot accept them. Conversely, a DTE subscribing to outgoing calls barred can receive calls, but cannot initiate them. Furthermore, once a call is established, for either option, the associated virtual circuit retains its full-duplex capability.

Closed User Group

The closed user group facility enables users to configure virtual networks within a large public network by allowing a DTE to belong to one or more closed groups. Within a closed group, a DTE is allowed to communicate with all members of the group, but is precluded from communicating with all the other DTEs. Consequently, the closed user group facility provides a highly desirable access restriction service to achieve security and privacy in an open network.

The limit on the number of closed user groups is network dependent. Typically, public data networks support up to 99 closed user groups for a single DTE. It is possible, however, to extend this number to 10,000 closed user groups. Once a DTE belongs to multiple closed user groups, a preferential group must be specified.

Fast Select

The fast-select facility allows the DTE to include up to 128 octets in the call setup and clear packets. A number of transaction-oriented applications can take advantage of the expanded call user data field to transfer a considerable amount of information about the nature of the transaction. The call user data field in the call request can readily contain information such as the name of the customer, the credit card number, the credit expiration date, and the total amount of the transaction. Conversely, the called DTE can use the expanded call user data field of the call accepted packet to transfer the transaction confirmation status back to the calling DTE. Carried over a "conventional" virtual call, this transaction would entail the transfer of eight control packets just to support the exchange of the data packets. The use of the fast-select facility, assuming that the calling DTE has subscribed to this option and the called DTE has subscribed to the fast-select acceptance option, minimizes the communication overhead, thereby reducing the overall transaction time to a minimum.

The fast-select facility offers the users with two options for fast data transfer. The first option, fast select with call setup, uses the same sequence of packets to establish a circuit, transfer data, and clear the circuit as conventional virtual calls. The established circuit remains active to support data packets if subsequently required until the call is cleared. The only advantage of fast select with call setup is the ability to include information in the expanded call user data field.

The second option, fast select with immediate clear, uses a streamlined call setup and clear sequence with no actual data-transfer phase. The called DTE responds to the call request, which may contain up to 128 octets of user data, by sending a clear request packet to which user data may be attached. The circuit is then immediately cleared.

Flow Control Parameter Negotiation

The flow control parameter negotiation facility allows the modification of the flow control parameters on a per-call basis. These parameters are the packet size and the window size at the DTE–DCE interface at each direction of data transmission.

The packet size refers to the maximum length, in octets, of the user data field of the data packets. The standard default packet size is 128 octets. This size, however, can be changed to 16, 32, 64, 256, 512, 1024, 2048, and 4096 octets. Other permissible values are network dependent.

The window size determines the maximum number of packets that can be transmitted before an acknowledgment must be received. The standard default window size is 2. It is possible, however, to negotiate other window sizes, including 1 and 3 to 7. The availability of these alternative sizes is dependent on the network service provider.

The negotiation of flow control parameters takes place between the calling and called DTEs, with bidding always toward the default value. The final negotiated values are returned by the called DTE in the call accepted packet. If the negotiation facility is not supported at either end of the connection, the default flow control parameters are used.

One-Way Logical Channels

The one-way logical channel facility is essentially a channel-selective version of the call barring facility discussed above. Unlike call barring, which designates all logical channels as either incoming or outgoing, the one-way logical channel facility provides the option of restricting network access to individual or grouped logical channels. If all the logical channels defined at the DTE–DCE interface are affected by the one-way logical channel access restrictions, the DTE is effectively call barring.

Two options are available under the one-way logical channel facility. The first option, one-way logical channel outgoing, restricts the channel use to originating outgoing virtual calls only. The channel is barred from receiving incoming calls. The second option, one-way logical channel incoming, restricts the channel use to receiving incoming virtual calls only. This option, however, is not an essential facility.

The one-way logical channel facility parameter is selected at subscription time for an agreed-on period of time. Furthermore, the restriction imposed by the facility option applies only at call setup time. Once the call is established, the transmission capabilities are full duplex.

Throughput Class Negotiation

The *throughput* of a virtual circuit refers to the maximum amount of information that can be transmitted through the network, over a defined time period, when the network is operating at saturation. At subscription time, each virtual circuit is assigned a default throughput class. The assigned value of the throughput class specifies the effective throughput, expressed in bits per second, of the virtual circuit. The throughput class value is based primarily on the speed of the physical link connecting the DTE to the network. Other factors may be considered, including the window size and the use of the D bit. The default throughput class value is agreed on by the user and the network service provider at subscription time for both directions of transmission.

The standard values for the throughput class are 3, 4, 5, 6, 7, 8, 9, 10, 11, 12, and 13. These values correspond to 75, 150, 300, 600, 2400, 4800, 9600, 19,200, 48,000, and 64,000 bits per second, respectively. Many network service providers, however, support higher throughput rates.

The throughput class negotiation facility allows the throughput rates to be negotiated on a per-call basis. This allows the called and carrying DTEs to adjust their throughputs to a mutually acceptable value. The facility, however, only allows a DTE to negotiate a lower throughput class value than the preassigned standard default value.

Transit Delay Selection and Indication

The transit delay selection and indication facility allow a DTE to select a desired maximum network transit delay for the duration of the call, thereby allowing the user control over the response time. The selection is established on a per-call basis.

When capable of providing this service, the network must assign appropriate resources to ensure that the actual transit delay applicable to the call does not exceed the desired transit delay. The network informs both DTEs about the transit delay selected for the call in the incoming call packet transmitted to the called DTE and the call connected packet transmitted to the calling DTE. Depending on the current network status, however, the indicated transit delay may be smaller than, equal to, or greater than the desired transit delay requested in the call request packet.

Additional User Facilities

Additional user facilities are only available in certain networks. Their support is not required internationally. The most commonly used additional facilities include bilateral closed user groups, call redirection, and call deflection, extended packet sequence numbering, hunt groups, reverse charging, and reverse charging acceptance, network user identification, Recognized Private Operating Agency (RPOA), and additional closed user group access.

Bilateral Closed User Groups

The bilateral closed user group facility is a selective version of the closed user group facility discussed above. Unlike the closed user group facility, however, the bilateral closed user group facility allows access restrictions to be applied to a pair of DTEs. This bilateral relationship allows the DTEs to communicate with each other, but precludes any communication with other DTEs in the open network.

A DTE can belong to multiple bilateral closed user groups. Furthermore, the DTE may subscribe to a bilateral closed user group with outgoing access. This last facility enables the DTE to subscribe to one or more bilateral closed user groups and to originate virtual calls to DTEs in the open network.

Call Redirection and Call Deflection

The call redirection facility allows a DCE to redirect calls destined for the "originally called DTE" to an "alternative DTE." The DCE does not forward any incoming call packet to the originally called DTE. Call redirection is usually performed when the originally called DTE is out of order or busy. Some network service providers, however, may offer systematic call redirection for other reasons specified by the user at subscription time. Furthermore, some network service providers allow the user to specify a list of alternative DTEs to which consecutive attempts of call redirection can be made, in the order described in the list, until the call is completed. A different type of service offers the capability of logically chaining redirections among DTEs so that, if the alternative DTE is busy or out of order, a redirection is attempted to its alternative DTE.

The call deflection facility is similar to the call redirection facility. The only difference is that, in call deflection, the originally called DTE receives an incoming call packet and then deflects the call to an alternative DTE.

Extended Packet Sequence Numbering

The extended packet sequence numbering facility extends the modulo of packet sequence numbering to 128 for all channels at the DTE–DCE interface. Consecutive packets are numbered from 0 to 127. In the absence of this facility, the standard packet sequence numbering restricts packet numbering to 0 to 7. Furthermore, the X.25 Recommendations do not allow a sequence number of a packet waiting acknowledgment to be reused until the acknowledgment is received.

The addition of the extended packet sequence numbering was needed to deal with long propagation time, observed when transmitting signals over satellite channels, and better accommodate transmission media with a very high transfer rate, such as optical fiber.

Hunt Groups

The hunt group facility allows incoming calls to be distributed to multiple DTE-DCE interfaces associated with the same hunt group address. Calls addressed to the hunt group can be received by any member of the group having at least one free logical channel, excluding one-way outgoing logical channels. The hunt group address is contained in the incoming call packet received by the called DTE. Other provisions, such as the assignment of a specific, unique address, can be made. Furthermore, permanent virtual circuits may be associated with a DTE-DCE interface that belongs to a hunt group. These circuits, however, operate independently of the hunt group operations. Figure 37 illustrates the basic operations of the hunt group facility. The label G represents the name of the group.

On establishment, the call is treated by the called and calling DTEs as a conventional call. Notice, also, that members of a hunt group may originate virtual calls. These calls are handled by the interface as conventional calls.

The DTEs attached to an interface that belongs to a hunt group may be physically attached to separate ports on the same device or spread across multiple devices. This feature is valuable to organizations with large computing facilities. It allows jobs to be directed to different resources, including front ends and computers.

Reverse Charging and Reverse Charging Acceptance

The reverse charging facility allows a subscribing DTE to request reverse charging in the call request packet transmitted to the DCE. If such calls are accepted by the network DTE, the remote DTE is charged for the network resources used during the call.

The reverse charging acceptance facility provides a subscribing DCE the ability to forward to the DTE incoming call packets that request reverse charging. The DTE may either accept these calls or reject them by issuing a clear packet. Acceptance of the calls results in having the called user subsequently charged by the network service provider for the resources used during the call.

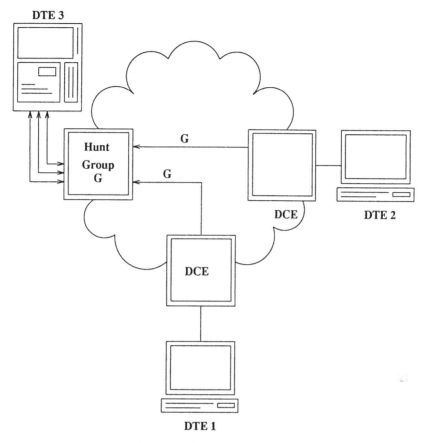

FIG. 37   Hunt group operation.

Network User Identification

The network user identification facility was incorporated in the 1984 X.25 Recommendations. The facility offers a standardized way for users to identify
themselves and provide information to the network for purposes of billing,
security, and network management on a per-call basis. Users can also invoke all
or parts of the subscribed facilities with each call. This allows the user to tailor
the interface to the specific call. The network user identification is composed of
three related facilities: the network user identification subscription, the network
user identification selection, and the network user identification override.

   The network user identification subscription facility is established for a period of time. This facility allows the subscribed DTE to provide the network
with the billing, security, and network management information. This information, carried in either the call request packet or call accepted packet, is conveyed
to the network using the network user identification selection facility.

   The network user identification override may be established for a period of
time. At subscription time, one or more user identifiers may be specified and

agreed on for a defined period of time. Associated with each network user identifier is a set of subscription time optional user facilities. The set of facilities may differ from one user identifier to another. When one of the network user identifiers is provided in a call request packet by means of the network user identification selection facility, the network user identification selection facility is used to override the set of subscription time facilities that apply to the interface by the set of subscription time facilities associated with the user identifier. The override remains in effect for the duration of the particular call to which it applies. The override, however, does not apply to other existing or subsequent calls on the interface.

Recognized Private Operating Agency Facility

The Recognized Private Operating Agency (RPOA) facility allows the subscribed DTE to specify one or more RPOAs as potential transit networks through which packets are routed. The RPOAs, typically international record carriers (IRCs), are packet network carriers that act as a transit network within one country or between different countries.

A subscribed DTE can use a call request packet to request use of the RPOA. The selected RPOA is identified by its data network identification code (DNIC). In response, the network attempts to route the call through the specified agency. If the call cannot be successfully routed, the network does not attempt to establish the call through an alternate route.

Additional Closed User Group Access

The closed user group facility offers more restrictive group-related access. Based on these additional closed user group options, access to a group may be restricted to receiving incoming calls from members of a closed user group or initiating outgoing calls to members of a closed user group. Furthermore, a DTE that subscribes to a closed user group with incoming access may receive calls both from DTEs that are members of the closed group and from DTEs belonging to the open part of the network. Conversely, a DTE that subscribes to a closed user group with outgoing access may initiate calls both to DTEs that are members of the closed group and to DTEs that belong to the open part of the network.

X.25 Terminal Access

In its standard configuration, X.25 allows a packet-mode DTE to interface and exchange packets with a PDN. In some environments, however, hosts may be required to interact with asynchronous, character-mode devices such as computer terminals, printers, and plotters. These devices, usually referred to as DTE-C terminals, are only equipped with a minimal set of functionalities to manipulate characters. Consequently, they are unable to handle all levels of the synchronous X.25 Protocol architecture. In order to accommodate DTE-Cs, a packet assembler/disassembler (PAD) must be used to provide an interface between the asynchronous, character-mode device and the X.25 synchronous protocol used to access the PDN.

The PAD performs several functions, including

- assembling the characters received from the DTE-C terminal into packets to be transmitted to the host over the X.25 PDN
- disassembling the packets received from the host into a stream of characters for transmission to the DTE-C terminal
- performing operations necessary to establish, reset, and clear a switched virtual circuit
- transmitting service signals to the terminal user to indicate operational status
- reacting to command signals
- controlling the DTE-C terminal functions

PADs are usually provided by the PDN administration. Access to the PAD may be leased or on a dial-up basis. Many companies, however, own private PADs. Private PADs reside in the X.25 user domain and are external to the pubic PDN. Owning private PADs offers several advantages, including a more cost-effective way of multiplexing several user connections over a single synchronous X.25 link. Furthermore, DTE-Cs connected to private PADs have the capability of both initiating and receiving calls, contrary to DTE-Cs connected to a public PDN that can only initiate calls, but not receive them.

The basic configuration of a PAD base X.25 network is depicted in Fig. 38. Initially, the conversation between the user and the PAD is local. The PAD presents to the user a command line prompt. The user issues the proper command to intruct the PAD to make a connection to the desired host. In response, the PAD issues a request to establish a virtual circuit with the remote host, assuming that no permanent virtual circuit exists between the user's terminal

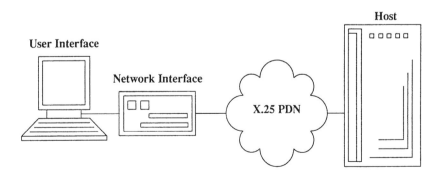

FIG. 38 Packet assembler/disassembler basic configuration.

and the host. On receiving the call request, the remote host determines if the user is allowed access and issues the appropriate "announcement" to the user in the same way that it would for a locally connected terminal. Immediately after the connection between the local user and the remote host is established, the PAD becomes essentially transparent: data from the user, received in the form of character streams, are assembled into packets and transmitted into the network, and packets received from the network are disassembled into character streams and presented to the user's terminal. At the end of the call, the virtual circuit is cleared, and the PAD normally issues a message to the user indicating call termination.

Recommendations Related to the Packet Assembler/Disassembler

Given the profusion of computing devices with a wide spectrum of characteristics, the CCITT defined three PAD-related recommendations that specify

1. How a terminal should be connected to a PAD
2. The range of user-definable services that should be offered by the PAD
3. How the host communicates with the PAD over the PDN

The three recommendations are identified as X.3, X.28, and X.29. The set of recommendations is commonly referred to as *triple X*.

The X.3 Recommendations define the PAD operating characteristics for a given DTE-C. These characteristics, collectively referred to as *the PAD parameters*, include flow control, operating speed, padding, editing, and echo of characters. These parameters are normally set by the user from the commands issued at the terminal or by the remote packet-mode DTE being accessed. The X.28 Recommendations define the procedures used by a DTE-C to communicate and control a PAD and the procedures used by the PAD to respond to these commands. The X.29 Recommendations, on the other hand, define the procedure used for the exchange of control information across the PDN between the PAD and the host DTE. Figure 39 illustrates a PAD-based connection to an X.25 network and identifies the boundaries of the various PAD-related CCITT recommendations. The combination of the X.3 parameters and the procedures offered by X.28 and X.29 to set and control these parameters provides a complete method of controlling a PAD-based connection to an X.25 PDN. The following sections provide a more detailed description of these protocols and their basic functionalities.

X.3 Protocol Specification

The CCITT X.3 Recommendations define the functions of a PAD and specify the parameters used to control its operations. The set of selected values for these parameters defines a PAD profile for a particular DTE-C. Since PADs may

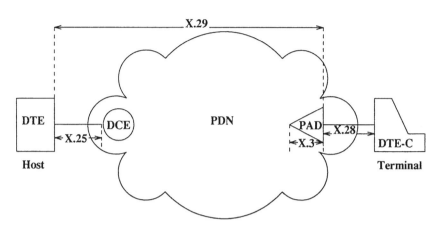

FIG. 39 International Telegraph and Telephone Consultative Committee (CCITT) recommendations related to the packet assembler/disassembler and their boundaries (DTE-C = character-mode data terminating equipment).

perform multiplexing of traffic to serve more than one terminal, a profile is maintained for each terminal.

The X.3 Recommendations define 22 parameters that determine the behavior and operating characteristics of the PAD. A list of these parameters is provided in Table 12.

For each parameter described, a list of a possible range of values is specified. To facilitate the use of the PAD, X.3 defines a standard profile that associates a default value to each parameter. Several alternative standard profiles have been defined for use with popular types of terminals. Users may either adopt the standard profile to set the PAD parameters to their default values based on the terminal type or selectively modify the values of the parameters to accommodate their own environment. The PAD profile is normally determined when the communications link between the terminal and the PAD is established. However, at any time during the call, any PAD parameter value except for the terminal speed parameter can be modified. The procedure used to define an initial PAD profile, and potentially modify the associated parameter values, is defined by the CCITT X.28 Recommendations.

X.28 Protocol Specification

The CCITT X.28 Recommendations define the protocol that governs the interaction between the PAD and the DTE-C asynchronous terminal. The protocol specifies the procedures used by the DTE-C to access the PAD, set the parameters to the required values, establish and clear virtual calls to remote packet-mode DTEs, and control the exchange of data between the terminal and the

TABLE 12 Packet Assembler/Disassembler Parameters

Number	Parameter	Function Description
1	Data transfer escape	To allow operator to escape from data transfer to PAD command state
2	Echo	To control PAD echoing back characters to terminal
3	Data forwarding signal	To trigger sending of partial or full packets
4	Idle timer	Time-out value to trigger sending partially filled packet by PAD
5	Ancillary device control	To control PAD exercising flow control (XON/OFF) over terminal output
6	PAD service signals	To control PAD sending control information to terminal
7	Procedure on break	To control PAD actions on receipt of break signal from terminal
8	Discard output	To control PAD discarding data intended for terminal
9	Carriage return padding	To determine number of padding characters inserted after carriage return
10	Line folding	To control PAD inserting characters to prevent terminal line overflow
11	DTE-C speed	To control terminal speed
12	Flow control by DTE-C	To control DTE-C exercising flow control over PAD, using XON/OFF
13	Line feed insertion	To control PAD inserting line feed after carriage return sent or echoed to terminal
14	Line feed padding	To determine number of padding characters inserted after line feed
15	Editing	To control PAD support for editing during data transfer
16	Character delete	Selected character for character delete
17	Line delete	Selected character for line delete
18	Line display	Selected character for line display
19	Terminal type	To control terminal type for editing PAD service signals
20	Nonechoed characters	To control set of characters not to be echoed by PAD to terminal when echo is enabled
21	Parity treatment	To control parity treatment of characters to and from terminal
22	Page wait line feed number	To control number of line feeds from PAD to signal page wait

PAD. The messages exchanged between the DTE-C and the PAD are referred to as *command signals* and *PAD service signals*. The CCITT X.28 Recommendations define the command signals that the PAD user can issue to control the PAD and the service signals that the PAD displays to the user's terminal to acknowledge the PAD command signals and to inform the terminal user of the identity or the internal state of the PAD in response to the command signals. The complete list of PAD command signals as defined by the CCITT X.28

TABLE 13   Packet Assembler/Disassembler Command Signals

PAD Command Signal Format	Function	PAD Service Signal Response to Terminal*
STAT	To request status information on a virtual call placed to a remote DTE	Free or engaged
CLR	To clear a virtual circuit	CLR CONF or CLR ERR
PAR? Params	To read the current values of the specified parameters	PAR with PAD Param reference numbers and their current values
SET? Params	To set and reset specific PAD Param values	PAR and PAD Param reference numbers and their current values
PROF Ident	To allocate a standard set of PAD parameters values	Acknowledgment
RESET	To reset a virtual circuit	Acknowledgment
INT	To transmit and interrupt	Acknowledgment
SET	To set all PAD parameters	Acknowledgment
SELECTION PAD COMMAND SIGNAL	To set up a virtual call with a remote DTE	Acknowledgment

*PAD service signals are disabled when Parameter 6 is set to 0.

Recommendations is given in Table 13. The availability of these commands, however, may depend on the type of PAD used.

Table 14 shows the complete list of PAD service signals as defined by the CCITT X.28 Recommendations. The availability of these signals may also depend on the type of PAD used. Furthermore, the acceptance of any PAD service signal is optional and can be suppressed by setting the value of Parameter 6 of the PAD to zero.

TABLE 14   X.28 Packet Assembler/Disassembler Service Signals

	Format	Function Description
COM		Indication of call connection
CLR	PAD command signal	Indication of call clearing
ENGAGED		Response to STAT PAD command signal
ERROR	Signal error	PAD command signal is in error
FREE		Response to STAT PAD command signal
PAD ID		Administration selected ID
PAR	Decimal value	Response to SET or SET and READ
RESET	DTE	Virtual circuit reset by remote DTE
	Local error	Reset due to local procedure error
	Network congestion	Reset due to network congestion

X.29 Protocol Specification

The CCITT X.29 Recommendations specify the interaction between the PAD and the remote packet-mode DTE. The protocol specifies the procedures used by the remote host DTE to change the PAD parameters of the DTE-C terminal attached to the PAD and exercises a degree of control over different PAD functions.

The CCITT X.29 Recommendations define eight control messages, referred to as *PAD messages*, to control the interaction between the remote packet-mode DTE and the local PAD. Table 15 summarizes these messages and briefly describes their purpose. The exchange of information between the remote packet-mode DTE and the PAD can occur at any time during a virtual call. The interaction, however, is usually defined in accordance with call establishment, data transfer, and call clearing.

Call establishment between the PAD and the packet-mode DTE follows the same procedure used by X.25. The PAD and the remote host are both packet-mode DTEs and can, therefore, exchange call setup packets across their respective DTE–DCE interfaces. Consequently, in order to establish a virtual circuit, the calling PAD issues a call request packet to the remote host. The X.29 call request packet is similar to the X.25 call request packet. The only difference is that the first four octets of the call user data field contain a protocol identifier field. The encoding of the protocol identifier of the X.29 call packet request informs the called packet-mode DTE that a PAD is calling. All other aspects of the X.29 call establishment procedure follow the X.25 Level 3 procedures.

On call establishment, data transfer may take place between the PAD and the remote packet-mode host. The procedure used for data transfer and flow

TABLE 15   X.29 Packet Assembler/Disassembler Message Types

PAD Message Type	Code	Direction	Message Purpose
Set PAD	0010	→ PAD	Set selected parameters to indicated values
Read PAD	0100	→ PAD	Read values of indicated parameters
Set & Read PAD	0110	→ PAD	Perform change of indicated parameter and require PAD to confirm change
Param indication	0000	← PAD	List of parameters and values in response to read commands
Invitation to clear	0001	→ PAD	Request for the PAD to clear the virtual call
Indication of break	0011	↔ PAD	Indication of a break from DTE or terminal
Error	0111	↔ PAD	Indicates error due to an invalid PAD message
PAD reselection	0101	→ PAD	Request for the PAD to clear the virtual call and establish a call to a different address

→ = to PAD; ← = from PAD; ↔ = to or from PAD

control is determined by the X.25 Level 3 Protocol. During data transfer, how-ever, the packet-mode DTE is able to communicate with the PAD by setting the Q bit in the data packet to inform the PAD that the packet is a PAD message. This indicates that the remaining information of the packet is intended for the use of the PAD and should not be disassembled and forwarded to the user. All PAD messages contain, in Octet 4, a specific message code to indicate the purpose of the message. Depending on the message type, the PAD takes the necessary actions.

The procedure used for clearing a virtual circuit between the PAD and a remote packet-mode DTE mainly follows the procedure specified by the X.25 Recommendations for call clearing. At any time, either side can transmit a clear request packet, thereby starting a call-clearing procedure. The only clearing procedure that is specific to a PAD and host interaction occurs when, under the instructions from a higher level process running on the host DTE, an "invitation to clear" the PAD message is issued to solicit the PAD to transmit a clear request message across its DTE–DCE interface. In response, the PAD indicates to the user that the virtual circuit has been cleared by sending the PAD service signal, CLR CONF, as defined by the CCITT X.28 Recommendations. In all other aspects, the interaction between the PAD and the remote packet-mode DTE to clear a call follows the X.25 Level 3 procedures.

Interconnections of X.25 Networks

The X.25 Recommendations define the interface used for connecting the DTE to the DCE associated with a packet-switched network. In large-scale applications, however, the associated DTEs may be connected through multiple data net-works. Consequently, a protocol must be defined to allow communication be-tween two DTEs connected to different networks. The protocol must allow separated virtual circuits to be set up across each intermediate network. More-over, in order to associate packets with a particular call, a virtual circuit must also be set up across the links that interconnect these networks together. These functionalities are provided by the X.75 Protocol.

The X.75 Model is based on the idea that an internetwork connection may be set up by concatenating a series of intranetwork and relay-to-relay virtual circuits. The relays, referred to as *signaling terminal exchanges* (STEs), are essentially half gateways. Each half is associated with its own network. Their basic functionalities include the establishment of the virtual circuits, both within and across networks, and the routing of the data packets to the DTEs at each end of the connection. Figure 40 illustrates the model used by X.75 to intercon-nect X.25 networks. Based on the X.75 Model, a DTE attempting to establish a connection to a host in a distant network proceeds the same way as in the conventional X.25 network setup. The calling and called DTEs both communi-cate with their respective networks using purely X.25 procedures, totally un-aware of any further complexity. Consequently, the presence of multiple virtual circuits associated with a call remains transparent to the DTEs. It is the responsi-

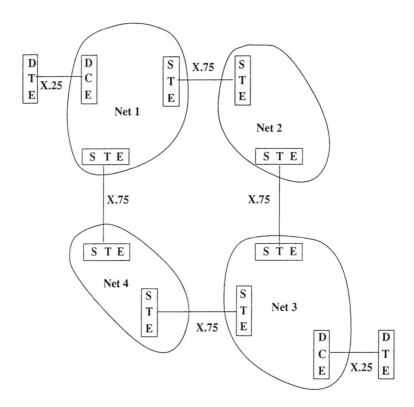

FIG. 40   The X.75 basic network configuration (STE = signaling terminal exchanges).

bility of the network, when it detects a remote destination address, to select an appropriate STE to which a virtual circuit is established. The selected STE records the existence of the virtual circuit in its routing tables and proceeds to establish a virtual circuit to the next STE. This procedure continues until the destination DTE is reached.

X.75 Protocol Architecture

The X.75 Protocol bears strong similarities to the X.25 Protocol. As in the case of X.25, three protocols are associated with the X.75 Standard: the Packet Layer Protocol, the Data-Link Layer Protocol, and the Physical Layer Protocol. This architecture is depicted in Fig. 41.

The physical links are expected to run at high-speed rates. A data rate of 64,000 bits per second is recommended, although higher speeds may be required in certain configurations to enhance the network performance. At the data-link layer, the multilink procedure (MLP) is recommended in order to enhance reliability and throughput.

The MLP was introduced by CCITT in its 1984 X.25 Recommendations. The objective of this procedure, which may be considered as a sublayer within

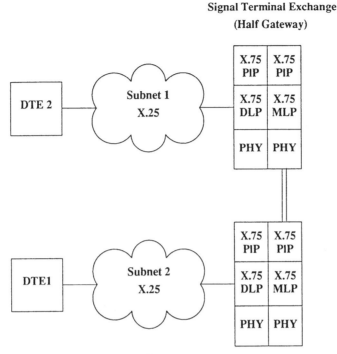

Signal Terminal Exchange
(Half Gateway)

PLP : Packet Layer Protocol
DLP : Data Link Layer Protocol
MLP : Multiple Link Protocol
(Used to enhance Reliability and Throughput)

FIG. 41   X.75 Protocol configuration.

the data-link layer, is to perform the function of distributing packets from the Packet Layer Protocol across multiple physical links, thereby increasing both the reliability and the throughput of the link. Each link is nothing more than a LAP-B link. The combined set of links is treated as a single entity when frames are being transmitted across such links. Consequently, when a frame is to be transmitted, any available link is selected, regardless of the logical channel identifier carried by the frame. The multilink control field in each frame is then used by the receiving end of the MLP to resequence frames before they are delivered to the Packet Layer Protocol.

The X.75 PLP is less complex than its X.25 PLP equivalent. This simplicity stems from the fact in X.75 communication takes place between two directly connected STEs, whereas in X.25 communication involves intermediate DCEs and possibly PADs. Consequently, the set of packets required to implement the functionalities of the protocol is reduced to call setup packets, call-clearing packets, data-transfer packets, flow control packets, and restart packets. The general format of these packets is depicted in Fig. 42. Figure 43 lists the main packets associated with the X.75 Protocol.

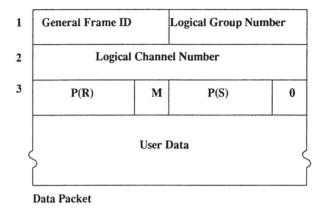

1	General Frame ID	Logical Group Number	
2	Logical Channel Number		
3	P(R)	M	P(S)	0
	User Data		

Data Packet

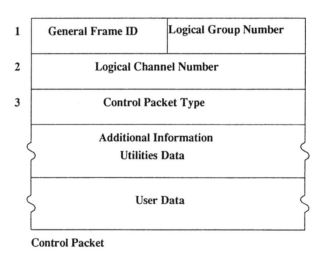

1	General Frame ID	Logical Group Number
2	Logical Channel Number	
3	Control Packet Type	
	Additional Information Utilities Data	
	User Data	

Control Packet

FIG. 42   The X.75 packet format.

Future of X.25

Although popular, especially in Europe and Asia, the X.25 network has several limitations. Its approach to error handling involves potential data loss. The X.25 Protocol does not provide a procedure guaranteeing delivery of data in transit when a virtual circuit is cleared or reset. Furthermore, the protocol does not state how many packets can be discarded in the process. When the D bit is set, up to a whole window size of packets may be lost. This complicates protocols at the transport layer that provide reliable service with X.25 at the network layer.

Another strong criticism of the X.254 network is that it was developed for use on public packet data networks using facilities that are rather suited for switched telephone networks. Consequently, the protocol was developed for a

Packet Types	Protocol Usage
Call Request Call Connected	Call Set-Up
Clear Request Clear Confirm	Call Clearing
Data Interrupt Interrupt Confirm	Data Transfer
Receiver Ready Receiver Not Ready Reset Confirm	Flow Control
Restart Restart Confirm	Restart

FIG. 43 The X.75 main packets.

communications environment in which transmission errors are frequent and the data rate is low compared to the speed achieved by current transmission facilities. These protocols and schemes built into packet switching were designed for functionality, not performance. These functionalities aimed to compensate for the errors introduced by the communications media and transmission facilities. As a consequence, the data-transmission schemes were complex and resulted in a large processing overhead. The overhead was required to process the information added for error detection, duplicate packet removal, and ordered delivery of these packets.

The current high-speed technology raises the average network speed beyond hundreds of megabits per second (Mb/s). Consequently, the bottleneck has shifted away from the network transmission speed to the processing power of the end stations and intermediate network nodes. Furthermore, the improved quality offered by the transmission facilities dramatically reduces the rate of errors. Consequently, the burden of error detection and correction is reduced, thereby allowing the design of simpler protocols to deal easily with the low rate of transmission errors at the end systems that operate above the level of packet-switching logic.

In order to take advantage of this new communications environment characterized by its high-speed data rate and low error rate, two new technologies, *frame relay* and *cell relay*, were developed. Frame relay is a form of statistical multiplexing designed to operate at user data rates of up to 2 Mb/s. This is achieved by reducing the functionalities of the intermediate nodes to a set of core functions and relying on the end systems to provide error recovery and flow control. From the user's perspective, a frame relay network can be viewed as a data line. This view is provided by an interface, through which the service can be accessed, and an architecture, which can provide the low functionality, streamlined services required by the concept of frame relay.

Cell relay can be viewed as a form of frame relay in which cells are short frames of constant lengths. Cells are usually associated with the primary switching fabric of the network. The design of cell relay networks is geared toward supporting different types of traffic, including data, voice, image, and video, at a very high speed. By using a fixed frame length, the processing overhead is further reduced to a minimum. Consequently, the technique offers several benefits, including low latency in the first node while the cell is accumulated, the possibility to allocate bandwidth in demand, and the potential for the design of simpler and faster switching nodes.

The new classes of protocols take advantage of the reliability and fidelity of modern digital facilities to provide faster packet switching. These protocols, such as frame relay, eliminate as much as possible the overhead inherent in X.25. Figure 44 illustrates the number of steps required by both X.25 and frame relay to establish a session between a source and a destination. Figure 45 compares the number of functionalities supported by X.25 and by frame relay in order to establish, manage, and maintain a connection over a link. Figure 46 compares the protocol architecture for both X.25 and frame relay. These figures confirm that the trend in frame relay is to streamline the protocol to a minimum to allow the network to do what it does best, "move packets." Protocols such as frame relay are finding wide application in public and private networks. For many, they are considered to be the "X.25 network of the future."

Summary

The main objective of this article was to provide a brief history of the development of X.25, discuss the basic concepts of packet-switched data networks, and examine the advantages and disadvantages of using X.25 as a common network interface protocol.

In the first part of the article, the general concept of the X.25 layered architecture was introduced, and the relationship of this protocol to the ISO Model was examined.

The second part of the article discussed the basic protocols and interfaces of the X.25 physical layer. The Physical Layer Protocol standards, recommended for use at the physical level by the X.25 source document specification, were described. Other protocols not cited in the X.25 source documents, but widely used with X.25, were also discussed.

The third part of the article was dedicated to discussing the issues related to

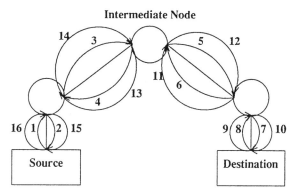

X.25 Packet Switching Network

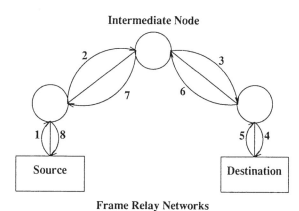

Frame Relay Networks

FIG. 44 The number of steps required by frame relay and X.25 to establish a session between a source and a destination.

the Data-Link Layer Protocol of X.25. A brief discussion of the HDLC Protocol family was provided. We then focused on Link Access Protocol–Balanced (LAP-B) and Link Access Protocol, D Channel (LAPD). The cases for which the logical link control (LLC) layer of LANs is used at the data-link layer were also examined.

The fourth part of the article focuses on the third layer of the X.25 Protocol architecture. The basic issues discussed include permanent virtual circuits, switched virtual calls, logical channel assignments, session management, and flow control. The packet structure of X.25 was also discussed, and an examination of each packet type was provided. Furthermore, the section discussed the addressing scheme used by X.25. This discussion included the X.25 and OSI addressing conventions, the international numbering plan for data networks (X.121), and examination of the suitability of these addressing plans to support internal network routing.

The fifth part of the article was dedicated to explaining how the facilities provided by X.25 operate and how these facilities can be used to enhance a user

Protocol Functions	X.25	Frame Relay
Generate/Recognize Flags	•	•
Transparency	•	•
Generate/Verify FCS	•	•
Recognize Invalid Frames	•	•
Discard Incorrect Frames	•	•
Translate Addresses	•	•
Fill Interframe Time	•	•
Multiplexing Logical Channels	•	•
Manage State Variables	•	
Buffer packets awaiting Acks	•	
Manage Retransmit Timers	•	
Acknowledge Received I-frames	•	
Check Received N(S) against V(R)	•	
Generate Rejection Messages	•	
Respond to Poll/Final bit	•	
Manage number of retransmissions	•	
Act on reception of REJ	•	
Respond to RNR	•	
Respond to RR	•	
Manage D,Q,M bits	•	
Detect out of sequence packets	•	
Manage Network Layer RR, RNR	•	

FIG. 45 Comparison of the number of functionalities supported by frame relay and X.25 to establish, manage, and maintain a connection over a link.

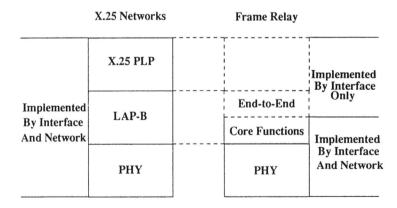

FIG. 46 Comparison of the protocol architectures of frame relay and X.25.

session. The facilities discussed include the international facilities described in Recommendation X.2, the CCITT-specified DTE facilities, the facilities offered by the originating public data network, and the facilities offered by the destination public data network.

The sixth part of the article provided an overview of the principal standards associated with X.25. These protocols are usually referred to as companion protocols of the X.25 Standard. Due to the complexity of these protocols, this section only provided a general tutorial of these standards. This tutorial can be viewed as a convenient reference guide for users of the X.25 network. Depending on the importance of these protocols and their relevance to current implementations of the X.25 network, a relatively in-depth analysis was provided. The protocols discussed include X.3, X.28, and X.29. There was a discussion of the CCITT packet assembler/disassembler (PAD) recommendations. An examination of some options for using these recommendations on an X.25 network was also provided.

The seventh section of the article focused on the principal internetworking recommendations and standards associated with X.25. This section provided a discussion of internetworking X.25-based networks using the X.75 Protocol. The X.75 packet types were examined, along with the X.75 utilities.

The last part of the article provided concluding remarks and discussed the future of X.25. With the advent of high-speed networks with relatively error-free links, it is becoming a common belief that X.25 is gradually getting outdated. The use of X.25, however, is still increasing.

Glossary

ACCESS METHOD. A set of rules for transferring data between hosts.

ACCESS CONTROL. Field used in token-ring MAC to control access to the channel according to the priority and reservation algorithm.

ACKNOWLEDGMENT. Message returned from receiver to the sender to indicate the successful reception of transmission.

ACKNOWLEDGE EVEN NUMBER MESSAGE (ACK0), ACKNOWLEDGE ODD NUMBER MESSAGE (ACK1). BISYNC acknowledgments for even and odd numbered messages, respectively.

ADDRESS. Data structure used to identify the destination of a message.

ADVANCED DATA COMMUNICATION CONTROL PROTOCOL. American National Standards Institute standard version of the bit-oriented Data-Link Control Protocol.

ALGORITHM. Method for solving a problem that is suited for computer implementation.

AMERICAN STANDARD CODE FOR INFORMATION INTERCHANGE (ASCII).

AMPLIFIER. Hardware device that may be inserted at various points of the communications system to impart a gain in signal strength.

AMPLITUDE MODULATION. Modulation technique that varies the amplitude of the carrier signal to convey information.

AMPLITUDE SHIFT KEYING. Modulation technique in which the two binary values are represented by two different amplitudes of the carrier frequency.

ANALOG SIGNAL. Continuously varying electromagnetic wave that may be propagated over a variety of media.

ANALOG TRANSMISSION. The transmission of analog signals regardless of their content.

APPLICATION LAYER. The seventh layer of the OSI Model.

ARPANET. One of the earliest and most influential computer networks in the United States.

ASYNCHRONOUS BALANCED MODE. HDLC mode for balanced configuration used only in point-to-point operation. The configuration consists of two combined stations and supports both full-duplex and half-duplex transmission.

ASYNCHRONOUS RESPONSE MODE. HDLC mode for unbalanced configuration used in point-to-point and multipoint operation. The configuration consists of one primary station and one or more secondary stations and supports both full-duplex and half-duplex transmission.

ASYNCHRONOUS TRANSMISSION. A type of transmission in which each character is transmitted independently of other characters. Each character is preceded by a start bit and ended by a stop bit. This type of transmission is often referred to as start-stop transmission.

ATTENUATION. The loss of power the signal suffers as it travels from the transmitting device to the receiving device.

AUTOMATIC REPEAT REQUEST (ARQ). Error-control technique in which the receiver detects packets with errors and requests their retransmission.

BACKOFF. Random delay before another attempt at transmitting is made by a colliding station in random-access medium protocols.

BALANCED CONFIGURATION. The configuration consists of two combined stations and supports both full-duplex and half-duplex transmission. The configuration is used only in point-to-point operation.

BALANCED TRANSMISSION. Mode of transmission in which the information is conveyed by comparing voltages in two circuits. For digital signals, this technique is known as differential signaling. It aims to minimize the effect of induced voltages.

BANDWIDTH. The difference between the two limiting frequencies of the spectrum. It is expressed in hertz.

BASEBAND. Transmission of signals without modulation. Since the entire scheme is occupied by one signal, this technique does not allow frequency division multiplexing.

BAUD RATE. Number of signal changes per second; used to convey information. Often misused for bit rate.

BINARY SYNCHRONOUS COMMUNICATION (BISYNC). A character-oriented data-link layer protocol for half-duplex applications.

BIT-ORIENTED PROTOCOLS. A class of Data-Link Layer Protocols based on the use of a flag character to delimit the frame.

BIT RATE. Number of bits per second transmitted or received.

BIT STUFFING. The insertion of extra bits into a data stream to avoid the occurrence of unintended control fields.

BLOCK CHECK CHARACTERS. Extra characters added to frames for error protection in the BISYNC Data-Link Layer Protocol.

BROADBAND. The use of coaxial cable for providing data transfer by means of analog signals. Modems are used to transmit digital signals over one of the frequency bands of the signal.

BROADCAST COMMUNICATION. A communications network in which transmission is received by all stations at the same time.

BUS. A conductor used for transmitting signals or power.

BYTE-COUNT-ORIENTED PROTOCOLS. A class of data-link layer protocols that rely on the byte count field in the frame to determine the length of the frame.

CARRIER. A continuous frequency that is impressed with a second signal to convey information.

CARRIER SENSE MULTIPLE ACCESS (CSMA).

CARRIER SENSE MULTIPLE ACCESS WITH COLLISION DETECTION (CSMA/CD).

CHARACTER-ORIENTED PROTOCOLS. A class of data-link layer protocols that uses standard alphanumeric control characters for communication.

CHECKSUM. An error-detection code based on a summation performed on the bits to be checked.

CLASS 0 TRANSPORT SERVICE. OSI transport protocol for Type A networks with error control, sequencing, and flow control. The transport service provides minimal connection establishment, data-transfer, and connection termination facilities.

CLASS 1 TRANSPORT SERVICE. OSI transport protocol for Type A networks with error control, sequencing, and flow control. The transport service provides minimal connection establishment, data-transfer, and connection termination facilities. In addition, it provides the capability to recover from network layer resets or disconnects and offers expedited data capabilities.

CLASS 2 TRANSPORT SERVICE. OSI transport protocol for Type A networks with error control, sequencing, and flow control. The transport service provides connection establishment, data-transfer, and connection facilities for multiplexing multiple transport layer connections over a network layer virtual circuit. The service uses credits to provide for transport layer flow control.

CLASS 3 TRANSPORT SERVICE. OSI transport protocol for Type B networks. The protocol adds the multiplexing and control functions of Class 2 to the error-handling functions of Class 1.

CLASS 4 TRANSPORT SERVICE. OSI transport protocol for Type B networks. The protocol is designed to handle worst-case networks. It provides comprehensive error-detection and -recovery mechanisms.

COAXIAL CABLE. Cable used for large-bandwidth electrical signals. It consists of an inner conductor, usually a small copper wire, a braided outer conductor, and a shield.

COLLISION. Result of simultaneous transmission of packets over the same medium.

COMBINED STATION. Combines the features of both primary and secondary HDLC stations. This type of station may issue both commands and responses. They are used in the asynchronous balanced mode.

CONNECTIONLESS DATA TRANSFER. Data transfer without prior establishment of a connection between the sender and the receiver. The exchanged units of information are usually referred to as datagrams.

CONNECTION-ORIENTED DATA TRANSFER. This mode requires the establishment of a connection before any exchange of data and the termination of the connection on completion.

CONTENTION. Attempt by two stations to use the same channel at the same time.

CONTINUOUS ARQ. Standard approach to ARQ for data-link control protocols that allows frames and acknowledgments to occur simultaneously. Depending on the implementation, the sender may be required either to retransmit packets with errors selectively or to retransmit the packet that was in error and all packets sent after it.

CONTROL CHARACTER. A character in a standard character set used to convey control information to a device.

CONTROL CIRCUIT. An X.21 circuit used to send control information from a DTE to a DCE.

CONTROL FIELD. Field in a frame containing control information.

CREDIT. Permit to transmit a number of messages as prescribed by the permit message. Used to allow variable-size windows.

CYCLIC REDUNDANCY CHECK. An error-detection code in which the code is the remainder resulting from the division of the bits to be checked by a predetermined binary number.

DATA CIRCUIT TERMINATING EQUIPMENT (DCE). Standards committee terminology for the attachment of user devices to the network. The data equipment takes on different forms for different network types.

DATA SET READY (DSR). An RS-232C interface circuit activated by the data set (modem) to indicate that it is powered up and ready to use.

DATA TERMINAL EQUIPMENT (DTE). A generic name for data-processing equipment connected to a data network, such as visual display units, computers, and workstations.

DATA TERMINAL READY (DTR). An RS-232C interface circuit activated by the data terminal equipment (DTE) to indicate to the data set (modem) that it is ready to send and receive data.

DATAGRAM. A self-contained packet of information that is sent over the network with minimum protocol overhead. The transmission of datagrams does not require the setting of a connection. Datagrams are handled independently of each other.

DATA-LINK ESCAPE (DLC). A communications control character used by BISYNC. The purpose of this character is to change the semantics of a limited number of contiguous characters that follow.

DATA-LINK LAYER. This layer corresponds to Layer 2 of the OSI Reference Model for Open System Interconnection and most other network architectures. The main functions implemented by this layer include the reliable transfer of data across the data link and the control of data flow between the sender and the receiver.

DEMAND ASSIGNMENT MULTIPLE ACCESS. A technique used to allocate in demand part of the capacity of the access medium to users.

DIGITAL DATA COMMUNICATION MESSAGE PROTOCOL (DDCMP). A byte-count-oriented data-link control protocol developed by Digital Equipment Corporation.

DIRECTORY SERVICE (DS). A protocol entity forming part of the application layer. The basic function of this protocol is to translate symbolic names as used by the application processes into fully quantified network addresses as used with the Open System Interconnection environment.

ENCAPSULATION. The addition of control information to the data obtained from the user.

END OF TEXT (ETX). A communications control character used in BISYNC to terminate the transmission of an entity.

END OF TRANSMISSION (EOT). A communications control character used to conclude a BISYNC transmission that may have contained one or more text messages and headings.

END OF TRANSMISSION BLOCK (EOB). A communications control character that indicates the end of a BISYNC transmission block.

ENQUIRY (ENQ). A control character used in standard alphanumeric codes and by some data-link layer protocols, such as BISYNC, for polling and selection or for assertion of the receiver status.

ETHERNET. A CSMA/CD-based local-area network.

EXPEDITED DATA. A short packet with high priority for treatment. It is usually used in emergency cases.

EXTENDED BINARY CODED DECIMAL INTERCHANGE CODE (EBCDIC). Eight-bit alphanumeric code used in many IBM products.

FAST SELECT. An option of the X.25 virtual call that allows the inclusion of data in the call setup and call-clearing packets.

FIBER OPTICS. A communications medium used for high-speed networks. It is usually composed of glass or similar fiber and associated circuits to transmit optical signals.

FILE TRANSFER, ACCESS, AND MANAGEMENT (FTAM). A protocol entity forming part of the application layer that allows the user application to access a distributed file system.

FLAG. A special configuration of bits used by data-link control protocols to achieve frame synchronization.

FLOW CONTROL. A technique used to regulate the traffic between the sender and the receiver. It aims to prevent the sender from overflowing the buffers of a slow receiver.

FRAGMENTATION. The need to break up the data into a smaller bounded size to accommodate a network with a short maximum packet length. Other motivations for fragmentation include more efficient error control, more equitable access to shared communications media, and better management of small-size buffers. The method may, however, incur more processing overhead and generate more interrupts to be served by the communications devices.

FRAME. The unit of information transferred over a data link. It is composed of control information and user data.

FRAME-CHECK SEQUENCE (FCS). A set of additional characters used for error detection.

FREQUENCY. Rate of signal oscillation in hertz.

FREQUENCY MODULATION. Modulation in which the frequency of an alternating current is varied.

FREQUENCY SHIFT KEYING. Modulation in which the two binary values are represented by two different frequencies near the carrier frequency.

FULL-DUPLEX TRANSMISSION. Data transmission in both directions at the same time.

GATEWAY. A device that connects two systems and allows data to be passed from one system to the other. When the two communicating systems operate with different protocols, the gateway performs protocol conversion.

GO-BACK-N ARQ. A variant of continuous ARQ. On detection of an erroneous frame, the receiver drops the frame in error and all succeeding frames. The receiver waits for the arrival of the correct version of the frame that was in error to resume receiving the succeeding frames.

HALF DUPLEX. An information exchange strategy in which data transmission may take place in either direction, one direction at a time.

HANDSHAKING. Sequence of messages exchanged to convey information and verify its receipt.

HEADER. Protocol-defined control information that precedes user data.

HIGH-LEVEL DATA-LINK CONTROL (HDLC). Standard data-link layer protocol used to regulate the data exchange between point-to-point data links or multidrop data links.

IMPULSE NOISE. A high-amplitude, short duration noise pulse.

IEEE 802. A set of standards for local-area networks being developed by the IEEE 802 Committee.

IEEE 802.2. Logical Link Control Protocol developed by the IEEE 802 Committee.

IEEE 802.3. CSMA/CD MAC Protocol developed by the IEEE 802 Committee.

IEEE 802.4. Token Bus MAC Protocol developed by the IEEE 802 Committee.

IEEE 802.5. Token-Ring MAC Protocol developed by the IEEE 802 Committee.

INSTITUTE OF ELECTRICAL AND ELECTRONICS ENGINEERS (IEEE).

INTERFACE. Boundary between equipment or between protocol layers.

INTERNATIONAL ORGANIZATION FOR STANDARDIZATION (ISO). International organization responsible for telecommunications networking standards, including the OSI Reference Model.

INTERNET CONTROL MESSAGE PROTOCOL (ICMP). An integral part of the Internet Protocol (IP) that handles errors and control messages. It allows hosts to interact with gateways and gateways to report problems about datagrams to the original source that sent the datagram. It also includes an echo request/reply used to test host reachability.

INTERNET PROTOCOL. An internetworking protocol that provides connectionless service across multiple packet-switched networks.

INTERNETWORKING. Communication among devices across several networks.

JOB TRANSFER AND MANIPULATION. A protocol entity forming part of the application layer. It is used by user application processes to transfer and manipulate documents describing job processing tasks.

LINK. Communications facility connecting nodes in a communications network.

LINK ACCESS PROTOCOL–BALANCED (LAP-B). CCITT X.25 version of bit-oriented Data-Link Control Protocol.

LOCAL-AREA NETWORK (LAN). A general-purpose network that spans a limited geographical area.

LOGICAL LINK CONTROL (LLC). Upper part of the data-link layer in the IEEE 802 architecture.

LOOPING. Routing that sends packets back to the source node. It is used for testing and checking.

MANCHESTER ENCODING. A digital encoding technique in which there is a transition in the middle of each bit time.

MEDIA ACCESS CONTROL (MAC). Lower part of the data-link layer in the IEEE 802 architecture.

MESSAGE HANDLING SERVICE (MHS). A protocol entity forming part of the application layer. It provides a generalized facility for exchanging electronic messages between systems.

MESSAGE SWITCHING. Technique used to transfer messages between nodes of a network.

MODEM. Acronym for modulator-demodulator. It translates a digital signal into a form suitable for transmission over an analog communications facility and translates back an analog signal into a digital form suitable for processing by a digital computer.

MULTIPLEXING. A technique used for sharing a resource among several contenders.

MULTIPOINT. A configuration in which more than two stations share a communications path.

NEGATIVE ACKNOWLEDGMENT (NAK). A message or a control character indicating to the sender a negative acknowledgment.

NETWORK LAYER. Layer 3 of the OSI Reference Model.

NODE. Data-processing equipment that is part of a telecommunications network.

NON-RETURN-TO-ZERO (NRZ). A digital signaling technique in which the signal is at a constant level for the duration of a bit time.

NORMAL RESPONSE MODE (NRM). Unbalanced configuration in the HDLC transfer modes of operation. In this mode, the primary may initiate a data transfer to a secondary, but a secondary may only transmit data in response to a poll from the primary.

NULL MODEM. An appropriate wiring of the RS-232C circuits that allows two devices to communicate directly with each other.

OPEN SYSTEM INTERCONNECTION REFERENCE MODEL. An architectural model of networking developed by ISO and adopted as an international standard.

OPTICAL FIBER. A transmission medium used for transmission of data in the form of light waves of pulses.

PACKET SWITCH. A mode of operation of a data communications network in which each message to be transmitted is first divided into packets. As they travel across the network, packets may be stored and later forwarded to the next node in the path.

PARITY BIT. Extra bit added to transmitted data to enhance error control.

PHASE MODULATION. A modulation technique that varies the characteristics of the carrier signal phase to convey information.

PHYSICAL LAYER. Layer 1 of the OSI Reference Model and most other network architectures.

PIGGYBACKING. A technique used to return acknowledgment information across a full-duplex data link without the use of special acknowledgment messages.

POINT TO POINT. A configuration in which two stations share a communications path.

PRESENTATION LAYER. Layer 6 of the OSI Reference Model.

PRIMARY STATION. The master station that is in charge of the HDLC link.

PROTOCOL. A set of rules and conventions that regulates the exchange of data between two communicating devices.

PULSE CODE MODULATION. A process in which the signal is sampled and the amplitude of each sample with respect to a fixed reference is quantized and converted by coding to a digital signal.

QUALITY OF SERVICE (QoS). Parameters that describe a quality of service as perceived by the user.

RANDOM ACCESS. Unscheduled access to the communications medium.

REMOTE-OPERATIONS SERVICE ELEMENT (ROSE). A protocol entity forming part of the application layer that provides facility for initiating and controlling operations remotely.

REQUEST TO SEND (RTS). The DTE activates this circuit in the RS-232C interface when it is ready to send data.

RING. Configuration of stations in the shape of a ring in which communication normally flows in one direction.

RING INDICATOR (R). The DCE activates this circuit in the RS-232C interface to indicate to the DTE the presence of a ringing signal due to an incoming call.

ROUTING. A technique used to determine the path of a message from the source to the destination.

RS-232-C/RS-449/422-A/423-A. Standards developed by the American Electronic Industries Association for a physical layer interface between communications devices.

SECONDARY STATION. A station in an HDLC configuration that is not primary.

SEGMENTATION. Fragmentation of a message or stream of data into small fragments.

SELECTIVE REJECT FRAME (SREJ). HDLC frame used as a negative acknowledgment of a frame. The frame requires the transmission of the erroneous frame only.

SERVICE ACCESS POINT (SAP). Used in the OSI Reference Model to define a point at which a service is provided.

SESSION CONTROL LAYER. Layer 5 of the OSI Reference Model.

SIMPLEX TRANSMISSION. A type of information exchange strategy between two communications devices by which transmission can only occur in one direction.

SLIDING WINDOW FLOW CONTROL. A method of flow control in which transmitting stations may send numbered packets within a dynamically changing window of numbers.

SPECIFIC-APPLICATION SERVICE ELEMENT (SASE). A collection of protocol entities forming part of the application layer responsible for providing various specific application services.

SPECTRUM. Refers to an absolute range of frequencies.

START OF HEADING (SOH). Communications control character used in BISYNC at the beginning of a sequence of characters that contain address or routing information.

START OF TEXT (STX). Communications control character used in BISYNC to indicate that the following sequence of characters, usually referred to as text, is to be treated as an entity.

STOP-AND-WAIT ARQ. A flow control strategy in which the sender sends a block of data and waits for an acknowledgment before proceeding with the following block.

SUPERVISORY FRAME. Frame used by data-link control protocols for supervision of the link.

SYNCHRONOUS DATA-LINK CONTROL (SDLC). A data-link layer protocol developed by IBM. It is the IBM version of HDLC.

SYNCHRONOUS TRANSMISSION. Data-transmission mode in which the time of occurrence of each signal representing a bit is related to a fixed time frame.

TIME DIVISION MULTIPLEXING. A technique used to share the channel capacity of a communications channel by carrying different signals at different time slots.

TOKEN BUS. Local-area network architecture using a coaxial cable as a physical transmission medium and a token to guarantee a regulated access to the medium.

TOKEN RING. Local-area network architecture in which all stations are connected in the form of a physical ring and messages are transmitted by allowing them to circulate around the ring. A regulated access to the medium is guaranteed by a token.

TRANSMISSION CONTROL PROTOCOL (TCP). The TCP/IP standard transport-level protocol that provides a reliable, full-duplex, stream service. It is a connection-oriented transmission protocol and relies on the Internet Protocol to transmit data across the physical network.

TRANSPORT LAYER. Layer 4 of the OSI Reference Model.

TWISTED PAIR. A transmission medium consisting of two insulated wires arranged in a regular spiral pattern.

UNBALANCED CONFIGURATION. An HDLC link configuration used in point-to-point and multipoint operations. The configuration consists of one primary and one or more secondary stations. It supports both full- and half-duplex transmission.

UNNUMBERED FRAME. HDLC frames to handle operations such as link start-up and shutdown, specifying modes, and other maintenance-related operations.

USER DATAGRAM PROTOCOL (UDP). The TCP/IP standard transport-level protocol that allows an application layer on one machine to send a data-

gram in another application layer in another machine. It is a connectionless protocol and relies on the Internet Protocol to deliver the datagrams
to the destination.

VIRTUAL CIRCUIT. A transmission path set up by the connection protocol from
one end to the other.

WACK. The response contains an acknowledgment and requests the sender to
pause for a time duration.

X.21. Physical layer interface included in the OSI Reference Model.

X.25. Standard defining the user interface to pubic data networks.

Bibliography

General References for Computer Communication

Bertsekas, D., and Gallagar, R., *Data Networks*, Prentice-Hall, Englewood Cliffs, NJ,
1992.

Black, U. D., *Data Communications and Distributed Networks*, 2d ed., Reston Publishers, Reston, VA, 1987.

Day, J. D., and Zimmermann, H., The OSI Reference Model, *Proc. IEEE*, 71:1334–
1340 (December 1983).

Folts, H. C., *Compilation of Data Communications Standards*, 2d rev. ed., McGraw-
Hill, New York, 1983.

Halsall, F., *Data Communications, Computer Networks and Open Systems*, 4th ed.,
Addison-Wesley, Reading, MA, 1995.

Henshall, J., and Shaw, A., *OSI Explained. End to End Communication Standards*,
Ellis Horwood, Chichester, England, 1988.

Linnington, P. F., Fundamentals of the Layer Service Definitions and Protocol Specifications, *Proc. IEEE*, 71:1341–1345 (December 1983).

McNamara, J. E., *Technical Aspects of Data Communication*, 3d ed., Digital Press,
Bedford, MA, 1988.

Martin, J., *Computer Networks and Distributed Processings: Software Techniques and
Architecture*, Prentice-Hall, Englewood Cliffs, NJ, 1981.

Martin, J., *Introduction to Teleprocessing*, Prentice-Hall, Englewood Cliffs, NJ, 1972.

Quarterman, J. S., and Hoskins, J. C., Notable Computer Networks, *Commun. ACM*,
29(10):932–971 (October 1986).

Schwartz, M., *Telecommunication Networks, Protocols, Modeling and Analysis*, Addison-Wesley, Reading, MA, 1987.

Seyer, M. D., *RS-232 Made Easy: Connecting Computers, Printers, Terminals and
Modems*, Prentice-Hall, Englewood Cliffs, NJ, 1982.

Sherman, K., *Data Communications, A User's Guide*, 3d ed., Prentice-Hall, Englewood
Cliffs, NJ, 1990.

Stallings, W., *Data and Computer Communications*, 4th ed., Macmillan, New York,
1994.

Tanenbaum, A. S., *Computer Networks*, 2d ed., Prentice-Hall, Englewood Cliffs, NJ,
1988.

Zimmermann, H., OSI Reference Model—The ISO Model of Architecture for
Open Systems Interconnection, *IEEE Trans. Commun.*, COM-28:425–432 (April
1980).

Physical Layer References

Bell Telephone Laboratory, *Transmission Systems for Communications*, Bell Telephone Laboratory, Murray Hill, NJ, 1982.

Bertine, H. U., Physical Level Protocols, *IEEE Trans. Commun.*, 28(4):443–444 (1980).

Bleazard, G. B., *Hand Book of Data Communications*, NCC Publications, 1982.

Chou, W., *Computer Communications, Vol. 1: Principles*, Prentice-Hall, Englewood Cliffs, NJ, 1983.

Cooper, E., *Network Technology*, Prentice-Hall, Englewood Cliffs, NJ, 1986.

Davies, D. W., *Communication Networks for Computers*, Wiley, New York, 1973.

Freeman, R., *Telecommunication Transmission Handbook*, Wiley, New York, 1981.

Held, G., *Data Communication Components*, Hayden, Rochelle Park, NJ, 1979.

Intel Corporation, *The Intel 8251A/S2657 Programmable Communications Interface*, 1978.

Luetchford, I. C., CCITT Recommendations Network Aspects of the ISDN, *J. Sel. Areas Commun.*, SAC-4:334–342 (May 1986).

McClelland, F. M., Services and Protocols of the Physical Layer, *Proc. IEEE*, 71:1372–1377 (December 1983).

Peterson, W. W., *Error Correcting Codes*, MIT Press, Cambridge, MA, 1981.

Yanoschak, V., Implementing X.21 Interface, *Data Communications* (February 1981).

Data-Link Layer References

Conard, J. W., Service and Protocols for the Data Link Layer, *Proc. IEEE*, 71:1378–1383 (December 1983).

Doll, D. R., *Data Communications: Facilities, Networks and System Design*, Wiley, New York, 1980.

Field, J. A., Efficient Computer-Computer Communication, *Proc. IEEE*, 23:756–760 (August 1976).

Fraser, A. G., Delay and Error Control in a Packet Switched Network, *Proc. ICC*, (22.4)121, 125 (1977).

Fraser, A. G., Towards a Universal Data Transport System. In *Advances in Local Area Networks* (K. Kummerle, F. Tobagi, and J. O. Limb, eds.), IEEE Press, New York, 1987.

Rudin, H., An Informal Overview of Protocol Specifications, *IEEE Commun. Mag.*, 23:46–52 (March 1985).

Media Access Control Sublayer References

Burg, F. M., Chen, C. T., and Folts, H. C., Of Local Networks, Protocols, and the OSI Reference Model, *Data Communications*, 129–150 (November 1984).

Bux, W., Closs, F., Janson, P. A., Kummerle, K., and Muller, H. R., A Reliable Token Ring System for Local Area Communication, *Proc. National Telecommunications Conference*, A2.2.1–A.2.2.6 (1981).

Carpenter, R., A Comparison of Two Guaranteed Local Network Access Methods, *Data Communications* (February 1984).

Clark, D., Program, K., and Reed, D., An Introduction to Local Area Networks, *Proc. IEEE '66*, 1497–1516 (November 1978).

Dixon, R. C., Strole, N. C., and Markov, J. D., A Token-Ring Network for Local Data Communications, *IBM Sys. J.*, 22:47–62 (January–February 1983).

Eswaran, K. P., Hamacher, V. C., and Shedler, G. S., Collision Free Access Control for Computer Communication Bus Network, *IEEE Trans. Software Eng.*, SE-7 (November 1981).

Fine, M., and Tobagi, F. A., Demand Assignment Multiple Access Schemes in Broadcast Bus Local Area Networks, *IEEE Trans. Computers*, C-33(12) (December 1984).

Fratta, L., Borgonovo, F., and Tobagi, F., The EXPRESS_NET: A Local Area Communication Network Integrating Voice and Data. In *Performance of Data Communication Systems* (G. Pujolle, ed.), Elsevier, North-Holland, New York, 1981, pp. 77–88.

Institute of Electrical and Electronics Engineers, *Carrier Sense Multiple Access with Collision Detection (CSMA/CD) Access Method and Physical Layer Specifications*, American National Standard ANSI/IEEE Standard 802.3-1985, New York, 1985.

Institute of Electrical and Electronics Engineers, *Logical Link Control*, American National Standard ANSI/IEEE Standard 802.2-1985, IEEE, New York, 1985.

Institute of Electrical and Electronics Engineers, *Token Ring Access Method and Physical Layer Specifications*, American National Standard ANSI/IEEE Standard 802.5-1985, New York, 1985.

Institute of Electrical and Electronics Engineers, *Token-Passing Bus Access Method and Physical Layer Specifications*, American National Standard ANSI/IEEE Standard 802.4-1985, New York, 1985.

Network Layer References

Bell, P. R., and Jabbour, K., Review of Point-to-Point Routing Algorithms, *IEEE Commun. Mag.*, 24:34–38 (January 1986).

Bertsekas, D., and Gallager, R., *Data Networks*, Prentice-Hall, Englewood Cliffs, NJ, 1987.

Chapin, A. L., Connections and Connectionless Data Transmission, *Proc. IEEE*, 71: 1365–1371 (December 1983).

Davies, D. W., Barber, D. L. A., Price, W. C., and Solmonides, C. M., *Computer Networks and Their Protocols*, Wiley, New York, 1979.

Deasington, R. J., *X.25 Explained: Protocols for Packet Switched Networks*, 2d ed., Ellis Horwood Series of Computer Applications, Chichester, England, 1988.

Dhas, C. R., and Konangi, V. K., X.25: An Interface to Public Packet Networks, *IEEE Commun. Mag.*, 10:74–84 (June 1977).

Green, P. E., Protocol Conversion, *IEEE Trans. Commun.*, COM-29:726–735 (May 1981).

Hawe, B., Kirby, A., and Stewart, B., Transparent Interconnection of Local Area Networks with Bridges, *J. Telecommunication Networks*, 3:116–130 (1984).

International Organization for Standardization, *Network Service Using X.25 and X.21*, ISO TC97/SC5/N2743, 1981.

International Telegraph and Telephone Consultative Committee, Data Communication Networks: Services and Facilities, Interfaces; Recommendation X.25, Interface Between Data Circuit Terminating Equipment (DCE) for Terminals Operating in the Packet Mode and Connected to Public Data Networks by Dedicated Circuit. In *Blue Book*, Vol. 8, Fascicle 8.2, International Telecommunication Union, Geneva, 1989.

Meijer, A., and Peters, P., *Computer Network Architectures*, Computer Science Press, Rockville, MD, 1982.

TAIEB F. ZNATI

Network Regulation
in the United States

Introduction

The current multiple-network environment in the United States is undergoing significant change as a result of the phenomena of network and media convergence[1] and privatization.[2] The speed with which various industry firms seek to make a joint venture or combine to enter new markets underscores this fact.[3] As this change occurs, one of the critical questions for First Amendment theorists and scholars concerned with mass media and telecommunications is what access and speech rights will network owners and users have after the convergence and privatization of the mass media and telecommunications networks?[4]

As the convergence phenomenon intersects with growing privatization, there is potential danger to access and broad-based speech opportunities. Privatization of the technologies and networks can lead to the concentration of control over content in the hands of private network owners. Because of the historical tendency to equate speech rights with ownership of the means of transmission, privatization of the merging of technology, network function, and information streams could effect a transfer of the current shared control over access and speech from the current public/private constitutional arrangement to private/contractual arrangements. Such a result could be detrimental to the potential speech and access opportunities of existing and future network users.[5] Private owners may not be motivated by public interest considerations of access and inclusion. Instead, their major motivation to provide access and speech to employees will be utilitarian, and their major motivation to serve a particular customer or group of customers will depend (in the most ideal sense) on the desirability of that individual or market as a customer base and their ability to pay. Because these decisions are private, there is arguably less opportunity to rest the justification for access and speech rights on constitutional grounds given the alleged absence of state action.[6]

Under such circumstances, it is reasonable to ask: Will privatization of postconvergence multifunctional, multimedia networks result in speech rights only for network owners and those they elect to employ or to serve under contract? In an era of privatization in the provision of network services, what, if any, affirmative responsibility will the government retain to ensure access and speech rights for the non-facilities-based public?

In the area of employee access and speech rights on employer networks, a pragmatic judicial and/or legislative balance must be struck between the network owners' legitimate business needs, employees' access, speech and privacy rights, and public policy concerns, including the public's right to know. Ultimately, employee speech rights should not depend on whether they are employed by public or private firms. Rather, at minimum, employee speech "rights

© 1994 by Allen S. Hammond IV, professor of law and director of the Communications Media Center at the New York Law School. Updates have been added for this publication.

should encompass concerns regarding wages and working conditions, as well as safety and product quality issues about which the public may have interest.[7]"

In the areas of nonaffiliated programmer and subscriber access and speech rights, the application and enforcement of libel, indecency, and obscenity laws can serve to encourage some network owners to relinquish editorial control over content in order to avoid liability. This assumes that the government relinquishes its strategy of imposing responsibility and liability on both the network owner and the subscriber. Ultimately, without the assertion of editorial control by the network provider, responsibility and liability for speech should reside with the speaker.

Similarly, the removal of government-sanctioned limitations on carrier tort liability would encourage some network owners to eschew private carriage for the protection that public common carriage affords. Tort liability under state law would attach whenever the private carrier negligently handled subscriber information. Private carrier and closed user group attempts to exempt themselves from such liability via exculpatory contract clauses or tariff language would be deemed unconscionable and unenforceable as a matter of law when it could be established that the subscriber does not possess equal bargaining power. Only carriers providing service to the general public or interconnection with public networks should enjoy the protection from tort liability. Like the imposition of libel and criminal liability, the application of tort liability would also serve as an incentive for private networks to eschew control over content or, at the very least, provide access via interconnection among other networks.

The government should continue a qualified reliance on the antitrust laws[8] and structural safeguards[9] to ensure access to network facilities. Other than the use of these strategies, the government may proactively encourage access and speech by creating regulatory policies and tax incentives that favor the building of open, switched, interconnected networks incorporating distributed intelligence. Such networks, whether public or private, would provide multimedia interactive capabilities to large numbers of users. They also provide a preferable alternative to multichannel, unidirectioned distribution systems in which network architecture and functionality preclude two-way, broadband interactive communications while facilitating the network owners' exercise of private editorial control.

When combined with the regulatory strategies outlined above, selection of a switched-network architecture would also ensure that neither network owners nor users forfeit meaningful access or speech rights. For, in a switched broadband interactive network environment, the notion of scarcity on which the constitutional and technical regulation of antecedent technologies is based, should no longer serve as a viable justification for limiting access and speech rights.[10]

Networks and Closed User Groups Defined

General Characteristics

In their most basic manifestation, networks are collections of interconnected users.[11] They can be defined in terms of numerous characteristics, including (1) technology (spectrum, wire, fiber), (2) information (video, voice, data), (3) ownership (facilities based or non-facilities based), (4) control of content (editor, hybrid, common carrier), and (5) control of network access and/or functionality.[12] This article discusses them primarily in terms of ownership of facilities, control of network access and/or functionality, and, ultimately, control of content.

The Network Lexicon

Public Switched Network: "Open Network"

The commercial common-carrier network owners own the network facilities and retain control of all levels of network functionality. There are no predetermined limits on who may or may not join the network. All who timely pay the subscription fee (tariff rate for the particular class and volume of service) may gain access and enjoy usage.[13] Because the network is available to virtually all potential users,[14] it may be defined as being *open*.[15] These networks are the long-distance, regional, and local public switched networks.

Virtual Private Networks

In contrast to the public switched networks, virtual private networks (VPNs), offer their customers access to reserved private-line capacity on the public switched telephone network (PSTN).[16] A VPN is essentially a long-distance service in the United States. In the case of VPNs, the user manages network applications, while the carrier manages all other levels of network functionality.[17] It is anticipated that by 1997, VPN services will account for 17% of the domestic service revenues of the three biggest U.S. long-distance carriers—AT&T, MCI, and U.S. Sprint.[18]

Private Networks

In the case of private networks, all telecommunications facilities are owned by an entity other than a government-certified commercial common carrier or the user leases dedicated lines from certified carriers but maintains control over both ends of the communications channel. In the last case, the user typically owns facilities on its premises (local-area networks or private branch exchanges [PBXs], i.e., "intrabuilding private networks") and leases from carriers anything that crosses public rights of way. For example, the company may lease a dedicated T1 between two privately owned PBXs. Network usage is confined to the owner and its affiliates and is not shared or aggregated on a commercial basis.[19]

A private network may be open or closed. However, most closed user group networks are based on privately owned or dedicated facilities. Thus, most closed user group networks are private.[20]

Hybrid Networks

A large number of U.S. users have now opted for hybrid networks combining leased lines between particular locations with VPNs. Users prefer to use private networks for sensitive business information (in the form of encrypted data) because these are considered more secure than VPNs.

Closed User Groups

Closed user groups are large-volume users that tend to communicate with each other "intensely." They combine to form alternative network associations for much of their communications needs. Associations' networks may have specialized performance attributes related to group needs.[21] User groups' networks may be closed for numerous reasons, including specialized equipment, specialized features, transmission speeds, security, service pricing, or speech-related restrictions.

Other Networks

The definitions above are admittedly limited to switched telecommunications networks in some way related to the public switched networks. In contrast, video distribution networks include traditional broadcasting and cable television networks that are nonswitched, essentially one-way distribution media that are not usually interactive.

Convergence and Metamorphosis: From Cable and Telephony to Broadband Networks

It is suggested likely that cable networks and local telephone networks will evolve into the switched broadband interactive networks of the near future.[22] Should this be true, the resulting networks could span the gamut from private networks, virtual private networks, and hybrid networks to public (common-carrier) networks and could incorporate a potential range of access options from nondiscriminatory access and mandated leased and/or tariffed public access to private access by negotiated contract or ownership. Speech options could range from owner control of a portion of available capacity with unrestricted user speech on the remaining portion to owner control of all capacity and ultimately all speech allowed on the network. This spectrum of alternative access and speech relationships is in essence the amalgam of access and speech relationships currently residing on cable and telecommunications networks.

In this context, FCC hearings regarding implementation of the cross-ownership and OVS provisions of the Telecommunications Act of 1996,[23] litigation challenging the must carry rules of the Cable Competition and Consumer Protection Act of 1992 and the telco-cable cross-ownership prohibition of the

Cable Communications Policy Act of 1984, as well as several electronic mail (E-mail) and electronic bulletin board service cases percolating through the judicial system, may establish much of the scope of access and speech rights network owner providers and users will have.

Current Scope of Network and User Group Access and Speech Restrictions

There are at least four levels at which a network owner or closed user group may control access and/or speech activities on their facilities. Control may be exercised over actual speech or communication (content), over access to the network as configured by the owner (network access), over the ability to reconfigure network functionality (network software intelligence), and over the ability to set equipment standards for network provisioning and interfacing with the network (equipment standards and network protocols). Current government policies affect the exercise of access and speech at each of the levels.

Legal sanction of the exercise of control varies depending on the manner in which the network is used. When the network is merely one of many tools or assets used by a firm to conduct its business, the "network owner" enjoys wide latitude over each of the four levels. When the network and its related functions and services are the product the private or public firm sells to customers, the network owner's ability to control access has been subject to greater government restraint, depending on architecture, market power, and traditional rights accorded networks having similar technologies and functions.

Owner-Imposed Limitations on Access: Subscriber/User-Initiated Access to Third Parties

Network Owner/Employer Restrictions on Outgoing Calls. Employers may be subscribers to the public switched networks, virtual private network subscribers, owners of their own networks, or a collection or association of users forming a closed user group. They may be private or public firms. In any event, because of the utility of long distance and E-mail, as well as the growth in availability of 800 and 900 services, employers often find it necessary to block access to certain networks and phone services to limit corporate expenses. Call blocking, for instance, is used to limit employee access to the above-mentioned services.[24] In the process, employees are denied access to the networks over which such services are provided and the information providers at the other end of the line. Federal and municipal government call-blocking restrictions on access to dial-a-porn and long-distance calls are well-known examples.[25] Employers also engage in call monitoring as a means of policing their restrictions on network usage.[26]

Employers' justifications for engaging in these practices include the need to manage or reduce costs or fraud involved in unauthorized 900 number and E-mail calls, which have cost companies hundreds of thousands of dollars. In addition, some companies use computers to monitor customer services employees' performance, such as keystrokes per minute, time between phone calls, length of breaks, and number of errors. [27]

There are potential dangers inherent in call blocking and monitoring that raise significant public policy issues. Call blocking has been argued to implicate First Amendment concerns because the employer's limitations on access to the network and hence those an employee might contact constitute limitations on potential speech activities in which the employee might otherwise engage. Call monitoring is said to raise issues of worker privacy as the employee's expectation of privacy is infringed by periodic monitoring. [28] Call-blocking and -monitoring activities raise nettlesome problems for the public's "right to know" as well for these practices may also be used to detect whistle-blowers instead of individuals calling dial-a-porn providers and other unauthorized users. [29] Thus, it is understandable that some observers argue that employer/network owner control of access or usage of the corporate telephone or network affects employee's constitutional speech and/or privacy rights. [30]

Third-Party Access to Private Network or Virtual Private Network Facilities. Firms also attempt to limit third-party access to their networks to protect against toll fraud. [31] While computerized telephone equipment such as voice mail often help companies conduct business with greater efficiency and lower cost, they also often provide access for electronic thieves who steal thousands of dollars of long-distance telephone service. [32] According to some experts, computer hacking may cost U.S. companies between $2.2 billion and $4 billion nationally. [33]

In a less arcane realm, E-mail, voice-mail, and telephony systems may be used by union organizers, law enforcement authorities, friends, or family members to communicate with employees. However, efforts of union organizers to make use of the employer-owned telecommunications systems and/or networks to communicate with employees under Section 7 of the National Labor Relations Act (NLRA) may prove unsuccessful. [34] As a practical matter, recent precedent supports the employer's right to bar union access absent a union showing that the union possesses no other reasonable means of communicating its organizational message to employees. [35]

Restriction on Membership in Closed User Groups: Control of the Jointly Owned Network Asset. Some firms engage in a joint venture to develop interorganizational network systems (IONSs). IONSs can increase the efficiency and competitiveness of their owners. [36] In the process, IONSs also can serve as a catalyst to realign the relative market position of nonowners by creating new barriers to market entry and exit, which are often controlled by the IONSs owners. [37] By establishing, maintaining, and changing its pricing structure, as well as network applications, standards, protocols, and internal control procedures, IONSs owners can raise barriers to system (and, often, market) access. [38]

Network Owners' Exercise of Content Controls

Public and private firms also use call monitoring to manage employee communications to the firm's customers. Telemarketing and travel reservations services are two common examples.[39] Computer networks such as Prodigy have asserted some control over bulletin board content in response to various user protests regarding speech on controversial subjects.[40]

Aside from firm business-oriented restrictions on access and in-house, on-line speech, there are the traditional and evolving limits on access and speech in the realm of broadcasting,[41] cable television,[42] and plain old telephone service (POTS).[43]

Long-distance and local network providers arguably are private speakers possessing the right to refuse carriage or billing services to subscribers seeking carriage of programming the carrier deems undesirable.[44] To date, the courts have not taken a similar view with regard to cable television access channels.[45] The District of Columbia Circuit distinguished the telephone and cable contexts, noting that, unlike the government-compelled offering of leased and public access channels in cable, the billing services provided by telephone companies were voluntary and therefore private.[46] The state action/private action distinction relied on by the circuit court in the Alliance case could have a profound effect on future regulation of customer access to the networks of foreborne carriers.[47] They, like the local telephone companies' offering of billing services, offer their common-carrier communications services on a voluntary basis. Thus, per the Alliance for Media analysis, they would be free to engage in discriminatory provision of services.[48]

Constitutional Dimension

The above, admittedly cursory, review leads to the conclusion that Network owner providers limit access and speech activities of employees, user/members, and third parties to accomplish numerous tasks, including protection of property, assets, costs, and market share, as well as to achieve competition advantage, limit or constrain dissemination of proprietary information, manage the communication of information, and/or discourage the procurement of sometimes illicit information.

Under such circumstances, when may we say that a user/member or a potential outside communicator is impermissibly constrained from gaining access or engaging in speech? Is it possible to distinguish between legitimate business needs and impermissible constraints on speech activities? At first blush, based on the discussion above, one could suggest that permissible firm needs include all of those listed above.[49] By the same token, others might argue that impermissible firm needs include many of the same goals articulated as legitimate.[50]

One possible way to answer this dilemma is to examine the manner in which the law has addressed and apportioned access and speech rights in the varying relationships between network owner providers and closed user groups on the one hand and network users and third parties on the other. A critical distinction in the manner in which such rights are apportioned is the relative status of the network. When the network is an asset established primarily for the internal use of the corporation or closed user group, the employees, closed user members, and third-party communicators have very limited access and speech rights. When the network is the product or service offered by the corporation or closed user group, subscribers and viewers have been accorded greater access and speech rights based on constitutional, economic, and other public policy principles.

Network as a Business Asset or Tool

Employer/Employee Relationships

As mentioned above, employers may be network owner providers, closed user groups, or simply network subscribers. The nature of their status as public or private institutions, however, has a significant effect on the scope and expectations an employee may have regarding constitutional protection of their arguable rights of access to company facilities and ability to speak on those facilities.

Recently, there have been a spate of law suits filed by employees alleging that their constitutional rights have been violated when employers monitored their conversations over E-mail or telephone networks.[51] While this new area of the law remains in a state of flux as cases are resolved, there are some indications of the extent of protection afforded employee speech.[52] For instance, most experts agree that while the Federal Electronic Communications Privacy Act (ECPA) of 1986 protects the privacy of electronic messages sent through public networks to which individuals or companies subscribe, it does not apply to internal E-mail.[53] Thus, to the extent that employees enjoy speech rights on company E-mail facilities, those rights are limited to communication over public E-mail systems. Employers retain the right to restrict access to and monitor internal E-mail.

Public Employer/Employee.   The courts have held that a public employee has First Amendment constitutional protection for speech about "matters" of public concern.[54] In cases in which the employee is acting as a "whistle-blower," public policy and legislation in an increasing number of jurisdictions support a public employee's right to speak.[55] It is clear that employees do not enjoy an unfettered right of speech, however. For instance, current cases allow the employer to deny such speech when it may disrupt the workplace.[56]

Private Employer/Employee.   Under the National Labor Relations Act, an employee has statutory protection for speech concerning work-related activities.[57] There are also whistle-blower statutes in many states that protect employee speech about company wrongdoing.[58] Otherwise, under the "work at will" doctrine, the employee ostensibly has no recognized speech rights in the face of legitimate company interests.[59] The scope of an employee's statutory license to use company E-mail and/or telecommunications facilities to realize their work-related speech right is not established, however.[60]

Closed User Group and Members (Actual and Potential)

When firms or users associate via network facilities that they have acquired, they may exercise control over member and nonmember network access and speech.[61] While the scope of a closed user group's (private network's) liability for actionable speech is unsettled, it seems intuitively appropriate that its liability track that of bulletin board system operators (sysops) such as CompuServe and Prodigy.[62] The more extensive its control over the communication of content is, the more extensive the liability for that content ought to be. At the same time, the more extensive the control of content is, the less extensive individual user speech rights will be.

Network as Product or Service: Relationships Between Network Owner Providers and Users

Network Owner Provider and Subscriber Users (Common Carriage)

The largest category of relationships between network owners and consumers exists in the provision of network transmission capacity and network-related services. Telecommunications network owners may provide transmission between two or more points at varying speeds with a variety of ways to manipulate the various types of transported information. Services range from the provision of transmission capability for private networks to virtual private and hybrid networks, 800 and 900 number services, and billing, to POTS. The provision of network transmission, switching, billing, and intelligence-based services may be accomplished pursuant to regulated tariff, by contract, or by a combination of the two.

As competition has increased, regulators have tended to afford network owner providers greater flexibility in providing services under contract.[63] Even when services are not provided pursuant to contract, network owner providers have been granted greater flexibility in providing many services under tariff.[64] When the services are offered on a common-carrier or quasi-common-carrier basis, the network provider has tended to limit network access based on the type and class of service, network integrity, security, and capacity.

Aside from government-mandated responsibilities to foreclose opportunities for harassing, indecent, or obscene speech to reach protected subscribers, carriers have tended to eschew control of information content, thereby foregoing liability for customer communication. This practice has been sanctioned by federal and many state regulatory bodies. Also, carriers have traditionally sought to limit their liability for loss or damage to customer communications.[65] Until now, these efforts have been successful.[66]

However, recent court decisions have held that telecommunications network owners have speech rights as extensive as cable video distribution network providers.[67] Assuming these cases are ultimately upheld on appeal, should the telecommunications network providers exercise their speech and editorial rights by limiting the access and speech of users, attempts to limit speech-related liabilities may, and increasingly should, prove less successful.[68]

For instance, in a related area, bulletin board/E-mail providers such as CompuServe and Prodigy have the ability to control access and screen speech content on their systems.[69] While CompuServe does not actively seek to control access or, more importantly, content, Prodigy does. As a result, while CompuServe has been successful in avoiding liability for libelous statements made by one of its users, it is not clear that Prodigy will as well.[70] Moreover, the decision to control content places the service provider in a difficult position when it either fails to prohibit offensive speech quickly or prosecutes other speech in a seemingly biased manner.[71]

Network Owner/Information Provider and Consumer/Subscribers

Telephony (Plain Old Telephone Service) and Telecommunications.   The need for interconnection and the economies of scale inherent in provision of local telephone service led in significant part to the creation of government-sanctioned telephone monopolies. Government then sought to ensure the public access to the monopoly provider by requiring that the provider not discriminate between customers on the basis of facilities or the price paid for the services provided.[72]

As a further means of ensuring nondiscrimination, the telephone company was not allowed any control over the content of information it transmitted. More recently, however, telephone companies have been allowed to deny billing and collection services to dial-a-porn providers deemed undesirable by the carriers.[73] Also, government requirements that long-distance common carriers may not engage in the provision of information services and local common carriers may not provide electronic video services within their local markets have been overturned. According to a recent district court opinion, local telcos now have video electronic speech rights.[74] Should the decision be upheld on appeal, there is still a question of how this newly articulated speech right will merge with the telco owner's property right vis-à-vis control of access and content.[75] Many potential competitors and customers of local carriers possessing essential facilities have voiced concern over the potential for unfair competition.[76]

In the area of switched, interactive telecommunications, the diverse set of relationships addressed above is expanding even farther as interactive video distribution capabilities come on line and user access to network functionalities increase via manipulation of network intelligence. It is here that a significant potential for new and increased access and electronic speech is to be found.[77]

As fiber-optic, computer, and switched telephony technologies merge, so do the heretofore separate network functions and information streams of telephony, broadcasting, cable, and print.[78] As this occurs, there is a potential danger that the network owner, as a potential speaker, may experience a conflict of interest with the provision of network-related services to users who, like the network owner, are also information providers. Newspaper publishers, cablecasters, and broadcasters have raised this potential for conflict of interest as a reason for continuing the prohibition against local telephone companies' entry into the information and video distribution markets.[79] While these arguments earlier found sympathetic ears in Congress, recently they have proved unpersuasive before the Federal Communications Commission (FCC) and the courts.[80]

Similar complaints have been raised in other instances in which access to transmission and owner speech merge. These instances concern the exercise of access and content control by bulletin board service providers and the provision of access and speech-related services by cable television media.[81]

Computer Networks. According to a number of legal commenters, individual subscribers to commercial or private computer bulletin board services have no access rights. Access is garnered by contract, and control of access (and ultimately speech) resides, in the first instance, with the service provider or the sysop. While there is very little information on the criteria employed for denying initial access, revocation of access is the ultimate sanction employed by sysops to discipline miscreant member users.[82] There are options short of denial of access that are also employed.

At base, the rationale for sysop control of access is ownership of the system facilities. With regard to sysop content control, the recent *Cubby versus CompuServe* decision provides some indication of the considerations militating against sysop exercise of content control.[83] The greater the discernible control the sysop exercises over access and content, the greater its potential liability to users and third parties for damage caused by the information's content.

In *Cubby*, CompuServe, an on-line information service provider, was sued, unsuccessfully, for the alleged libel of a third-party competitor of a bulletin board provided on the service provider's system. In determining that CompuServe was not liable for the alleged libel, the court established by implication that heightened control of the communicated content would have resulted in liability.[84] In another libel action ultimately settled out of court, Prodigy, another sysop, was sued for an alleged libel of a third party by a Prodigy subscriber.[85] Unlike CompuServe and many other sysops, however, Prodigy has publicly distinguished itself based on the extent of control it exercises over transmitted content.[86] As a consequence, there was speculation that Prodigy might not have easily extricated itself from liability.[87] In Stratton Oakmont, Inc. v. Prodigy Servs. Co., 23 Media L. Rep. 1794 (1995), a subsequent case, Prodigy was held to be liable in part because of its publicly stated policy of maintaining control over content.

Constitutionally Based Access and Speech Rights in Traditional Media: Broadcasting and Cable Television. Historically, market entry and technological considerations have affected the apportionment of access and speech rights between media owner providers and the public. While, as a practical matter, electronic speech has been protected under the constitution regardless of whether it is in print,[88] voice,[89] or video[90] format, traditional media owners in each industry have been accorded different First Amendment rights vis-à-vis users based on differing assessments of the ease of economic and technological entry into each market.

Broadcasting. The initial scarcity of broadcast frequencies relative to public demand for access resulted in the requirement that the broadcast licensee share the frequency with the public.[91] With FCC-engineered deregulation of broadcasting, the fairness doctrine, community ascertainment regulations, and programming guidelines were abolished or seriously compromised.[92] Subsequent to deregulation and the abolition of the fairness doctrine, the scope of access sharing was ultimately limited to candidates for political office.[93] Even before the fairness doctrine was "abolished," its potential power to require access had been significantly limited by judicial decisions.[94]

The current scope of government-exercised content control over the broadcast licensee extends to the prohibition of speech that is libelous, indecent, or obscene.[95] Users have a right to diverse information, but no right to speak as individuals or information providers without owner permission.

Cable Television.   According to several courts and legal scholars, government regulation of access to cable channels is justified because franchises are scarce due to the physical limits inherent in the use of public rights of way.[96] The physical scarcity is further exacerbated by the economies of scale inherent in the provision of cable service.[97] For these reasons, the cable franchisee is required to share the franchisee's channels of communication with the public and other information providers. Concerns about the continued availability of local news and public affairs programming, as well as economic market and anticompetitive constraints alleged to have been imposed by cable firms, have been used to justify limits on the control cable franchisees may exercise over broadcaster access to the cable network.[98] The leased access, must carry, and public access channels are an attempt by Congress to ensure third-party access to cable networks.[99] According to some scholars, the leased access rules have proved only moderately successful. And, due to recent litigation, the must carry requirements are under a potential constitutional cloud.[100]

The cable franchisee's control of communicated content is constrained by legal sanctions that may be imposed for indecent or obscene speech.[101] In part due to the necessity to avoid government-imposed sanctions, cable franchisees may be compelled to exert editorial control over matter provided by third-party information providers that may be deemed indecent or obscene.[102]

Regulatory Shifts in the Age of Convergence and Privatization: Some Preliminary Answers

Network as Asset

Private Firms and Closed User Groups

When the network is the private asset of the firm, employee and third-party efforts to assert First Amendment rights of access or speech over internal communications systems will be limited.[103] In the case of employees of private firms, the National Labor Relations Act may allow them to negotiate for speech rights provided the rights are exercised for the protest or discussion of working conditions.[104] All arguments for fairness and ethics aside, beyond the narrow entitlement of the NLRA, employees of private firms enjoy little real access or speech rights to corporate network assets. Ultimately, the company network owner may limit and/or control access and speech. [105]

Public Firms

Employees of public (government) firms are similarly limited. The First Amendment has been interpreted to afford such employees the right to speak on matters of public interest. [106] They, like their private brethren, also receive some protection from a variety of state whistle-blower statutes. Aside from these protections, however, public employees have no rights of access or speech to internal communications systems. At least one commenter has forcefully argued that private and public employees should enjoy the same scope of speech rights encompassing comment on work- and product-quality-related matters. [107]

Access to the networks of closed user groups is also limited. Here, without a showing that the network asset is being used to restrain competition unlawfully, the user group may exercise control over access and/or speech on virtually all aspects of the network. However, the exercise of control over access and speech carries a certain level of responsibility for actionable speech violations. The precise level of responsibility has yet to be measured, however, and may ultimately depend on the technology and the circumstances of each case. [108]

Network as Product or Service: The Evolution of Speech Regulation in Traditional Media and Telecommunications

While the traditional regulatory apportionment of network provider and user access and speech remains virtually intact in broadcasting, it is under challenge in cable and telephony. Congress's decision to impose must carry requirements on cable franchisees has been upheld for the moment. [109] Congress's prohibition against local telco ownership of cable facilities in its service area and provision of video programming has been challenged and overturned in several courts. [110] The challenges to the must carry provisions and the telephone–cable cross-ownership ban are significant because they provide two of the judicial pillars on which regulation of the future electronic broadband networks will be built. This follows because the cases address the regulation of speech in cable and telephone, the two industries from which much of the broadband infrastructure is likely to emerge. [111]

It is clear that the cable television and regional telephone industries are in the process of merging at a rapid pace, following hard upon technology's lead. [112] Most regulators and industry analysts expect this merger of industries and technologies to result in the provision of interactive, broadband, multimedia services. Thus, judicial pronouncements on the relative rights of network owners to provide information over their networks and to determine who may speak over their facilities, other than themselves, are critical to the evolution of speech rights on the new and evolving infrastructure.

A decision overturning the must carry rules is possible. Majorities in both the district court and Supreme Court concluded the rules are content neutral and the government's interests compelling.[113] However, a new majority of the Supreme Court still could conclude otherwise.[114] Justice O'Connor, like Judge Williams below, concluded that Congress rested a significant portion of the justification for its must carry rules on its desire to ensure the continued provision of local news and public affairs, as well as educational programming by broadcasters.[115] If so, Congress's reasons for adopting the must carry rules may be argued to rest in part on the content of broadcasters' speech and may be deemed impermissible.[116] Also, on remand, evidence of economic harm may, on closer analysis and examination of prior history, prove insufficient to establish a sufficient threat to the government's interest in the retention of iable broadcast stations.[117] Thus, it is possible that the rules may be overturned under the reading of the law as espoused by Supreme Court dissenters in *Turner*.[118]

Meanwhile, numerous courts have held that the congressional prohibition against telco provision of video services is unconstitutional.[119] According to the district court in *C & P et al. versus United States*, the availability of workable regulatory alternatives that do not deny local video speech to the entire class of telephone company speakers renders the prohibition unconstitutionally overbroad. The C & P suit is not the first effort undertaken by the Regional Bell Operating Companies (RBOCs) to defeat the cross-ownership ban on constitutional grounds.[120]

Under such circumstances, the user control over access and speech on cable television and local telephone switched networks will be revised to accommodate increased network owner control. The scope of user access and speech rights most likely would be established by contract and reflect the relative bargaining power of the parties.[121] In such a scenario, in the absence of state action,[122] small users and individuals would have access and speech rights solely at the sufferance of the network provider/owner and the specter of private censorship unmediated by government becomes quite real.

Should the must carry rules be upheld based on economic market and antitrust regulation and the telco prohibition be overturned, at minimum, opportunities for access and speech would continue to incorporate the current statutory delineations of common carriage, leased access, public access, and network owner access.[123] Opportunities for speech would be broadened to include telephone network owner/speakers and cable network speakers, as well as the merged cable–telco network owner speaker, and would continue to include unaffiliated information provider "speech" and user subscriber speech. Under this set of outcomes, the focus of access and speech policy arguably shifts to a government-mediated inquiry into the extent and the manner in which the owner provider may limit or prohibit the exercise of access and speech rights by potential and actual user/subscribers. So long as owner providers and network users retain access and speech rights, the First Amendment is likely to be better served. For reasons stated above, the more probable outcome will be modification, if not outright repeal, of the must carry rules and repeal of the telco–cable cross-ownership prohibition.

Conclusions

The convergence and privatization of telecommunications networks will continue as a market and technological reality and as a preferred regulatory tool. While the outcome of the Turner and C & P cases will affect the scope of network owner control over access and content, questions regarding the scope of access to private networks and the extent of private network control over content will remain. For instance, in the event *Turner* is overturned and *C & P* upheld, how may telephone and cable network owners establish criteria for access and speech on their networks? In the context of private action by network owners, how might the government seek to ensure affirmatively subscriber/user access and relatively unfettered speech, while avoiding inappropriate regulation of network owner speech?

Access presents a particularly interesting set of problems. For instance, to the extent government regulation of network owner control over access is based on technical scarcity, we may be approaching a time when technical scarcity will cease to be a credible concern.[124] Admittedly, however, an abundance of technical channel or switching capacity does not ensure access to all potential users. Marketplace failures due to wealth distribution, limited network infrastructure availability, and selective market competition still will play a significant role.

These questions of speech and access are doubly critical given the current proposed mergers. To the extent that large telephone and cable corporations are allowed to merge, economic scarcity will remain a valid policy concern. Aside from reducing potential competitors while driving up the price for market entry, the types of services made available and the manner in which they are priced by the merged firms will affect who will have access to network functionalities. If the postmerger economics follow the same trends as prior periods of merger in related media industries such as broadcasting, debt service demands will ultimately force the merged firms to cut costs, serve more lucrative markets, and raise prices.[125] In such an event, some market segments may receive less service while other segments pay more. Such developments would certainly affect the cost of access, may preclude significant segments of the market from having meaningful access, and will affect the speech activities of those who acquire access.

Some scholars have argued that the nation's constitutional laws be changed to reflect the growth of speech-related activity engendered by the convergence of computer, network switching, and fiber-optic technologies. For instance, at least one eminent constitutional scholar has argued for an amendment to the First Amendment to protect speech activities conducted over computers.[126] Other scholars have argued that the First Amendment in its current form may be interpreted to protect access and speech activities conducted over computer-augmented broadband interactive switched networks.[127]

Short of constitutional solutions, however, the government retains other regulatory tools for ensuring "universal" access and relatively unfettered speech for network owners and users. Government may address these problems reactively (antitrust, liability for content, and tort liability) or proactively (setting minimum technical parameters for networks that favor distributed intelligence and switched interactive network technologies) and a universal service requirement to ensure access and speech in the face of the above-mentioned market failures.

Control of employee and subscriber speech is problematic for all the reasons mentioned above. There clearly are legitimate and compelling reasons for employer and/or network owner limits on employee or subscriber speech in some instances. However, the potential for private censorship remains great, and its negative impact is no less devastating to the individual than when engaged in by the government.

The impact of these risks can be positively affected by the exercise of regulatory choices that federal and state governments now have before them. These regulatory choices affect the exercise of access and editorial control at the content, network configuration, and equipment levels—the same levels at which network owners exert control.

Macrocontrols

As for macrocontrols, for instance, given the extensive cost of deploying fiber optics to the home, federal and state regulators could allow private industry to continue to build network information delivery systems composed of one-way, compressed-channel technology (cable and open video systems) rather than switched, two-way, interactive technology (Integrated Services Digital Network [ISDN]/broadband). While this approach may be favored by portions of the industry, there is a significant danger that such a solution would postpone the advent of switched, interactive, multimedia communications. More important, however, it replicates the current regulatory difficulties that accrue when the government cedes control over distinct, clearly discernible transmission paths to network owners and then imposes liability for speech.

Microcontrols

For microcontrols, at the network intelligence and applications level, the government has initiated regulatory proceedings aimed at equalizing user interconnection to the local monopoly, public, switched-network architecture and increasing network service offerings by enhancing network flexibility through distributed network intelligence. These proceedings have yet to be concluded. A resolution that favors distributed intelligence and shared user/network control over network functionalities would maximize speaker control over the process by which information is communicated.

Market Regulation

When the network owner exercises control over the network via access or content control to deter or forestall competition, the antitrust laws regarding market regulation also should be applicable.[128]

Control and Liability for Speech

As privatization continues, the lessons learned in the CompuServe case and alluded to in the Medphone case concerning Prodigy, as well as other recent cases regarding publisher liability, should give would-be private network editors pause. When a network owner exercises control over access and content, they may not be able to avoid liability for that content when it is harmful to the public. [129] Similarly, network owners may be held liable for negligent or careless manipulation and control of subscriber/user information when such action results in injury to the user or to third parties. [130] Certainly, the libel, obscenity, and indecency laws will remain making control of content a cause for liability. Thus, even when network owners seek to eschew all content regulation, they are likely to be no more successful than telephone common carriers and cable operators, which by statute must exert some control over obscene or indecent subscriber speech. Ultimately, self-preservation and protection of the bottom line may motivate firm efforts to curb libelous speech. But, forgoing editorial control of content would remove a downside cost of doing business that may be preferable to the cost of maintaining the monitoring of subscriber and programmer speech and the potential liability the exercise of editorial control brings.

Tort Liability

There is another way in which network owner control of speech may be tempered by government sanction: the imposition of tort liability. The exercise of control over access and content necessarily invites expectations that the network owner, in the exercise of its editorial discretion, has reviewed and sanctioned all information it transmits. Moreover, should the network owner lose or damage customer information in storage, manipulation, or transmission or negligently preclude the transmission of customer information entrusted to its care, it is reasonable to require that the owner compensate the customer to the extent of its legally recognized tort damages. A recent case in Illinois addressed this issue. [131]

At least one commenter has noted that, in an era of deregulation, the reasons for continuing to limit the tort liability of nondominant telecommunications common carriers cease to be applicable. [132] At least three reasons have been used by the courts to justify the continuation of exculpation clauses limiting common-carrier liability. First, federal and state regulators may be held to possess the regulatory authority to establish such limits. Second, without such limits, judgments paid by monopoly carriers would be passed on to subscribers having no alternative service providers. Third, limited liability provisions preserve national uniformity in the provision of services and avoid discrimination between like situated but geographically dispersed subscribers.

Today, however, such reasons retain little credibility. First, the Communications Act of 1934 does not authorize federal regulators to preempt state law tort remedies then existing at common law or by statute. Rather, such remedies as the act provides are in addition to existing state remedies. [133] In addition, courts have not automatically granted primary jurisdiction over state tort liability claims to regulatory agencies, but often have found such claims to be within the purview of the courts.[134] Second, in an era of convergence and expanding competition at all market levels in the telecommunications and, ultimately, multimedia marketplace, many subscribers increasingly have and will continue to have alternative sources of service. Finally, in an era of privatization in which a substantial portion of the existing telecommunications infrastructure is owned by a growing disparate number of private owners serving distinct "high-end user" submarkets rather than the larger local or national markets of various subscribers, national uniformity appears to be less a function of government action and more a function of the relative market power of the service provider, the purchaser, and market demand. For these reasons, government-sanctioned, carrier-initiated limitations on tort liability should be abolished except when a carrier elects to serve all classes of customers via public switched multimedia networks.

A decision to remove the tort liability limitation except when applied to carriers serving the majority of all classes of users via a public switched multimedia network or providing significant interconnection between public switched networks would serve as a financial incentive for some carriers to maintain service to a broad subscriber base, to expand their service offerings to include other consumer groups, or, at the very least, ensure sufficient compatible interconnection.

The Concept of Unconscionability

Whether the nondominant network provider or providers resort to contracts or tariffs as the vehicles for the offering of services to subscribers, there may be instances in which the doctrine of contract unconscionability may be invoked. If the network owner, as provider of scarce network resources, leveraged its economic position by employing form contract language to limit its tort liability, its attempt to enforce such restrictions might be denied by the courts on the grounds of unconscionability.[135] Moreover, as one public service commission has observed, given the increasing complexity of tariffs it would be " 'unconscionable' to assume that any telephone subscriber had consented either impliedly or expressly to broad liability waivers."[136]

Each of the above-mentioned policies affects the incentive structure under which information carriers would exercise control of access and speech on their networks. None of the proposed policies is inconsistent with established constitutional law. They do not preclude network owner exercise of control over access and speech. They merely remove liability protections enjoyed by public common carriers, expand technical opportunities for user access and speech, and continue preexisting economic regulation. As such, they should be adopted as regulatory policy regardless of whether constitutional law is changed.

Acknowledgments: Special thanks are due to the Board of Trustees and Administration of the New York Law School, which provided a summer research grant to support the work on this article. Thanks also are due to my colleague Michael Botein, who provided valuable editing insights, as well as my research assistants Anjali Singhal and John McGill and especially Camille Brousard, my library liaison, all of whom provided excellent support.

Notes

1. The convergence phenomenon can be seen in the merger of fiber-optic, telephone, and computer technologies into broadband telecommunications networks technology and the consequent disintegration of the distinctions between video distribution and public switched networks. See, generally, Joshua Quittner, Online to a Revolution: The Amazing—and Some Say Ominous—New World of TV, Telephone and Computer Is Heading Your Way, *Newsday*, News section, p. 4 (July 18, 1993); and Electronic Media Regulation and the First Amendment: Future Perspective, *Data Channels, Incl. Computer Digest* (February 3, 1992).

2. By most accounts, the privatization of the public network increases daily, fueled by the growing demand for specialized communications services. It is estimated that as much as a third of the nation's total yearly telecommunications investment is channeled into private networks, virtual private networks, and related hybrid services. Privatization is also the chosen vehicle for building the broadband/electronic/superhighway, which many believe will be the logical evolution of the public switched network and its more private analogs. See Information Infrastructure Task Force, *The National Information Infrastructure: Agenda for Action*, September 15, 1993, pp. 1–2, 4–16. See also John Holliman, Vice President Gore Press Conference on Info Highways, Transcript #267-1, live report, *CNN News Show*, 1:00 P.M. and 6:30 P.M. eastern time, December 21, 1993.

3. See Alan Deutschman and Joyce E. Davis, The Next Big Info Tech Battle, *Fortune*, 39 (November 29, 1993); John Huey, Andrew Kupfer, Jane Furth, and John Wyatt, What That Merger Means for You, *Fortune*, Domestic edition, 82 (November 15, 1993); Jolie Solomon, Daniel Pedersen, et al., Big Brother's Holding Company, *Newsweek*, 38 (October 25, 1993); John Greenwald, John F. Dickerson, Thomas McCarroll, et al., Wired! Bell Atlantic's Bid for Cable Giant TCI Is the Biggest Media Deal in History, *Time*, 50 (October 25, 1993); and Sandra Sugawara and Paul Farhi, Merger to Create a Media Giant; $26 Billion Bell Atlantic-TCI Deal Is a Vision of TV's Future, *Washington Post*, p. A1 (October 14, 1993).

4. Some scholars have begun to address this question. See Special Report: Universal Telephone Service; Ready for the 21st Century? 1991 Annual Review of the Institute for Information Studies. A Joint Program of Northern Telecom and the Aspen Institute, *Edge*, 175(6) (December 2, 1991).

5. Users may be divided into two major groups composed of facilities-based and non-facilities-based users. The vast majority of users are non-facilities based. These individuals, firms, or groups have no ownership of the networks and services they use. They are most often semipassive recipients of information transmitted one way over other networks (broadcasting and cable). The communications needs of these users vary substantially and are evolving at different speeds and in multiple directions. For instance, many businesses already have significant needs for high-speed, high-capacity broadband communications networks. See Dertouzos, Communications, Computers and Networks, *Scientific American*, 265:62, 64 (September 1991); Gore, Infrastructure for the Global Village, Computers, Networks and Public Policy, *Scientific American*, 265:150, 152 (September, 1991); Dertouzos, Building the Information Marketplace, *Technology Review* 94:28, 31–32 (January 1991). By comparison, the general public has not yet generated needs sufficient to precipitate demands for greater network speeds and capacities. See Gary Yaquinto, The Information Superhighway; Construction Ahead: What Regulators Should Ask About the Information Superhighway, *Public Utilities Fortnightly*, 31 (June 15, 1994).

6. Several scholars have criticized the current state/private dichotomy established by the Supreme Court in light of the continuing trend toward privatization in American life. See Rodney A. Smolla, The Bill of Rights at 200 Years: Bicentennial Perspective: Preserving the Bill of Rights in the Modern Administrative-Industrial State, *Wm. and Mary L. Rev.*, 31:321 (1990); and Clyde Summers, The Privatization of Personal Freedoms and the Enrichment of Democracy: Some Lessons from Labor Law, *Univ. Ill. L. Rev.*, 1986:689 (1989).

   Generally, without a showing of an independent nexus of involvement by the state, however, neither the chartering, funding, licensing, regulating, or tax exemption of a corporation by the government constitutes state action. See *Cohen versus Illinois Institute of Technology*, 524 F. 2d 818 (7th Cir. 1975) cert. denied 425 U.S. 943 (1976) and *Sament versus Hahnemann Medical College and Hospital of Philadelphia*, 413 F. Supp. 434 (E.D. Pa. 1976), affirmed mem. 547 F. 2d 1164 (3rd Cir. 1977) (charter); *Aasum versus Good Samaritan Hospital*, 542 F. 2d 792 (9th Cir. 1976); *Trageser versus Libbie Rehabilitation Center*, 590 F. 2d 87 (4th Cir. 1978), cert. denied 442 S. 947 (1979) (funding); *Moose Lodge No. 107 versus Innis*, 407 U.S. 163 (1972) (licensing); *Jackson versus Metropolitan Edison*, 419 U.S. 345 (1974) (regulation). When the private entity exercises powers traditionally reserved to the state, state action may be found; see *Nixon versus Condon*, 286 U.S. 73 (1932) (election); *Marsh versus Alabama*, 326 U.S. 501 (1946) (company town); and *Evans versus Newton*, 382 U.S. 296 (1966) (municipal park).

7. See, generally, Cynthia Estlund, What Workers Want? Employee Interests, Public Interests, and Freedom of Expression, *U. Pa. L. Rev.*, 140:921 at 925, 935–936, 960–964.

8. Information providers that find access to the network or communication over the network constrained by the network owner may be able to establish that an antitrust violation has occurred. If an information provider can successfully establish that the network provider either possesses and seeks to maintain monopoly power, *United States versus Grinnell Corp.*, 384 U.S. 563, 570–571 (1966), or owns essential facilities, or is attempting to monopolize a market segment, the network provider's activity may be prohibited.

   Most industry observers, scholars, and commenters have suggested that some portion of the future broadband network infrastructure may be composed of essential facilities. An antitrust violation will arise when such facilities are (1) extremely difficult if not impracticable for competitors to duplicate, (2) owned by one or a group of firms, and (3) not made available to competitors of the network facilities' owner without an appropriate business justification or apparent efficiency, especially when the network owner is also an information provider. See *Vill. L. Rev.*, 38:571, 575, 584–585 (1993), which cites *City of Anaheim versus Southern California Edison Co.*, 955 F.2d 1373, 1380 (9th Cir. 1992) and *MCI Communications Corp. versus AT&T*, 708 F.2D 1081, 1123–1133 (7th Cir. 1983).

   Aside from the significant cost of litigation, the difficulty in establishing the relevant market at a time of fluctuating market boundaries is substantial.

9. In telecommunications, the term *structural safeguards* refers to the separation of a vertically integrated firm into corporate segments based on whether they provided basic network services or enhanced services. The FCC defined enhanced services as services "which employ computer processing applications that act on the format, content, code, protocol or similar aspect of the subscriber's transmitted information; provide the subscriber additional, different, or restructured information; or involve subscriber interaction with stored information" (47 C.F.R. @ 64.702(a), 1990). Enhanced services include data-processing services, as well as

videotext, audiotext, database retrieval, and other computer and communications technologies applications.

See, generally, *Second Computer Inquiry*, 77 F.C.C.2d 384, recon., 84 F.C.C.2d 50 (1980), clarified on further recon., 88 F.C.C.2d 512 (1981), aff'd sub nom.; *Computer and Communications Indus. Assoc. versus FCC*, 693 F.2d 198 (D.C. Cir. 1982), cert. denied, 461 U.S. 938 (1983); *BOC Separation Order*, 95 F.C.C.2d 1117 (1983), aff'd sub nom.; *Illinois Bell Tel. Co. versus FCC*, 740 F.2d 465 (7th Cir. 1984), recon. denied, 49 Fed. Reg. 26,056 (June 26, 1984), aff'd sub nom.; *North American Telecommunications Assoc. versus FCC*, 772 F.2d 1282 (7th Cir. 1985). In addition, see Robert J. Butler, In the Aftermath of *California v. FCC*: Computer III Remand Proceedings Pose Difficult Policy Choices for the Enhanced Services Industry, *Computer Lawyer*, 24 (May 1991) and Barbara D. Khait and Andrew S. Elston, RHCs and Information Services: Gateways to Opportunity? *Online*, 27 (September 1989).

The FCC determined that it would be sufficient for the RBOCs to provide enhanced services as integrated entities and offer their "unbundled" basic network functions to other enhanced service providers on a tariffed, nondiscriminatory basis. See *Computer III*, 104 F.C.C.2d at 964-65, 1063-66. See *Filing and Review of Open Network Architecture Plans*, 4 FCC Rcd. 1 (1988), recon., 5 FCC Rcd. 3084, amended plans conditionally approved, 5 FCC Rcd. 3103 (1990). The FCC concluded that requirement to unbundle the network functions, combined with accounting and other nonstructural safeguards, would obviate the need to rely on the separate subsidiary requirement to prevent the RBOCs from engaging in access discrimination and anticompetitive cross-subsidization that would favor their enhanced service operations. *Computer III*, 104 F.C.C.2d at 1007-12, 2 FCC Rcd. at 3039. Also see Public Service Commission Paper Attacks Computer III Ruling, *Worldwide Videotex Tele-Service News*, 5(7) (July 1993); RHCS, ESPS, States Form Familiar Lines in Computer III Remand, *BOC Week*, Vol. 8, Section 11 (March 18, 1991); *Computer III Remand Proceedings*, 5 FCC Rcd. 7719 (1990) and 6 FCC Rcd. 174 (1990).

At least one court was highly skeptical of the Open Network Architecture (ONA) plan's efficacy whether in its Computer II form or its subsequent Computer III form. See *U.S. versus Western Electric et al.*, 673 F. Supp. 525 (1987). Ironically, the same mechanisms of the ONA plans that Judge Harold Greene found so ineffective in 1987 are the very mechanisms the FCC proposes to implement under its Computer III regulations modified after *California versus FCC*.

10. See Note 124.
11. For the purposes of the article, *networks* are defined as collections of interconnected users (National Telecommunications and Information Administration, *NTIA Infrastructure Report: Telecommunications in the Age of Information*, October 1991, pp. 13-20, 92). The type of transmission and the receive/send machinery employed varies. These points may or may not be capable of engaging in interactive communication. This definition acknowledges that cable and broadcast television systems may be deemed to be networks just as the public switched interexchange and local-exchange systems constitute networks. This definition also facilitates the exploration of the broader array of access solutions presently employed and likely to be employed in the regulation of future networks.
12. Network functionality is determined by hardware and software architecture and specified protocols. See, generally, Peter Fetterolf, Connectivity: The Sum of Its Parts, *Byte*, 197 (November 1991).
13. A *tariff* is a published set of rates charged and conditions under which various classes of service are offered by common and private carriers.

14. The notion of network and service availability is subsumed within the definition of universal service. *Universal service* is a government and industry policy that encouraged AT&T (then a monopoly) to make telephones and service available to the majority of the American public at reasonable rates. Subsidies of less profitable (or unprofitable) provision of service to rural and poorer areas were built into the business and long-distance charges. David Coursey, Battle of the Bandwidth, *Infoworld*, Perspectives, 34 (January 14, 1991). The traditional goal of universal service was to ensure that "all but the poorest Americans could afford to make and receive telephone calls, even if they lived in remote, expensive to serve areas." From Special Report: Universal Telephone Service; Ready for the 21st Century? *Edge*, 6(175), forward (December 2, 1991). As such, universal service operated as a kind of equality in access and likeness in service offerings. In the current era of increased competition and privatization, however, universal service may no longer mean likeness (or comparability) of service or equality in technical access. Idem.

15. The term *open* as used here in the general sense to describe technical and price-related nondiscriminatory access to communicate on the network should not be confused with the FCC's Open Network Architecture (ONA) policy. The FCC's policy is an attempt to provide enhanced service providers (a specific class of users) with fair, nondiscriminatory access to local telecommunications networks. See Dawn Bushaus, Enhanced Services—ONA and AIN on a Collision Course, *Communications Week*, 32L, 6 (June 17, 1991).

16. See, generally, VPNs Set to Challenge Private Networks and PSTN During Nineties, *Fintech Telecom Markets*, 31 (April 15, 1992); Mark Luczak, Tapping the Hidden Savings in Virtual Networks; Hybrid Networks, *Telecommun.*, 25(3):45 (March 1991); and Robert Violino, A Network of Their Own, *Information Week* (January 14, 1991).

17. See Note 12.

18. VPNs Set to Challenge Private Networks and PSTN During Nineties, *Fintech Telecom Markets*, 31 (April 15, 1992).

19. The networks typically are created to meet the needs of their respective users for transmission of high-speed data, information processing, voice traffic, and/or security. Consequently, they serve closed sets of users with relatively cohesive sets of needs, as well as eligibility, procurement, and financing criteria. See, generally, James I. Cash, Jr., and Benn R. Konsynski, IS Redraws Competitive Boundaries, *Harv. Bus. Rev.*, 134 (March 1985/April 1985); Venkatraman, IT-Enabled Business Transformation: From Automation to Business Scope Redefinition, *Sloan Mgmt. Rev.*, 73n (January 1994).

20. This definition does not include closed networks established without the use of private or dedicated facilities.

21. These user groups may be local, regional, national, or international in scope. Examples of such user groups include ad agencies, media firms, printers, insurance agencies, hospitals, record rooms, police, automobile manufacturers, parts suppliers, dealers, financiers, and computer networks. See, generally, James I. Cash, Jr., and Benn R. Konsynski, IS Redraws Competitive Boundaries, *Harv. Bus. Rev.*, 134 (March 1985/April 1985); and Venkatraman, IT-Enabled Business Transformation: From Automation to Business Scope Redefinition, *Sloan Mgmt. Rev.*, 73N (January 1994). Also see John Helliwell, Networks Provide a Critical Competitive Edge for Airlines: Reservation-System Leaders Reap Diverse Benefits, *PC Week*, C1 (January 19, 1988); and Salvatore Salamone, Airline Reservation Network Flies into New Age of LANs; Software Front End, QIK RES, Speeds Terminal Termination at American Airlines, *Network World*, 34 (Novem-

ber 26, 1990) (regarding airline closed user networks and usage) and More Shared Networks Approved Under @4(c)(8), *Banking Expansion Reporter*, 11 (August 1, 1983) (regarding banking). Also see Rita Marie Emmer, Chuck Tauck, Scott Wilkinson, and Richard G. Moore, Marketing Hotels Using Global Distribution Systems; Use of Electronic Listings and Computers in the Hospitality Industry, *Cornell Hotel and Restaurant Administration Quarterly*, 34(80):80 (December 1993) (regarding hotels).

22. See Alan Deutschman and Joyce E. Davis, The Next Big Info Tech Battle, *Fortune*, 39 (November 29, 1993); Multimedia: The Tangled Webs They Weave, *Economist*, Special Section, p. 21 (October 16, 1993); and Sandra Sugawara and Paul Farhi, Merger to Create a Media Giant; $26 Billion Bell Atlantic–TCI Deal Is a Vision of TV's Future, *Washington Post*, p. A1 (October 14, 1993). For an explanation of some of the strategic market and technical reasons for the merger of telephone and cable networks, see S. Ronald Foster, CATV Systems Are Evolving to Support a Wide Range of Services, *Telecommun.*, 95 (January, 1994); Dave Schriftgiesser, Key Trends in Broadband Communications: The Next Five Years, *Telecommun.*, 101 (January 1994); Rick Pinkham, Combining Apples and Oranges; Part 1; Telecommunications and CATV Companies Merge to Form Full Service Hybrid Networks, *Telephony*, 32 (January 24, 1994); and Alan Stewart, Classless Cables; Common Networks for Televisions and Telephones, *Commun. Intl.*, 8 (October 1993).

23. Section 302 of the Telecommunications Act of 1996 established new sections 651 through 653 amending Title VI of the Communications Act of 1934. The sections provide alternatives for telephone company entry into video programming markets. One option for entry is to provide cable service over an open video system. The FCC, which is charged with establishing rules to implement the new statutory provisions, adopted a Report and Order and Notice of Proposed Rulemaking (NPRM) in which it sought comment on how to implement the provisions of the Telecommunications Act of 1996 governing open video systems. The Commission also eliminated its video dialtone rules and policies and the applicability of Section 214 to a telephone company's video programming delivery systems. See *In the Matter of Implementation of Section 302 of the Telecommunications Act of 1996 Open Video Systems*; *In the Matter of Telephone Company–Cable Television Cross-Ownership Rules, Sections 63.54–63.58* (FCC Release No. 96–99), 1996 FCC LEXIS 1206 (1996). Also see, *Telco Business Report*, 9:13 (April 22, 1996); LEC-Cable TV Battle over OVS Shows no Signs of Easing; *Interactive Video News*, 8:4 (April 15, 1996); Barrie Tabin, Coalition Spells out Video System Concerns, Open Video Rules in Watched Closely Telcos Hail OVS as Successor to Burdensome Video Dialtone; *Nation's Cities Weekly*, 19:15; 1 (April 15, 1996).

24. Carl Warren, Abuse of Company Facilities for E-Mail Must Be Curbed, *Network World*, Top News section, E-Mail, p. 25 (March 30, 1992).

25. For instance, many New York city agencies have configured their phones to prevent city workers from dialing long distance and calling specialty phone services such as dial-a-porn and sports information lines. See Jennifer Preston, It's OK, as Long as It's a Local Call, *Newsday*, City edition, News section, p. 5 (October 26, 1989).

26. Ronald E. Roel, Advances in the Campaign for Workers' Rights; Laws, Court Rulings Offer Protection in Areas Where Bill of Rights Doesn't, *Newsday*, Nassau and Suffolk edition, Business section, p. 84 (January 10, 1988); also see Plan to Monitor Calls Made by Civil Servants Attacked, *Los Angeles Times*, Home edition, Part 1, p. 11, column 1, National Desk (March 10, 1985).

27. Roel, Note 25.

28. Idem. An examination of employee privacy rights is beyond the scope of this article.

29. See Plan to Monitor Calls Made by Civil Servants Attacked, *Los Angeles Times*, Home edition, Part 1, p. 11, column 1 (March 10, 1985). The laws protecting corporate employees are inconsistent at the state level. See Tom Devine, A Whistleblower's Checklist, *Chemical Engineering*, 207 (November 1991).

30. Carol Wolinsky and James Sylvester, Privacy in the Telecommunications Age, *Assoc. for Computing Machinery*, 23 (February 1992). Such arguments have met with only limited success to date. Even though E-mail and intelligent network technologies provide new opportunities for speech activities, the articulation of these activities as rights squarely pits them against the heretofore established and legally recognized property rights of the employer/network owner.

31. Susan E. Kinsman, Toll Fraud on Rise, SNET Says; SNET Warns Businesses to Guard Against Toll Fraud, *Hartford Courant*, p. B1 (July 29, 1992). Also see Annabel Dodd, When Going the Extra Mile Is Not Enough, *Network World*, p. 49 (April 12, 1993).

32. Unauthorized entry can be accomplished by calling a company's toll free 800 number or a voice mailbox and using a computer with an automatic dialer to break the security code and gain access to the company's telephone system and outgoing lines. See Kinsman, Note 31. Also, electronic bulletin boards are sometimes used to exchange generic passwords that provide access to company maintenance ports, exchange programming instructions for various systems, or procure programming manuals for voice systems enabling unauthorized parties to gain operational control, including the ability to unblock restrictions on international dialing and turn off on-site call accounting equipment. Annabel Dodd, When Going the Extra Mile Is Not Enough, *Network World*, 49 (April 12, 1993).

33. See Kinsman, Note 31.

34. 29 U.S.C. sections 7, as amended, 29 U.S.C.A. section 157.

35. For instance, without a showing by union organizers that (1) they possess no other reasonable alternative means of communication to reach nonunion employees or (2) that the employer is discriminating against the union by denying access to facilities the employer otherwise makes available, the courts are unlikely to afford the organizers access to an employer's private E-mail or telecommunications facilities. See *Lechmere, Inc., versus NLRB*, 112 S. Ct. 841, 848 (1992), quoting and affirming *NLRB versus Babcock and Wilcox Co.*, 351 U.S. 105, 112 (1956). Given the circumstances cited by the *Lechmere* court as justifying a conclusion that no other reasonable access to communication existed (logging and mining camps or remote resort hotels; *Lechmere* at 849), one commenter has concluded that organizers must establish "employee isolation" in order to prove the absence of reasonable alternative means of communication; Michael L. Stevens, The Conflict Between Union Access and Private Property Rights: *Lechmere, Inc., v. NLRB* and the Question of Accommodation, 41 Emory L.J. 1317, *1335 (1992) (arguing that the Supreme Court's "reasonable alternative means of communication" standard first announced in *Babcock* and later affirmed in *Lechmere* is the appropriate standard.) cf. Peter J. Ford, The NLRB, Jean Country, and Access to Private Property: A Reasonable Alternative to Alternative Means of Communication Under Fairmont Hotel, *Geo. Mason U.L. Rev.* 13:683 (1991) (arguing that the *Jean Country* decision strikes a more appropriate balance between employer property rights and employee access to information under Section 7 of the NLRA).

36. See Cash and Konsynski, Note 21. There are three levels at which companies can participate in an IONS: information entry and receipt (content), software development and maintenance (network intelligence), and network (network as configured) and processing management. Idem.

37. Idem.

38. For instance, it is alleged that owners of major airline reservation systems have acted in anticompetitive ways by using their systems to minimize the bookings of competing nonowners. See John Helliwell, Networks Provide a Critical Competitive Edge for Airlines: Reservation-System Leaders Reap Diverse Benefits Networks Provide a Critical Competitive Edge for Airlines, *PC Week*, C1 (January 19, 1988). While the actual use and impact of the networks are still debated, it is clear that, in the airlines, travel, hotel, and vacation/leisure businesses, the use of interorganizational networks has often led to a significant competitive edge in the market. See Salvatore Salamone, Airline Reservation Network Flies into New Age of LANs, *Network World*, 34 (November 26, 1990) and Helliwell.

39. There are also numerous content/subject matter restrictions as well. See Wolinsky and Sylvester, Note 30.

40. Prodigy, an information services company, has been involved in a number of controversies regarding the content of messages transmitted over its facilities in part because until recently it has sought to maintain editorial control. See *Stratton Oakmont, Inc. v. Prodigy Servs. Co.*, *Media L. Rep.* 23:1794 (1995). Also see Alison Frankel, On-Line, On The Hook, *The American Lawyer*, (October, 1995), 58; *Technology Review*, 98:7, 22; Edwin Diamond and Stephen Bates, Law and Order Comes to Cyberspace (October, 1995); Michael Dunne and Elizabeth Barba, The Evolving Rules of Cyber-Libel, *New Jersey Law Journal*, Supplement, Intellectual Property Law (July 24, 1995), p. 3; and Richard Raysman and Peter Brown, On-Line Services and Defamation, *New York Law Journal* (July 11, 1995), p. 3. W. John Moore, Taming Cyberspace, *National J.*, 745 (March 28, 1992) and Felicity Barringer, The Nation: Electronic Bulletin Boards Need Editing. No They Don't, *New York Times*, p. 4, column 1 (March 11, 1990). Some other bulletin board providers have no policy regarding what may or may not be said over their facilities. For them, the communicator of the information bears the ultimate responsibility for the content. Their position has met with judicial approval in one instance. See *Cubby versus CompuServe Inc.*, 776 F. Supp. 135 (S.D.N.Y. 1991).

41. The First Amendment protects the exercise of speech and editorial control over programming decisions and transmissions by broadcast licensees. *Syracuse Peace Council*, 2 FCC Rcd. 5043, reconsideration denied 3 FCC Rcd. 2035, affirmed, *Syracuse Peace Council versus FCC*, 867 F.2d 654.

42. Cable television operators enjoy significant protection as well. The First Amendment protects the exercise of speech and editorial discretion by cable television operators. *Leathers versus Medlock*, 499 U.S. 439,444 (1991); *Los Angeles versus Preferred Communications, Inc.*, 476 U.S. 484, 494 (1986).

43. The First Amendment has been held to protect voice communications over the telephone (*Sable Communications of California, Inc., versus FCC*, 492 U.S. 115 (1989)). Recently, the First Amendment has been held to accord local-exchange network operators the right to engage video communication to their service area subscribers. *Chesapeake and Potomac Tel. Co. of Va. versus United States*, 830 F. Supp. 909 (E.D. Va. 1993); affirmed, 42 F.3d 181 (1994). One other circuit court and four other district courts have also found the cable–telephone cross-ownership ban unconstitutional on First Amendment grounds. See *U.S. West, Inc. versus United States*, 855 F. Supp. 1184, June 15, 1994, affirmed, *U.S. West, Inc. versus United States*, 1994 U.S. App. Lexis 36775; 95 Cal. Daily Op. Service 15, December 30, 1994; *GTE California, Inc. versus Federal Communications Commission*; United States of America, 39 F.3d 940, October 31, 1994; *Ameritech Corporation et al. versus United States*, 867 F. Supp. 721, October 28, 1994; *Bellsouth Corporation versus United States*, 868 F. Supp. 1335, September 23,

1994; USTA, OPATSCO, NTCA Win Lawsuit to Lift Cable-Phone Ownership Ban, *BNA Mgmt. Briefing* (January 30, 1995). Also see 47 U.S.C. @ 533 (b), *The Cable Communications Policy Act of 1984*, prohibiting local telephone companies from providing video programming to potential viewers in their service area directly or indirectly through an entity owned by the telephone company or under its common control.

In another proceeding, Judge Harold Greene ruled that Bell Atlantic and Pacific Telesis may offer video services throughout the United States because such action would not violate the Modification of Final Judgment prohibition against the provision of long-distance telephone service. *United States versus Western Electric Co., Inc.*, D.C. No. 82-0192 (HHG), March 16, 1995. For a report of the decision, see Two Regional Bell Operating Companies May Offer and Deliver Video Programming, *Antitrust and Trade Regulation Report (BNA)*, Vol. 68, No. 1705, p. 387, March 23, 1995.

44. Sable Communications, Note 43, concurring opinion of Justice Scalia; *Dial Info. Servs. Corp. of New York versus Thornburg*, 938 F. 2d 1291 (2nd Cir. 1991); *Carlin Communication, Inc. versus Mountain States Telephone and Tel. Co.*, 827 F. 2d 1291 (9th Cir. 1987); and *Information Providers Coalition for the Defense of the First Amendment versus FCC et al.*, 928 F. 2d 866 (9th Cir. 1991) (information providers).

    In each of the cases, the information providers were allowed to provide messages, but the telephone companies refused to provide billing services for the transmitted messages. The difficulties associated with collections without the assistance of the phone companies rendered the information providers' businesses marginal at best.

    In *Information Providers*, the court inter alia considered the petitioners' assertion that FCC regulations requiring the individual wishing to receive dial-a-porn messages notify the carrier in writing constituted a prior restraint. The court concluded that no prior restraint was involved because there was no government action to enjoin speech, require advanced governmental approval for speech, censor or license speech. Instead, the court found that only the telephone companies are involved. And, as they are private actors, they are constitutionally free to ban dial-a-porn from their networks and/or refuse to make available billing services to dial-a-porn information providers (922 F. 2d at 877). Similar conclusions were reached in the other cases. See Carlin, 827 F. 2d at 1293, 1295, 1297 n10 and Dial Info., 938 F. 2d 1543.

45. See *Alliance for Community Media versus FCC*, 10 F.3rd 812 (D.C. Cir. 1993) vacated upon the granting of request for rehearing 15 F.3rd 186 (D.C. Cir. 1994), rev'd, 56 F.3rd 105 (D.C. Cir. 1995), cert. granted, 64 U.S.L.W. 3070 (US Nov. 13, 1995).

46. Idem. at 821 and accompanying notes. Without the voluntary billing services distinction, it can be argued that there is no reasonable distinction between the telephone and cable contexts as addressed by the courts. Access to cable is mandated by statute, while access to telephone carriers is ostensibly by election of the carrier. However, given the long-standing state and federal regulatory policies regarding universal access, nondiscriminatory service for like customers ordering like service, and carrier content neutrality, an explanation of the Ninth and District of Columbia Circuit decisions based on the above-referenced distinction between cable and telephony is not likely to maintain credibility. Regardless of Congress's disinclination to label cable a common carrier, even with reference to access channels, given the first-come, first-served nondiscriminatory operation of access channels, it is difficult to make a relevant distinction between cable access

channels and basic telephone service under tariff without the "voluntary billing" distinction.

47. See *In the Matter of Tariff Filing Requirements for Nondominant Common Carriers*, 8 FCC Rcd 6752, August 18, 1993; *In the Matter of Policy and Rules Concerning Rates for Dominant Carriers*, Part 1 of 3, 4 FCC Rcd 2873, April 17, 1989; and *MCI Telecommunications Corporation, Petitioner versus Federal Communications Commission et al.*, 765 F.2d 1186, July 9, 1985.

48. Under a narrow reading of the applicable precedent, they arguably would be free to ban dial-a-porn from their networks by refusing to offer billing services to dial-a-porn information providers. See Note 44.

49. While the Electronic Communications Privacy Act protects users of E-mail and bulletin boards against the intentional monitoring of their messages by third parties, employers seeking to protect company information and assets can monitor employee messages on internal E-mail systems. Julie Bennett, Firms' Rights Protected by Electronic Mail Laws, *Crain's New York Business*, Takeout, Telecommunications, p. 28 (October 8, 1990). The Electronic Communications Privacy Act of 1986 also allows employers to read employee E-mail messages situated on company computer systems that permit third-party access, provided the employee gives permission. Rosalind Resnick, The Outer Limits, *Natl. Law J.*, 1 (September 16, 1991).

50. Some commenters take the position that any monitoring of E-mail or searching through personal employee files is ethically wrong regardless of the law. See Glenn Rifkin, Do Employees Have a Right to Electronic Privacy? *New York Times*, late edition–final, section 3, p. 8, column 1, Financial Desk (December 8, 1991). Aside from questions of ethics, commenters have argued that the use of monitoring is demoralizing to employees and therefore counter productive. Glenn Rifkin, The Ethics Gap; Despite Growing Attention, Many IS Managers Say, "It's Not My Job," *Computerworld*, Executive Report, IS Ethics, 83 (October 14, 1991).

51. See Linda Wilson, Addressing E-Mail Rights, *Information Week*, Supplement, Enterprise Computing, 54 (February 15, 1993); Electronic Mail Raises Issues About Privacy, Experts Say, *BNA Daily Labor Report*, Current Developments section (November 17, 1992); More E-Mail Legal Actions, *Computer Fraud & Security Bulletin* (February 1992); Rifkin, Note 50, Do Employees Have; Alice Kahn, Careful — The Boss Might Be Reading Your Electronic Mail, *San Francisco Chronicle*, Metro edition, Variety, 3E (November 20, 1991); Bennett, Note 49; and Resnick, Note 49.

52. Victoria Slind-Flor, What Is E-Mail Exactly? *Natl. Law J.*, 3 (November 25, 1991). Kevin McDermott, Labor Targets 'Eavesdropping' Law; AFL-CIO Is Confident it Can Win Suit Despite Early Ruling by Judge, *St. Louis Post-Dispatch*, January 1, 1996, Monday, p. 1; Donald H. Seifman and Craig W. Trepanier, E-mail and Voicemail Systems. Evolution of the Paperless Office: Legal Issues Arising out of Technology in the Workplace, part 1, *Employee Relations Law Journal*, 21:3, 5 (December 22, 1995); Communications; Tampering With E-mail: Proprietary Rights and Privacy Issues, *Law Practice Management*, 21:8, 36 (November 1995/December 1995); Laura B. Pincus and Clayton Trotter, The Disparity Between Public and Private Sector Employee Privacy Protections: A Call for Legitimate Privacy Rights for Private Sector Workers, *American Business Law Journal*, 33:1, 51 (September 22, 1995).

53. See Brian D. Pedrow and Debra E. Kohn, Tampering With E-mail: Proprietary Rights and Privacy Issues, *Law Practice Management*, 21:8, 36 (November 1995/December 1995); and Laura B. Pincus, Clayton Trotter, The Disparity Between

Public and Private Sector Employee Privacy Protections: A Call for Legitimate Privacy Rights for Private Sector Workers, *American Business Law Journal*, 33: 1, 51 (September 22, 195); Electronic Media Regulation, Note 1. See, generally, Wilson, Note 51, and Rifkin, Note 50, The Ethics Gap.

Also, to the extent that state constitutions afford an employee a right of privacy or speech, they may not be precluded by the ECPA. For instance, a recent attempt to argue federal preemption failed in California. See Slind-Flor, Note 52, discussing *Alana Shores versus Epson America, Inc.*, SWC112749 and *Flanagan versus Epson America*, BC007036.

54. *Rankin versus McPherson*, 483 U.S. 378 (1987) (public employees may not be fired for making statements about matters of public concern). See, generally, Estlund, Note 7, pp. 921, 923–924. However, the question of whether employees can make such statements over the company's E-mail and/or telephone systems has not been addressed to date.

55. See, Estlund, Note 7, at p. 924, n8. Also see Matthew W. Finkin et al., Legal Protection for the Individual Employee, 284–286 (1989).

56. Whether a public employee's speech concerns a matter of public interest is determined by the content, form, and context of the statement, gleaned from the entire record before the court. See *Connick versus Meyers*, 461 U.S. 138, 147–148 (1983).

57. These include Section 7, concerted activities for the purposes of mutual aid, such as union organizing, and striking to improve working conditions. They arguably also include protests and advocacy that predate cognizable collective efforts to organize. See Estlund, Note 7, at 924, n8, and Charles Morris, NLRB Protection in the Nonunion Workplace: A Glimpse at a General Theory of Section 7 Conduct, *U. Pa. L. Rev.*, 137:1673, 1677 (1989).

58. See Estlund, Note 7, at 924. Also see Finkin, Note 55.

59. One expert has argued that, despite the fact that free speech is a constitutional right outside the workplace, speech can be regulated in the workplace so long as there are legitimate business reasons for doing so. Also, there should be a clear corporate policy enunciated that sets forth the reasons for the restrictions. See Electronic Mail Raises, Note 51.

The arguable absence of legally sanctioned speech rights has not deterred those who view employee speech as a right. See Rifkin, Note 50, The Ethics Gap.

To date, businesses have not authored many guidelines for internal corporate E-mail networks. There are, however, as many as 200 state statutes covering E-mail-related issues. See Electronic Mail Raises, Note 51.

60. At least one scholar argues that employers are free to invade employee privacy on E-mail as well. Steven B. Winters, Do Not Fold, Spindle or Mutilate: An Examination of Workplace Privacy in Electronic Mail, *S. Cal. Interdisciplinary L. J.*, 1: 85 (1992).

61. Other forms of control are used as well. For instance, on the Internet, an amalgam of research-oriented networks moving toward commercialization, group users sometimes "gang up on abuses [by a particular user] in a form of citizens' arrests [sic] in which abusers are asked to stop disrespectful behavior" (J. A. Savage and Gary H. Anthes, Internet Privatization Adrift, *Computerworld*, p. 1, November 26, 1990). Also see, generally, Moore, Note 39.

62. While it is possible that a sysop may be held responsible for libelous information residing on its bulletin board systems, the current law is unsettled as to the scope of such liability or the circumstances under which such liability would attach. See Robert Charles, Note: Computer Bulletin Boards and Defamation: Who Should Be Liable? Under What Standard? *J. L. and Tech.*, 2(121): 134 (1993). Also see, generally, David J. Conner, *Cubby v. CompuServe*, Defamation Law on the

Electronic Frontier, *Geo. Mason Ind. L. Rev.*, 2:227 (1993); and David R. Johnson and Kevin A. Marks, Mapping Electronic Data Communications onto Existing Legal Metaphors: Should We Let Our Conscience (and Our Contracts) Be Our Guide? *Vill. L. Rev.*, 38:487 (1993). However, it is reasonably argued that if a sysop knows the statement to be false, or should have known, or if the sysop fails to delete libelous information once notified by the injured party, the sysop may be sued for publication of libel. See Charles, pp. 147–148, and Johnson and Marks, p. 497. Also see *Cubby, Inc., et al. versus CompuServe et al.*, 776 F Supp. 135 (1991); William Jackson, CompuServe Picked Its Fight in Libel Case, *Business First — Columbus*, 8(11), section 1, p. 4 (November 18, 1991); Brock N. Meeks, As BBSes Mature, Liability Becomes an Issue, *Infoworld*, S14 (January 22, 1990); Kahn, Note 50; and Geoffrey Stone, The First Amendment Is Safe at Prodigy, *New York Times*, section 3, p. 13 (Dec. 16, 1990).

63. *Memorandum Opinion and Order in the Matter of AT&T Communications, Contract Tariff F.C.C. No.[s] 2[-13], and AT&T Communications, Contract Tariff F.C.C. No. 15, et al.*, 1994 FCC Lexis 242, Release-Number: FCC 93-537, January 19, 1994 (affirming the FCC decision to allow AT&T to offer business services under contract tariffs). The commission has permitted AT&T to offer services under tariff via individually negotiated contracts provided the contract tariffs are made generally available to similarly situated customers under substantially similar circumstances. See *Competition in the Interstate Interexchange Marketplace*, CC Docket No. 90-132, Report and Order, 6 FCC Rcd 5880, 5896–97, reconsidered in part, Memorandum Opinion and Order, 6 FCC Rcd 7569 (1991), further recon., Memorandum Opinion and Order, 7 FCC Rcd 2677 (1992) (interexchange order).

64. *In the Matter of Tariff Filing Requirements for Non-Dominant Common Carriers*, 8 FCC Rcd 6752, August 16, 1993. While the FCC has substantially deregulated the telecommunications industry, it cannot compel carriers to eschew the filing of tariffs if they so desire. *MCI Telecommunications Corporation, Petitioner, versus Federal Communications Commission et al.*, 765 F.2d 1186 (July 9, 1985).

The commission's permissive tariffing policy, which allowed nondominant carriers to elect not to file tariffs, was recently overturned by the Circuit Court of Appeals for the District of Columbia. This result was much to the disagreement of at least one former chair of the FCC. See Sikes in Parting Shot to Congress Wants Forbearance Restored, *Report on AT&T*, January 18, 1993. Nevertheless, shortly after the circuit court's decision, the commission approached the line of absolute deregulation by allowing nondominant carriers to file tariffs on one day's notice under the rationale that they do not possess sufficient market power to set rates for competitive service offerings. See *In the Matter of Tariff Filing Requirements for Nondominant Common Carriers*, 8 FCC Rcd 6752; 1993 FCC Lexis 4285; 73 Rad. Reg. 2d (P & F) 849, August 18, 1993.

65. See Phillip S. Cross, Utility Liability Waivers: New Rules for New Technologies, *Public Utilities Fortnightly*, 129(12):34 (1992) and James Brook, Contractual Disclaimer and Limitation of Liability Under the Law of New York, *Brooklyn L. Rev.*, 49:1, 22 (1982). Also see, generally, *Liability of Telegraph or Telephone Company for Transmitting or Permitting Transmission of Libelous or Slanderous Messages*, 91 ALR3rd 1015 (1993). Telephone companies retain the right to refuse service when a subscriber uses obscene or profane speech. See Allan L. Schwartz, Right of Telephone Company to Refuse, or Discontinue, Service Because of Use of Improper Language, 32 ALR3rd 1041 (1993).

66. Carriers are often successful in limiting their liability for provision of service. *M.R.C.S., Inc. versus MCI*, 1987 WL 12813 (E.D.La.) (claims against carrier for

poor quality transmission are limited to the terms of the tariff.). Also see Brook, Note 65. However, there are numerous instances in which the courts have refused to allow exculpatory language in carrier tariffs to limit carriers' liability. See *In Re Illinois Bell Switching Station Litigation*, 1993 WL 323120 (Sup. Crt. Ill.) (1993) (carrier's exculpatory tariff language limiting liability for consequential damages is not controlling in the face of willful violation of a state statute and regulations requiring utility to provide adequate and efficient, just, and reasonable facilities); *Source Assoc., Inc. versus MCI*, 1989 WL 134580 (1989) (tariff does not limit liability for willful misconduct); *D. Clarico et al. versus Southwestern Bell Telephone Co.*, 725 S.W. 2d 304 (1986) (reasonableness of public utility's tariff limitation becomes an issue of fact when utility can, but does not, timely remedy customer's problem, resulting in a loss that exceeds tariff limitation on liability); and *Lahke et al. versus Cincinnati Bell, Inc.*, 439 N.E. 2d 928 (1981) (carrier's exculpatory tariff language is not controlling in the face of violation of a state statute requiring utility to provide necessary and adequate facilities).

67. See *Chesapeake and Potomac Telephone Co.*, Note 43, holding that telco–cable cross-ownership regulation that prohibited telephone companies from providing video programming to subscribers in the telephone companies' service areas contravenes the First Amendment.

68. There are numerous instances in which the courts have refused to allow exculpatory language in carrier tariffs to limit carriers' liability. See *In Re Illinois Bell*, Note 66; *Source Assoc. Inc.*, Note 66; *D. Clarico et al.*, Note 66; and *Lahke et al.*, Note 66.

    There are also a growing number of cases extending tort liability to providers of goods and services generated via the use of computer and information technologies. See, generally, Barry B. Sookman, The Liability of Information Providers in Negligence, *Computer Law and Practice*, 5:141 (1989).

69. "Sysops have the right to run their systems any way they see fit. They have no 'common carrier' obligations, as do the telephone companies, to transmit everyone's messages" (Meeks, Note 62, p. S14). According to some, a sysop is a publisher with the corresponding right to edit or shape the bulletin board's message traffic as they see fit. Idem.

70. A sysop may be held responsible for libelous information residing on its bulletin board systems. If they know the statement to be false, or should have known, or they fail to delete libelous information once notified by the injured party, they may be sued for publication of libel (Meeks, Note 62). Because CompuServe exercised no editorial control over information on one of its bulletin board services, it avoided potential liability for libel. See *Cubby, Inc., et al.*, Note 62. Also see Jackson, Note 62.

71. See Stuart Silverstein, Prodigy Services' Fee Set Up Under Probe, *Los Angeles Times*, p. D1 (April 16, 1991); Stone, Note 62.

72. Warren G. Lavey, The Public Policies that Changed the Telephone Industry into Regulated Monopolies: Lesson from Around 1915, *Fed. Com. L. J.*, 39:171 (1989).

73. *Carlin Communication, Inc.*, Note 44.

74. See *Chesapeake and Potomac Telephone Co.*, Note 43.

75. See Edmund Andrews, Ruling Frees Phone Concerns to Offer Cable Programming, *New York Times*, Al, column 2, continued on D5, column 1 (August 25, 1993) (announcing the decision of the U.S. District Court overturning the telephone–cable television cross-ownership ban at 47 U.S.C. @ 533 (b), *The Cable Communications Policy Act of 1984*). Also see, 47 U.S.C. @ 533 (b), *The Cable Communications Policy Act of 1984*, prohibiting local telephone compa-

nies from providing video programming to potential viewers in its service area directly or indirectly through an entity owned by the telephone company or under its common control. Waivers have been granted under statue and FCC rule. See FCC Upholds GTE Cerritos Waiver, Grants Another, *Broadcasting Magazine*, 136 (May 1, 1989).

76. See Edmund L. Andrews, A Communications Free-for-All, *New York Times*, section D, p. 1, column 3 (February 4, 1994); *Communications Daily*, p. 8 (February 11, 1993) (cable TV opposition); and John Aloysius Farrell, Newspapers Roll Out Lobbyists in Electronic Information Fight, *Chicago Tribune*, p. 5 (October 27, 1991) (newspaper publishers).

77. See Note 1.

78. See Allen S. Hammond IV, Regulating Broadband Communications Networks, *Yale J. on Reg.*, 9:181, 183–191 (1992).

79. See Hammond, Note 78, at pp. 196–198 and accompanying notes.

80. See *Chesapeake and Potomac Telephone Co.*, Note 43.

81. Regarding restrictions on access to on-line data systems, see Silverstein, Note 71, p. D2; Michael Schuyler, Systems Librarian and Automation Review: Rights of Computer On Line Service Users, *Small Computers in Libraries*, 41 (December 1990); and Stone, Note 62.

    With regard to discriminatory provision of access to cable television, see Donna N. Lampert, Cable Television: Does Leased Access Mean Least Access? In *Cable Television Leased Access: A Report of the Annenberg Washington Program in Communications Policy Studies* (Northwest U., ed.), 1991, pp. 10–12, 15–16; Henry Gilgof, Report Card on Cablevision: Mixed Signals: Programs Praised, Fees Criticized, *Newsday*, 2 (September 10, 1990); and Chuck Stogel, Amid Cable TV Tangle, Is Viewer Being Served, *Sporting News*, 45 (August 27, 1990). The more recent leased access provisions have been upheld as constitutional. See Most Provisions of 1992 Cable Act Survive First Amendment Challenge, *BNA Antitrust and Trade Regulation Report*, 65(1632):387, News and Comment Section (September 23, 1993).

82. It may be that, due to the relative newness of these services and the necessity to have access to the appropriate telephony and computer equipment, access is controlled by economic and market demand factors.

83. *Cubby, Inc., et al.*, Note 62.

84. CompuServe was deemed a distributor rather than a publisher based on several factors. Based on its determination that CompuServe was a distributor, the court held that CompuServe would have had to have knowledge or reason to know that the remarks of the Journalism Forum were allegedly defamatory. See *Cubby, Inc., et al.*, Note 62.

85. Medphone Corporation, a small New Jersey company, sued Peter DeNigris, a 41-year-old Long Island, New York, elections forms processor and amateur stock investor, in federal court in New Jersey. Medphone alleged that DeNigris' comments on Money Talk, a bulletin board service operated by Prodigy, helped cause an almost 50% decline in the company's stock in the summer of 1992. Medphone also alleged that DeNigris engaged in libel and securities fraud. Amy Harmon, New Legal Frontier: Cyberspace; Millions of Americans Swap Information — and Barbs — on Computer Bulletin Boards. But the Laws Governing Free Speech Are Making an Abrupt Entry into This Spaceless, Timeless World, *Los Angeles Times*, Home edition, Part A, p. 1, column 1, National Desk (March 19, 1993).

86. Prodigy is not named as a defendant in the Medphone suit. However, its insistence on screening all electronic messages on its system has led some to argue it is

a publisher and therefore should have some liability for libelous statements made over its facilities. Harmon, Note 85.

The $40 million suit filed against DeNigris was settled for $1 in late November 1993. Fred Volgelstein, Computer Libel Suit Settled, But the Issue Isn't, *Newsday*, p. 7 (December 28, 1993); Kurt Eichenwald, Medphone Blames Messenger for Its Stock Price Troubles, *New York Times*, Section D, p. 8 (December 28, 1993).

87. Harmon, Note 85.

88. *Miami Herald Publishing Co. versus Tornillo*, 418 U.S. 241 (1974).

89. *Sable Communications*, Note 43.

90. *Leathers*, Note 41 (cable) and *Red Lion Broadcasting versus FCC*, 395 U.S. 367 (1969) (broadcasting).

91. See *Metro Broadcasting versus FCC*, 497 U.S. 547 (1990) and *Red Lion Broadcasting*, Note 90. In *Red Lion*, the Supreme Court recognized broadcasters as having a qualified constitutional speech right. However, broadcasters' editorial speech rights were held secondary to the rights of listeners and viewers to receive diverse information and ideas (*Red Lion*, Note 90, at pp. 389–390).

92. See *Columbia Broadcasting System versus Democratic National Committee*, 412 U.S. 94 (1973) (broadcasters could not be compelled to accept editorial advertisements covering controversial issues) and *Syracuse Peace Council*, Note 41.

93. Over time, the broadcast licensee's speech right has been expanded. Broadcasters may not be compelled to accept editorial advertisements for broadcast when they are already adhering to an obligation to present controversial issues of public importance fairly. They retain the right to decide what controversial "issues are to be discussed and by whom, and when"; see *Columbia Broadcasting System*, Note 92. Most recently, the Federal Communications Commission was upheld when it abolished the Fairness Doctrine because the doctrine allegedly "chilled" broadcasters' exercise of their editorial discretion; see *Syracuse Peace Council*, Note 41.

94. *Columbia Broadcasting System*, Note 92 (broadcasters could not be compelled to accept editorial advertisements covering controversial issues).

95. *Action for Children's Television versus FCC*, 932 F. 2d 1504 (1991), cert. denied 112 S. Ct. 1282 (1992) and *FCC versus Pacifica Foundation*, 438 U.S. 726 (1978).

96. See Michael Meyerson, The First Amendment and the Cable Operator: An Unprotected Shield Against Public Access Requirements, *Comment* 4:1 (1981).

97. See *Berkshire Cablevision of Rhode Island, Inc. versus Burke*, 571 F. Supp. 976 (1983), vacated as moot, 773 F.2d 382 (1st Cir. 1985) (holding that mandatory cable channel access rules are constitutional based on theory of economic scarcity). Cf. *Preferred Communications versus City of Los Angeles*, 754 F.2d 1396 (9th Cir. 1985), aff'd 476 U.S. 488 (1986) (requiring cable operator to set aside mandatory and leased access channels diminishes the operator's freedom of expression).

98. See *Turner Broadcasting System, Inc. versus FCC*, 1994 WL 279691 (U.S. Dist. Col.), June 27, 1994 at 5 and 6, citing congressional fact-finding hearings regarding the Cable Television Consumer Protection and Competition Act of 1992, summarized in S. Rep. No. 102-92, pp. 3–4 (1991) and H. R. Rep. No. 102-628, p. 74 (1992), as well as the conclusions Congress reached that are recited in Sections 2(a)(1) through (21) of the act.

99. See *The Cable Communications Policy Act of 1984*, Note 75, 47 U.S.C. Sections 531 (public, educational, and governmental access channels) and 532 (leased access channels).

100. The Supreme Court has unanimously affirmed that cable operators engage in constitutionally protected speech activities. See Turner Broadcasting System,

Note 98, citing *Leathers*, Note 42. While the court held that the government possessed a compelling interest in enacting the must carry rules, it also concluded that the government had failed to provide sufficient evidence to establish that broadcast stations are in economic jeopardy and that the must carry rules will actually advance the government's interests by materially alleviating the economic harm (Turner, Note 98, pp. 20–22). The court then remanded the case for a determination of whether broadcasters were indeed adversely affect by competition from cable television. Idem. See Ana Puga, Congress Upheld on Cable Rule; But Court Orders Review of First Amendment Issues, *Boston Globe*, p. 35 (June 28, 1994); More Evidence Needed; U.S. Supreme Court Vacates Must-Carry Decision, Remands It to Lower Court, *Communications Daily*, p. 1 (June 28, 1994); John Lippman, For Now, TV Viewers Are Spared Another Juggling of the Channels; Cable: A Final High Court Ruling on the "Must-Carry" Statute Could Be Years Away, *Los Angeles Times*, Part D, p. 5, column 3 (June 28, 1994); David G. Savage, High Court OKs Congress' Right to Regulate Cable TV, *Los Angeles Times*, Part A, p. 1, column 1 (June 28, 1994); Linda Greenhouse, The Media Business: Justices Back Cable Regulation, *New York Times*, Section D, p. 1, column 6 (June 28, 1994); and Joan Biskupic, Supreme Court Connects Cable TV to Free Speech Protections of Press, *Washington Post*, p. A1 (June 28, 1994). For a synopsis of many of the key holdings of the case, see Case Digests: Federal Cases; United States Supreme Court; First Amendment—Communications, *New Jersey Law J.*, 76 (July 4, 1994).

101. *Alliance for Media et al.*, Note 44. Also see *First Report and Order in the Matter of Implementation of Section 10 of the Cable Consumer Protection and Competition Act of 1992*, FCC Rcd 998, 1993 Lexis 3144 (February 3, 1993).

   For commentary and limited analysis of Section 10, see Timothy B. Dyk and Sarah L. Wanner, Developments in Communications Law: The FCC's Indecency Proposals Under Fire, *Legal Times*, 25 (May 17 1993) and Bruce Fein, Cable Discretion and the First Amendment, *Washington Times*, G1 (December 2, 1992) (arguing that the Cable Television Consumer Protection Act of 1992 properly permits cable operators to eschew carriage on access channels of programming the operators deem to be obscene or sexually explicit).

102. Idem.

103. See Electronic Media Regulation, Note 1. See, generally, Wilson, Note 51, and Rifkin, Note 50.

104. See Note 54.

105. The monitoring of employee E-mail is justified as a legitimate method of protecting business assets and perogatives. See Rifkin, Do Employees Have, Note 50.

106. A government employee cannot be fired for nondisruptive exercise of the First Amendment right to speak on matters of public concern (*Connick*, Note 56) provided, however, that the employer does not possess an interest in "effective and efficient fulfillment of its responsibilities to the public [which] outweigh the employee's interest in speaking" (*Connick*, Note 56, pp. 150–151).

107. Estlund, Note 7, pp. 921, 923–924.

108. See, generally, Henry H. Perrit, Jr., The Congress, The Courts and Computer Based Communications Networks: Answering Questions About Access and Content Control, Introduction, *Vill. L. Rev.*, 38:319 (1993).

109. Sections 4 and 5 of the Cable Consumer Protection and Policy Act require cable systems of a certain size to carry, on broadcaster request, the signals of certain licensed commercial and noncommercial broadcast stations in the cable operator's market. Prior to enacting the must carry rules as part of the Cable Communications and Consumer Protection Act of 1992, Congress determined that cable

operators often enjoy monopoly status as the only multichannel provider in their respective markets.

According to the Senate, Congress enacted the regulations to ensure that cable operators do not exercise their control over their distribution facilities in a manner that discriminated against broadcasters (*Senate Report*, 138 Cong. Rec. 1133). The Senate also stated that the signal carriage provisions are "not at all based on the content of those signals, but instead . . . counterbalance cable systems' commercial or economic incentive to exclude . . . [broadcast signals]" (*Senate Report No. 102-92*, 138 Cong. Rec. 1133, 1189, 1992).

Several cable operators, among others, challenged the constitutionality of the rules. *Turner Broadcasting Systems, Inc., et al. versus Federal Communications Commission et al.*, 1993 U.S. Lexis 4339, *10-*11.

The district court panel ruled 2 to 1 that the provisions are constitutional. According to the court, the provisions are "essentially economic regulation designed to create competitive balance in the video industry as a whole, and to redress the effects of cable operators' anti-competitive practices" (Turner Broadcasting, pp. *19-*20).

110. The 4th Circuit decision was appealed to the Supreme Court, which granted certiorari. However, the Telecommunications Act of 1996 repealed a significant portion of the telco-cable cross-ownership provision by allowing qualified telephone company entry into video distribution markets coextensive with their telephone subscriber markets. The Supreme Court subsequently remanded the C&P case back to the 4th Circuit to address the issue of mootness.

     One other circuit court and five other district courts have found the cable-telephone cross-ownership ban unconstitutional on First Amendment grounds. See, *US West Inc., versus United States*, 855 F. Supp. 1184, June 15, 1994, affirmed, *US West, Inc. versus United States*, 1994 U.S. App. Lexis 36775; 95 Cal. Daily Op. Service 15, December 30, 1994; *GTE California Inc. versus Federal Communications Commission; United States of America*, 39 F.3d 940; October 31, 1994: *Ameritech Corporation et al. versus United States*, 867 F. Supp. 721, October 28, 1994; *Bellsouth Corporation versus United States*, 868 F. Supp. 1335, September 23, 1994; USTA, OPATSCO, NTCA Win Lawsuit to Lift Cable-Phone Ownership Ban, BNA Management Briefing, Jan. 30, 1995.

111. Deutschman and Davis, Note 22; Editorial, Policing the Information Highway, *Chicago Tribune*, North Sports, Final edition, p. 30, Zone N (November 26, 1993); Huey et al., Note 44; Solomon et al., Note 44; Greenwald et al., Note 44; and Sugawara and Farhi, Note 44.

112. Idem.

113. The district court panel ruled two to one that the must carry rules are constitutional. *Turner Broadcasting Sys. Inc. versus FCC*, 819 F. Supp. 32 (D.D.C. 1993). While the Supreme Court majority affirmed the constitutionality of the rules in theory, the Court questioned whether the government had established that its compelling interest in protecting the access of the 40% of Americans to broadcast television was actually threatened by cable television competition. Consequently it remanded the case to the district court on the issue of economic harm. *Turner Broadcasting Sys., Inc. versus FCC*, 114 S. Ct. 2445, 2469-70 (1994).

114. The conclusion that the must carry rules pass constitutional muster may be in jeopardy. The court has changed with Justice Blackmun's retirement and Justice Breyer's succession. Consequently, there is no majority for much of the *Turner* opinion as currently written, if the case returns to the court subsequent to a full hearing in the district court. It is not possible to know with any assurance how Justice Breyer would vote if the case returns to the court. Moreover, while it is unclear what further evidence the parties will submit in the district court, it is

likely that a subsequent Supreme Court will have new information before it. Thus, there are opportunities for a new majority to emerge.

115. Both the Supreme Court and district court dissents came to this conclusion. Justice O'Connor, in her dissent joined by Justices Ginsburg, Scalia, and Thomas, concluded that Congress's reasons for adopting the rules certainly made significant reference to the content of information to be provided by broadcasters. As such, strict scrutiny is required even when the government's goals may be laudable, *Turner Broadcasting Sys., Inc. versus FCC*, 114 S. Ct. 2445, 2476-78 (1994) (O'Connor).

Judge Stephen Williams's dissent in the district court also took considerable exception to the impact of the must carry rules on cable operators and programmers, *Turner Broadcasting Sys., Inc. versus FCC*, 819 F. Supp. 32, 60 (D.D.C. 1993) (Williams).

The Supreme Court majority's conclusion that the must carry rules are unrelated to the content of the messages that the respective broadcasters and cablecasters carry is questionable. It ignores a major thrust of congressional and FCC broadcast policy stretching back more than 30 years. See *Memorandum Opinion and Order in the Matter of Formulation of Rules and Policies Relating to the Renewal of Broadcast Licenses*, 44 F.C.C.2d 405, 29 Rad. Reg. 2d (P & F) 1 (December 12, 1973); *Memorandum Opinion and Order on Reconsideration of the Fairness Report in the Matter of the Handling of Public Issues Under the Fairness Doctrine and the Public Interest Standards of the Communications Act*, 58 F.C.C.2d 691, 36 Rad. Reg. 2d (P & F) 1021 (March 19, 1976); *Report and Order in the Matter of Primer on Ascertainment of Community Problems by Broadcast Applicants*, 27 F.C.C.2d 650, 21 Rad. Reg. 2d (P & F) 1507 (February 18, 1971); *Memorandum Opinion and Order in the Matter of Petition for Rulemaking to Require Broadcast Licensees to Maintain Certain Program Records*, 43 F.C.C.2d 680 (October 3, 1973); *Report and Order in the Matter of Amendment of the Primers on Ascertainment of Community Problems by Commercial Broadcast Renewal Applicants and Noncommercial Educational Broadcast Applicants, Permittees and Licensees*, 76 F.C.C.2d 401, 47 Rad. Reg. 2d (P & F) 189 (March 12, 1980); *Memorandum Opinion and Order in the Matter of Amendment of the Commission's Rules Concerning Program Definitions for Commercial Broadcast Stations by Adding a New Program Type, "Community Service" Program, and Expanding the "Public Affairs" Program Category and Other Related Matters*, 88 F.C.C.2d 1188, 1982 FCC Lexis 721, 50 Rad. Reg. 2d (P & F) 1245 (January 13, 1982); *Report and Order in the Matter of the Revision of Programming and Commercialization Policies, Ascertainment Requirements, and Program Log Requirements for Commercial Television Stations*, 98 F.C.C.2d 1076, 56 Rad. Reg. 2d (P & F) 1005 (June 27, 1984); *Deregulation of Radio*, 84 FCC 2d at 988; *Commercial Television Stations*, 98 FCC 2d at 1096; and *Report and Order in the Matter of Policies Regarding Detrimental Effects of Proposed New Broadcast Stations on Existing Stations*, 3 FCC Rcd 638, 64 Rad. Reg. 2d (P & F) 583 (November 24, 1987).

116. For both dissents, strict scrutiny is triggered by the rules' impact. First, the must carry rules mandate speech that the cable operators would not otherwise make and prohibit cable operators from programming a portion of their channels as they might otherwise have done. (Turner, 114 S. Ct. at 2749; Turner, 819 F. Supp. at 59). Second, the rules do so in a manner that directly burdens the cable operators' exercise of editorial control and speech. (Turner, 114 S. Ct. at 2749; Turner, 819 F. Supp. at 59-60). As a consequence of the cable operators' loss of control over their channels of transmission, they suffer a direct, palpable, diminution of speech.

The dissenting opinions diverged once they concluded that the rules' impact triggered strict scrutiny. Justice O'Connor concluded that the government's interests were not sufficiently compelling to justify content-based speech restrictions (Turner, 114 S. Ct. at 2748). Unfortunately, she does not state what would constitute such a compelling interest. Judge Williams concluded that, while the government's interest in diversity was sufficiently compelling, its means of achieving its goal was not sufficiently narrowly tailored to accomplish the government's purpose. First, there is insufficient proof that requiring carriage of broadcasters will increase diversity (Turner, 819 F. Supp. at 61–62). Second, there are less burdensome alternatives, such as the leased access channel provision, that would accomplish Congress's purpose (Id.). While not conceding that the government possessed a compelling interest, Justice O'Connor too concluded that the rules are insufficiently tailored to achieve the government's stated goals (Turner, 114 S. Ct. at 2749).

117.	The dissent acknowledged as compelling the government's goal of ensuring access to television for Americans financially disinclined or incapable of subscribing to cable, as well as Americans who remain geographically remote from areas where broadcasting is offered. However, the available evidence does not support the congressional finding that broadcasting is being economically threatened by cable. On review of the proffered evidence, the dissent concluded that

> The legislative findings do not support the inferences needed to sustain must-carry . . . (1) there is no finding of any present or imminent harm; (2) the evidence of some dropping of some local broadcast channels in itself fails to show any widespread problem; (3) the proliferation of local broadcast stations since the end of the FCC's must carry rules undermines any inference of a problem; (4) the findings as to structure and incentives, taken together with the evidence of cable's dependence on broadcasting, fail to raise the concern beyond the level of speculation; and (5) even if the hazard were perceptible, the record does not address the less intrusive alternatives (Turner, 819 F. Supp. at 65).

Moreover, Congress's efforts to establish evidence of broadcasting's economic demise prove no less effective than prior efforts by the FCC and broadcasters. The must carry question is not the first instance in which economic harm to existing broadcast stations has been raised against new competitors.

In the broadcast economic injury cases, the courts and the Federal Communications Commission concluded that an existing broadcaster could prevent the entry of a new broadcast competitor based on pleading economic harm unless its allegations of economic injury were supported by proof of a significant loss in news and public affairs programming occasioned by a loss of advertising revenues. It also had to establish that this loss in news and public affairs programming would not be alleviated by the new entrant. After years of litigation, the FCC concluded that no broadcaster had been able to meet the public interest burden successfully. See *Report and Order in the Matter of Policies Regarding Detrimental Effects*, Note 115. The consistent inability of broadcasters to meet the burden of proof necessary to establish an actionable economic detriment lead to the abolition of the economic injury objection in comparative proceedings. The FCC, in abolishing the objection, stated that

> By this action, the Commission abolishes certain policies that address the issue of economic injury to existing broadcast stations. Our decision is based on our experience in implementing these policies and the intervening growth of the electronic media which lead us to conclude that the public interest is no longer served by their retention.

> Our review of more than 80 cases indicates that, although parties may have routinely pleaded [economic injury issues], they have been unable to demonstrate

sufficient evidence to warrant a finding of harm that would result in a net loss of service to the [public.]

We also conclude that the underlying premise of the Carroll doctrine, the theory of ruinous competition, i.e., that increased competition in broadcasting can be destructive to the public interest, is not valid in the broadcast field. The court, in Carroll, conceded that "private economic injury is by no means always, or even usually, reflected in public detriment. Competitors may severely injure each other to the great benefit of the public."

Consideration of allegations of economic injury to determine whether they will lead to an overall derogation of service to the public is like looking for the proverbial "needle in a haystack." On this basis, we will no longer entertain claims of Carroll injury. *In re Policies Regarding Detrimental Effects of Proposed New Broadcast Stations on Existing Stations*, 3 F.C.C. 638 at 640–641 (1987).

In coming to its conclusion, the FCC also cited congressional determinations that pleadings of economic injury would be insufficient to preclude competitive market entry. See, 2 FCC Rcd 3134, March 26, 1987, *Notice of Inquiry.*

The FCC's conclusions regarding existing broadcasters' inability to prove economic injury over many years in numerous licensing proceedings bear a significant similarity to the conclusions reached by the Circuit Court for the District of Columbia in *Quincy Cable TV, Inc. versus FCC*, 768 F.2d. 1434 (D.C. Cir. 1985) and those reached by the dissent in *Turner*. In each instance, the evidence proffered is too speculative. In *Quincy*, the court concluded that, irrespective of the ultimate constitutionality of the rules,

The Commission had not adequately substantiated its assertion that a substantial governmental interest existed . . . the problem the must carry rules purported to prevent—the destruction of free, local television—was merely a "fanciful threat" unsubstantiated by the record or by two decades of experience with cable tv. (*Century Communications versus FCC*, 835 F.2d 292, 295 (D.C. Cir 1987) quoting Quincy, 768 F.2d at 1457).

Indeed, in its subsequent effort to justify the must carry rules, the FCC did not even advance an economic harm argument, a fact that the court noted, deeming that argument "foreclosed by Quincy Cable TV" (Id. at 299 n. 4).

118. As stated in Note 104, the court's majority remanded the case to the district court, requiring the court to hear further evidence because the government had failed to provide sufficient evidence to establish that broadcast stations are in economic jeopardy and that the must carry rules will actually advance the government's interests by materially alleviating the economic harm (Turner, 114 S. Ct. at 2445). Only Justice Stevens would have voted for affirmance of the district court opinion on which the appeal was based (Turner, 114 S. Ct. at 2473).

119. See note 110.

Regulated local-exchange carriers (LECs) have been prohibited from providing video distribution in their local markets since 1970. At that time, the FCC issued a rule prohibiting a telephone company from owning a cable concern in the same market. See 47 CFR 63.54(a) and (b) and Note l(a). The rule was promulgated to prevent anticompetitive activities of some LECs that sought to control the entry of cable into their markets by restricting or controlling cable operator access to telephone facilities and pole attachments. See, generally, *Applications of Telephone Common Carriers for Section 214 Certificates for Channel Facilities Furnished to Affiliated Community Antenna Television Systems (Final Report and Order)*, 21 FCC 2d 307, recon. in part, 22 FCC 2d 746 (1970), aff'd *General Telephone Co. of S.W. versus United States*, 449 F.2d 846 (5th Cir. 1971).

In 1984, Congress codified the FCC's telco–cable cross-ownership rules in the Cable Communications Act of 1984. See the *Cable Communications Policy Act of 1984* (84 Cable Act), 47 U.S.C. @ 613(b). The legislative history of Section 613(b) indicates that it was intended to codify the then-current FCC telco–cable cross-ownership rules prohibiting telephone companies from directly providing video programming to subscribers in their telephone markets. See H.R. Rep. No. 934, 98th Cong., 2d sess. at 56, and 130 Cong. Rec. H 10,444 (daily, October 1, 1984).

The FCC subsequently reversed its earlier decision and concluded that the public interest would be better served by partially lifting the cross-ownership ban. See *Further Notice of Inquiry and Notice Proposed Rulemaking*, 3 FCC Rcd 5849, CC Docket 87-266, Telephone Company–Cable Television Cross Ownership Rules, Sections 63.54–63.58, FCC 88-249 (released September 22, 1988). The commission concluded that, subject to safeguards, the public would receive significant benefits if telephone companies were allowed to provide cable television service. It tentatively concluded that "construction and operation of technologically advanced, integrated broadband networks by carriers for the purpose of providing video programming and other services [would] constitute good cause for a waiver of the prohibition" (see 3 FCC 5849 (1988), citing 69 FCC 2d 1110). However, Congress did not repeal its law.

In light of Congress's refusal to remove the prohibition, Bell Atlantic filed suit alleging that the 1984 Cable Act prohibition violates the First and Fifth Amendments' rights of local-exchange carriers, as well as the First Amendment rights of subscribers.

120. Prior to the C & P (Bell Atlantic) law suit, the government addressed the constitutionality of restricting telephone company access to information markets in three separate proceedings, occurring in three respective venues. The issue was first raised in the Modification of Final Judgment (MFJ) proceedings in 1982 and 1987. In 1982, District Court Judge Harold Green issued an order that inter alia precluded the soon-to-be-divested Regional Bell Operating Companies from entering the information services market as "electronic publishers." See *United States versus AT&T*, 552 F. Supp. 131, 227 (D.D.C. 1982), aff'd sub nom. *Maryland versus United States.*

In 1987, during the first triennial review of the decree, the Justice Department and the RBOCs recommended that the ban on entry into the information services market be lifted because market conditions had changed significantly and entry would have positive effects on the industry. Further, some RBOCs argued the ban should be removed because it infringed the RBOCs' First Amendment rights. In response, the court ruled that the RBOCs could enter the information services market as transmission providers, but not as information providers.

Upon RBOC appeal of the court's decision, the circuit court ruled that the district court was required to apply a different standard to assess the advisability of removing the restrictions as the parties recommended. The circuit court did not reach the First Amendment issue, however. In the subsequent remand by the Circuit Court of Appeals, the district court reluctantly removed the restriction.

From 1988 to 1992, the RBOCs also sought relief from the cross-ownership ban through a variety of bills proposing to allow telephone company entry into the video transmission and provision market. See H.R. 1504, 103rd Cong., 1st sess. (1993); S. 1200, 102nd Cong., 1st sess. (1991); H.R. 2546, 102nd Cong., 1st sess. (1991) (companion legislation to H.R. 2546); 2. 2800, 101st Cong., 2nd sess. (1990); S. 1068, 101st sess. (1989); and H.R. 2437, 101st Cong., 1st sess. (1989) (companion legislation to S. 1068). Yet, in all the legislation that has addressed the issue of telco ownership of video distribution facilities, despite efforts by telephone companies to raise the constitutional speech issue before the FCC and the MFJ court, to date there are only two references to First Amendment speech issues in previously unsuccessful legislation. See H.R. Rep. No. 934, 98th Cong., 2nd sess. (1984), reprinted in 1984 U.S.C.C.A.N. 4655, 4668, citing *Associated Press versus U.S.*, 326 U.S. 1, 20 (1945), explaining the relationship between the telephone–cable cross-ownership prohibition and the First Amendment goal of diversity of ownership and viewpoint. Also see S. 1200, *Communication Competition and Infrastructure Modernization Act of (1991)*, Title I, Section 101 (14), which states

The Congress makes the following findings: (14) A broadband communications infrastructure will be every American's tool of personal emancipation; will generate a quantum increase in Americans' freedom of speech.

The issue was addressed a second time when Congress determined that its codification of the FCC's telephone–cable television cross-ownership rules was constitutional.

Finally, the issue was addressed in the context of FCC deliberations concerning its telephone company–cable television cross-ownership restriction.

121. See generally *In re Competition in Interstate Interexchange Marketplace Petitions for Modification of Fresh Look Policy*, 8 F.C.C.R. 5046 (1993); and *In re Competition in Interstate Interexchange Marketplace*, 6 F.C.C.R. 7569 (1991).

122. See note 6.

123. The statutory requirement that cable operators provide leased, educational, and governmental access channels was upheld recently. See *Daniels Cablevision, Inc. versus United States*, 835 F. Supp. 1 (D.D.C. 1993).

124. At least one communications expert asserts that there is no shortage of available spectrum, only a shortage of current human ingenuity to harness it. See George Gilder, What Spectrum Shortage? *Forbes*, On the Cover section, Computers/Communications, p. 324 (May 27, 1991). Similarly, digital, switched interactive telecommunications networks can provide another source of increasing capacity for the transmission of information to the home. Consequently, they too reduce scarcity. See comments of Mitch Kapor in Rockley L. Miller, Digital World Future Systems, Inc., *Multimedia and Videodisc Monitor*, 11(7) (July 1993).

125. Andrea Adelson, The Media Business; Radio Station Consolidation Threatens Small Operators, *New York Times*, Late edition–Final, Section D, p. 1, column 1 (April 19, 1993); Edmund L. Andrews, The Media Business; Plan to Ease Rule on Buying Radio Stations, *New York Times*, Late edition–Final, Section D, p. 1, column 3 (February 27, 1992); and 1985: A Year Like No Other for the Fifth Estate; Changes in the Broadcasting Industry, *Broadcasting*, 109:38 (December 30, 1985).

126. Speaking at a recent conference in San Francisco on computers, freedom, and privacy, Laurence H. Tribe, a professor of constitutional law at Harvard Law School, called for an amendment to the U.S. Constitution that would protect privacy, speech, and other constitutional rights made possible in part, but now threatened by, computer technology.

The Tribe amendment reads, in full:

This Constitution's protections for the freedoms of speech, press, petition and assembly, and its protection against unreasonable searches and seizures and the deprivation of life, liberty or property without due process of law, shall be construed as fully applicable without regard to the technological method or medium through which information content is generated, stored, altered, transmitted or controlled.

In Professor Tribe's view, the current constitutional amendments do not protect the rights of computer users adequately. See, Resnick, Note 49.

127. See, generally, Hammond, Note 78 and Perrit, Note 108, pp. 334–35.

128. *Associated Press versus United States*, 326 U.S. 1, 19–20 (1945).

129. For instance, *Soldier of Fortune* magazine was recently held liable for an advertisement it published that the court interpreted as inter alia soliciting contract killing jobs. See Ronald Smothers, Soldier of Fortune Magazine Held Liable for Killer's Ad, *New York Times*, Late edition, Section A, p. 18, column 5 (August 19, 1992).

130. See Sookman, Note 68.

131. At least one state has limited the applicability of telephone carriers exculpatory language to ordinary negligence and does not allow disclaimers for acts of gross negligence, willful neglect, or misconduct. See State OKs Liability Disclaimers for Telcos, *Public Utilities Fortnightly*, 130:42 (December 15, 1992) (discussing *Re Inclusion of Liability Limitations*, Case Nos. 90-774-T-GI et al., October 30, 1992 (W.Va.P.S.C.)).

The Supreme Court for the State of Illinois recently reached the opposite conclusion. It determined that parties suffering economic injury totaling millions of dollars as a result of a severe fire at an Illinois Bell switch could not recover. The court held that the parties' statutory claims for economic losses were not recoverable in a tort action and that the exculpatory language in Illinois Bell's tariff properly limited claims for disruption of service to compensation for the cost of the calls. See *In re Illinois Bell Switching Station Litigation*, Docket No. 73999, Supreme Court of Illinois, 1994 Ill. Lexis 97, July 28, 1994, Filed (decision is not final until expiration of the 21-day petition for rehearing period).

For a summary of the impact of the Hinsdale switch fire on subscribers, see, generally, Art Barnum, Bell Fire Fiasco Rings in Memory Five Years Later, *Chicago Tribune*, Du Page section, p. 2 (April 30, 1993). Also see Central Office Fiascos Expose User Vulnerability, *Network World*, 34 (July 18, 1988); Barton Crockett, Citing Negligence, Users File Suits Against Ill. Bell, *Network World*, 1 (June 6, 1988) (discussing the law suits against Illinois Bell alleging millions of dollars in damages due to a fire in the Hinsdale central-office switch); and Barton Crockett, Users Seek Damages from Illinois Bell; Companies Charge BOC with Inadequate Fire Prevention in Wake of Hinsdale CO Disaster, *Network World*, 11 (May 23, 1988).

132. Christy Cornell Kunin, Unilateral Tariff Exculpation in the Era of Competitive Telecommunications, *Cath. L. Rev.*, 41:907 (1992).

133. 47 USC section 414 (1988).

134. Kunin, Note 132, pp. 914–915 and 926.

135. It has been aptly observed that "the law, by protecting freedom to contract does nothing to prevent freedom of contract from becoming a one-sided privilege. Society, by proclaiming freedom of contract, guarantees that it will not interfere with the exercise of power by contract. Freedom of contract enable enterprisers to legislate by contract and, what is even more important, to legislate in a substantially authoritarian manner without using the appearance of authoritarian forms"; see Kessler, Contracts of Adhesion—Some Thoughts About Freedom of Contract, *Colum. L. Rev.*, 43:629, 640–641 (1943), cited in Summers and Hillman, *Contract and Related Obligation: Theory, Doctrine and Practice*, West, St. Paul, MN, p. 585 (1987).

136. *Re Equicom Communications, Inc.*, 109 Pur 4th 540 (1990), cited in Phillip S. Cross, Utility Liability Waivers: New Rules for New Technologies, *Public Utilities Fortnightly*, 129(12):34 (1992).

ALLEN S. HAMMOND, IV

Network Reliability
(see Communications and Information Network Service Assurance)

Network Routing and Flow Control
(see Flow and Congestion Control)

Network Security   (see Information and Network Security)

Network Signaling   (see Common Channel Signaling)

Network Simulation   (see Modeling and Simulation of Communications Systems)

Network Software Reliability and Quality

Introduction

The subject of network reliability in North America must be approached by first examining quality and reliability assurances, without which no adequate means of quantifying and qualifying network reliability exists. Even more significantly, proactive, preventive measures must be considered. This article pays particular attention to software; many of the issues surrounding hardware reliability have been adequately addressed, while software reliability is in many ways a less exact and less understood discipline. Specifically, customer objectives and expectations — crucial to the success of network operation — are taken into account, as are service quality models, which are useful in providing benchmarks for software development efforts. Determining how to measure software quality and reliability is as critical as the process of development itself; this topic is covered in the final part of the article. It is important to note that each of the areas mentioned interact in a dynamic pattern that constantly reinforces the process and (it is hoped) the end result of network quality attainment: measurements strengthen assurances, customer feedback modifies both assurances and measurements, and service quality models evolve as the process itself is better understood.

Any discussion of network reliability must begin by making a distinction between hardware and software. The former is becoming more and more of a commodity in the marketplace as extremely sophisticated and reliable systems are deployed throughout the world with great success. New SONET (Synchronous Optical Network) fiber-ring networks, for example, are designed with self-healing capabilities that automatically reroute calls if a break occurs in the network. It is also relatively easy to partition network elements, building "fire walls" to contain problems to the smallest possible area. Software, on the other hand, is becoming increasingly complex, and the lines of available code are growing at a furious rate. As the importance of software grows, so does the need to pursue methods of measuring and ensuring its reliability. Difficult to self-heal or partition, software by its nature presents its own set of challenges if we are to measure and control its reliability.

Telecommunications network architecture is inherently modular, hierarchical, and distributed. This is relatively easy to see in hardware configurations, as central-office switches feed remotes and other peripheral units, while connecting private branch exchanges (PBXs), interexchange switches, and other central offices through trunks. It is useful to view software architecture, however, in terms of four main subsystems: support system software, call-processing software, maintenance software, and administrative software (billing, etc.). The main task of call-processing software is to identify an incoming call, establish and maintain a communications path, and disconnect the call. Its architecture, therefore, must meet strict reliability standards for both the core system and individual features. Failure of a feature means loss of the use of that feature; system failure may mean network downtime (1).

Among the many obvious negative effects of network outages, three may be identified as having a particularly strong impact on users. One is the loss of revenues, to both the telephone company involved and any customers who rely on telephone service for their business. Stock brokerages, for example, can lose transactions worth millions of dollars from just minutes of downtime; similarly, retail mail order houses stand to lose sales. Another impact of outages is the loss of service itself, which can yield a variety of consequences, from unanswered 911 calls, to the loss of credibility, to the mere inconvenience of being without telephone service. The third impact is the potential threat to national security; because of the degree to which networks are interconnected, and their critical role in supporting communications, any outage that cascades from its point of origin and compromises other networks could seriously impair the ability of a government to operate efficiently.

Outages, unfortunately, are not hypothetical: there have been notable disruptions of service that serve as examples of what can happen if the telecommunications network on a local, regional, or national level is lost. In 1991, a cable cut in Newark, New Jersey, affected long-distance telephone service in New York City for several hours, causing the shutdown of the New York Stock Exchange. During the same year, a software error played havoc with the Signaling System No. 7 (SS7) network in several regions of the United States. Another software error, in 1990, caused a sharp reduction in the capacity of AT&T's long-distance network. In 1988, a central-office fire in Chicago's Hinsdale office caused a severe service disruption; a similar fire in New York City in 1975 affected traffic in Manhattan for a month.

Network Quality from the Customer's Perspective

There is a growing need to understand postrelease software quality not only in its own right, but from the customer's (in this case, the service provider's) perspective (2). Customers are demanding an increasing level of quality in products and services. If a company is perceived as offering greater quality than its competitors, it will have a competitive advantage. Surveys and research of customers of switching suppliers have identified 39 key elements of software quality that customers consider when rating a software load as good or bad. Listed below are the top 14 elements in order of criticality.

1. Amount of downtime
2. Supplier's willingness to solve problems
3. Reliability for end users
4. Accuracy of documentation
5. Adequacy of supplier prerelease testing
6. Availability of technical support
7. Timeliness of response to problems
8. Timeliness of repair

9. Severity of problems
10. Technical competence of support people
11. Completeness of documentation
12. Accuracy of repair
13. New features meet needs
14. Number of problems

Five of these factors relate to problems in the software (software reliability) and six items relate to the supplier's ability to solve problems after they are reported (service quality). Service quality, therefore, has become a powerful competitive weapon in most service organizations (3). The message is clear: To increase customer satisfaction, a field service organization must be in a position to resolve problems within the customer's time frame (2).

Achieving excellence in software-related services, however, provides unique challenges for which the right measurements are essential. As the dependence on software by central-office switching equipment increases, those services related to software quickly become the focus of the customer-service organizations among telecommunications systems manufacturers. These organizations are assuming the responsibility for the manufacture, delivery, and performance of software products. The number of installed software applications is growing so rapidly that software maintenance and support will become a major industry in its own right by the close of the century, if not sooner (4).

Telephone operating companies spend approximately $6 per digital line per year due to failures related to software services. This creates a nearly $300 million per year opportunity for suppliers to reduce telephone operating company costs by improving software service quality. High service quality keeps customers loyal to a supplier, while low quality drives them to the competition. In a survey of 2374 customers from 14 organizations, Forum Corporation found that more than 40% listed poor service as the number one reason for switching product loyalties (5). Another study concluded that "service related attributes comprise 49.5% of the weight" when business telecommunications end users select a supplier (6). Management of processes to improve service quality necessitates measuring both customer perception and the quality of supplier deliveries.

Businesses such as banking and airlines tend to have a high degree of uniformity, so product or technology differences do not provide much of a competitive edge. This point has also been reached in the central-office switching segment of the telecommunications equipment market. While a supplier may hold a short-term edge in the speed with which any new feature is brought to market, most products are similar in function and capability (7). Software service quality thus may be the new primary differentiator between competitors. If a company exceeds the quality of its competitors in aspects important to its customers, it has a strategic competitive advantage.

Service quality has become, therefore, the most powerful competitive weapon of most service organizations (3). It has been repeatedly demonstrated that excellence in quality translates into increased market share and return on investment, as well as lower manufacturing costs and improved productivity

(8). Because service quality produces measurable benefits in profit, cost savings, and market share, understanding service quality and how it is achieved must be a high priority (9). Further, because telecommunications information systems are critical to the success of most companies, end users of central-office-based services are demanding an optimum balance among cost, timely availability of new features, and quality level as experienced in field operation (10). The postrelease deliverables in the software life-cycle are key elements in this optimization.

Service Definition

Service—that ephemeral part of customer interaction that all businesses like to think they provide—has been described as intangible, heterogeneous, and perishable (11). Service transactions themselves can be further classified as routine or nonroutine (12). Routine events include processes such as applying new software releases. Nonroutine service transactions in the switching environment include replacing a switch that has been destroyed by a catastrophic event or helping a telephone operating company recover from a major software failure that caused system downtime.

Servicing software in the switching environment is highly intangible. Because of the inherent differences in each service event, there is a degree of heterogeneity, due largely to the human aspect of the process. In most software services, the customer witnesses the performance of the service, as when a problem is solved or a software release is delivered. Unlike physical products, however, a software failure's exact cause may be hidden from view. The user's ability to describe what is going wrong is often marginal because of the abstract nature of many software products. This makes software one of the most troublesome products to maintain and support (4).

Quality Definition

In regard to software, *quality* means conforming not only to customers' technical specifications, but also to their implied specifications—that is, their expectations about delivered services. It is the customer's perception of quality that matters most, not the supplier's interpretation of quality control data or self-image. Perceived service quality is a global judgment or attitude about a service, whereas satisfaction is related to a specific transaction (11). Many individual cases of customer satisfaction over time lead to a perception of overall service quality.

Measurement uses symbols to represent the properties of objects (12) or services. It maps a set of objects and services onto a set of symbols. For example, failures, time, and rating are characteristics that can be measured. Examples would include using failures to evaluate a software development process, using time to explain why a customer may be unhappy even after a problem has been solved, or using a rating to describe the customer's perception of software delivery reliability.

Software Service Dynamics

Dynamic Interactions

The environment in which telecommunications suppliers operate is dynamic; this is particularly evident with software. This dynamic environment has four facets. The first is a complex model that relates customer perception and expectation of software service quality to what the supplier actually delivers. The second facet is composed of primary events in the supplier's service delivery process that result in customer satisfaction or dissatisfaction. The third facet involves elements that make up perceived customer satisfaction and the elements a customer considers in the satisfaction decision process. The fourth facet concerns environmental dynamics that are influenced by forces that cause customers to change their definitions of satisfaction over time, usually in a direction that makes this goal more difficult for the supplier to achieve.

This environment can be conceptualized as two parallel planes (Fig. 1), with the lower plane indicating service events and the upper plane representing the threshold of customer satisfaction. This threshold is constantly moving upward because of competition, information, changing technology, changing customer requests, and the promotional efforts of equipment suppliers and telephone operating companies (12). The space between the planes represents

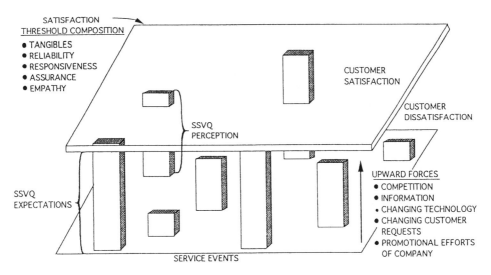

FIG. 1 Software service environment dynamics (SSVQ = software service quality).

areas of customer dissatisfaction, and the distance between the planes symbolizes software service quality expectations. Each of the major service delivery areas has many individual service events, represented by the vertical blocks; the height of the vertical blocks represents perception of software service quality for each event. The height of the customer satisfaction plane may be different among various telephone operating companies and between the telephone operating companies and the end users of telecommunications services.

Each service event may or may not cross the threshold into customer satisfaction. A service event will have an impact on customer satisfaction if the customer's expectations are met in regard to tangibles, reliability, responsiveness, assurance, and empathy (3). Many events of customer satisfaction, over time, will generate a customer perception of the service excellence of the supplier.

The Satisfaction Threshold

For the satisfaction threshold, there are five areas that customers consider when deciding how a service event compares with their expectations. These are listed below in increasing order of their evaluation difficulty.

1. Tangibles. This area includes the physical aspects of the service — facilities, equipment, and appearances, including the appearance of service personnel, the "high-tech" look of the information systems to which customers have access, the design of workstations used to deliver new software releases, and the appearance of the training facility. Tangibles are primarily search properties, those factors that are important to a customer looking for a supplier during the early stages of a customer–supplier relationship.
2. Empathy. This area concerns the provision of caring, individualized attention, such as a quick and helpful response to a fire, storm, or other disaster. Empathy includes experience properties, such as how easily the customer can contact the supplier and the supplier's understanding of the customer.
3. Responsiveness. This consideration involves the willingness to provide prompt service that helps the customer, such as rapidly solving software problems discovered and reported by the customer.
4. Reliability. Reliability relates to performing a service dependably, accurately, and consistently. This can include solving customer-reported problems without requiring recalls, delivering software releases 100% on time with no customer issues, or applying software patches that provide the required increases in functionality without causing other problems. Responsiveness and reliability become most important as the customer–supplier relationship matures.
5. Assurance. The most difficult area to evaluate, assurance involves the supplier's knowledge, courtesy, and ability to engender trust and confidence. Assurance includes search, experience, and credence factors. However, in a mature relationship such as a strategic partnership, it is the credence properties of security and competence that dominate (13).

Importance of Satisfaction Components

The above five satisfaction component areas are not equally important to customers. Measurements should be taken to determine their ranking and identify areas needing attention. Typical rankings of relative importance for two customer types in a mature customer–supplier relationship are shown below.

	Telephone End Users	*Maintenance and Support*
Reliability	60.5%	57.3%
Responsiveness	16.0%	19.9%
Assurance	12.6%	12.0%
Empathy	10.3%	9.6%
Tangibles	0.3%	1.2%

Factors that Raise the Satisfaction Threshold

There are five attributes that signal change over time in the threshold of perceived customer satisfaction (14) that apply to the switching-system software environment:

1. Competition. Since deregulation, competition in the telecommunications industry has intensified, causing both suppliers and service providers to differentiate themselves by providing increased levels of service. This influences customer experiences and, consequently, their expectations.
2. Information. With the advent of the information age, customers are influenced by greater volumes of information distributed at faster speeds. This directly affects their attitudinal changes.
3. Changing technology. Expectations increase as technology advances. For the customer, technology frequently implies better, faster service.
4. Changing customer tastes. As customer tastes and priorities change, so will their expectations.
5. Promotional efforts of companies. What suppliers say about their services influences what the telephone operating company or the end user expects.

These attributes must be considered when making decisions about the content and timing of measurements. If the software service environment changes significantly, then customer expectations must be measured frequently—at least every six months. However, the telephone operating company's and end user's expectations may change at different rates and may be affected by these five factors in different ways.

Process Assurances

As we have seen, the reliability of computer-controlled network software is of critical importance. Unfortunately, creating zero-defect software is not a simple task for a number of reasons. First, the sheer volume of software contained in a large system can exceed millions of lines of code, making it unwieldy. The resulting complexities can lead to all sorts of problems. Because software is interconnected, small changes do not always have small effects; some problems can even be intractable. In addition, computer programs and large collections of interconnected communications systems usually involve interactive and concurrent processes, which can make potential errors subtle and hard to locate. Finally, each new feature set requires new software modifications or installations, which can have an impact on an entire interconnected system (15).

In spite of these limitations, progress has been made in developing assurances that software products developed for telecommunications networks adhere to reliability and quality standards. Before examining several of these, it is advantageous to define what we mean by "software reliability."

Software Reliability

Software reliability is an evolving science (or art, depending on your point of view) that can be discussed on a number of levels: safety, testing, debugging, measures, statistical measurement techniques, visualization, environments, and formal methods.

Software Safety.   Although there are relatively few researchers in the area of software safety (reducing the risk of using software), N. G. Leveson has done some pioneering work in the techniques of fault-tree analysis and publicized the importance of computer safety.

Software Testing.   Two major types of testing are used today. One, dubbed the *black box method*, correlates selected combinations of inputs with outputs and compares results to design or specification documents. The other, which can be called the *glass box method*, attempts to test all paths through the software code. Used as an element of the design stage, glass box testing prepares and updates a test program on completion of coding of each module. These test programs can be used as tools throughout the design process.

Software Debugging.   There has been steady progress in the area of software debugging. For example, "slicing" tools allow users to backtrack from checkpoints without reexecuting a program to reach recent prior states. Current debugging tools, however, still depend on the experience and judgment of the person doing the testing.

Software Measures.   While a number of statistical and graph-theoretical software metrics have been proposed, their relevance has not yet been established. Few examples of software reliability measurement technology have been developed that can prove the achievement of high reliability and safety; the ones that do exist, however, can demonstrate that high reliability and safety have not been achieved, and thus have value. In addition, Carnegie Mellon University's Software Engineering Institute's (SEI) Capability Maturity Model, Bell Canada's Trillium, and Bellcore's Customer Supplier Quality Program, have each constructed a "software maturity" model that can measure the process of developing software, ranging from chaotic to well organized.

Statistical Measurement Techniques. Although the most commonly used software reliability metric, statistical techniques largely ignore the structure and complexity of the software, as well as the nature and quality of previous testing. They are useful, however, in estimating failure probability by monitoring the process and applying data-analysis techniques.

Visualization. Visualization techniques can be helpful in monitoring and analyzing data, or in statistical process control and network performance analysis.

Disciplined and Controlled Environments. Typical techniques for disciplined and controlled environments include requirements for design, walk-through, specifications of design techniques or coding style, and enforced sign-off criteria for life-cycle phases.

Formal Methods. The formal approach to rigorous verification relies on logic and views programs as symbolic manipulators (15).

While current technologies are far from adequate in dealing with network reliability issues, seven areas that postulate a framework for establishing effective methods of reliable software and communication can be identified.

1. Establish the underlying principles of reliable software. Testing and verification in particular depend heavily on quantitative parameters.
2. Cope with complexity. The use of theoretical studies may provide some insight into the problems created by the staggering complexity that characterizes current telecommunications software systems.
3. Explore software architectures that promote reliability. Answers may lie in combining knowledge in logical design and performance analysis.
4. Discover ways to model and engineer concurrent and real-time systems. This is essential for developing machine-processable specifications and standardized tools for automating the design and implementation of distributed systems.
5. Define "reliable" systems and ways to test them. The profusion of varying "flavors" of technologies such as SS7 create fundamental problems in establishing baseline criteria for system/software reliability.
6. Develop a methodology for measuring and verifying the reliability of a design. Partitioning a large problem into a smaller one, for example, can make a complex task more manageable.
7. Explore the question as to whether a systematic approach can even be

developed to ensure software reliability. It is hoped a coherent systematic approach can be developed from the presently used glass box, monitoring/visualization, formal, and model checking methods (15).

Software Architecture Reviews

Software Architecture Reviews (SARs) were established by Bellcore to ensure the quality of a Bellcore client company's network elements and network switching systems. The SAR considers software as a collection of interrelated components. The term *software architecture* defines the functions and interfaces of the software components, their relationships, and the run-time behavior of software components in the system. Knowledge gained from the software architecture is used to improve analysis of capability evolution, fault management, and system robustness issues (16).

The reviews assess three aspects of software reliability and quality: the impact of new requirements on existing software architectures, the fault tolerance and robustness of software, and the fault-prevention methods used before product deployment. It is particularly suitable to employ the review processes in today's networks, which are becoming increasingly complex as new services— common channel signaling, Integrated Services Digital Network (ISDN), advanced intelligent network (AIN)—are deployed to provide ever more sophisticated services to end users (16).

System robustness is determined through various failure-management techniques. Failure detection identifies faults once they have occurred, failure recovery determines how a system recovers from a failure, failure containment prevents faults from spreading, and failure reporting provides information about the failure and the recovery actions taken. Fault prevention, which uses formal techniques for program verification, prevents failures from occurring in the first place (16).

During an SAR, specific items are listed to check against the software being tested. These items fall into two basic checklists: one for evolution and one for dependability. Large systems are expensive and meant to last a number of years. Accordingly, they must be easy to modify (evolve) in order to add new functionality as it becomes available. Systems are therefore evaluated for intellectual manageability, or the flexibility and maintainability of a system. Data independence is also checked: What are the shared data structures in the system? How are they represented? Finally, the interoperability of releases is evaluated to determine what mechanisms provide interoperability within the system and what mechanisms are provided for updating individual modules (16).

High dependability of a switching system ensures continuity of service, avoids catastrophic consequences on the environment, prevents unauthorized access or handling of information, survives catastrophic events, and is modifiable at a reasonable cost. The checklist includes general information on fault tolerance, fault prevention, failure detection, failure recovery, failure containment, and failure reporting (16).

TABLE 1 Benefits versus Activities Matrix

Activity	Benefits								
	1	2	3	4	5	6	7	8	9
Certified code reuse	X	X	X	X	X				
Complexity analysis	X	X	X	X	X	X	X	X	X
Customer surveys	X	X		X					X
Estimation: code size, cost, defect, schedule	X	X	X	X	X	X	X	X	
Fail modes, effects, criticality analysis	X	X	X	X					
Fault-tree analysis	X	X	X	X					
Fault insertion	X	X	X	X			X		
Field performance measurements	X	X		X					
Field testing	X	X		X					
Incremental development	X	X	X	X	X	X	X	X	X
Inspections and reviews	X	X	X	X	X	X	X	X	X
Operational profiles	X	X	X	X	X	X	X	X	X
Process assessments				X	X	X	X	X	X
Procurement control	X	X	X	X	X	X	X		
Reliability and reliability growth models	X	X	X	X	X	X	X	X	
Reliability data and reporting	X	X	X	X			X	X	X
Reliability objectives							X	X	
Reliability prediction				X	X	X	X	X	
Reliability tests	X	X	X	X	X	X	X	X	X
Requirements traceability	X	X		X			X	X	
Robustness strategy	X	X	X	X					X
Root cause analysis	X	X	X	X	X	X	X	X	X
Sequential software testing	X	X	X	X					
Service planning	X	X		X					
Statistical process control							X	X	X
System architecture analysis	X	X	X	X	X	X			X
Training and education	X	X		X				X	

Benefits: 1 = reduced outage and downtime experienced in the field; 2 = reduced Field Failure Rate and customer reported problems; 3 = reduced number of patches; 4 = improved customer perception of software quality/reliability; 5 = shortened development time; 6 = shortened test time; 7 = better, more informed gate and release decisions; 8 = improved process management, control, and predictability; 9 = increased speed to an industry leading position.

Benefits/Activities Matrix

A benefits/activities matrix provides an effective guide to establishing and maintaining network reliability. Activities that contribute to network quality are assessed and their benefits listed in an easy-to-read matrix (Table 1). The following 27 activities are included.

1. Certified code reuse. It is possible to estimate system reliability based on reuse of a specific collection of components, assuming that the reliability of the code is known.

2. Complexity analysis. This is the degree to which a system or component has a design or implementation that is difficult to understand and verify.

3. Customer surveys. Periodic surveys of existing customers can help identify such important issues as product failure rate and downtime.

4. Estimation of code size, cost, defect, and schedule. Accepted tools can provide project and program managers with estimates of these project components.

5. Failure modes, effects, and criticality analysis (FMECA). FMECA is a form of bottom-up analysis of an existing or proposed design in which malfunctions are postulated and their effects on the system are traced through the design.

6. Fault-tree analysis. In this technique, the known unacceptable situations or states of the system are each analyzed to determine how they can be prevented.

7. Fault insertion. Also known as fault or error seeding, fault insertion is the process of intentionally adding known faults to those already in a computer program in order to monitor the rate of detection and removal and to estimate the number of remaining faults.

8. Field performance measurements. Analysis of field performance data can be used to improve development and testing processes.

9. Field testing. This category includes any testing done in the field on software running in a live, in-service switch.

10. Incremental development. This is a software development technique in which requirements' definition, design, implementation, and testing occur in an overlapping, iterative (nonsequential) manner, resulting in incremental completion of the product.

11. Inspections and reviews. Product inspections and reviews during software development help to improve quality by detecting and removing defects.

12. Operational profiles. These provide a description of the expected use of a system in its intended environment. Using operational profiles while testing is the most efficient method of testing in that this increases reliability per unit execution time more rapidly than any other approach.

13. Process assessments. These activities emphasize meeting common quality and reliability improvement needs through quality surveys, audits, and analysis.

14. Procurement control. Although important to understand third-party vendor reliability controls, using such software can reduce costs and schedules of a software development cycle.

15. Reliability and reliability growth models. Reliability is the ability of a system or component to perform its required functions under stated conditions for a specified period of time. Reliability growth is the improvement in reliability that results from correction of faults.

16. Reliability data and reporting. Goals can be set – and data gathered – for reliability metrics that can assure significant improvement from a previous release.

17. Reliability objectives. Specific reliability objectives should be defined for each phase of development.

18. Reliability prediction. This exercise evaluates the ability of new software

to meet reliability requirements, extrapolates existing reliability data, and derives reliability data for a software-deliverable release unit.

19. Reliability tests. Testing should be defined for each stage of development to estimate initial failure intensities and other measurements.
20. Requirements' traceability. This is a set of mappings from the original requirements to the products of each phase in the development cycle.
21. Robustness strategy. A robustness strategy seeks to improve a system's functioning in the presence of invalid inputs or stressful environmental conditions.
22. Root cause analysis. A key defect-prevention process, root cause analysis provides a means to identify process improvement opportunities.
23. Sequential software testing. Testing with a preestablished sequence allows for testing to cease when product reliability either meets the target or is worse than expectations.
24. Service planning. Product cost and schedule estimates should include plans for preventive and corrective service of the product.
25. Statistical process control. As a process with a measured performance that is within statistically acceptable variations of a standard, statistical process control requires process goals, measures, and standards.
26. System architecture analysis. This calls for the development of a functional block diagram and system architecture in order to design in reliability and detect defects early.
27. Training and education. Any software reliability engineering effort should be supported by an extensive education and training program.

Protocols

Protocols provide the rules of interaction between different parts of a large system or across systems. As such, they are important elements of network function, and, ultimately, its reliability. Some current developments are specification languages, protocol validation, and conformance testing.

Specification Languages

Three specification languages have been developed by the International Telegraph and Telephone Consultative Committee (CCITT) and International Organization for Standardization (ISO). The System Description Language (SDL) was developed by CCITT and finalized in 1985. The Language of Temporal Ordering Specification (LOTOS) and Estelle were developed by ISO in 1989.

Protocol Validation

Validation in this sense means getting a "yes" or "no" answer. Model checking is more practical in that it is built in and posts and answers queries automatically. A large number of answers, therefore, can either provide confidence in the protocol or be used to construct counterexamples.

Conformance Testing

Conformance testing checks whether a given protocol implementation meets all requirements in its specification. There has been steady progress in the systematic generation of test sequences (15).

Toward A Customer Software Quality Model

In order to make concrete some of the ideas contained in this article, it is useful to take a brief look at software quality models developed for the system test, verification, and deployment phases of software development (2). These models are of importance on two counts: they were developed by one of the world's largest telecommunications software manufacturers, and they come from an industry that has written and deployed more critical software than any other. If there is a single industry that should be concerned with software reliability, it is telecommunications.

Software Reliability Model

The Software Reliability Model can be used to calculate the cumulative number of problems for any future in-service time. Problems can be forecasted for software loads with greater than 5% of their total expected in-service time using the logarithmic Poisson execution time (LPET) model, as follows:

$$\text{Cumulative Problems } (t) = \mu(t) = \beta_0 \times \text{Ln}((\beta_1 \times \text{Time}) + 1)$$

where $\beta_0 = 1/\text{slope}$ and $\beta_1 = \text{slope} \times \exp(\text{intercept})$.

The model was applied to forecast future regional problem arrivals based on two options: current field performance and future software loads with no actual field performance. The model application is driven by

- the failure intensity of the software at the time of release
- the expected peak deployment
- a marketing deployment forecast
- typical software product deployment profiles

The use of the Software Reliability Model is invaluable in making informed decisions as to when to release new software. The failure intensity is a measure used to forecast problem arrivals, and it can be used at the release decision point (as opposed to the Software Releasability Model, discussed below, which is used earlier in the process). Failure intensity values must meet a preestablished target. Metrics on key elements of the service delivery process are analyzed to help pinpoint difficulties. Good metrics can predict the customer perception-expectation gap to some degree (17). Forecasting problems for future software loads using the failure intensity at release is done using the same model (LPET).

As a result of the Software Reliability Model, the failure intensity metric has been used independently as an indicator of the releasability of the software.

Objectives defined for the failure intensity value are now part of the software release criteria.

Software Releasability Model

Once these objectives were established, then predicting the failure intensity at the planned release, or predicting when the target failure intensity would be reached, became a primary concern. This led to the development of the Software Releasability Model, which is used during the software verification phase. An effort to give some predictability to this metric produced excellent results. The failure intensity at release is one variable used to predict customer problems. One way to provide more control over resulting customer perception is to predict failure intensity at release during the testing process and thus predict customer problems by using the failure intensity value in the Software Reliability Model. The objective of the Software Releasability Model is to project final failure intensity and customer problems while testing is in progress.

Service Planning Model

The Service Planning Model has shown that required support levels can be forecast within acceptable accuracy ranges. This methodology could be easily expanded to other functions, products, and markets that have workloads on which software quality and deployment volumes have an impact. It can also be used to create a unified and balanced staffing plan to assure that customer-expected levels of service for the installed base of switching systems can be met.

The effort began by developing a process flow of software from design to the field. From this analysis, model elements were identified that would statistically project future requirements for service organization support. As a result of this effort, a sophisticated service planning model prototype was developed. This planning tool forecasts future problem arrivals based on software reliability delivered to the verification organization from design, forecasted number of features in each software load, and the number of deployed systems.

Any methodology developed to improve software quality must include the following benefits:

- increased product reliability based on improving release decisions
- improved integrated product verification, operations, and support planning
- greater control, measured product verification, and service support
- continuous, closed-loop resource planning
- reduced product risk to switching supplier and customer
- ability to measure productivity
- improved proactive and reactive decision making

Software Quality Technology Engineering

Bellcore has established the Software Quality Technology Engineering (SQTE) organization, which works with suppliers to ensure the long-term reliability and quality of software. At stake is not only the integrity of the network, but the bottom line of the telephone service providers. With the size of telephone switching software releases growing from 1 to 1.5 million lines, and with quality levels at approximately 2 faults per 1000 lines of code, the Bell Operating Companies are likely to see 1000 new problems, which could cost them about $6 million (18).

SQTE builds quality into the software system with a process approach, with three primary focuses. Analyzing the supplier's software quality program through a process called quality systems analysis establishes a baseline from which quality processes may be refined and developed. Implementing and using a quality measurement program allows objectives to be established, such as executing all test cases and passing 95% of them, demonstrating a decreasing rate of failure, and resolving all open major trouble reports. Stimulating and monitoring quality improvements helps to classify defects by where they originate and are detected. Software surveillance programs thus help to put effective quality systems in place while inspiring quality improvements, which in turn can yield dramatic financial savings to telephone companies (18).

Solutions for Network Reliability

Reliability in the Network at Large

One reason that networks are so vulnerable to outages is their interconnected nature. Although there are hundreds of telephone service providers, networks are so tightly interwoven that they essentially result in a single, integrated system. This fact underscores the need for reliability, both within an individual system and on a networkwide level, since the potential for wider and more serious outages increases as a network becomes more connected (19).

New network components are based on a set of criteria that reflect this virtually seamless national and international network. These are (19)

- the ability to introduce new capabilities and services over large geographic areas
- modular construction allowing network components to evolve cost effectively at different rates
- flexible architecture prolonging the network's life and avoiding technical dead ends
- the ability to share resources among many network nodes
- the highest achievable level of reliability

Since digital fiber systems provide virtually unlimited capacity, the emphasis on network components to direct calls and identify high-usage routes has diminished. Today, the costs of installing equipment and switching calls are more important than straight equipment costs. Common channel signaling — the ability of network nodes to communicate over dedicated channels without the inefficiencies of in-band communications — has led to the development of intelligent networks (INs), which require several nodes acting in concert to complete rudimentary call-processing functions. The resulting network architecture consists of a single computer with different subsystems performing specialized functions (19).

Not unlike the computer industry, however, telephone networks must look beyond a network topology that centralizes intelligence and becomes vulnerable to unreliability. Today, it is unthinkable to have a business run all of its word-processing systems from one central mainframe; if one system fails, everyone's terminal goes down. As a result, the computer industry has concluded that, in general, multiple microprocessors are better than a single central processing unit (CPU) for reliability. The telephone industry is putting Open Network Architecture (ONA) in place, which requires even simple functions to have a number of network nodes to complete a process, and Bellcore has identified the importance of operational support systems in its advanced intelligent network (AIN) plan (19).

Addressing Network Reliability

Two steps have been proposed toward achieving network reliability. The first is to develop a network interconnection plan that ensures enough physical paths to avoid network congestion and potential outages due to system overload. The second step uses the screening capability inherent in SS7 nodes to build "fire walls" to prevent individual problems from rippling though the network at large. By challenging fundamental network planning philosophies, however, there are a number of ways that telephone networks can regain network reliability (19):

- treat each node as a foreign element with the potential of letting errors cascade through the network
- ensure diverse physical and logical routes to each node (fiber-ring structures are good examples)
- require survivability mechanisms as part of the network evolution plan
- avoid having too many generations of technology in a single network (lower costs and improved switching and transmission capabilities enhance options)
- avoid using the lack of industry standards as an excuse for poor reliability performance

Bidirectional SONET rings also provide inherent network survivability. If a fiber line on a SONET ring is severed, traffic is automatically rerouted to reach all central offices on the ring through alternate paths. To enhance survivability further, telephone companies can duplicate links between adjoining SONET rings. Two central offices can serve as junctions between the rings; if there is a failure in one central office, traffic between the rings can automatically be redirected through the second. SONET provides flexible, dynamic bandwidth management, enabling the efficient transmission of services with bandwidth requirements ranging from narrowband to broadband. SONET technology also permits operation, administration, maintenance, and provisioning (OAM&P) to be extensively automated and integrated (20).

The Telecommunications Management Network

A vital element in ensuring reliability across telecommunications networks is OAM&P. The Telecommunications Management Network (TMN) is the internationally accepted concept for managing the dynamic networks that will provide new advanced services. A key component of the TMN architecture, and the focus of extensive work in industry forums and standards bodies, is the element management layer (EML). This layer manages subnetworks of network elements and the particular technologies within those elements (20).

The EML mediates between subnetworks and the high-level operations systems that manage services and network resources on a networkwide basis. The layer shields the high-level operations systems from the necessity of managing the technology details of network elements. It also masks from the operations systems differences in technologies from various equipment suppliers. As well, the EML consolidates and analyzes operational data and then communicates streamlined information to the operations systems. Relieved of the task of managing implementation details within network elements, the operations systems can concentrate on the effective and responsive management of the network as a whole, as well as the services it provides to users (21).

The TMN architecture enables new services to be introduced rapidly. Services are no longer tightly coupled to the technologies that implement them. High-level operations systems deal with generic network resources, combining these resources as needed to create new services. TMN's streamlining of OAM&P-related functions and communications reduces the possibility for incongruities between databases, increasing service reliability. Finally, TMN's distributed OAM&P functionality opens the potential for users to change and manipulate their service portfolios directly (20).

Logic in the Intelligent Network

The hallmark of the intelligent network is the rapid, low-cost introduction and customization of services. This requires service functionality that can readily and reliably be reconfigured to create advanced services. Flexible logic has been implemented within the intelligent network concepts by international standards bodies. Data-driven logic reduces the time frame within which services can be created from service-independent building blocks, the basic units of functionality that are combined to build IN services. Flexible logic, in conjunction with IN service-creation concepts, enables telecommunications operating companies to deploy differentiated, competitive services rapidly, reliably, and at lower cost. Companies can also readily customize services to the needs and preferences of particular subscribers (20).

The service-rich era that is unfolding will impose sophisticated challenges for switching software. The software must readily accommodate new and evolving services. As well, it must enable the user to invoke multiple services during a single call. A generic services framework enables the ongoing and rapid deployment of new intelligent services and features while ensuring that services perform reliably for users. This framework speeds service development though software reuse, and automated tools within it recognize, catalog, and manage reusable software elements. Other tools aid designers in obtaining software from inventory and integrating it to create new services and features rapidly. As well, a prototyping tool lets designers quickly verify designs (20).

Network Congestion Control and Reliability

Network congestion control and reliability are important issues in real-time communications networks. Functions that restrict new traffic from flowing into an already congested area, reroute new calls, and provide information to callers about the nature of the congestion must be built into the design of the system. Running concurrent software programs—the equivalent of duplicating hardware components—does little to ensure the reliability of the program itself. The program is either reliable or it is not. Unreliable software—software that produces errors—often results in network congestion. Congestion produced by hardware is due to the failure of a hardware component, which reduces the number of resources in a network. Software congestion, on the other hand, is caused by a decreased efficiency with which network resources are used. Congestion-controlling mechanisms have proven effective in dealing with hardware-related congestion; software is another matter (22).

There has been some progress made in the containing of software congestion within packet-switched networks. In low-speed packet networks, for example, flow control mechanisms control the amount of information entering the network by setting limits on the amount of information that can be sent from a source to a given destination at any one time. In high-speed networks, which differ from low-speed networks in having minimum bandwidth requirements and higher speeds, flow control is impractical. Rather, algorithms have been developed that proactively allocate resources rather than reactively throttling the sources sending the data (22).

Sometimes, however, despite the best efforts to control congestion, its cause can be traced to software problems. Auditing software through automatic control of software errors may enable the identification of an error before it causes congestion or system collapse. While still more art than science, audits play a crucial role in monitoring and discovering potentially catastrophic software errors. As a last resort, the best way to deal with an outage may simply be to reduce its duration and reinitialize the network manually and quickly (22).

Hardware reliability may be increased by deploying duplicate components throughout the network. Duplicating software, however, would also cause duplication of software faults. One approach toward software reliability, then, is to develop multiple versions of a software system. Applying diversity as early as possible in the software life-cycle reduces the probability of faults and errors. Another approach is through the partition analysis method, which increases program testing and verification effectiveness by using smaller pieces of code.

In addition, reliable software can be achieved through careful architectural design and elimination of faults in the components that have the most significant impact on the system reliability—those used most frequently or those connected architecturally to many other components, thus affecting their reliability (1).

Network Reliability Council

The Network Reliability Council (NRC) meets on a regular basis to review overall network performance of the public telecommunications network and recommends policies and procedures to ensure its integrity. The council appoints task groups to investigate technological advances, business/customer demands, interconnectivity issues, and rising competition, all of which can affect telecommunications. The NRC presently has five task groups; these track network reliability performance, increased interconnection, reliability concerns resulting from changing technologies, essential communications during emergencies, and telecommuting as a backup as a result of disasters.

The NRC reports to the Network Reliability Steering Committee (NRSC), which is sponsored by the Alliance for Telecommunications Industry Solutions, Incorporated (ATIS). The NRSC is a concensus-based industry committee composed of exchange carriers, interexchange carriers, manufacturers, major users, the Communications Workers of America (CWA) union, the National Association of Regulatory Utility Commissioners (NARUC), cellular service providers, and alternative access providers. The NRSC analyzes reports of network outages to identify trends and distributes its findings throughout the telecommunications industry.

Service Quality Measurement

Software service quality measurements quantify customer perceptions of and expectations about delivered services during the postdevelopment phase of the software life-cycle (23). The supplier provides software services, and the customer then uses the services and develops a perception of their quality. Measurements can be made of customer experiences and perceptions and of supplier deliveries. This information can be fed back into the process to improve it. Understanding why the measurement is to be made and how the information will be used is critical to identifying the causes of, and improving, service quality problems.

The software service quality measurement process is, therefore, a management tool that can be used to improve postdevelopment software quality from the customer's perspective. It requires management and corporate commitment: "Materially improving service quality is a long term, multifaceted task. To approach this task on a short-term unidimensional basis is to invite failure" (24). To be effective, a good quality measurement system must be simple, accurate, easy to use (25), make sense in the application, and be mathematically sound. Customer expectations and the software services delivered should be well defined to allow measurement of the key gaps in the service delivery process that make up customer satisfaction. The measurement process should be sensitive to the forces that cause customer expectations to change over time. The implementers of the measurement process should realize that each contact with a customer is an opportunity for that customer to experience the quality of the service and an opportunity for the supplier to create repeat business (26). Software service quality measurement can be viewed as a way to predict how much repeat business a supplier is likely to receive.

Measurement Points in Switching-System Software Service

Conceptual Model of Service Quality

The conceptual model of service quality starts with customer expectations about the service. These expectations are influenced by the customer's past experiences, personal needs, and word-of-mouth communications with other customers. The supplier's process of service delivery to meet these expectations starts with management's perceptions of customer needs. Measuring Gap 1 identifies the difference between management's perception of customer expectations and actual customer expectations. The process then moves to the marketer's translation of these perceptions into service quality specifications and, finally, to the service delivery itself. Measuring Gaps 2 and 3 results in a representation of the difference between management perception of customer expectations and the service delivered. What the marketer communicates to the customer about the service delivery is also a part of this process. Measurement at Gap 4 characterizes the difference between the service quality delivered and what the marketer communicates to the customer about the delivered service. The "moment of truth" occurs because, when the service is delivered, the customer sees its delivery. Gap 5 shows the difference between customer expectations and perceptions.

Switching Software Service Quality Measurement Model

The switching software service quality measurement model, as applied to the switching software environment, is shown in Fig. 2. The supplier side remains basically the same as depicted in the conceptual model, but the customer side divides into two sections, one for the telephone operating company and one for end users (in this context, large organizations that contract with the telephone operating company to use software-based features). The telephone operating company's expectations are influenced by previous experiences and corporate or personal needs. These expectations are also influenced by communications with other telephone operating companies in industry forums, with Bellcore, and with regulatory bodies (such as the U.S. Federal Communications Commission [FCC] and public utilities commissions). Furthermore, the fact that telephone operating companies do business with more than one major supplier and are

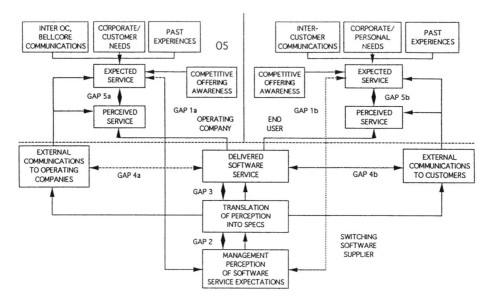

FIG. 2 Switching software service quality model (OC = operating company).

quite aware of what the competition offers is an additional, and quite significant, factor in their service expectations. Similarly, end-user expectations are
influenced, though to a lesser extent, by intercustomer communications and
awareness of competitive offerings. The formation and growth of end-user
forums, such as the National Centrex Users Group, will inevitably increase the
influence of intercustomer communications.

Although Fig. 2 shows the customer as a single box, this does not imply
similarity in expectations among customers. Customers are not a homogeneous
group. Telecommunications customers' expectations do not change significantly
by any demographic grouping, including geographic region, industry, or internal staffing characteristics (20). However, research shows that end-user expectations change dramatically according to their business situation. Five very different groupings have been identified (6):

1. Not critical. Businesses involving primarily voice-only users.
2. Critical. Businesses in which telecommunications is perceived as a competi-
    tive advantage and both voice and data service are required.
3. Need help. Companies in their start-up stages that need outside support and
    expertise on basics.
4. Early adapters. Companies with in-house expertise and that have a high
    degree of maturity.
5. Technology. Typically, established companies in highly competitive mar-
    kets that look for advanced and innovative suppliers.

The study discussed above concluded that differences among customers are
closely related to factors such as maturity and rate of growth of the company,

the nature and competitiveness of their business, and the amount of in-house expertise they had in the area of telecommunications.

Figure 2 identifies the "gaps" to be measured, as in the conceptual model. Gaps 1a and 1b measure the difference between telephone operating company/end-user expectations and the supplier's perception of these expectations. Gap 2 measures how well the supplier translates management's perceptions of software service expectations into service specifications. Gap 3 measures the difference between these specifications and the service actually delivered. Gaps 4a and 4b represent the difference between the quality of the software services delivered and what the supplier communicates to the telephone operating companies and end users about service delivery. The paramount measures, 5a and 5b, are the deltas between customer perceptions and expectations. These gaps depend entirely on Gaps 1–4. As in the conceptual model, Gap 5 is the difference between consumer expectations and perceptions, which in turn depends on the size and direction of the four gaps associated with the delivery of service on the marketers side (27). In short, what is measured is the size and direction of Gaps 1–5.

The term *gap costs*, then, refers to the costs associated with using a product of benchmark reliability as opposed to one of lesser reliability. After these costs have been calculated for all products in a given category, standardization managers can analyze the report and make their decisions by (28)

- weighing long-term maintenance costs associated with product selection
- choosing optimum product functionality versus costs for a given application
- visualizing the long-term costs of underbuilding or overbuilding a system
- determining which suppliers should be retained as primary and secondary product sources
- determining when it is economical to replace installed products with newer equipment
- deciding if positive aspects of a product such as system availability, shipment timeliness, or repair turnaround offset higher annual maintenance costs, rather than using another product altogether

Software Services Delivered

Software Services

Specific examples of a customer-service focus include the following:

1. Delivery and application of new software releases to provide the customer with a new software generic that contains the base program, new features, and central-office engineering parameters that are unique to each installation.
2. Delivery and application of software fixes to provide rapid and accurate resolution of customer software problems.
3. Early in-service performance to offer the correct performance of software and the close attention to the occurrence and severity of software problems from the first instant of operation following installation.

4. Handling problems in the field by ensuring technical assistance is designed to resolve quickly all issues and problems encountered with the software.
5. Training and documentation to teach customers about the operation, maintenance, and support of the software in the supplier's training facility or at the customer's site.
6. Customer testing and technical interactions to provide customer access to testing at the supplier's site and to conduct formal technical information exchanges, such as joint testing efforts, technical services symposiums, and information exchange forums.
7. Customer demonstrations and inspections to demonstrate to customers how new software features work by allowing them to examine and tour the facilities where software is generated and software services are managed.

These are the major points of customer encounters during the service portion of the software life-cycle. Within each one of these there are thousands of "moments of truth," occasions for the customer to experience and witness the supplier's software service quality. The measurement program focuses on these service delivery processes, and achieving excellence in each area is necessary if a supplier wishes to claim total service excellence.

Software Services Reliability Model

Software service quality should be perceived as a chain of customer encounters in which the quality of the supplier's services will be seen as no better than the worst event. Software service quality measurement summaries should consider this. This is an area for which averages can give supplier management a false sense of security. If a supplier strives to achieve total service excellence, each element must achieve total service excellence and the measurement summary statistics should be consistent with this idea. As an example, a supplier may excel in 8 of 10 major customer encounters by doing exactly what the customer expects. However, in 2 of 10 service encounters, the supplier is not doing what the customer expects and, yet, because the average is maintained, no cause for alarm or action is signaled. Measurement sensitivity is a serious consideration to prevent inappropriate action.

Assessment Methods

Many measurements are needed to capture continuously a complete and accurate picture of customer expectations and perceptions. When assessing gaps involving different units and scales, it is necessary to use the most appropriate method. The use of multiple approaches to measure service quality is advocated by Berry: "A portfolio of research approaches compensates for the limitations of any one methodology and offers richer and more comprehensive insights into what is really going on—and why" (29). Some methods include

1. Surveys of telephone operating company and end-user expectations and perceptions. These measure customer expectations of the service and what they perceive it to be presently; they must be ongoing and in real time. The method includes focus group and executive interviews. Used to measure Gap 5 and contributes to measurement of Gap 1.

2. Telecommunications market surveys. Surveys measure expectations and perceptions of noncustomers, perception of the competition, and contribute to measurement of Gap 1.

3. Problem solicitation. Problems reported by telephone operating companies and end users help identify areas needing improvement and contribute to measurement of Gaps 1 and 4.

4. Employee surveys. Employee input is invaluable for identifying areas requiring improvement. Each customer-service employee participated in information gathering for measuring Gaps 1, 2, 3, and 4.

5. Management–customer encounters. Supplier management meetings with customers provide an opportunity to gain information that can be used to improve the software service delivery process. Used to measure Gaps 1, 4, and 5.

6. Internal process metrics. Metrics on key elements of the focused service delivery processes are analyzed to help pinpoint difficulties. Used to measure Gaps 2 and 3.

7. Customer report cards. Ratings by customers help to identify the degree to which expectations have been met. Used to measure Gap 5 and contributes to measurement of Gap 1.

Management and Measurement

For measurement processes to thrive, they should be part of a proactive view of software service that creates an environment in which customer input and opportunities to improve are solicited. A customer-oriented measurement philosophy demanded by top management is an important element of the software service quality measurement process. Quality improvement efforts can succeed only when the direction is clear, consistent, and supported by the actions of top management (30). This is an example of a proactive management–customer encounter method of measurement.

Perceived customer satisfaction can only be measured by interviewing the customer, which measures Gap 5. Since it is impractical to do this for every service delivery, internal metrics are used to predict customer satisfaction, which measures Gap 3. As a result of customer survey analyses and management–customer encounters, a management goal and measurement philosophy is specified for each of the service delivery areas. These goals are set to achieve customer satisfaction.

Suppliers use myriad software service quality measures to try to predict customer satisfaction. Some of these are defined in the Bell Communications Research document, *Reliability and Quality Measurements for Telecommunications Systems* (*RQMS*), (GR-929-CORE) (31).

With end users requiring optimized system costs, rapid availability of new software features, and a high operational quality level, understanding and measuring software service quality may be a prerequisite for the success of telecommunications suppliers. A software quality metrics revolution is just beginning. This revolution will see emphasis in software engineering shift from "developer-oriented quality measures to customer-oriented or user-oriented ones" (10). The concepts of software service quality measurement will have an important role to play in this revolution.

Establishing Metrics For Determining Reliability

All telecommunications network switching system suppliers attempt, with varying degrees of success, to provide excellent software and excellent customer service for that software. To achieve this, a measurement process must be in place to gauge progress toward a zero-defect goal and to indicate when corrective action must be taken to keep the process on track. A proactive measurement strategy integrated into the service delivery process provides many benefits beyond the mere collection and analysis of data. In the telecommunications switching environment, quality measurement is a strategic tool that can help increase a supplier's competitiveness. It can be used to measure gaps in the software service delivery process and explain environment dynamics and the composition of customer satisfaction.

While there are hundreds of quality and performance metrics that can be used to track and report the performance of a network, it is clearly not feasible, if not impossible, to use every one of them. A set of "managed metrics" has been identified, chosen for their importance to customers. To qualify, a metric must have been selected by customers through one of the following mechanisms:

- National Electronic Systems Assistance Center (NESAC) managers focus set of RQMS metrics
- Software Quality Conference (SQC) (subcommittee of the Reliability Review Forum) set of focus metrics
- customer-specific "executive" metrics

Table 2 below lists the managed metrics and the reasons for their selection.

Bellcore has defined the following field performance metrics and requirements. These metrics are part of various customer performance monitoring vehicles including RQMS, Local Switching System Generic Requirements (LSSGR), Software Quality Conference, National ESAC, and Cost of Poor Quality. In some cases, specific metrics may be reported in more than one of these vehicles.

Bellcore's Reliability and Quality Measurements for Telecommunications Systems (RQMS) defines 32 metrics that are used to track system performance, hardware performance, software performance, firmware performance, and compliance with RQMS.

As to system performance measurements, system outage performance is

TABLE 2 Managed Metrics

Metric	RQMS	NESAC	SQC	Customer Specific
Host total outage frequency	X	X	X	X
Host total outage downtime	X	X	X	X
Host partial outage frequency	X			X
Host partial outage downtime	X			X
FCC-reportable outages			X	
Outage distribution	X	X		
Patch volume	X	X	X	
Defective patches	X	X	X	
Patch propagation delays	X	X	X	
Software faults	X	X	X	
Problem reports by severity	X	X	X	X
Fix response time	X	X	X	X
Fault fix history	X	X		
Retrofit application aborts	X	X	X	X
Application reports	X	X	X	
Product predictability				X
Development interval				X
Transparency metric				X

measured in terms of frequency (incidents per system per year) and duration (minutes per system per year). Total outage frequency is the number of incidents per system per year for total (100%) call-processing outages. Only outages affecting 100% of call processing for the entire system are counted in this metric. Outages are categorized as follows:

- host total outage downtime — number of outage minutes per system per year for call-processing outages (100% lines or 100%) greater than 30 seconds
- host total outage frequency — number of outage minutes per system per year for partial call-processing outages
- host partial outage frequency — number of incidents per system per year for partial call-processing outages
- host partial outage downtime — number of outage minutes per system per year for partial call-processing outages
- remote total system outage performance — measured in terms of frequency (incidents per system per year) and duration (minutes per system per year) of outages in remote systems
- remote partial system outage performance — measured in terms of frequency (incidents per system per year) and duration (minutes per system per year) of outages in remote systems

Table 3 shows the expected LSSGR requirements for suppliers to meet by the fourth quarter of 1995.

TABLE 3   LSSGR Outage Requirements to be Met by the End of 1995

		Frequency	Downtime
Host			
Total outage			
Scheduled	Supplier attr.	1.0/sys/yr	1.0 min/sys/yr
Unscheduled	Overall	0.1/sys/yr	2.0 min/sys/yr
	Supplier attr.	.075/sys/yr	1.0 min/sys/yr
Partial outage	Overall	0.4/sys/yr	3.0 min/sys/yr
	Supplier attr.	0.3/sys/yr	1.5 min/sys/yr
CCS Isolation	Overall	0.1/sys/yr	3.0 min/sys/yr
	Supplier attr.	.075/sys/yr	1.5 min/sys/yr
Remote			
Total outage			
Scheduled	Supplier attr.	1.0/sys/yr	1.0 min/sys/yr
Unscheduled	Overall	0.1/sys/yr	2.0 min/sys/yr
	Supplier attr.	.075/sys/yr	1.0 min/sys/yr
Partial outage	Overall	0.4/sys/yr	3.0 min/sys/yr
	Supplier attr.	0.3/sys/yr	1.5 min/sys/yr
Other			
Total outage duration (100% + recovery)		Graph below	
Total outage recovery		<20 minutes	
Total ISDN circuit switch capability		1.0 min/sys/yr	
Total ISDN packet switch capability		1.0 min/sys/yr	
Control capability		3.0 min/sys/yr	
Visibility and diagnostic capability		3.0 min/sys/yr	

The LSSGR supplier-attributable scheduled outage requirements will tighten even more through the end of the century (Table 4).

Release application performance measures any service degradation or abort occurring during an upgrade of an existing system to a new software release. For each release, the percentage by month is computed on a cumulative basis over the data for all previous months.

- Retrofit application abort—defined as an application attempt that does not switch activity to and remain on the new software load

TABLE 4   Supplier-Attributable Scheduled Outage Requirements (LSSGR)

	Frequency	Downtime
Year end 1995	1.0 inc/sys/yr	1.0 min/sys/yr
Year end 1997	0.0 inc/sys/yr	0.0 min/sys/yr

- Retrofit application problems — defined as any service degradation or abort occurring during an upgrade of an existing system to a new software release

Table 5 shows the RQMS requirements for release application performance. Problem reports are received from in-service sites and are measured in number of reports per system per month.

- Total problem reports — as defined above
- Problem reports by severity level — include all problems regardless of cause (including duplicates) received from a customer

Product change notices (PCNs) are measured as the number of product change notices (PCNs) released along with sites and circuit packs affected.

Customer-service response times define the fix response times for software problem reports and are measured as the time from the opening of a problem report by the customer until delivery of a software fix. Fix response time is provided separately for critical, major, and minor problems.

Measurement of patches includes

- Released patches are measured in terms of the number delivered to the vendor cumulatively over the life of the release.
- Defective patches are measured as the percentage of patches that were released and later found to be defective.
- Additional patch measurements include the number of patches not propagated within two weeks and the percentage of patches that introduce new features.

As to software faults,

- Actual faults and predictions are measured as the number of software code corrections, including patches, applicable to the market. Fault prediction is done using a software reliability model. The prediction error rate should be less than 10%.
- Software faults by severity level are calculated based on problem priority.

TABLE 5 Reliability and Quality Measurements for Telecommunications Systems Release Application Requirements

		Year End 1995
Software application	Problems	5.0%
	Aborts	1.0%

TABLE 6   Other Reliability and Quality Measurements
              for Telecommunications Systems
              Software Requirements

Metric		Requirement
Problem reports	Critical	0.03/sys/mo
	Major	0.05/sys/mo
	Minor	0.14/sys/mo
	Total	0.22/sys/mo
Software faults	Critical	30/rel
	Major	200/rel
	Minor	570/rel
Patches	Volume	600/rel
	Defective	0.5%/rel
Problem response time	Critical	24 hours
	Major	30 days
	Minor	180 days

- Cumulative fault density data is normalized as faults per million new and changed noncommentary source lines of code.

Table 6 shows the RQMS requirements for the remaining software metrics.

Hardware performance measurements include circuit pack returns, which are measured as the volume of returns per thousand equipped lines and trunks for in-service sites. Firmware performance measurements include firmware changes, which are measured by the number of packs with executable firmware requiring physical replacement or update and the percentage of codes changed when upgrading software.

Cost of poor quality (COPQ) measurements standardize the data-collection and analysis methods for determining ongoing product ownership costs. One of the COPQ systems is derived from RQMS to quantify—in monetary terms—the negative impact a customer suffers from less-than-perfect quality. Achieving favorable COPQ ratings, therefore, provides not only more reliable networks, but greater incentives for vendors to improve the quality of their products.

Software COPQ components include

- problem discovery—the cost of identifying software faults
- unisolated faults—The cost to clear errors caused by residual software faults
- patching—the cost to administer and apply patches to correct software faults
- defective patch—the cost to remove bad patches
- outages—the cost associated with lost revenue and system recovery due to outages
- load failure—the cost to recover from a failed software application

- image failure — the cost to redo tape image of office database
- problem reporting — the cost of handling subscriber reports

Hardware COPQ components include

- hardware diagnosis — the cost of identifying hardware faults
- handling and shipping — the labor costs to replace defective hardware
- material repair — the unit repair cost to repair defective hardware
- maintenance spares — the cost to maintain a ready supply of replacement hardware
- change notice — the cost associated with the administration, analysis, and implementation of hardware change notifications
- software change — the cost to implement a software change
- routine maintenance — the costs to perform ongoing administrative and routine maintenance on the system

The switching-system life-cycle costs (SSLCC) model tracks five phases in the life-cycle of a switching system: acquisition costs, installation costs, operation costs, growth costs, and retirement costs. SSLCC is a new metric that is still evolving.

FCC metrics define an upper limit of 0.05 FCC-reportable outages per system per year. An FCC reportable outage is one that lasts 30 minutes or longer, and one that affects 30,000 or more lines.

Software Quality Conference (SQC) measurements focus on nine metrics and have been identified by the SQC, which is a subcommittee of the Reliability Review Forum. All except the first are based on RQMS definitions:

- FCC reportable outages
- outage downtime
- outage frequency
- application aborts
- problem report fix response time
- problem report volume
- defective patches
- software faults
- patch volume

National ESAC Conference measurements focus on the following 13 RQMS metrics:

1. Problem reports
2. Outage downtime
3. Outage frequency
4. Customer-service response time

5. Patches
6. Defective patches
7. Patches not propagated within two weeks
8. Application aborts
9. Application problems
10. Fault fix history
11. Software faults
12. Fix response time
13. Outage distribution

Customer-service metrics include other quality performance metrics that are important to customer satisfaction. These are

- Application out-of-sync time
- Application time
- Peripheral loading interval
- Recovery time
- Obsolete patch volume
- Disruptive patch volume

Conclusion

Reliability is the probability of failure-free operation for a specified period in a defined operational environment. Software reliability engineering is the applied science of planning, predicting, measuring, and managing the reliability of software-based systems to maximize customer satisfaction. Software reliability engineering helps a supplier gain a competitive edge by satisfying customer needs more precisely and thus more efficiently. Implementing a software reliability program is also a way to increase return on investment for a corporation.

In order to increase network reliability, a product must be managed for reliability. Good reliability management requires a plan that clearly defines the reliability objectives, activities to be performed, and measurement and performance expectations. Reliability performance requirements should be spelled out in detail, and specific reliability-related activities designated to the appropriate parties. A good software reliability program takes a holistic life-cycle approach to reliability management. It stresses reliability in the design of the product and places an emphasis on analysis activities early in the life-cycle. It maximizes the use of known and proven techniques, technologies, and tools, provides the framework to manage a product for reliability, and is quantitative and auditable. In short, it takes advantage of every known means to design reliability into the product and anticipate problems before they have the opportunity to become manifest in the finished product.

A well-planned and well-implemented reliability program will yield a number of significant benefits, which can include reduced outage and downtime in the field, reduced field failure rate and customer-reported problems, and improved customer perception of software quality and reliability. Reduced development time, test time, and text/fix intervals are also results often realized from a quality program. In general, companies can expect more informed gate and release decisions, and improved process management, control, and predictability when they make quality and reliability inherent aspects of the product planning, development, and manufacturing processes.

As the hardware and architectural infrastructures of telecommunications networks become increasingly reliable, it will be up to software engineers and quality control professionals to devise ever more stringent and ingenious methods by which the development and implementation of software will yield a high-quality, reliable product. While there is not a single answer for software reliability problems, continued work in all of the areas just mentioned will clarify the issue and, if nothing else, increase the efficacy of current methodologies.

References

1. Hac, A., Improving Reliability Through Architecture Partitioning in Telecommunications Networks, *IEEE J. Sel. Areas Commun.*, 12(1):193–204 (January 1994).
2. Hudepohl, J., Snipes, W., Hollack, T., and Jones, W., A Methodology to Improve Switching System Software Service Quality and Reliability, *IEEE Global Telecommunications Conference* (December 1992).
3. Berry, L. L., Parasuraman, A., and Zeithami, V. A., The Service Quality Puzzle, *Business Horizons*, 31(5) (September–October 1988).
4. Falkner, R. G., Get The Bugs Out, *Service News* (November 1989).
5. Blume, E. R., Giving Companies the Competitive Edge, *Training and Development J.* (September 1988).
6. Bultman, C., How to Define Customer Needs and Expectations, paper presented at Customer Satisfaction Measurement Conference, Atlanta, GA, February 1989.
7. Eastern Management Group, *Central Office Switching and Central Office Adjuncts*, Eastern Management Group, Market Report, Parsippany, NJ, May 1988.
8. Parasuraman, A., Zeithami, V. A., and Berry, L. L., A Conceptual Model of Service Quality and Its Implications for Future Research, *J. Marketing*, 49 (Fall 1985).
9. Buzzell, R. D., and Gale, B. T., *The PIMS Principles: Linking Strategy to Performance*, Free Press, New York, 1989.
10. Musa, J. D., Faults, Failures, and a Metrics Revolution, *IEEE Software* (March 1989).
11. Parasuraman, A., Zeithami, V. A., and Berry, L. L., SERVQUAL: A Multiple-Item Scale for Measuring Consumer Perceptions of Service Quality, *J. Retailing*, 64(1) (Spring 1988).
12. LoSardo, M. M., Measuring the Quality of Services: One Company's Experience, paper presented at Measurement Research and Operations Symposium, Nashville, TN, September 1988.
13. Zeithami, V. A., Service Quality Management, course presented at Northern Telecom, RTP, NC, November 1989.

14. Zeithami, V. A., The Concept of Perceived Quality, paper presented at Measurement Research and Operations Symposium, Nashville, TN, September 1988.

15. Chung, F. R. K., Reliable Software and Communication I: An Overview, *IEEE J. Sel. Areas Commun.*, 12(1):193–204 (January 1994).

16. Erickson, R. L., Griffith, N. D., Yai, M. L., and Wang, S. Y., Software Architecture Review for Telecommunications Software Improvement, IEEE 0-7803-0950, 1993.

17. Hudepohl, J., Measurement of Software Service Quality for Large Telecommunications Systems, *IEEE J. Sel. Areas Commun.*, 8(2) (February 1990).

18. Pence, J. L., and Hon, S. E., III, Building Software Quality into Telecommunications Network Systems, *Quality Progress* (October 1993).

19. Pecoraro, A. V., and Rendall, D. S., Progress and Network Reliability Present Growing Concern, *Telephone Engineer and Management*, 97(9):54J–54O (May 1, 1993).

20. Slack, D. (ed.), Planning for Advanced Services, *FocusOn*, internally generated BNR customer publication, 1995.

21. Anderson, E. C., Goett, J., Evans, D., Mydosh, R., and Aidarous, S., The Role of the Element Management Layer in Network Management, paper presented at the Network Operations and Management Symposium '94, Orlando, FL, February 14–17, 1994.

22. Coan, B. A., and Heyman, D., Reliable Software and Communication III: Congestion Control and Network Reliability, IEEE 0733-8716, 1994.

23. Hudepohl, J., Software Service Quality Measurement for Switching Systems, 13th Intl. Switching Symposium Proc., 4(C5) (1990).

24. Berry, L. L., Zeithami, V. A., and Parasuraman, A., Quality Counts in Services Too, *Business Horizons*, 44–52 (May–June 1985).

25. Brown, M. G., Bridging the Quality Gap, *Quality* (June 1987).

26. Wagel, W. H., Feldman, D., Fritz, N. R., and Blocklyn, P. L., Quality—The Bottom Line, *Personnel* (July 1988).

27. Zeithaml, V. A., Berry, L. L., and Parasuraman, A., Communication and Control Processes in the Delivery of Service Quality, *J. Marketing*, 51:35–48 (April 1988).

28. Scott, T., Assessing Network Product Reliability: The Reliability Gap Cost, *Telephony* (April 19, 1993).

29. Berry, L. L., Multiple Measures Reflect True Quality of Service, *American Banker*, 153(48) (March 10, 1988).

30. Leonard, F. S., The Case of the Quality Crusader, *Harvard Business Review*, 12–20 (May–June 1988).

31. Bellcore, *Reliability and Quality Measurements for Telecommunications Systems (RQMS)*, GR-929-CORE, Bellcore, Piscataway, NJ, 1994.

JOHN P. HUDEPOHL

Networked Distributed Virtual Reality: Applications for Education, Entertainment, and Industry

Introduction

Case was twenty-four. At twenty-two, he'd been a cowboy, a rustler, one of the best in the Sprawl. He'd been trained by the best. . . . He'd operated on an almost permanent adrenaline high, a byproduct of youth and proficiency, jacked into a custom cyberspace deck that projected his disembodied consciousness into the consensual hallucination that was the matrix.[1]

Cyberspace is a globally networked, computer sustained, computer generated, multidimensional, artificial, or virtual reality. In this world, onto which every computer screen is a window, actual, geographical distance is irrelevant. Objects seen or heard are neither physical, nor, necessarily, representations of physical objects, but are rather—in form, character, and action—made up of data, of pure information. This information is derived in part from the operations of the natural, physical world, but is derived primarily from the immense traffic of symbolic information, images, sounds, and people, that constitute human enterprise in science, art, business, and culture.[2]

We foresee that computing environments in the next decade will be very widely distributed, ubiquitous, open-ended, and ever changing. All the computers in the world will be mutually connected. New services will be added from time to time, while old services will be replaced. New computers will be connected, and the network topology and capacity will be changing almost continually. Users will demand the same interface to the environment regardless of login sites. Users will move with computers and will move even while using them. Users will also demand much better user interfaces, so that they will be able to communicate with computers as if they are communicating with humans.[3]

This article addresses virtual reality (VR) and how can it be networked to support multiple-user immersion environments, joined over long distances. The sites are networked using modem-to-modem telephone lines, the Internet, and high-bandwidth telecommunications. The major contribution is a discussion of the networked virtual reality projects produced at the STUDIO for Creative Inquiry, Carnegie Mellon University (CMU; Pittsburgh, PA). The project team has designed and constructed the Networked Virtual Art Museum, an art museum that joins telecommunications and virtual reality. Other networked virtual reality applications are in various stages of construction, including a virtual city, a virtual design and teleconferencing station, the virtual showroom, and a virtual test track, among other educational and industrial projects. Included is an international survey of networked virtual reality developers and their most noted applications. The conclusion forecasts a not-so-distant future in which education, entertainment, and industry will employ networked immersion environments.

The promise of virtual reality[4] has captured our imagination; networks will render it accessible. There can be little doubt that networked immersion environments, cyberspace, artificial or virtual reality, or whatever you want to call it will evolve into one of the greatest ventures ever to come forward. Virtual reality will draw from and affect the entire spectrum of culture, science, and commerce, including education, entertainment, and industry. It will be multinational and will introduce new hybrids of experience for which descriptors presently do not exist.

Gibson, Benedikt, and Tokoro are cited above. At first reading, they might appear to be divergent tracks, but welding them together contributes significantly toward the framing of a "matrix," a "computer-sustained, computer-generated, multi-dimensional, artificial, or virtual reality" that is "widely distributed, ubiquitous, open-ended, and ever changing." They also suggest three essential areas of recent cultural and technical development.

First is the formation of a cyberculture, which includes individuals who prefer to inhabit the domain of distributed digital media—electronic bulletin boards, databases, and multiuser simulation environments, including virtual reality. These inhabitants more or less live in such domains; the majority of their time is occupied within them. There they can alter their identities, their manner of social interaction, and their relationship with society. They become virtual beings in a virtual place. By living in such domains, a society becomes established, and a morality may emerge. What kind of morality will this be? Will it be governed? By whom and for what? This line of questioning becomes even more involved when one considers distributed virtual reality as a three-dimensional (3-D) environment that may contain private spaces or residences, which contain personal objects and possessions.

Second, the emulation of the physical world and private spaces may have doors, closets, and windows that look out onto multidimensional vistas. Toolkits allow the transformation of the world, and extensions of it consist of a never-ending field of pure data. The field of data can include all walks of commerce and produce worlds that do not fit our present descriptors. Some experiences will be familiar, like going shopping or going to a concert. Other things will be unusual, like going to an ancient place or another planet.

Third, the pervasiveness of the data field is everywhere, and people are moving with computer devices. Interfaces become intuitive. Guides or agents co-inhabit the domains. Agents acquire knowledge, become familiar, and grow old with us.

While this could read as science fiction, extensive research is already being conducted in networked or distributed virtual reality. It currently constitutes a very small industry, but one with great potential for growth. Our research at the STUDIO for Creative Inquiry at Carnegie Mellon University investigates networked virtual reality and relevant applications within it.

Survey of Networked Virtual Reality

The evolution of the VR industry is a fascinating study. Howard Rheingold, author of *Virtual Reality*, is the leading expert in this area.[5] It is helpful, however, at this point to survey examples of networked virtual reality, applying the term broadly.

Literature

The novelist William Gibson is often cited as one who sparked the imagination of legions who are waiting for their moment to jack into the Sendai Deck.[6] Their efforts have spawned a myriad of Gibson-like MultiUser Dimension (MUD) and Multiuser simulation environment (MUSE) systems that catapult a user into a spiraling on-line matrix of experiential role playing and other amusement. Here, users employ text to describe the world, themselves, and their interaction with other users.

MUD systems have been influential in the development at Carnegie Mellon University of the Oz Project, conceived by Joe Bates.[7] Predominantly a text-based narrative environment, the project explores the creation of the personality of artificial characters, which ultimately will be expressed as three-dimensional animations. Bates recently premiered the Woggles, four bloblike creatures programmed with apparent emotional behavior. The Woggles exhibit the appearance of personalities and act independently. Bates argues that most virtual worlds express the concerns of replicating physically and functions of objects, but that other directions must be informed by cinematic conventions: character, plot, and story. If the entertainment of the future will be networked virtual reality, an essential study is mimesis formed by cinematic or narrative-based interaction.[8]

Virtual Communities

One can also point toward growing numbers of on-line virtual communities. These communities are the subject of the book, *Virtual Communities*, by Howard Rheingold.[9] They introduce social, if not cinematic, interactions in a virtual place.

The Whole Earth 'Lectronic Link (WELL) is an exemplary on-line service, offering an expanded electronic bulletin board service (BBS). One example of an interdisciplinary virtual crossroads on the WELL is the Art Com Electronic Network (ACEN), which was founded in 1986 by myself and Fred Truck.[10] There exist definite feelings of community and familiar social interaction among the user base. Sometimes entire stories are written by the users, and they assume the role of distinct characters and unfold a plot. ACEN focuses on the interface of contemporary art and communications technologies—and offers a discussion board, electronic art galleries, and even an electronic shopping mall for art-related books, software, and video.

The overall shape and flavor of on-line services are staggering in variety, and they are not just to be found in the West. In Japan, the Coara network, based in Oita Prefecture, has been operating for several years. The Oita local government is highly supportive and encourages the growth of the network. In Korea, the Association for Computer Communication Amateur Network functions as a service group for emerging grassroots networking efforts in Southeast Asia. There are myriad other examples. A global perspective on the massive growth of computer networks is provided by John Quarterman in *The Matrix*.[11]

Entertainment

It is not by chance that Fujitsu Limited of Japan has supported both the work of Joe Bates, investigating interactive drama, and Habitat (a commercial on-line service in Japan), originated by Randall Farmer and Chip Morningstar at LucasFilms.[12] The eventual merging of these projects, or projects similar to their concerns, indicates the advancing edge of dramatic entertainment, ultimately to be accessed through networked immersion environments.

Farmer has demonstrated that the argument for connectivity and virtual reality is not about bandwidth; Habitat employs simple, low-speed, 300-baud modems. For display, it employs a simple monitor. It is not an immersion environment. The core requirements for networked virtual environments are

- A processor with an operating program
- Display and navigation devices
- Distribution, using computer networks, television, or telephone
- Thoughtful world design, employing cinematic conventions

The addition of a head-mounted display (HMD) and other immersion devices is a relatively simple matter that does not add considerably to the complexity of the operating program. The difficulty is economic—few of the users of Habitat, a consumer-based network, could afford an HMD next to their personal computer. However, within the coming years, possessing a consumer-grade HMD will become an option.

In addition to drama, there is music. Jerry Garcia of the infamous rock band, the Grateful Dead, was reported to desire interactive concerts at which the "audience could participate in creating the performance." The recent presentation at SIGGRAPH by Jaron Lanier, formerly of VPL, Incorporated, was a music performance incorporating immersive virtual reality.[13] Move over MTV (Music Television), here comes MVR (music virtual reality).

Education

The field of education offers an opportunity for applications to investigate that which is otherwise inaccessible and to present challenges for the representation of and interaction with information. Scientific visualization is one area noted for advances in imaging applications. For example, Warren Robinett at the University of North Carolina (UNC) has introduced the Flying Through Protein Molecules project, which allows for interaction with molecular models scaled to gigantic proportions. Robinett, who began at Atari and went to the National Aeronautics and Space Administration (NASA) before the UNC, has produced a number of advanced immersion applications.[14]

Another project is ExploreNet, developed by Charles Hughes and Michael Moshell at the University of Central Florida. It is a simulation network that assists in teaching core disciplines in elementary schools. Inspired by Habitat, Hughes and Moshell have produced two prototypes and have simulated Caruba, an entire wild world with caves, jungles, deserts, waterfalls, and mountains. It addresses immersion experience as a psychological phenomenon for which a head-mounted display is less important than compelling, dramatic interaction.[15]

The cognitive and behavioral sciences suggest that the capacity for learning is greatly enhanced when sight, sound, and touch augment the presentation of text or numerical data. Sound is especially important as it naturally directs our eyes. From this perspective, we can recognize the flight simulator and the computer game as advanced educational tools. Virtual reality—and eventually virtual reality networked over long distances—will be employed to conduct a multitude of education and training projects, ranging from grade school classes to corporate seminar applications and beyond. Long-distance education will benefit tremendously, and students will have the opportunity to visualize and even touch that which is otherwise inaccessible.

The Workplace

Yet another direction is represented by the TeamWorkStation, a project in search of the paperless office, by Hiroshi Ishii of Nippon Telephone and Telegraph (NTT), Japan, which consists of linked multimedia workstations.[16] There are also a number of other networked workstation applications under development.

Telepresence can also be considered a networked extension of virtual reality, especially with regard to operators manipulating robots located in remote factories or perhaps in outer space. In telepresence, force feedback can be an essential aspect.

The Military

The origins of virtual reality are largely to be found in the military, as recounted in Rheingold's *Virtual Reality*. One current focus is SIMNET, a nonimmersive virtual environment project involving armored tank simulation with remote nodes linked in real time.[17] SIMNET was used to train personnel for Operation Desert Storm in the Middle East. The experience is reported to evoke strong physical and emotional reactions among the trainees. SIMNET originated at the University of Central Florida along with a number of other networked projects such as Caruba, discussed above. The Department of Computer Science at the Naval Postgraduate School has recently launched NPSNET, with the goal to establish a government-owned, workstation-based network of visual simulators that employ SIMNET databases.

As this brief survey shows, networked virtual reality exists. It has a history and has been given different forms shaped by varied intentions. The projects discussed above range from desktop applications to immersion environments. What unifies them are their networking applications.

The Networked Virtual Art Museum

Perhaps it is useful to report on one project at the STUDIO in greater detail. The project is the Networked Virtual Art Museum, which joins telecommunications and virtual reality through the design and development of multiple-user immersion environments networked over long distances.

The essential areas investigated through the project include visual art and architecture, world-building software, telecommunications, artificial intelligence, communications protocol, cost analysis, and computer programming and human interface design.

Visual Art and Architecture.   The fusion of disciplines is the basis for collaborative authorship of virtual worlds. The construction of the virtual museum involves the participation of visual artists, architects, computer-aided design

FIG. 1   Networked Virtual Art Museum, SIMLAB, NASA/Robotics Engineering Consortium, Carnegie Mellon University.

teams, computer programmers, musicians, and recording specialists, as well as other disciplines.

World Building.   The project serves as a testing site for world-building software and associated hardware.[18] The programming teams have added considerably to the functions of the software tested. Public releases are planned.

Telecommunications.   Critical to the project is the development and implementation of networking approaches, including modem-to-modem, server, and high-bandwidth connectivity. Telecommunications specialists collaborate with the design team to resolve problems of connectivity in immersion environments. Project achievements in this area are discussed in greater detail below.

Artificial Intelligence.   The application of artificial intelligence, in the form of agents (or guides) and smart objects, is an essential area of development. The inclusion of investigators in the areas of interface design, smart objects, and artificial intelligence is a major component.

Groupware and Communications Protocol.   The project documents multiuser interaction and groupware performance, establishes protocols within networked immersion environments, and suggests standards. The contribution of communications specialists addresses aspects of documentation and standardization.

Cost Analysis.   Another planned study will address the practical nature of networked immersion environments, investigate the effectiveness of information access for the end user, and profile the end-user experience. The project involves the participation of cost analysis specialists and formulates a practical cost basis for networked immersion environments.

The project team has designed and constructed a multinational art museum in immersion-based virtual reality. The construction of the museum involves a developing grid of participants located in remote geographical locations. The nodes are networked using modem-to-modem telephone lines, the Internet, and, eventually, high-bandwidth telecommunications.

Each participating node will have the option to interact with the virtual environment and contribute to its shape and content. Participants are invited to create additions or galleries, install works, or commission researchers and artists to originate new works for the museum. Tool rooms will be available, so one can construct additional objects and functions for existing worlds or build entire new ones. Further, guest curators will have the opportunity to organize special exhibitions, explore advanced concepts, and investigate critical theory pertaining to virtual reality and cultural expression.

The museum also functions as a stand-alone installation and is easily transportable for presentation in cultural, educational, or industrial venues.

The Museum

The design of the museum centers on a main lobby from which one can access adjoining wings or galleries. Several exhibitions are completed, while others are under construction. The first exhibition to be completed is called Fun House and is based on the traditional fun house found in amusement parks. The museum also contains the Archaeopteryx, conceived by Fred Truck and based on the ornithopter, a flying machine designed by Leonardo da Vinci.[19] Imagine flying a machine designed by one of the world's greatest inventors. The team is also collaborating with Lynn Holden, a specialist in Egyptian culture, to complete Virtual Ancient Egypt, an educational application based on classic temples mapped to scale. The gallery exhibitions mentioned are being constructed at CMU. However, we are anticipating other additions conceived and constructed by participating nodes in Australia, Canada, Japan, and Scandinavia.

Now that the framework of the museum project has been described, perhaps it is useful to discuss the essential points of one application: the Fun House.

Fun house: a building in an amusement park that contains various devices designed to startle or amuse. (20)

For the first installation in the Networked Virtual Art Museum, a fun house was designed. While making metaphorical reference to the "fun house" found

FIG. 2   Lobby, Networked Virtual Art Museum, SIMLAB, NASA/Robotics Engineering Consortium, Carnegie Mellon University.

FIG. 3   Virtual Ancient Egypt: Temple of Horus, SIMLAB, NASA/Robotics Engineering Consortium, Carnegie Mellon University.

throughout traditional amusement parks, the application is an investigation of interaction and perception employing networked, immersion-based virtual reality. It was this world that was utilized during the first long-range demonstration conducted between Carnegie Mellon University and Munich, Germany, in September 1992. A more recent demonstration, featuring a different virtual world, was conducted between CMU and Tokyo, sponsored by the International Conference on Artificial Reality and Tele-existence, July 1993, Japan.

The fun house metaphor is particularly applicable as a container for virtual experience. On entering a fun house, one is acutely aware of being cast into a different world; one's senses are amused and assaulted by a number of devices—trick mirrors, fantasy characters, manipulation of gravity, spatial disorientation, mazes, and sound, for example. In the virtual Fun House, various traditional devices are adapted and some new ones are offered.

Key attributes to be found in the Fun House include

- Objectification of "self" within an immersion environment. Users can select their image from a library that includes Frankenstein, Dracula, and a doctor, among others; the Cookie Man has proved to be a favorite. When entering the Fun House, users can see their image reflected in real time in a mirror. They can also see the images of other users. Users can extend their hands and wave at each other—a basic and highly communicative form of human expression.
- Interaction with a client (or agent) that has an artificial intelligence. When

FIG. 4   Temple Virtual Cat, SIMLAB, NASA/Robotics Engineering Consortium, Carnegie Mellon University.

entering the Fun House, a client greets you and speaks. It has a polite behavior and is programmed to face you, follow at a certain distance, and to stay out of your way. After a while, it stops following and says good-bye. Smart objects are also incorporated; touching them brings up events within the program.

- Interaction with multiple users in real time. Networked telecommunications allows for the simultaneous support of multiple users within the Fun House. For the demonstration between CMU and Munich, the users selected the Dracula and Cookie Man personas from the library. Each user could see the other, had an independent point of view, and could move objects.
- Links to moving objects. The Fun House features a merry-go-round; users can grab hold and catch a ride while music plays.
- Objects attach themselves to users. The Fun House features a flying saucer ride by which users are transported into the spacecraft, and they can pilot its flight. The event brings about the "beaming up" of the user and a whirring sound associated with flying saucers.
- Attributes of physical laws. The Fun House features a ball game in which users pick up a ball and throw it at targets. The ball falls, bounces, and loses velocity. Thus, gravity, velocity, and friction are articulated. The motion of the ball is sound intensive.

FIG. 5   Temple Agent, SIMLAB, NASA/Robotics Engineering Consortium, Carnegie Mellon University.

New Directions

Following the Fun House, a number of applications for Ford Motor Company were designed; these are discussed in detail in the section, "Teleconferencing and Design." However, while working on the applications, the project team became increasingly interested in approaching the immersion environment as a site at which things can be constructed or created. For Ford, virtual reality was to be utilized as a virtual design studio, but what are other approaches?

Currently under design are three applications for a public institution or educational setting; they are also network capable: the Music Room, Construction Room, and Painting Room. These are friendly and intuitive environments that require only a short learning curve to utilize. Instructions are bilingual (Spanish and English), and the environment is co-habited by small agents or "beasons"[21] that are programmed with a low-level artificial intelligence. The beasons are guides to the various interactive objects and demonstrate how the environment functions by literal illustration. In the Music Room, they run around, make contact with the instruments, and thus play music. Users can see which instruments produce what sounds and how to perform with them. There are a number of other intuitive controls for navigating around in the room.

The Music Room contains three basic instruments. The largest is a six-note

FIG. 6   Sanctuary, Temple of Horus, SIMLAB, NASA/Robotics Engineering Consortium, Carnegie Mellon University.

keyboard attached to a wall. This instrument plays a pentatonic scale in three voices—orchestra, choir, and percussion. It is performed by touching the keys or by simply waving a hand within close proximity; this requires the use of a glove-type input device. The other instruments are a drum and shakers.

The Construction Room is designed for young children. It contains building blocks that can be assembled to construct objects, sort of a virtual Lego set. In this case, there is a nearly infinite supply of blocks, and one can enter into some of the objects created, such as a house. The beasons co-inhabit this site as well, bounce merrily on the blocks, and illustrate how to stack them.

The Painting Studio is quite literal—a site for making paintings or graffiti. The user can select from a number of "brush" effects and choose colors.

As of this writing, negotiations are under way to present these three applications in a public science museum setting and to set up a network to connect the facility with yet other facilities so people from different locations can simultaneously interact. They are also designed for use by school districts and other institutions.

The Virtual City

Another application currently under design is a virtual city. Inspired in part by applications like the Music Room, the city is an actual city inhabited by a

FIG. 7 Virtual Polis, SIMLAB, NASA/Robotics Engineering Consortium, Carnegie Mellon University.

multitude of participants, each with their own purposes. Imagine a virtual city complete with private spaces or domiciles, parks, stores, and entertainment centers. As much as a grand social experiment, it also is a far-reaching graphical user interface (GUI) for electronic home shopping and entertainment. The precedent for such an application as the city is Habitat, the commercial on-line service available from Fujitsu in Japan. As mentioned above, it features two-dimensional applications and currently has 10,000 subscribers.

The concept of a city was previously discussed by the team. The conception of it accelerated when the team was approached by a film production company to produce a set for a commercially distributed motion picture. The set would, of course, be a city, and from that point the virtual city formulated quite quickly.

The salient points of the virtual city include

- A distributed, three-dimensional inhabitable environment
- Investigation of teleexistence in a distributed virtual construct
- Capability of supporting potentially unlimited participants
- Private spaces, property, and a moral code
- Exploration of tools to alter the environment while inhabiting it
- Interface (GUI) for home shopping and entertainment

The idea is of a distributed application that is based on the notion of an inhabited city. Traversing the city and encountering other inhabitants will be a startling experience.

Teleconferencing and Design

The last area of investigation includes distributed virtual reality as an interface for teleconferencing and design. The range of possible applications is broad, but education and industry are obvious examples.

Education can benefit with regard to long-distance learning, and industry can gain from a higher level of videoteleconferencing. This raises the question: What is the advantage over videoteleconferencing? The answer is, the relationship to the subject.

In a teleconferencing session employing distributed virtual reality, multiple participants can share a dynamic relationship with their subject. For example, imagine a team of automobile designers discussing options via videoteleconferencing. To look at the subject, they might program a prerecorded videotape highlighting the desired aspects.

In contrast, for the Ford Motor Company, I designed a Virtual Showroom and Virtual Test Track. During the teleconference, the participants can actually walk around the car—open doors, hood, and trunk, test drive it. In detailed applications, participants can examine specific, even minute, parts in detail.

While the Ford showroom and driving applications were produced from a consumer's standpoint, it has become increasingly apparent that these applications could benefit teleconferencing sessions. Their key advantage is the dynamic relationship with the subject. Of course, the subject need not be automobiles, and applications can be developed to serve a wide range of interests.

We also produced a prototype Virtual Design Station for Ford that demonstrates the value of virtual reality as a site to build things. (Design is used broadly here, including design of other multimedia applications.)

The following are salient aspects of a design studio that would employ distributed virtual reality:

- The virtual environment supports interaction among networked remote production teams for the purpose of industrial design and teleconferencing.
- The application establishes a relationship between an actual workstation and a virtual workstation; operators have the capability of switching back and forth, employing a windowing method.
- Teleexistence is made evident through synchronous voice communication and the capability of each member to "see" the other members in the virtual environment.
- Stations consist of display, stylus, keyboard and mouse; additional input devices include head trackers, data gloves, and machine vision for voice and gesture recognition; additional output devices include head-mounted displays.
- Design teams utilize proprietary software for the purpose of producing wire frame geometric models, painting, and mapping.

The advantages of distributed design stations are numerous, but an essential point is economy. Teams can be located in remote sites and benefit from a collective design experience.

Connectivity

The basis for distributed virtual reality is a function of telecommunications. Two or more sites are joined by an operating system that can employ a number of telecommunications delivery services:

- Direct dial-up lines
- Internet computer network
- High-band networks
- Cable television
- Wireless

The project is at the forefront of the investigation of connected immersion environments. Presently, a single point-to-point link employing low-band (9600 baud) modems is supported. The demonstrations between CMU, Munich, and Tokyo mentioned above proved successful, and the update delay between each pair of cities was a barely perceptible 100 milliseconds or less. Servers and broadband telecommunications are in the planning stages, as discussed below.

Select functions for point-to-point connectivity include

- Providing all functions of the virtual world-building software in a distributed manner. The world and its attributes are distributed to each node, and one node is specified as the controller.
- Providing constant views and updates of each user's object manipulations. Users can move objects, including themselves, and updates to position and change occur with imperceptible update cycles.
- Writing and saving files to record the manipulations of objects. Users can change worlds and carry an object with them into another world.

As of this writing, the programming of a distributed client-server code has begun. Here, multiple users can share an immersion environment by interfacing with a node. The following are key points of investigation:

- The formation of a client-server model by which multiple users can simultaneously share immersion environments.
- The servers or nodes are to be located across the world and can be updated automatically via the Internet.
- The nodes will offer sites for access and distribution of virtual reality software.
- The nodes will support multiple platforms.

The next phase of the project will support broadband connectivity. To facilitate this phase, a consortium has recently been developed with associates in Japan, Scandinavia, and the United States. Members will conduct technology transfers and test-bed projects. Actual applications are scheduled for development. Members of the consortium will collaborate to produce basic elements of connectivity for distributed simulations from client-server systems over the Internet to broadband ATM (asynchronous transfer mode), cross-platform graphical user interfaces, and executables.

Conclusions

The server model points toward the matrix that Gibson so often refers to in *Neuromancer*. In addition to various worlds, the server will contain rooms with tools. Here, users can construct other virtual objects, including entire other worlds. Imagine a networked immersion environment capable of supporting multiple users, with each having the possibility of changing the existing virtual world or constructing a new one. Imagine the bulletin board model applied to networked virtual reality. This could make a continuously changing reality, which may produce some anarchy. The promise of constructing virtual worlds while within an immersion environment is open ended. This approach can be extended to any number of educational, entertainment, and industrial purposes.

Distributed virtual reality has a history and has been given many different forms, shaped by varied intentions. It also has the promise of a future, marking the advancing edge of a new industry.

Acknowledgments: Gratefully acknowledged for their support and interest are STUDIO for Creative Inquiry, Carnegie Mellon University; Ascension Technology Corporation; Sense8 Corporation; and Virtual Research, Incorporated.

Notes

1. Gibson, W., *Neuromancer*, Ace, New York, 1984, p. 5.
2. Benedikt, M., *Collected Abstracts for the First Conference on Cyberspace*, University of Texas, Austin, 1990, p. i.
3. Tokoro, M., *Toward Computing Systems for the 2000s*, Technical Report SCSL-TR-91-005, Sony Computer Science Laboratory, Tokyo, 1991, pp. 17–23.
4. The term *virtual reality* can be applied to a broad range of applications. The fullest virtual reality is immersion based. This typically means to put on a head-mounted display (HMD) and navigate in three-dimensional space with a hand-directed input device, or perhaps in addition wearing a kind of body-tracking suit. This allows a user to be deeply immersed in the world, and the world is created and updated in real time from the data received from the input devices. For example, when one moves his or her head from left to right, one sees the world in the HMD being updated from a left-to-right orientation. With the addition of more input

devices and 3-D sound, the immersion environment becomes a convincing simulation.

Perhaps the most common applications of virtual reality are desktop models. These consist of traditional computer graphics that one can manipulate, if not navigate. Computer games made by Nintendo or Sega are prime examples. Computer networks distributing electronic mail or bulletin boards are also a type of virtual reality.

One reason for such a broad range of interpretation of virtual reality is the conflict inherent in the following definitions from *Webster's Dictionary* (Ottenheimer, 1977):

* Virtual, being so in effect or essence, although not in actual fact
* Reality, 1. the quality or state of existing or happening as or in fact, 2. actual, true, objectively so, etc., 3. not merely seeming, pretended, imagined, fictitious, nominal, or ostensible.

To combine the "the quality or state of existing" and "although not in actual fact" opens up an enormous range of possibilities, involving both hardware and software, not to mention intention. This suggests options that range from electronic mail to the Holodeck in Star Trek, which remains the property of Paramount script writers for the time being. The advance of virtual reality results from a merging of previously disconnected disciplines. It is intermedia. One major premise of virtual reality is an experiential relationship to data or information. As suggested above, immersion applications provide an entirely different type of experience than a desktop application. The difference is the distinction between looking at a blue lagoon and swimming, if not going for a scuba dive.

5. Rheingold, H., *Virtual Reality*, Simon and Schuster, New York, 1991.
6. The Sendai Deck is an electronic interface device often mentioned in *Neuromancer*; one would jack into it, like plugging into a device. Once jacked in, one was connected to the cyberspace matrix.
7. See Bates' contribution to this collection, "The Nature of Characters in Interactive Worlds and the Oz Project."
8. For further information, see B. Laurel, *The Art of Computer Interface Design*, Addison-Wesley, New York, 1990.
9. For further information, see H. Rheingold, *The Virtual Community*, Addison-Wesley, Reading, MA, 1993.
10. See C. E. Loeffler, The Art Com Electronic Network, *Leonardo*, 21(3):320–321 (1988).
11. Quarterman, J. S., *The Matrix: Computer Networks and Conferencing Systems Worldwide*, Digital Press, Bedford, MA, 1990.
12. See Farmer's contribution to this collection, "Social Dimensions of Habitat's Citizenry."
13. See J. Lanier, Music from Inside Virtual Reality, The Sound of One Hand, *Whole Earth Review*, 320–334 (Summer 1993).
14. See W. Robinett, Electronic Expansion of Human Perception, *Whole Earth Review* (Fall 1991).
15. See the contribution by Hughes and Moshell, "ExploreNet," in this collection.
16. See Ishii and Miyake, Toward an Open Shared Workspace: Computer and Video Fusion Approach of TeamWorkStation, *Commun. ACM*, 34(12):37–50 (December 1991).
17. See Rheingold's *Virtual Reality*, pp. 360–361.

18. For further information, see the *WorldToolKit Manual*, Sense8 Corporation, Sausalito, CA. WorldToolKit is a computer program for the creation of virtual reality applications.

19. See Truck's contribution to this collection, "Leonardo's Flying Machine and Archaeopteryx"; for a fuller account, see his *Archaeopteryx*, Fred Truck, Des Moines, IA, 1992.

20. *Webster's Third New International Dictionary*, G. and C. Merriam, Springfield, MA, 1981.

21. The beasons are named after Curtis Beason, a programmer associated with the virtual reality project at Carnegie Mellon University.

CARL EUGENE LOEFFLER

Networking for Computers (see Introduction to Computer Networking)

Networks for Local-Area Communications—An Overview of LANs

Introduction

The local-area network (LAN) is used to communicate information among computers and other devices over a limited region. Although the devices interconnected by a LAN are most commonly computers, they are not limited to any class of equipment and include video display units, printers, storage devices, sensors and monitoring devices, and facsimile machines. The geographical span of LANs can range from a single room to a collection of buildings; typically, the LAN provides coverage for a region of diameter less than 10 kilometers. Unlike a public communications network, the LAN is usually installed and operated by a single enterprise or organization.

Since its first appearance in the 1970s (nearly simultaneously with the advent of the single-user workstation), the LAN has grown steadily in popularity. Users have realized significant benefits from sharing resources accessible over LANs, and new applications, such as distributed file systems, electronic mail, windows-based graphical user interfaces, hypermedia browsers, and desktop videoconferencing, have spread primarily through the LAN environment. Indeed, the entire area of client–server computing has come about largely because of the LAN.

Today, nearly every networked computer has a LAN as its primary network interface. The LANs are then interconnected to form a global internetwork, such as the well-known Internet, which now encompasses over 30,000 LANs worldwide. (The Internet is a large, global, packet-switching network of computers that use the Internet Protocol [IP]. It is the direct descendent of the ARPANET, the world's first packet-switching network.)

This article describes the different services provided by LANs, surveys the technologies used in their implementation, and explains the principles of operation behind many LANs in use today.

LANs transport blocks of data, which are referred to by different names (e.g., packets, cells, frames, segments, slots, and worms). To prevent confusion, here the basic unit of transfer in the LAN is called a *message*.

Service Requirements

LAN users have different requirements. In this section, the most common services required of LANs are reviewed. The services provided by a LAN can include transmission of different traffic classes, broadcast and multicast transmission, error detection and recovery, prioritized transmission, and connection-oriented and connectionless transmission.

333

Communications network traffic is often divided into three separate classes: asynchronous, synchronous, and isochronous traffic.

Asynchronous traffic consists of blocks of data that may occur at any time. There is, in general, no significant timing relationship among the bits of information in an asynchronous traffic stream. Representative of this class is so-called bursty traffic, which is composed of data blocks of arbitrary length that can arrive at any given instant. Many computer applications generate asynchronous traffic, including file transfers, electronic mail exchanges, transaction processing, and human–computer interactions. Asynchronous traffic is a very general class and encompasses nearly any type of traffic imaginable. By its nature, asynchronous traffic tolerates high variation in delay and even message loss, which can be compensated for by retransmitting the lost message.

Traffic that must be delivered before a specified deadline is called *synchronous traffic*. Although this traffic may arrive at arbitrary time instants, the data rate of the stream does not exceed a specified maximum. Thus, parameters relevant to a synchronous traffic flow specification consist of a maximum access time and required bandwidth. Examples of synchronous traffic are real-time signals, such as event interrupts and interactive graphics, in which the user expects nearly instantaneous response time. Synchronous traffic can be viewed as a restricted subset of asynchronous traffic; any LAN that supports synchronous traffic supports asynchronous traffic by default (though not necessarily efficiently).

While asynchronous and synchronous traffic admit nondeterministic arrival characteristics, *isochronous traffic* arrives at regular time instants and in fixed-size blocks of data. The prototypes for isochronous traffic are sound and video streams, which can be viewed as fixed-sized blocks of data that occur at multiples of a fundamental time scale. For instance, the human voice is often transmitted as a single octet (eight bits) sampled every 125 microseconds (μs), whereas full-motion video can be represented as a sequence of frames that occur every 16.67 milliseconds (i.e., at the rate of 60 hertz). Isochronous traffic is a subset of synchronous traffic, and support for isochronous traffic implies support for synchronous traffic.

To support a given traffic class, a LAN must guarantee that the essential properties of the incoming stream will be preserved at the receiving end. Clearly, asynchronous services are easily provided, and all LANs provide this type of service. Synchronous traffic requires bounded access delays and bandwidth allocation mechanisms; a few LANs can support this service. Because isochronous service provides the strongest delay and bandwidth guarantees to the LAN user, its implementation is generally more complex than that of the others. Consequently, only a few advanced LANs provide isochronous service. Given the complexity of supporting synchronous and isochronous traffic, as well as current users' limited need for these traffic classes, most LANs do not provide support for all three traffic classes.

The LAN provides other services as well. In the connection-oriented service model the exchange of information among nodes is preceded by the establishment of a logical connection and followed by the termination of a connection. On the other hand, the connectionless service model requires no logical connection for the exchange of information. The connection-oriented model generally implies a reliable delivery service, which ensures that all the sender's information arrives at the receiver in the order sent and without loss. The connectionless model provides no such guarantee, and the parts of the sender's information might be lost or arrive at the receiver out of order. Connectionless service is sometimes referred to as best-effort or datagram service. Although most LANs accommodate the connectionless model, a few are based on the connection-oriented model.

It has become desirable for computers to use broadcast or multicast transmission (in which all or a specified subset of LAN nodes are sent data) in support of many applications. Realizing the importance of broadcast and multicast in computer applications, most LAN designers provide this feature.

It is sometimes useful for some messages to take precedence over others. It is possible to assign priorities to a message so that messages of higher priority will be given preferential treatment in comparison to messages of lower priority. Such prioritized service is essential in many applications, such as error-recovery functions and network management. It might also be desirable to allow paying users a choice of priority if their willingness to accept a lower quality of service is rewarded by lower service charges. Several LANs support prioritized service.

Often overlooked or taken for granted, network management is an integral part of any LAN, although it is usually provided in different measure by different LANs. Within the scope of network management can be included prosaic functions like node insertion and removal, address resolution, network initialization, and connection management. At a higher level, one also finds error, configuration, security, performance, and accounting management. Furthermore, many LAN protocols rely on the use of an ensemble of parameters that govern the protocols' behavior and performance. The networkwide collection of these parameters forms a unified data structure called the *management information base*. An important management function, therefore, is the monitoring and updating of the management information base.

Implementation Technologies

As a system, the LAN is based on many technologies. Electronic, optical, and software technologies are all used in LANs. Given that the LAN interconnects digital computers, it is not surprising that electronic technology is used widely in the implementation of the LAN. The node's interface to the LAN is usually a printed circuit board that hosts the integrated circuits, processor, and software to translate the node's data into the digital signals that traverse the LAN's transmission medium. This interface also performs other functions, such as media access control, which provides several users with orderly access to a shared medium.

A wide range of media is used in the LAN. For data rates up to about 150 megabits per second (Mb/s), electrical transmission over bit-serial wires and cables is most common. For higher data rates, both bit-parallel electrical wires and cables and optical-fiber waveguides are used; the upper limits of these transmission media are in the range of many gigabits per second (Gb/s). It is also possible to use airwaves as the transmission medium. To accommodate the widespread choices in media, it is commonplace to offer a selection of media types that can be used in a particular LAN; thus, the medium can be changed independently of the network interface electronics.

Twisted pairs of 16–36-gauge metallic wires (principally copper) have been used as electrical transmission media in the local loop of the telecommunications network for a long time. Similarly, the LAN can use both shielded and un-shielded twisted-pair media for transmission over a limited distance. Baseband modulation is most often found in twisted-pair media. To insure that a clock can be recovered by the receiver, data can be encoded so that a sufficient number of line-state transitions are observed in the incoming signal. Many coding schemes are used for this purpose, including Manchester, which guarantees a line-state change in every bit time regardless of which data pattern is transmitted, or 4B/5B, which encodes four-bit data nibbles as five-bit code-words in such a way as to ensure that no sequence of codewords contains more than three consecutive zeros. (The terminology nx/my, where n and m are positive integers and x and y represent the bit or trit designators B or T, respectively, will be used to denote a code that translates n x-symbols [bits or trits] into m y-symbols [bits or trits].) To reduce the effects of electromagnetic interference, especially in high-speed transmission over an unshielded twisted pair, tristate multilevel transmission codes are used in which three distinct line states are used to encode the 1 bit as a (rising or falling) line-state transition and the 0 bit as the absence of a line-state transition. Further reduction of electromagnetic interference can also be accomplished by random scrambling of the line states.

When properly engineered and manufactured, twisted pairs currently support LAN data rates up to 150 Mb/s. The other electrical transmission media commonly found in the LAN are 50- and 75-ohm coaxial cables. In addition to baseband modulation, other modulation schemes, such as frequency-, amplitude-, and phase-shift keying (FSK, ASK, and PSK, respectively), are used with coaxial cable. These so-called broadband coaxial systems often consist of multiple 6-megahertz–spaced carriers in the 5–300-megahertz (MHz) spectrum, to which a node can tune and modulate. Coaxial systems can support aggregate data rates of more than 200 megabits per second.

Most often, electrical media carry bit-serial data, but parallel runs of wire or cable can also carry bit-parallel data. Parallel media, however, cannot be used over appreciable distances because it is difficult to maintain synchronization of the individual bits. By choosing a wide enough data path, a bit-parallel system can achieve a data rate comparable to an optical-fiber system, but the transmission distance is much greater in the optical-fiber system.

The tremendous bandwidth of the glass-core optical fiber in the 0.8-, 1.3-, and 1.5-micrometer (μm) wavelength, low-loss windows enables a very high transmission rate. The superior attenuation characteristics of optical fiber, as small as 0.2 decibel per kilometer, permit low-loss, highly reliable data transmission. The speed–distance product of both the multimode optical fiber, with a core of about 50 μm, and the single-mode optical fiber, with a core of about 10 μm, is high enough to allow the construction of very high speed LANs. Both step- and graded-index optical fibers are used in the LAN. The optical LAN modulates an optical signal produced by either a light-emitting diode or a laser diode. Optical signals are usually encoded as 4B/5B symbols, arbitrary sequences of which are guaranteed to contain fewer than three consecutive low-power line states (i.e., the "dark" pulse). This facilitates clock recovery. Some optical LANs, however, transmit data in the simple nonreturn-to-zero (NRZ) code, which can produce arbitrary runs of nontransitioning line states, and must therefore use a scrambling technique to ensure that line-state transitions occur frequently enough to permit receiver synchronization. In this scheme, data bits are scrambled by modulo-2 addition with the output of a properly chosen and seeded linear feedback shift register that produces a pseudorandom sequence of bits. After transmission, descrambling is performed at the receiver by adding (modulo 2) the output of a similar shifter-register sequence to the incoming stream of bits in order to recover the original data.

Despite the promise of coherent detection (by which the frequency, amplitude, and phase of light can be modulated) for future optical LANs, all major optical LANs today use direct detection of baseband signals. Two kinds of light sources are used in LANs, the light-emitting diode and the laser diode. The light-emitting diode, which generates light in the 0.8- and 1.3-μm wavelengths, is inexpensive and entirely suitable for low-to-medium–speed optical LANs. Its output power levels are also satisfactory for the geographical limits of LANs. Based on gallium arsenide (GaAs, AlGaAS, InGaAs, and InGaAsP) edge- and surface-emitting *pn*-junction semiconductors, light-emitting diodes produce a wide spread of wavelengths that can be directly launched into a multimode optical fiber. While the light-emitting diode operates in the regime of spontaneous emission, it is possible to increase the input current so that it becomes a laser diode, which operates in the regime of stimulated emission. The laser diode, therefore, can radiate coherent, narrow linewidth light (less than 5 nanometers in wavelength) in all three windows. Considerably more costly than the light-emitting diode, the laser diode is used with single-mode optical fibers at the 1.3- and 1.5-μm wavelengths. Intended for operation at very high speed and over a long distance, single-mode laser diodes are rarely used in the LAN.

A light detector is used to convert received lightwave signals into electronic signals. The *p*-intrinsic-*n* (PIN) photodiode consists of three layers of doped semiconductor material (such as germanium). Light in the 1.3-μm range that falls on the intrinsic layer is then converted into photocurrent. Like the PIN device, the avalanche photodiode uses a three-layer diode structure, but it is heavily voltage biased to produce a stronger photocurrent. The avalanche photodiode is used mainly with lightwaves at the 0.8-μm wavelength. Although the avalanche photodiode has much higher responsivity than the PIN photodiode, its high voltage bias and operating wavelength (0.8 μm) make it less suitable for use in optical LANs.

An important consideration in installing an optical LAN is the engineering of the optical cable plant so that adequate signal distribution can be achieved. Optical fibers are joined by splices, couplers, and taps to build complex distribution networks. The splice simply joins two optical fibers by mating them end to end in a low-loss connection. The fused biconical-taper coupler is manufactured by fusing lengthwise the claddings of two single-mode fibers so that light traveling through one fiber will couple into the other. Different splitting and combining ratios can be determined at the time of manufacture. The biconical-taper process can also be used to construct optical star couplers of higher fan-in and fan-out, or devices with very high fan-in and fan-out can be fabricated from planar waveguides. A tap, which allows a tributary fiber to join with a main fiber, can be constructed from simple, fused, biconical-taper couplers. Components used to build an optical distribution system are passive, requiring no electrical power and performing no conversion of the optical signal to or from the electrical domain. However, the emergence of the erbium-doped fiber amplifier makes possible the amplification of lightwave signals at arbitrary points in the distribution network. Pumped by a laser, the erbium-doped fiber amplifier requires an electrical power supply, but performs no optoelectronic conversion. It is available for use only in the 1.5-μm window. The components discussed above can be used to implement a wide range of distribution networks. Ring, bus, star, tree, and perfect-shuffle–based structures are all possible.

The wireless LAN relies on atmospheric propagation to transport signals. Wireless transmission in both the microwave and infrared regions of the electromagnetic spectrum are used in LANs. Infrared signals are both line of sight and omnidirectional, but are confined to a single room since the signals do not penetrate solid barriers such as walls. Infrared signals can be generated by a light-emitting diode and received by a photodetector and often use FSK as their modulation scheme. In the United States, the portions of the spectrum at 0.902–0.928, 2.400–2.483, and 5.725–5.850 gigahertz (GHz) have been reserved for unlicensed industrial, scientific, and medical (ISM) use, and wireless LANs are commonly found in these bands.

The recommended modulation scheme for ISM applications is spread spectrum, in which a symbol time is divided into a number of smaller time units called *chips*. In the frequency-hopping approach, a pseudorandom sequence of carrier frequencies is generated at the chipping rate by the sender and is modulated by the stream of data symbols to be sent. In the direct-sequence approach, the sender uses a code to generate a pseudorandom sequence of chipwide pulses of a single carrier frequency and modulates it with the stream of data symbols to be sent. A result of the higher chipping rate is to spread the modulated signal's power spectral density over a wider range than the signal's inherent spectrum. If the pseudorandom codes are chosen properly, then several transmitters can occupy the spectrum without interfering with each other; thus, spread spectrum allows multiple users access to a channel. Table 1 shows the characteristics of infrared and radio-frequency transmission systems.

Local-Area Network Characteristics

LANs share many fundamental features and characteristics; these are discussed next. LAN standards play a central role in promoting the goal of universal connectivity among a community of users. The standardization of communications services and protocols allows all conforming implementations to exchange information and discourages proprietary solutions that tend to limit competition

TABLE 1 Wireless Transmission System Characteristics

	Infrared		Industrial, Scientific, Medical		Radio Frequency
	LOS	Omni	DSSS	FHSS	
Maximum Range (m)	60	30	240	909	40
Frequency	0.8–0.9 μm	0.8–0.9 μm	902–928 MHz, 2400–2483 MHz, 5725–5850 MHz	902–928 MHz, 2400–2483 MHz, 5725–5850 MHz	18 GHz
Maximum data rate (Mb/s)	4	10	10	20	3
Modulation technique	OOK	OOK	FSK, QPSK	QPSK	FSK
Mobility	Mobile	Stationary	Mobile	Mobile	Mobile

LOS = line of sight; DSSS = direct sequence spread spectrum; FHSS = frequency hopping spread spectrum; OOK = on-off keying; FSK = frequency-shift keying; QPSK = quaternary phase-shift keying

and keep equipment prices high. Consequently, the importance of LAN standards has grown steadily. Currently, several LAN standards have been established to support the different communication requirements of users. As new applications and requirements arise, new standards will undoubtedly emerge.

The first LANs—developed in the early 1970s—were proprietary products meant to interconnect one vendor's computer products. By 1980, however, Project 802 of the Institute of Electrical and Electronics Engineers (IEEE) had recognized the need for publicly disseminated LAN standards and eventually published a specification of the carrier sense multiple access with collision detection (CSMA/CD) protocol, which provides enough detail for vendors to implement interoperable equipment. Furthermore, the definition of the standard was sanctioned by companies that participated in the IEEE committee's balloting process, so that the standard was viewed as an open, nonproprietary solution. The IEEE sponsors several committees that work on defining LANs to meet several needs. Some of the IEEE's committees and their charters are

- IEEE 802.3: Carrier sense multiple access with collision detection LANs
- IEEE 802.4: Token bus LANs
- IEEE 802.5: Token ring LANs
- IEEE 802.6: Distributed queue dual bus (DQDB) LANs
- IEEE 802.11: Wireless LANs
- IEEE 802.12: Demand priority LANs

Several other IEEE 802 committees have been established to work on issues common to all LANs (e.g., interconnection of LANs, management of LANs, higher layer link protocols, and LAN security). The IEEE is today the preeminent organization in LAN standardization.

The American National Standards Institute (ANSI), which represents the

United States of America in the International Organization for Standardization/International Electrotechnical Committee (ISO/IEC), has developed or cross-adopted LAN standards. ANSI's X3T9.5 and X3T9.3 Committees are responsible for the fiber distributed data interface (FDDI) and the high-performance parallel interface (HIPPI) LAN standards, respectively.

The Telecommunications Standardization Sector (TSS) of the International Telecommunication Union (ITU), which was formerly called the International Telegraph and Telephone Consultative Committee (CCITT), produces the body of standards that define the broadband integrated services digital network (BISDN), of which the asynchronous transfer mode (ATM) is a fundamental component. Although its focus is essentially on wide-area telecommunications, the ITU-TSS's work often embraces LANs.

Not strictly in the LAN business, the Telecommunications Industry Association (TIA) of the Electronic Industries Association (EIA) plays a crucial role in the LAN arena. The EIA produces and publishes standards for wiring buildings, which is a key concern of any user installing a LAN. These standards influence how a LAN's cable plant will be laid out and expanded. The EIA wiring standards recommend an approach to installing shielded and unshielded twisted-pair, coaxial, and optical-fiber cables in commercial buildings. The recommended topology is essentially a tree, by which the offices of users on a floor of a building are wired to a local telecommunications closet. The telecommunications closets are then wired to the building's intermediate cross-connect equipment room, which is often located in the building's basement. Intermediate cross-connect equipment rooms can be attached hierarchically to a designated main cross-connect equipment room in another building to interconnect all the campus's users. In the telecommunications closets and cross-connect rooms, all connections are made passively, without electrically active equipment (e.g., by patch panels). The distances between cross-connect equipment range from 500 to 2000 meters, depending on the type of media used. Although not optimal from the point of view of cable cost, this structured approach ensures flexibility for the LAN operator by greatly simplifying LAN installation and maintenance.

The Internet Society is an informal, user-supported body that manages the Internet protocols. This society—especially through its Internet Engineering Task Force (IETF)—is responsible for developing and maintaining Internet standards. Because the Internet is largely a collection of LANs, the IETF sponsors the development of LAN-related protocols. Unlike other standards-making bodies, however, the IETF concerns itself with formulating standards that facilitate the use of Internet protocols (e.g., the internet protocol, the transmission control protocol, and the user datagram protocol) in the LAN environment.

Another user group that is active in defining and promoting ATM technology and standards is the ATM Forum. The ATM Forum, an industry group with a charter to promote the development and acceptance of ATM networks, concentrates on filling in the gaps left by formal standards.

The LAN consists of switching nodes and links that connect the nodes. The spatial relationship among the nodes and links of a LAN is called the *topology* of the LAN. There are five basic topologies used in the LAN: the linear bus, the ring, the star, the tree, and the arbitrary mesh. The linear bus is a single transmission line to which nodes are connected by taps (see Fig. 1-*a*). In the ring,

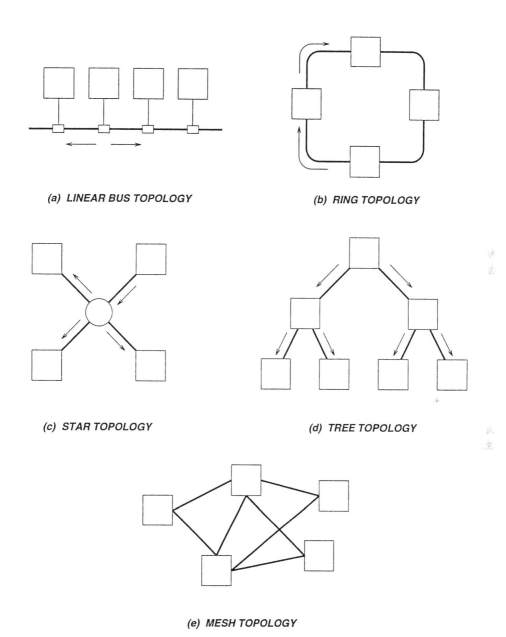

(a) LINEAR BUS TOPOLOGY

(b) RING TOPOLOGY

(c) STAR TOPOLOGY

(d) TREE TOPOLOGY

(e) MESH TOPOLOGY

FIG. 1   Examples of local-area network (LAN) topologies.

each node is connected to a successor node until the last node is connected to the first node to form a cyclic structure (Fig. 1-*b*). In the star, each node is connected to a central switching node (Fig. 1-*c*). The tree topology places nodes in the vertices of a tree (Fig. 1-*d*). The arbitrary mesh is an unstructured topology in which nodes may be interconnected in any way (e.g., as in Fig. 1-*e*).

The topologies discussed above can be augmented for the purpose of tolerat-

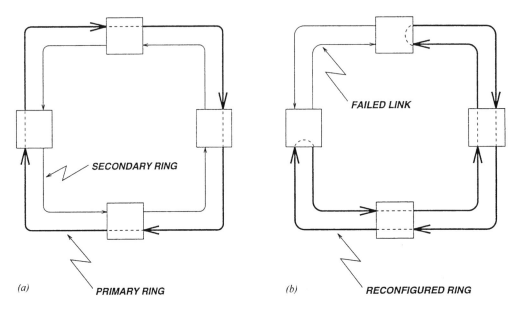

FIG. 2   Fault-tolerant dual ring: *a*, dual counterrotating ring; *b*, reconfigured nodes following a
failure.

ing specific classes of faults (e.g., media, node, and transceiver failures). A
commonly encountered, fault-tolerant topological structure is the dual counterrotating ring (Fig. 2-*a*). Nodes are attached to the dual ring by a pair of transmitters and a pair of receivers. One set of transceivers is the primary interface,
while the other is the secondary interface. Upon detection of a failure in the
medium, the nodes on either side of the failure reconfigure themselves so that a
new cyclic path is formed to replace the old one. The reconfigured ring is
illustrated in Fig. 2-*b*.

Local-Area Network Protocols

The LAN operates by means of *protocols*, which are rules that govern how two
or more nodes interact with each other. The protocols of interest in the LAN
are physical layer protocols, which specify how signals are encoded and transported on the medium; link layer protocols, which specify how nodes share the
medium to transmit blocks of information; and management protocols, which
specify how to perform essential actions (e.g., LAN initialization, node addition, node removal, and recovery from media faults). Several link layer protocols are used in the LAN. The type of link layer protocol used is heavily influenced by the topology of the LAN.

LANs that use broadcast media (i.e., media in which all nodes receive the
transmissions of any other node) must coordinate transmissions by means of a

multiple-access protocol. Broadcast media can be implemented in bus, ring, star, and tree topologies.

The linear bus topology often employs the carrier sense multiple access protocol, in which a node may transmit if it senses no other transmission in progress. Should two nodes, sensing the medium idle, simultaneously transmit messages, the messages will collide and be unintelligible to their intended recipients. Thus, the carrier sense multiple access protocol is often supplemented with a collision-detection capability, which permits curtailing the colliding transmissions and rescheduling them to be sent later. Carrier sense multiple access with collision detection provides nearly instantaneous media access at low traffic loading, but can become inefficient at high traffic loads since frequent collisions waste bandwidth and introduce additional delay on account of the rescheduling of messages. A crucial parameter in the design of such systems is the worst-case propagation delay between nodes, as this defines the time window of vulnerability during which, if two nodes initiate transmissions, then those transmissions will collide. Thus, all conditions being equal, the larger the worst-case propagation delay is, the higher will be the collision rate. Furthermore, since the rescheduling mechanism for collided messages introduces an essentially random rescheduling delay, there is no guarantee of deterministic service. Thus, the protocol is useful for the transmission of only asynchronous traffic. CSMA/CD may be used in other broadcast media, including the star and the tree. Broadcasting and multicasting are therefore accomplished in a natural way. True priorities are not easily implemented in this protocol.

Token passing is another widely used multiple-access technique. In token passing, a special message (the token) is passed from one node to its successor node. If a node receives the token and is prepared to transmit, then it temporarily removes the token from circulation and transmits its queued message. When it has finished transmitting, the node reintroduces the token, passing it to its successor node. The maximum length of time that the node may transmit is governed by the protocol's token-holding policy. The node may transmit no more than one queued message (one-shot service); it may transmit all messages queued at the time the node removed the token (gated service); it may transmit all messages, releasing the token only when its queue is empty (exhaustive service); or it may employ timers to dictate the token-holding time (timed service).

Because token passing realizes round-robin polling of nodes, token latency can prevent timely access to the medium. Even in a lightly loaded LAN, a node with a queued message must wait, on the average, for half the other nodes to examine and use the token, which imposes a relatively large delay before the waiting node has the opportunity to transmit its queued message. On the other hand, this latency is bounded, if the token-holding time at each node is bounded, and the scheme is capable of handling heavy traffic loads without suffering excessive delay, saturating only when the load approaches the LAN's capacity.

Asynchronous traffic is easily supported in token-passing LANs, but the support of synchronous traffic involves the use of sophisticated mechanisms to ensure that token-holding times are carefully controlled by special timers and that each node is explicitly allocated a specific share of the medium's bandwidth.

A well-known method for providing synchronous service in a token-passing LAN is the timed-token scheme, described below.

The most natural topology for token passing is the ring since it defines inherently a sequence of successor nodes among which the token is to be efficiently relayed. However, any broadcast medium easily supports token passing by including in the token's header the successor node's address, although the overhead of extra header fields and the possibility that a successor node might be relatively distant from its predecessor node introduce inefficiency. When token passing is used in the ring topology, it is called the token ring protocol; when it is used in the linear bus topology, it is called the token bus protocol. Broadcasting and multicasting are straightforward in both token ring and token bus LANs. Priorities can be provided in token ring and token bus LANs, either by demanding that queued, prioritized messages be transmitted in the order of their priorities during the time that the node holds the token or by the more sophisticated priority reservation scheme, in which a reservation to acquire the token at a higher priority is made by changing a field in the header of an in-transit message, so that when the token is released, it may be seized only by high-priority users.

Slotted multiple-access protocols rely on the fact that a node "sees" an uninterrupted stream of slots (i.e., short, fixed-length messages), which the node can read and modify on the fly. A slot is seen by a sequence of nodes in a specific order so that a linearly ordered topology (e.g., the bus or ring) must be used. In the bus, a dual bus must be provided so that a node can receive slots from nodes to its left and its right. A passing slot may be empty or occupied, as indicated by a "busy" bit in its header. An empty slot is available to the node, which can write data into appropriate fields of the slot, setting its busy bit so that it cannot be overwritten by downstream nodes. Slots are generated by a single node and transmitted back to back so that there are no gaps between them; consequently, the slots define a synchronized reference stream that can be used to transport asynchronous, synchronous, and isochronous traffic. To provide isochronous service, a fraction of the slots is dedicated exclusively for use by preassigned nodes. These dedicated slots can thus pass by the owning node on a periodic basis so that the sending and receiving nodes are synchronized and enjoy a fixed portion of the LAN's bandwidth. Slotted multiple-access protocols can provide low access delay under light-to-moderate traffic loads, with saturation occurring only when the load approaches the LAN's capacity.

A disadvantage imposed by the slot structure is that the transmitting node might need to segment a large message into many small slots, which would then be reassembled by the receiving node. Slotted LANs unfortunately suffer from unfairness when some nodes—because of their advantageous location on the medium—are able to "hog" an unfair portion of the passing slots, thereby depriving downstream nodes of empty slots. Mechanisms must be added to the protocol to prevent this from happening. Therefore, mechanisms—such as bandwidth balancing, discussed below—must often be introduced to ensure that all nodes enjoy fair access to the medium. Broadcasting and multicasting are straightforward since all nodes see every slot. Priorities can be implemented in slotted systems much as in token-passing systems.

Essential tasks, such as the addition and removal of nodes, are implemented by a management protocol to relieve the LAN operations staff of the burden of manually performing these tasks. In token-passing protocols, activities like node addition and removal can cause the token to be lost. In such cases, a new token must be issued. When timers are involved in multiple-access protocols, as in the timed-token scheme, such service disruptions might also invalidate the relationships among nodes' timers, so a protocol for dealing with out-of-bounds timers is also necessary. Similarly, the detection of a media fault and the actions to work around the fault are performed automatically by management protocols. Such a management protocol generally works on a link-by-link basis to deduce when transmissions have ceased on a specific link. After detection, the protocol activates redundant links and deactivates failed links, effecting a new signal path.

Select Local-Area Networks

In the sections below, several LANs in use today are described. Four classes of LANs are discussed: broadcast, switched, bit-parallel, and wireless LANs.

Broadcast LANs

In the broadcast LAN, every node receives every transmission. However, a node ignores any message not addressed to it. Broadcast LAN technology is cost effective and easy to implement. It avoids the use of complex switching and other centralized equipment. This appeals to personal computer and workstation users, who are unwilling to share common costs for centralized equipment—the broadcast LAN allows each user to attach by merely purchasing a host interface. Today, most client–server computing systems use broadcast LANs. Ethernet is the principal broadcast LAN.

Carrier Sense Multiple Access LANs

IEEE 802.3 Ethernet.   By far the most successful LAN in use is Ethernet, which is essentially the LAN defined in IEEE Standard 802.3. Rich in the options it offers, Ethernet at the physical layer is built on a number of different technologies, including cable, twisted-pair, and optical media. The primary standard, called 10BASE5, uses a 50-ohm coaxial cable with a diameter of 10 millimeters to transmit Manchester-encoded baseband signals at 10 Mb/s. The 10BASE2 option is similar to 10BASE5, but uses a less expensive, more flexible coaxial cable of 5-millimeter diameter and is sometimes referred to as "Cheapernet." Transceivers in 10BASE2 are also integrated into the node, which further lowers the cost of the LAN. The 10BASE-T option uses Category 3 unshielded twisted pairs for the transmission of Manchester-encoded baseband signals at 10 mb/s.

Similar to 10BASE-T, the newer 100BASE-TX option uses 100-ohm Category 5 unshielded twisted pairs or 150-ohm shielded twisted pairs for the transmission of 4B/5B-encoded baseband signals at the rate of 100 Mb/s; to meet legislated requirements for electromagnetic interference, scrambling is used so that signal energy is spread uniformly over a range of frequencies. 100BASE-TX Ethernet transmits a three-level code in which a line-state change represents the 1 bit and the absence of a line-state change represents the 0 bit. 100BASE-FX is similar to 100BASE-TX, except that it uses multimode fiber with a 62.5-μm core and a 125-μm cladding. 100BASE-FX Ethernet also uses the 4B/5B encoding, but transmits bits in a nonreturn-to-zero-inverted (NRZI) code. The final 100-Mb/s Ethernet is 100BASE-T4, which uses two twisted pairs to provide three data paths in the send and receive directions. An 8B/6T encoding is used to multiplex ternary sextets (six trits) onto the three data paths. Unlike the rest of the nBASE-T family, 100BASE-T4 does not have full-duplex transmit and receive capability.

The nBASE-T options now dominate the market because their twisted-pair media are compatible with EIA-recommended wiring approaches for commercial buildings, and the technology is essentially identical to telephone wiring. The 10BASE-F option transmits Manchester-encoded baseband signals at the 0.85-μm wavelength over multimode fiber with a 62.5-μm core and a 125-μm cladding; the data rate is 10 Mb/s. Intended to be operated as a single 10-Mb/s channel of a broadband 75-ohm coaxial cable, the 10BROAD36 option uses FSK or ASK over a 36-MHz bandwidth; this approach would allow multiple Ethernet channels to coexist in one cable plant.

The 10BASE5 and 10BASE2 Ethernet topologies are logical buses, which are made up of segments of cable interconnected by repeaters. Each 10BASE5 segment is less than 500 meters long and provides attachment for no more than 100 nodes, and each 10BASE2 segment is less than 200 meters long and provides attachment for no more than 30 nodes. The maximum transmission path in a 10BASE5 Ethernet is 5 segments. Nodes are attached by a transceiver, which usually connects to the cable by means of a piercing "vampire" tap. Between each node and its transceiver is a cable that runs no more than 50 meters. The maximum end-to-end lengths of 10BASE5 and 10BASE2 LANs are 2500 and 925 meters, respectively.

Ethernet uses CSMA/CD as its multiple-access protocol. If a node does not sense the presence of a carrier on the medium, it may transmit a message with a length that ranges from 64 to 1518 octets. If two nodes transmit nearly simultaneously, then their transmissions will collide with each other. The collision-detection feature of the protocol requires that transmitting nodes recognize when collisions occur. In the baseband versions of Ethernet, colliding signals produce an out-of-bounds direct-current component, which is easily detected by a node's transceiver. In the optical-fiber version, received signals are finely sampled for violations of the Manchester encoding (i.e., multiple levels within a half-bit time), which indicates the presence of two superimposed Manchester-encoded bit streams. When there is a collision on the medium, all nodes must abort their transmissions and refrain from retransmitting for a short time.

To reschedule a collided transmission, the node implements the so-called truncated binary exponential backoff algorithm, which injects a random delay between the time of collision and the retransmission of the message. If the message has already had k collisions, then the algorithm randomly chooses an integer r in the range $0 < r < 2^k$ and waits r time units (where a time unit is defined as the worst-case propagation delay) to retransmit; the value of r is limited to 1023, and the maximum number of retries is 16. This algorithm relieves temporary congestion by spreading the transmission times of contending messages over a successively wider range.

Token-Passing LANs

Among the token-passing LANs, the best known are those conforming to ANSI X3T9.5, IEEE 802.4, and IEEE 802.5 specifications, referred to as FDDI, token bus, and token ring, respectively.

ANSI FDDI. The FDDI is a high-speed ring based on optical-fiber technology. It is used in applications that require high-speed data transfer between high-performance computers (e.g., graphics and multimedia applications) and as a backbone for interconnecting LANs. However, the FDDI charter has been expanded to include electrical media as well; the wire-based LAN is sometimes referred to as the copper distributed data interface (CDDI).

The choice of media is wide: multimode, graded-index optical fiber; single-mode optical fiber; and unshielded twisted pair. All media support a data rate of 100 Mb/s and use the 4B/5B encoding. FDDI primarily uses multimode optical fiber of 62.5-μm core and 125-μm cladding, but other core/cladding dimensions are permitted (e.g., 50/125, 85/125, and 100/140 μm).

With multimode optical fiber, economical light-emitting diodes are used to generate lightwaves in the 1.3-μm window, and either PIN or avalanche photodiodes may be used as light detectors, but PIN photodiodes are preferred for their lower cost. The single-mode, optical-fiber option for FDDI specifies an 8.7-μm core in a 125-μm cladding. Laser diodes in the 1.3-μm window supply light for the single-mode optical fiber, and detection usually employs PIN photodiodes. It is even permitted to mix fiber dimensions on a LAN, as long as there is adherence to optical power budgets.

FDDI carries both asynchronous and synchronous traffic. Each node specifies its synchronous traffic requirement as a fraction of the LAN's capacity (i.e., as a fraction of the 100-Mb/s transmission rate). Obviously, the sum of all nodes' synchronous traffic requirements should be less than 1. When a node is passed a free token by the previous token holder, it may hold the token (and transmit messages) so long as its token-holding timer (THT) has not expired. FDDI makes use of the timed-token protocol to control the token-holding time at each node. In this protocol, which is essentially token passing with a sophisticated token-holding scheme, there is a LANwide target token rotation time (TTRT), which is agreed on by all nodes. The protocol guarantees that the elapsed time between successive visits of a free token to a specific node is on the average less than or equal to the value of the TTRT. The protocol also guarantees that this elapsed time will never exceed twice the value of the TTRT. Each node has a token rotation timer (TRT) that it sets to the value of the TTRT whenever the token is passed to it. (As soon as a timer is set to a value, it begins counting down at the rate of once per clock tick, expiring when its value reaches 0). When the token returns to the node, it sets the THT to the current value of the TRT and resets the TRT to the value of the TTRT, thus allotting excess token rotation time for use by the node if the THT has a positive value. Regardless of the value of the THT, the node may transmit any synchronous messages, and thereafter it may transmit its asynchronous messages as long as the THT has a positive value. Since each node has a contract to use no more bandwidth than is warranted by its synchronous traffic requirement, the mean token rotation time is less than the TTRT, and there should always be some excess capacity for asynchronous traffic.

Because transmissions in progress may continue even after THT has reached 0, the token might be late in returning to a node (as indicated by a value of 0 for the TRT). The late arrival of a token will not preempt synchronous transmissions, only asynchronous transmissions. The net effect is that, even when the token is late, its tardiness is bounded by the TTRT (i.e., the TRT will never exceed $2 \times$ TTRT). It has been proved mathematically that under this protocol the mean value of the THT is less than or equal to the TTRT and the maximum value of the THT is less than or equal to $2 \times$ TTRT.

Synchronous traffic is distinguished by eight priority levels, each with its own token rotation time requirement. The FDDI protocol guarantees that during each token rotation the bandwidth requirement of a higher priority message is satisfied before that of a lower priority message. The bandwidth not utilized by synchronous traffic is available to asynchronous traffic, which does not require the bounded access delay and guaranteed bandwidth of synchronous traffic. The net effect is that in FDDI low-priority messages are transmitted only if the LAN's utilization is low.

Another type of asynchronous traffic, called *restricted traffic*, is allowed. Specific nodes can be given exclusive access to the LAN for a short period of time (e.g., to download a large block of data or to back up a disk). Reminiscent of the token ring's priority scheme, nodes can enter into a restricted dialog when one of them sets a special bit in the token header that indicates the token may be used only by a restricted set of nodes. This gives the restricted set exclusive access to the ring for a limited period of time, as the remaining nodes are prevented from seizing the token and transmitting messages. So that the remaining nodes are not preempted forever, the node that originally converted the token from the unrestricted to the restricted format is responsible for reconverting it back to the unrestricted format. If the token remains in the restricted format beyond a specific length of time (e.g., because the converting node has failed), then error-recovery mechanisms automatically reconvert it to the unrestricted format. Restricted traffic is rarely supported in FDDI LANs.

Each message transmitted on the FDDI ring may contain a maximum of 4500 octets. Every message is protected by a cyclic redundancy code capable of detecting all single-bit errors.

The FDDI provides for fault tolerance through its dual counterrotating ring topology. No single failure of the media or a node can prevent full connectivity (see Fig. 2). If nodes are attached to the ring by bypass switches, then multiple node failures can be tolerated.

An FDDI LAN can include up to 500 nodes and cover a ring circumference of 200 kilometers.

IEEE 802.4 Token Bus. The IEEE 802.4 token bus was originally intended to be used on the factory floor. Its timed-token protocol is essentially identical to FDDI's and provides for synchronous data transfer, which is required for the real-time applications found in manufacturing and factory automation systems (e.g., robotic manipulation and process control). The chosen physical medium (75-ohm broadband coaxial cable) is well suited to the factory environment because of its immunity to the electromagnetic interference generated by factory equipment. Moreover, with frequency division multiplexing, the cable may be shared with other, non-LAN users, such as closed-circuit video and voice communication. The multichannel token bus usually employs a bidirectional midsplit cable in which signals in the spectrum from 5 to 108 MHz travel in the reverse direction from the sending nodes to a head-end, where they are translated (and possibly remodulated) to the spectrum from 162 to 265 MHz and sent in the forward direction to receiving nodes; it is also possible to use separate reverse and forward cables so that frequency translation at the head-end is not required. It is also possible to dispense with the head-end and operate the LAN as a single-channel, omnidirectional bus. The nodes connect to the trunk cable by means of short, 35-ohm to 50-ohm drop cables. Signals are either FSK (in single-channel LANs) or amplitude-modulated PSK (in multichannel LANs), with carrier spacings of 1.5, 6, and 12 MHz for data rates of 1, 5, and 10 Mb/s, respectively.

Token bus recently offered the option of using multimode optical-fiber media at data rates of 5, 10, and 20 Mb/s. The broadcast media are based on the optical star, over which Manchester-encoded baseband signals are transmitted. Light-emitting diodes are used to produce signals in the 0.8-μm window. To limit electromagnetic emissions when coaxial cable is used, before modulation the baseband data stream is scrambled according to the irreducible generating polynomial $1 + x^6 + x^7$.

Like FDDI, token bus uses the timed-token protocol. Token bus allows up to four priority levels — called access classes — for messages. The highest priority access class, which is reserved for ring-maintenance messages and other high-priority traffic, is analogous to the synchronous class in FDDI. The three lower priority access classes correspond to the top three asynchronous priority classes of FDDI. Token bus also permits a transmitting node to delegate the right to transmit to a node that is not its logical successor, creating a service that is comparable to FDDI's restricted-token service.

The addition and removal of nodes from the LAN is an important administrative function in token bus. Each node is obliged periodically to solicit a new node to join the logical ring. This is accomplished by the node issuing a solicit-successor message and waiting for a responder to reply with a set-successor message; insertion is completed when the soliciting node and the joining node establish new successor nodes. If more than one node responds, then a contention resolution process ensues to insert a single new node into the ring.

IEEE 802.5 Token Ring.   After Ethernet, the IEEE 802.5 token ring LAN is the most widely deployed. Token ring offers deterministic performance at a cost comparable to Ethernet. Consequently, it is most often used to interconnect personal computers in a client–server LAN. At the physical level, token ring uses shielded or unshielded twisted-pair media for a data rate of 4 Mb/s and shielded twisted-pair media for data rates of 4 or 16 Mb/s. Transmissions are differentially Manchester encoded. The topology is a physical ring in which each node is an active repeater. A maximum of 250 active nodes on the ring is specified. The ring is implemented in accordance with the EIA wiring guidelines by running a twisted pair from a group of nodes to a wiring concentrator. When the node is unpowered, it activates a bypass relay in the wiring concentrator to exclude the unpowered node from the ring.

Token ring uses straightforward token passing with a fixed token-holding time at each node. However, many implementations enforced the one-shot service discipline, thereby simplifying the design of the media access controller. The original protocol required that a node could not release the token until it had received and stripped from the ring the last message it had transmitted. This rule militates against efficient utilization of the medium, especially when the ring latency (time for a bit to rotate around the ring) is high. The protocol has been amended to allow for the early release of the token, which means that the token may be forwarded without the releasing node waiting to receive and strip its last transmitted message.

Token ring supports eight LANwide priority classes. The token is labeled with a priority, and nodes may seize the token for transmitting messages of equal or higher priority than the token. Furthermore, every message header is labeled with a priority. A node can make a reservation at a higher priority level by changing the priority label in the header of an in-transit message so that when the message is stripped from the ring by its sender, the token is released at the higher priority level. This gives the reserving node preferred access since only nodes with higher priority messages are now allowed to seize the token. After seizing the token and transmitting its higher priority message, the reserving node is then responsible for releasing the token at the priority it possessed when the reservation was made. Nested reservations can be used to realize the successive preemption of a lower priority level by a higher one. Token ring's approach to priorities is noteworthy in that it implements a global scheme that orders the transmission of all waiting messages in the LAN.

Slotted LANs

The slotted LANs are designed to provide isochronous data-transfer services. Each node receives data in short, fixed-length slots that periodically traverse the medium at a constant rate. The principal slotted LANs are ANSI's FDDI II, an enhanced version of the original FDDI LAN, and the IEEE 802.6 DQDB LAN. These LANs have not yet been widely deployed.

ANSI FDDI II.   FDDI's lack of support for isochronous traffic has led ANSI to define the FDDI II standard, which is compatible with the original FDDI standard but provides isochronous service as well. FDDI II uses the same ring topology, media, and transmission rate as FDDI. Every FDDI II LAN has a specially designated node called the cycle master that is capable of generating and stripping 125-μm time slots that contain fixed-length messages. A fixed-length message is made up of 1560 octets followed by a 20-bit spacer gap that precedes the next slot. Of those 1560 octets, 1536 octets are arranged in 16 wideband channels of 96 octets each. Thus, a wideband channel of 96 octets is repeated every 125 μs, achieving an effective data rate of 6.144 Mb/s. These wideband channels can be dedicated to isochronous users.

FDDI II accommodates two operating modes: the basic mode, in which the FDDI token-passing protocol is overlaid on top of all wideband channels at the rate of 98.384 Mb/s, and the hybrid mode, in which the FDDI protocol is overlaid on top of fewer than 16 wideband channels and the remaining wideband channels are dedicated to the transport of isochronous traffic. As wideband channels are allocated for synchronous use, the effective bandwidth usable by the FDDI token-passing protocol diminishes. In addition, there are 12 special octets in every slot that must be used to carry the FDDI token-passing protocol messages; this is necessary to ensure that a minimum of 768 kilobits per second are assigned to carry ring-monitoring and management information. Thus, the maximum speed at which FDDI messages can be transmitted is 99.072 Mb/s. In the hybrid mode, the wideband channels dedicated to isochronous traffic can be assigned in whichever manner is desired. In the 12-octet header are two octets that are reserved as a maintenance channel; this channel provides isochronous service at 128 kilobits per second and could be used for voice communication among operations and maintenance personnel.

One wrinkle in operating the FDDI token-passing protocol over wideband channels is that the data rate might vary. It is essential, then, that the parameters of the timed-token protocol be carefully chosen to account for such variation. In particular, the target token rotation time should be large enough so that a decrease in the effective transmission rate does not overly retard the rate of token rotation.

IEEE 802.6 DQDB. DQDB is intended to be used as a metropolitan-area network (MAN), which would usually be operated as a public utility. Since DQDB would be operated over public telecommunications transmission facilities, it adopts the physical protocols of these networks, which provide for data rates of 44.736, 155.520, and 622.080 Mb/s over electrical and optical media in accordance with the synchronous digital hierarchy (SDH) of the ITU-TSS. DQDB uses a dual-bus topology (Fig. 3). For added reliability, the end nodes of the bus can be connected to each other to form a looped-bus variant of the dual-bus topology.

The looped-bus DQDB is a dual-ring topology similar to that of FDDI. Each end node is a slot generator for a particular direction of the bus. A transmitting node sends to a destination node either on Bus A or Bus B, depending on their relative positions. The slots are 53 octets in length and contain a field to indicate whether the slot is occupied and a request field, the purpose of which is explained below. A fixed fraction of the slots may be reserved for use by isochro-

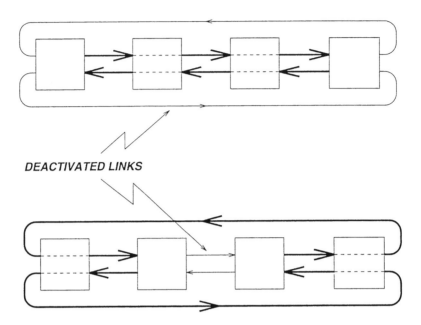

DEACTIVATED LINKS

FIG. 3   Distributed queueing dual bus (DQDB).

nous traffic in a manner similar to FDDI II's wideband channels. The remaining slots are available for use by nonisochronous traffic. If a node has a message (which is small enough to fit within a slot) to send to another node on Bus A, then it makes a request to send by marking the request field of a slot passing by on Bus B. This indicates to all upstream nodes that the marking node has made a reservation to transmit on Bus A. Each node counts how many requests have been made by downstream nodes, decrementing this count when it observes an unoccupied nonisochronous slot passing by on Bus A, as this slot will be used to satisfy one of the pending requests. When the request count is 0, then the node may use the next free slot on Bus A to transmit its message. The request mechanism in essence implements a LANwide queue of messages waiting to transmit on a particular bus.

It was observed that the original DQDB access algorithm under heavy load would unfairly favor nodes located near the end points of the bus by granting to them a greater share of slots than to nodes near the midpoint of the bus. Nodes at the end points are able to monopolize request slots (which are generated by the nodes at each end point) before they reach the nodes near the midpoint of the bus. The original algorithm was therefore enhanced to improve its fairness. The enhancement, called *bandwidth balancing*, forces a node to forfeit a fraction of its bandwidth by periodically foregoing the use of a free slot, even when it has the right to do so. The scheme wastes a small amount of bandwidth, but ensures that a group of N nodes transmitting at peak rate will all receive slightly less than $1/N$ of the total bandwidth.

DQDB supports priorities by maintaining a separate distributed queue for each of its three priority levels. A request of a given priority level is placed in the distributed queue of the same priority level. All messages in a higher priority queue are transmitted before messages in a lower priority queue. Within a queue, however, messages are served in the order of their arrival.

Switched LANs

In contrast to the shared-media broadcast LAN, the switched LAN incorporates switches connected by point-to-point links. A switching LAN is not subject to the bandwidth limitations of a broadcast LAN, in which the aggregate throughput cannot exceed the media speed. Rather, a switching LAN can use its links in parallel to achieve an aggregate throughput that surpasses the media speed. The concept of switched versus broadcast LANs is illustrated in Fig. 4.

Switched Ethernet

Switched Ethernet is an approach in which Ethernet segments are attached to a central high-speed switch. Switched Ethernet products have become very popu-

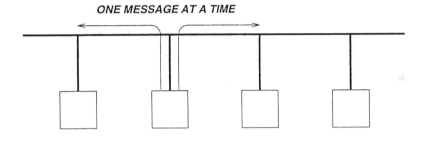

BROADCAST LAN

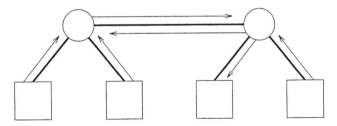

SWITCHED LAN

FIG. 4 Switched versus broadcast local-area networks.

lar, allowing LAN operators to relieve their heavily utilized Ethernet LANs and to extend a LAN's life by several more years. The principal cost of such an upgrade is the price of the switch—no host interfaces or cabling need to be replaced. Several vendors supply products, some of which can switch up to 64 segments. Switching is very attractive in a client–server computing environment because high-performance servers can be placed on individual segments so that their transmissions do not compete with the traffic generated by clients' requests. A node uses the Ethernet CSMA/CD protocol to place a message on its home segment. When the message arrives at the central switch, it is switched to the segment of the node to which the message is addressed. Buffering of the message may occur at either the input or output of the switch. When the switch gains access to the outgoing segment, the message is transmitted to its destination node.

Ethernet switches commonly support both the 10BASE-T and 100BASE-T families of media. Because the upper-layer protocols in some LANs operate in a request–response mode that permits no more than one message per session to be unacknowledged at any given time, there is the concern that extra hops through intermediate switches could contribute unacceptable delay to the message. To reduce the latency of messages, Ethernet switches commonly use cut-through routing rather than store-and-forward routing. While store-and-forward routing requires a message to be buffered in its entirety during switching, cut-through routing allows the switch to begin sending the head of a message before it has received its tail. Thus, the message can be received and acknowledged by the destination with minimum delay.

ATM LANs

Asynchronous transfer mode technology is intended for local-, metropolitan-, and wide-area network (WAN) deployment. Currently, however, most ATM equipment is used in LANs. ATM has the benefits of supporting all traffic classes; providing a common, seamless interface between local- and wide-area networks; affording modular growth with its switch-based implementation; and offering a variety of media types that may be mixed in one network.

The ATM LAN operates at a variety of data rates, including 44.736, 100, 155.520, and 622.080 Mb/s over electrical and optical media. The 100-Mb/s speed uses the transparent asynchronous transmitter/receiver interface (TAXI) link protocol, and the other speeds use the SDH link protocol. The TAXI protocol is run over multimode optical fiber at the 1.3-μm wavelength. It uses the 4B/5B encoding. The SDH protocol is run over Category 5 unshielded twisted pair as well as multimode and single-mode optical fiber at the 1.3- and 1.5-μm wavelengths. It uses NRZ encoding, which requires that the bits be scrambled to avoid long runs of zeros or ones. The irreducible generator polynomial is $1 + x^6 + x^7$, as with DQDB. In addition, when ATM messages are mapped to SDH formats, another scrambler, based on the generator polynomial $1 + x^{43}$, is applied to the data to reduce the chance that an ATM message would produce a sequence of bits that correlates with the output of the SDH scrambler.

The ATM LAN consists of two different interfaces. The interface between a host and a switch is called a private user–network interface, while the interface between connected switches is called a private network–network interface. The interface between a public ATM network and an ATM LAN—whether through a switch or a host—is called the public user–network interface.

The ATM LAN carries messages of 53 octets. The ATM LAN is a connection-oriented network, so each message is associated with a connection between a source node and a destination node. Connections may be established permanently or on an as-needed or switched basis. A signaling protocol is needed to support the setup and teardown of switched connections.

In order to provide different services to higher layer protocols, the ATM layer services must be enhanced by a so-called ATM adaptation layer (AAL). Different AALs have been defined for different applications. The Type 1 AAL provides the services needed by constant-bit-rate isochronous traffic. The Type 2 AAL provides the services needed by variable-bit-rate isochronous traffic. The Type 3 AAL provides the services needed by variable-bit-rate connection-oriented nonisochronous traffic. The Type 4 AAL provides the services needed by variable-bit-rate connectionless nonisochronous traffic. The Type 5 AAL is a streamlined, performance-tuned version of the Type 4 AAL.

To enforce the rate of flow in a connection, the preferred method is the so-called leaky bucket algorithm.

Most ATM switches provide explicit support for multicasting. The signaling protocol supports the management of multicast groups by growing and shrinking connection trees in response to leaf-initiated requests to join or depart from the group.

Bit-Parallel LANs

Most LANs transmit data in a serial stream of symbols. To achieve very high data rates, a small number of LANs transmit data over bit-parallel electrical wires. In this sense, these LANs are somewhat like the extended backplane buses that are found inside digital computers. Consequently, they tend to have a very short span. Two bit-parallel LANs in use today are the Myrinet and HIPPI LANs.

Myrinet

The proprietary Myrinet LAN is manufactured and marketed by Myricom and has its roots in the multicomputer world, in which it was used as the interconnection network for a prototype parallel computer. It uses eight-bit-wide data paths between LAN elements, operating at a data rate of 640 Mb/s. In reality, the data path is nine bits wide, one bit indicating whether the other eight bits are to be interpreted as a data or a control symbol—this is essentially a type of 8B/9B encoding. Thus, in addition to data octets, several other nondata symbols are possible. Symbols are transmitted in an NRZI encoding, so that the 1 bit is always encoded as a transition. The physical medium is a flexible shielded cable measuring 12 millimeters in diameter and containing 20 twisted pairs of copper wire. The data channel is a full-duplex, point-to-point link from a host interface to a switching node or between switching nodes. The topology is arbitrary, being any configuration of interconnected host interfaces and switching nodes. Each switching node can accept several links, and nodes of up to 16 ports are manufactured. The Myrinet LAN has a limited spatial coverage since the maximum link length is 25 meters.

The Myrinet switches are simple, blocking switches that make switching decisions for a message by examining the source-supplied routing information in the message's header. The complete route from the source node to a destination node is supplied to the requesting source node by a special route-manager software entity. Myrinet uses a form of cut-through routing called *wormhole routing*, in which the head of the message may arrive at its destination node before the tail has even left the source node. This keeps latency very low. If an in-transit message is blocked at a switch, then the progress of the entire message is halted by backpressure; that is, the blocked switch sends control messages back toward the source, notifying it to cease transmission temporarily. To reduce message latency, these switches can switch a message in less than 600 nanoseconds.

The full-duplex channels use symbol-by-symbol, stop-and-go flow control. Special STOP, GO, and IDLE symbols are available for controlling the flow of messages. Every Myrinet host and switch interface has a so-called slack buffer that holds a small number of in-transit symbols. The size of the slack buffer is enough to hold twice as many symbols as can propagate simultaneously on a maximum-length link (27 symbols). When the receiving slack buffer has filled beyond a threshold (e.g., because an output port of the switch is blocked by the transmission of another message), the receiver sends a STOP symbol to halt the incoming flow. Since it could take up to a full link propagation delay for the STOP to arrive, the buffer must be able to absorb at least a link's worth of symbols beyond the threshold. When the sender receives the STOP, it immediately throttles its flow, sending IDLE symbols instead. The IDLE symbols are merely discarded by the receiver, so they consume no buffer space. When the switch's port unblocks and the slack buffer has drained below the threshold, the receiver sends a GO symbol to restart the flow.

A Myrinet message consists of a maximum of 5.6 million symbols.

Multicast is not supported in Myrinet. A source node would have to transmit to each destination node a copy of the multicast message.

High-Performance Parallel Interface

The HIPPI LAN standard is controlled by the ANSI X3T9.3 committee, whose responsibility is the computer–peripheral interface. HIPPI uses shielded cables that contain either 32 or 64 parallel data paths. The 32-bit-wide LAN is run at 800 Mb/s, and the 64-bit-wide LAN is run at 1.6 Gb/s. The 32-bit-wide version of HIPPI has a cable of 50 twisted pairs, of which 43 are active signal wires; the 64-bit-wide version uses two 32-bit-wide cables. HIPPI's links are point to point and simplex; thus, two cables are required for a full-duplex channel. A full-duplex channel is not mandatory, however, and some applications (e.g., frame buffers) only need a simplex channel. The cable has a maximum length of 25 meters. The data is transmitted in an NRZ encoding, and there is a parity bit transmitted for each octet of the dataword. Like Myrinet, HIPPI consists of host interfaces and switches.

HIPPI is a connection-oriented LAN, and a connection from the source node to the destination node must be set up before a message can be transmitted. It is normal for implementations to tear down the connection as soon as a group of back-to-back messages has been transmitted.

The maximum message size is 1024 data octets in the 800-Mb/s version and 2048 data octets in the 1.6-Gb/s version. Thus, 6.4 μs are consumed in transmitting a maximum-size message in both versions of the HIPPI LAN.

Multicast is not supported in HIPPI. A source node would have to transmit to each destination node a copy of the multicast message, opening a new connection for each transmission.

Wireless LANs

The wireless LAN fits into many niches that are inappropriate for tethered LANs. Yet, the wireless environment is usually harsher than that found in wireline systems, because of interference, multipath fading, and noise; thus, special measures must be taken in developing wireless LANs. Probably the most important justification for installing a wireless LAN is to enable roaming. This is essential in applications that require mobility, especially when the emphasis is on customer-oriented sales and service (e.g., point-of-sale terminals, customer service, and guide services). The wireless LAN is also found outdoors or wherever permanent infrastructure is not accessible. At times, it may also be desirable to turn to a wireless LAN for esthetics reasons (e.g., in architectural landmarks where the installation of wires or cables would not be advisable). Currently, wireless LANs do not command a significant share of the LAN market.

The IEEE 802.11 Committee has begun work on defining a wireless LAN standard. The 802.11 standard specifies direct-sequence and frequency-hopping spread-spectrum transmission in the 2.4-GHz ISM and 0.8-μm infrared bands. Transmit power is in the range of 10 to 1000 milliwatts. A basic data rate of 1 Mb/s and an enhanced data rate of 2 Mb/s are supported. A variant of CSMA/ CD—called carrier sense multiple access with collision avoidance and acknowledgment, in which a node requests from its base station permission to send a message and receives acknowledgment of its receipt—is the prescribed media access method. PSK and Gaussian minimum shift keying are employed in the radio-frequency band, and pulse position modulation is used in the infrared band. Many wireless LANs are marketed, but only four are discussed below: the InfraLAN token ring, the NCR WaveLAN, the Motorola ALTAIR LAN, and the Windata Freeport LAN.

InfraLAN Token Ring

The InfraLAN token ring is compatible with the IEEE 802.5 token ring. Employing light-emitting diodes, it uses line-of-sight infrared signals to achieve data rates of 4 and 16 megabits per second. Computers are attached to base units that present an electrical, wire-based token ring interface on one side and an infrared interface on the other side. The base units are connected together to form a larger ring.

AT&T's WaveLAN

AT&T's WaveLAN product uses radio-frequency signals in the ISM band from 902 to 928 MHz. Transmissions are in a direct-sequence spread-spectrum format at the rate of 2 Mb/s. Nodes use omnidirectional antennas over a range of about 240 meters. For added security, the node interfaces incorporate hardware to implement the digital encryption standard.

Motorola's ALTAIR LAN

Motorola's ALTAIR LAN transmits radio-frequency signals in the 18-GHz band using quaternary FSK. An area of more than 4600 square meters (equivalent to a commercial building) can be serviced by the LAN. The LAN is compatible with 10BASE2, 10BASE5, and 10BASE-T Ethernet but uses a data rate of 5.7 Mb/s. Up to 50 nodes can be accommodated. The signal may optionally be scrambled to enhance privacy, since the signals are vulnerable to interception.

Windata Freeport LAN

Like the ALTAIR LAN, the Windata Freeport LAN is a radio-frequency, Ethernet-compatible, wireless LAN with a 5.7-Mb/s data rate. The Freeport LAN, however, uses direct sequence spread spectrum in the ISM bands from 2.400 to 2.485 and from 5.745 to 5.830 GHz. The different frequency bands are used for transmission in different directions. Modulation is by a 16-level, PSK trellis code. The LAN consists of Ethernet-connected hubs that communicate wirelessly with host-attached transceivers, allowing hosts to roam from one location to another. The maximum separation distance between a hub and a transceiver is 80 meters. To reduce multipath fading, the receiver incorporates correlators and sums their outputs to minimize multipath interfence. Up to 256 nodes and 62 wireless transceivers are permitted on the LAN.

The Future of Local-Area Networks

The future of the LAN is not certain. Clearly, LANs occupy an important place in computer technology, but it is unclear which LANs will dominate over the long run. Some trends are distinguishable, however.

The distinctions between the computer bus, the LAN, the MAN, and the WAN have become blurred. Several LANs have crossed over into the territories formerly reserved for buses or MANs. The emergence of seamless ATM networks also eliminates many traditional differences between the LAN and the WAN. Although the tendency has been for LANs to cover greater geographical regions, LANs such as HIPPI and Myrinet are designed for limited coverage, relying on bridges and routers to interconnect individual LANs. The wireless world is in a state of flux, as wireless, cellular telecommunications networks are being planned or deployed to provide global coverage for mobile and stationary computers.

There is now a large set of LAN standards, as well as a number of proprietary LANs, and their numbers keep increasing. There is little differentiation among several of these classes of LANs. Because of the intense competition in the LAN market, most observers agree that many standards will never be deployed and most proprietary solutions will fail. In the marketplace today, many different LANs coexist, and this situation is expected to continue. The greatest challenge to LAN diversity now comes from the ATM arena, which makes promises of an integrated services network centered about the 53-octet cell, regardless of whether it is traveling over the local, metropolitan, or wide area. ATM equipment is being deployed rapidly, propelled by its support for several types of media and data rates. On the other hand, ATM is immature, and many details remain to be worked out before widespread acceptance occurs. Furthermore, ATM's connection-oriented style is at odds with the connectionless style used in nearly all LANs.

Upper layer protocols, such as the TCP/IP and the network operating systems commonly found in personal computer LANs, are the soul of the LAN, accounting for much of its intelligence and complexity. Implemented in software, upper layer protocols are often performance bottlenecks, rarely being able to match the peak data rate of the LAN. Furthermore, upper layer protocols are commonly embedded in the computer's operating system, which discourages frequent performance tuning by vendors or users. Nevertheless, upper layer

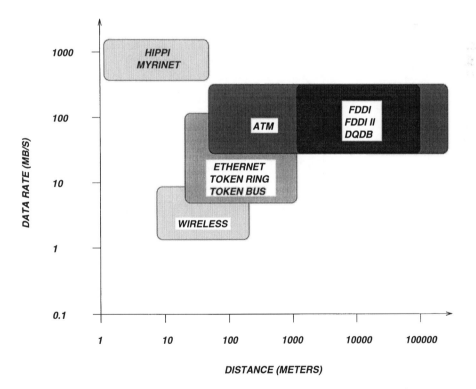

FIG. 5   Distance–speed characteristics of local-area networks (HIPPI = high-performance parallel interface; ATM = asynchronous transfer mode; FDDI = fiber distributed data interface).

TABLE 2 Local-Area Network Characteristics

LAN	Media Access Protocol	Data Rate (Mb/s)	Media Types	Modulation	Applicable Standards	Traffic Support
Ethernet	CSMA/CD	10–100	Coax, UTP, MMF	Manchester, 4B/5B, 8B/6T	IEEE 802.3	Async
Token ring	Token passing	4–16	UTP, STP	Manchester	IEEE 802.5	Async
FDDI	Timed-token passing	100	MMF, SMF, UTP	4B/5B	ANSI X3T9.5	Async, sync
Token bus	Timed-token passing	2–20	Coax	FSK, AM/PSK	IEEE 802.4	Async, sync
FDDI II	Slotted ring	100	MMF, SMF	4B/5B	ANSI X3T9.5	Async, sync, iso
DQDB	Slotted bus	44–622	MMF, SMF, UTP	NRZ	IEEE 802.6	Async, sync, iso
ATM	Circuit switching	44–622	MMF, SMF, UTP	NRZ, 8B/10B	ITU I series	Async, sync, iso
Myrinet	Wormhole routing	640	STP	NRZI, 8B/9B	Proprietary	Async
HIPPI	Circuit switching	800–1200	STP	NRZ	ANSI X3T9.3	Async
InfraLAN	Token passing	4–16	IR	Manchester	IEEE 802.5, proprietary	Async
WaveLAN	DSSS, CSMA	2	ISM		Proprietary	Async
ALTAIR	CSMA/CD	5.7	18 GHz	FSK	IEEE 802.3, proprietary	Async
Freeport	FHSS	5.7	ISM	PSK	IEEE 802.3, proprietary	Async

UTP = unshielded twisted pair; MMF = multimode fiber; STP = shielded twisted pair; SMF = single-mode fiber; IR = infrared; DSSS = direct sequence spread spectrum; FHSS = frequency hopping spread spectrum.

protocol performance must improve if users are to see the benefits of evolving LAN technology. This is even more critical when one considers that new LAN technologies are being asked to support more sophisticated services, many of which will be implemented primarily in software. An important area of research and development, therefore, is how to achieve scaleable performance in upper layer protocols.

The rapid spread of LANs has exposed networked computers to malicious attacks by miscreants. This unacceptable situation has led to the development of security measures to protect LAN-based computers, which are often connected to WANs that provide a portal for intrusions. The widespread use of firewalls (or security guards) and intrusion-detection software prevents unauthorized access by external users. These tools are now indispensable, but not foolproof. Therefore, LAN administrators are demanding additional protection, especially in the areas of privacy, authentication, and auditing. These features will become even more important as departmental LANs expand to encompass entire organizations.

Historically, the major LANs have all used broadcast media (e.g., Ethernet, token ring). These broadcast LANs have the advantage of simplicity, requiring only wires and an interface card. Most of the intelligence is outside the network. As the number of nodes on a LAN and their geographical dispersion have increased, the limitations of the broadcast media approach have become obvious: users get a decreasing fraction of the total bandwidth, and extending the media becomes more challenging. Moreover, because the network itself has so little intelligence, it is difficult to support network management. Some of these symptoms have been partially alleviated in Ethernet by the proliferation of hubs, which provide some network-management capabilities and make it possible to segment the media for more efficient use of bandwidth. Hub-based LANs have paved the way for the introduction of switched LANs such as switched Ethernet. Nearly all LAN standards currently in preparation are switch based. There is essentially no limit to the throughput of a switched LAN, so that growth in node populations and traffic load may be accommodated by merely adding switches. The management of these LANs is simplified because the switches contain processors that can be used to implement sophisticated management functions.

Yet, many challenges remain. Particularly troublesome are the rising costs of LAN operations and support. Significant effort is needed for an organization to plan, install, and maintain a LAN. While network-management systems help with these activities, they are only partial solutions. Better tools to support LAN administration are needed.

Conclusion

The LAN is an important element of current computer systems. A wide range of LANs has been developed to meet the various needs of computer users. A distance–speed plot of several operational LANs is shown in the graph of Fig. 5. Table 2 summarizes the characteristics of the major LANs.

Glossary

ALTAIR.   A wireless LAN produced by Motorola.

AMERICAN NATIONAL STANDARDS INSTITUTE (ANSI).   Principal standardization body in the United States of America.

ASYNCHRONOUS TRAFFIC.   Traffic in which there is no requirement for a timing relationship between the sender and the receiver.

ASYNCHRONOUS TRANSFER MODE (ATM).   A form of cell relay that forms the basis for the Broadband Integrated Services Digital Network.

ATM FORUM.   A standardization body focusing on ATM networking.

BANDWIDTH BALANCING.   A mechanism used to ensure fairness in DQDB.

CABLE PLANT.   The physical structure of a LAN.

CARRIER SENSE MULTIPLE ACCESS WITH COLLISION DETECTION (CSMA/CD).   A protocol in which the transmitter may send a message if it senses an idle carrier, but must abort transmission if it detects a collision.

CONNECTIONLESS SERVICE.   A service model in which nodes communicate on a "best effort" basis.

CONNECTION-ORIENTED SERVICE.   A service model in which communicating pairs appear to have a dedicated channel between them.

CUT-THROUGH ROUTING.   A form of switching that permits the head of a message to be forwarded before its tail has been received.

DIRECT DETECTION.   A method to demodulate wideband lightwave signals by using a photodetector to measure the energy of the incoming signal.

DISTRIBUTED QUEUING DUAL BUS (DQDB).   IEEE 802.6 MAN.

ELECTRONICS INDUSTRIES ASSOCIATION (EIA).   A group that produces building wiring standards relevant to LANs.

ETHERNET.   The colloquial name for the IEEE 802.3 LAN that uses carrier sense multiple access with collision detection.

FDDI II.   A slotted ring adaptation of FDDI that supports isochronous traffic.

FIBER DISTRIBUTED DATA INTERFACE (FDDI).   A timed-token ring that supports synchronous traffic.

FREEPORT.   A wireless LAN produced by Windata.

HIGH-PERFORMANCE PARALLEL INTERFACE (HIPPI).   A fast LAN used primarily in supercomputer installations.

INDUSTRIAL, SCIENTIFIC, AND MEDICAL (ISM).   A part of the radio-frequency spectrum reserved for unlicensed industrial, scientific, and medical use.

INFRALAN.   A wireless LAN produced by InfraLAN Technologies that uses infrared signals.

INSTITUTE OF ELECTRICAL AND ELECTRONICS ENGINEERS (IEEE).   A professional organization sponsoring Project 802 that produces and maintains many of the most important LAN standards.

INTERNET ENGINEERING TASK FORCE (IETF).   A group that controls the Internet standards.

ISOCHRONOUS TRAFFIC.   Traffic in which there is a strong timing relationship between the sender and receiver.

LIGHT DETECTOR.   A device that converts lightwave signals to electrical signals.

LIGHT SOURCES.   A device that converts electrical signals to lightwave signals.

LOCAL-AREA NETWORK (LAN). A network that interconnects an enterprise's computer equipment within a limited geographical region.

MANAGEMENT INFORMATION BASE. A collection of all managed attributes of a network.

MEDIA ACCESS CONTROL. The protocol that allows several nodes to transmit messages without interfering with each other.

MULTICAST TRANSMISSION. Transmission of a message that is received at more than one destination.

MULTIMODE OPTICAL FIBER. Large-core optical fiber that propagates many modes of a lightwave.

MYRINET. A high-speed proprietary LAN produced by Myricom.

NETWORK MANAGEMENT. The activity of monitoring, controlling, and administering a network.

PRIORITIES. An attribute of a message that determines the order in which it will be served.

SCRAMBLING. The act of transmitting the symbols of a message in permuted order to reduce electromagnetic emissions.

SINGLE-MODE OPTICAL FIBER. A small-core optical fiber that propagates only one mode of a lightwave.

SLOTTED. Refers to LANs in which a fixed-length frame structure is repeatedly broadcast or transmitted.

SPREAD SPECTRUM. A form of code division multiple access that uses a wideband signal to allow multiple transmissions to occupy a medium simultaneously.

SWITCHED ETHERNET. The use of switches to connect multiple Ethernet segments.

SYNCHRONOUS TRAFFIC. Traffic in which messages have delivery deadlines.

TELECOMMUNICATIONS STANDARDIZATION SECTOR (TSS). The principal international standardization body for telecommunications.

TOKEN BUS. The colloquial name for the IEEE 802.4 token-passing bus.

TOKEN PASSING. A multiple-access protocol in which transmission rights are granted by circulating a token among nodes.

TOKEN RING. The colloquial name for the IEEE 802.5 token-passing ring.

TOPOLOGY. The way in which nodes and links of a network are connected.

TRUNCATED BINARY EXPONENTIAL BACKOFF. The algorithm that is used in carrier sense multiple access with collision detection LANs to resolve collisions by inserting a random delay between retransmission attempts.

WAVELAN. A wireless LAN produced by AT&T.

WIRELESS. Refers to LANs that use atmospheric propagation of electromagnetic signals.

Bibliography

Stallings, W., *Local Networks*, 3rd ed., Macmillian, New York, 1990.

    A popular, frequently updated textbook on LANs.

Jain, R., *FDDI Handbook*, Addison-Wesley, Reading, MA, 1994.

    An excellent source of information about FDDI.

Johnson, H. W., *Fast Ethernet: Dawn of a New Network*, Prentice-Hall, Upper Saddle River, NJ, 1996.

A thorough compendium of Ethernet technology and standards.

Leading journals that publish research and survey articles on LANs include

- *IEEE/ACM Transactions on Networking*
- *Computer Networks and ISDN Systems*
- *IEEE Network*
- *IEEE Communications*

Four annual conferences that cover the topic of LANs are

- The IEEE INFOCOM Conference on Computer Communications
- The IEEE Conference on Local Computer Networks
- The ACM SIGCOMM Conference on Communications Architectures and Protocols
- The EFOC/LAN European Fiber Optic Communications and Local-Area Networks Conference

LAN standards may be ordered from the organizations that produced them. A list of these organizations and their addresses follows.

- American National Standards Institute (ANSI)
  ANSI Sales Department
  11 West 42nd Street
  New York, NY 10036
- ATM Forum
  480 San Antonio Road, Suite 100
  Mountain View, CA 94040-1219
- Electronic Industries Association (EIA)
  EIA Engineering Publications Office
  2001 Pennsylvania Avenue, N.W.
  Washington, DC 20006
- Institute of Electrical and Electronics Engineers, Inc. (IEEE)
  IEEE Customer Service
  445 Hoes Lane
  P.O. Box 1331
  Piscataway, NJ 08844-1331
- International Organization for Standardization (ISO)
  1 Rue de Varembé
  Case Postale 56
  CH-1211, Genève 29, Switzerland
- International Telecommunication Union (ITU)
  ITU Sales Section
  Place de Nations
  CH-1211, Genève 29, Switzerland

Alternatively, there are companies that specialize in selling copies of these standards.
Many Internet LAN standards are available as Requests for Comments (RFCs). A central Internet repository for these RFCs is found at ds.internic.net and several other sites on the Internet.

JOSEPH BANNISTER
MARIO GERLA

Networks for Wireless Digital Communications

Introduction

In the past several years, wireless communications have seen an explosive growth in the available number of services and types of technologies. Cellular telephony, radio paging, and cordless telephony have become commonplace and the demand for enhanced performance and capacity is growing. It is expected that, in the near future, wireless data and multimedia services will also become popular and that there will be an increasing need for spectrum-efficient radio networks to support such services.

Contrary to the common belief held in the late 1970s, radio communications have not been killed by fiber-optic communications. First, satellite communications proved to be a flexible way to support special end users by small maritime and land mobile terminals. The most conspicuous example was probably the small dish immediately unfolding in all the foreign locations visited by the roving U.S. Secretary of State Henry Kissinger. Such new, individual access links were different from the original multiplexed satellite trunk connections, which optical cables gradually have replaced since the 1980s. But, satellites would not be able to support the millions of subscribers that public wireless systems have today. A land-based system concept was developed to use the radio spectrum as efficiently as possible to provide seamless mobile access to the plain old telephone service (POTS). A cellular pattern of regular frequency reuse was introduced and became such a success that radiotelephony is called the *cellular* system by the American public at large. Figure 1 illustrates this concept. A wireless access network can connect a new subscriber in most urban and rural areas to the public telephone network at a lower investment cost than any local loop requiring digging.

In the 1970s, the principal design problem encountered in wireless or mobile radio was how to overcome the distortion of received signals by a time-varying and frequency-selective propagation path. Radio waves near the ground do not travel over a single, well-defined radio path as in free space; they are scattered against reflecting obstacles in the vicinity of the mobile antenna. Reflected waves may add destructively, causing the received signal to disappear or become heavily attenuated at certain locations. A moving user, as in vehicular telephony, receives a resulting signal that is rapidly varying in time. This effect is called *fading*. Moreover, waves excessively delayed by remote reflections cause distortion of the shape of transmitted waveforms. Modern (digital) signal processing (DSP) techniques can mitigate these effects to a great extent. As a result, now the most quoted critical issues of wireless systems are no longer directly related to multipath fading, but are

1. The scarcity of radio spectrum and the resulting mutual interference among users

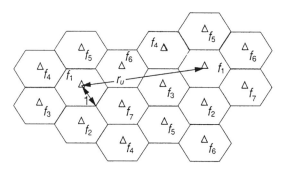

FIG. 1   Hexagonal cellular network with reuse of seven radio channels. Adjacent "cells" use different channels, but channels can be reused, observing a certain minimum separation distance r_u.

2.   The power consumption of portable terminals and the inadequacy of existing battery and other energy storage technologies
3.   The complexity of the software needed to support user mobility (e.g., from cell to cell or from operator to operator)

One of the earliest systems providing automatic radiotelephony, the Nordic Mobile Telephone (NMT) cellular system, was based on a standard developed in close cooperation between the five different operators (post, telegraph, and telecommunications [PTTs] administrations) and competing manufacturers in Denmark, Finland, Iceland, Norway, and Sweden. The NMT demonstrated the joint drive in European countries toward (inter)national cellular networks. Initially, the divided and divesting United States looked less able to develop and follow a common policy for mobile networking, even though the Bell Laboratories had played a leading role in the development of the novel cellular technology in the early 1970s (1). The United States eventually saw fit to agree to a common standard, the analog Advanced Mobile Phone System (AMPS), which subsequently captured about two-thirds of all mobile subscribers worldwide.

In 1992, the new standard for Pan-European digital cellular telephony known as GSM saw its first operational successes (2). The name GSM originated early in the 1980s as the French acronym for Groupe Speciale Mobile, the international working group most European PTT administrations gave the task of developing a common standard for cellular networks. A joint standard allows international roaming across the many European borders, until then only realized on a regional scale by the analog NMT standard mentioned above.

The main advantages of a digital system are a larger user capacity per unit of spectrum, ease of implementation of sophisticated encryption, authentication, and other security features, and robustness against radio-channel imperfections. A Pan-European standard further provides economies of scale in mass production of hand-held and car terminals, which would never have been achieved in the fragmented national markets in Europe. In the early 1980s, these two objectives were seen as the most critical success factors for achieving a much larger penetration of mobile telephone services in Europe. A third factor has been the introduction of competition in the monopolistic European markets—a cultural import from the United States.

In the past 10 years, the general emphasis of radio communications designs has shifted away from maximizing the capacity of individual links, limited by Gaussian noise and available bandwidth, to optimizing the capabilities of multiuser networks. The decisive interference now seldom comes from outside (as in military systems), but is produced by authorized users of the very same wireless network. Users thus share an interest in developing and adhering to the best possible protocols and standards for allocating the joint network resources. Accordingly, wireless systems engineering is developing into a more conscious search for the best common culture for multiple users in a real environment. The traffic capacity, spectrum efficiency, and cost effectiveness of modern radio networks are no longer won from nature (or an adversary) in a classical pursuit of individual gain.

Therefore, it has also become more important to consider the different requirements and cultures of new communities using wireless networks, including their needs for innovation of regulatory conditions and standards. This development has been commercially reinforced by the recent business trend in the information technology and defense sectors away from "selling the high-tech products that you can make" toward "making high-tech products that you can sell."

The Present Status of European Cellular Telephony

In the autumn of 1994, 102 operators in 60 countries were committed to GSM (see Table 1). The networks already in operation had 3 million subscribers, almost half of them in Germany. The commitment to GSM is formally expressed by a Memorandum of Understanding (MoU). In Europe, this is arranged by the Commission of the European Communities (CEC), which took great pains to ensure that common frequency allocations were made in the

TABLE 1 Global System for Mobile Communication Memorandum of Understanding Members (October 1994)

Region	Signatories	Countries
Europe	57	29
Arab States	10	10
Asia Pacific	29	17
Africa	6	4
Total	102	60

900-megahertz (MHz) band in the (then) 12 member states and in many other European countries. Later, the GSM standard was also adopted outside Europe, notably in important Asia–Pacific countries such as Singapore, Malaysia, India, Hong Kong, and Australia, and in the Middle East and Africa. Taiwan, Thailand, and New Zealand imposed it for at least one of their competing cellular networks. Accordingly, the old acronym GSM is now taken to mean Global System for Mobile Communication.

The general advantages of international standards, MoUs, and mutual roaming agreements for public mobile telephony are increasingly appreciated by different operators and markets. When the inception and implementation of such conventions lag behind, operators or their subscribers are denied some of the following mutual benefits:

1. Interoperability with the national public switched telephone network(s), including the Integrated Services Digital Network (ISDN)
2. Connectivity to all mobile users in an operator's service area, stimulating cooperation between adjacent carriers
3. Limiting investments in inflexible cable infrastructure to the backbone network
4. Low cost of introducing service area coverage, that is, proportional to the initial peak network use (erlang/unit area)
5. Simple upgrading of network capacity, when and where economically justified, by reducing cell size (cell splitting)
6. By virtue of 3, 4, and 5, fewer economies of scale than in hard-wired networks and, hence, less basis for "natural monopoly" arguments and for regulation against competition between local network operators (3)
7. Ability to locate vehicles and roaming user terminals, allowing automatic billing of users away from their home location or own operator
8. International portability of a subscriber identity module (SIM) smart card, authorizing personal and customized log in from compatible foreign terminals and competing networks by inserting the SIM card (2)

Such benefits have caused the annual growth of mobile communications markets to exceed 60% in many countries. No other telecommunications sector can boast similar growth rates at present. The infrastructure supply market is dominated by the few international manufacturers that combine expertise in both radio transmission and national switching systems. One of these manufacturers, L. M. Ericsson of Sweden, has some 40% of the cellular network market in the world, including a major proportion of the older networks using the American analog standard AMPS. The terminal market, on the other hand, is subject to the typical supply principles and economics of consumer electronics: short product development times and a "killing" competition due to the eroding profit margins for microelectronics commodity products since the late 1980s. The resulting drop in consumer prices assists in developing the service market much faster than in the past. Several operators even provide a terminal for free as part of a subscription for a certain minimum period of time.

At January 1, 1995, commercial GSM telephone service was already offered in 16 countries. The coverage area rolled out in Europe at that time is shown in Fig. 2. The major transit routes in Europe had been reasonably covered within one year after the first GSM service went on air in 1992, despite initial delays in acceptance testing of the first telephone handsets with their complicated software. Some services are restricted to voice-type (i.e., circuit-switched) con-

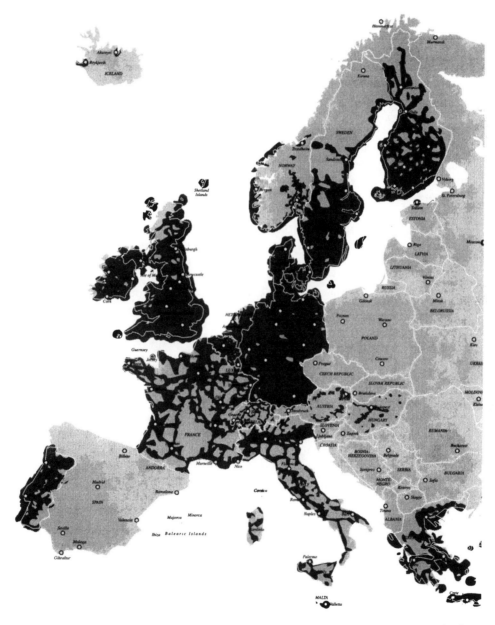

FIG. 2 Service area for airphones of European Global System for Mobile Communication operators by January 1995.

nections. A simple (packet-oriented) service allowing transfer of short data or alphanumeric (i.e., paging-type) messages is also available. Moreover, a set of data rates has been defined as bearer services in GSM (4) and are supported at a later phase of the market introduction of GSM.

Strikingly, the fastest adoption of GSM has been in Germany, where the national economy has been strained in the wake of the unification of East and West. After only six months of operation in December 1992, Mannesmann Mobilfunk announced the 100,000th subscriber to its GSM network, known as D2. In May 1993, 220,000 users subscribed to D2. Mannesmann's competitor in Germany, Deutsche Telekom, still holds the national monopoly for both public analog mobile telephony and the fixed telephone network. Some 90% of all telephone calls to or from mobile users (in general) originate or terminate in the fixed telephone network, and Telekom's analog cellular network, Cl, is already virtually saturated by its 800,000 subscribers. Nevertheless, Telekom's own GSM network D1 initially attracted fewer subscribers than the competing D2 network. The successful marketing of the D2 network services was designed by the American shareholder, AirTouch, a daughter company of Pacific Telesis. The early explosion of the GSM market in Germany caused a shortage of GSM terminals, especially portables, in other European countries, resulting in initial waiting lists and loss of network revenues.

The German case illustrates the general trend toward increased mobile network competition on the European continent, often spurred by more experienced business partners from the less monopolistic Anglo-American shores of the Atlantic Ocean. Clearly, the international GSM standard and MoU, with strict technical interface specifications and interworking requirements, paved the way not only to international cooperation and roaming, but also to local rivalry between competing cellular network operators. Introduction of network competition is indeed one of the GSM policy objectives of the European Union. This regulatory change is the third key factor in the dynamics of European mobile communications, next to the more technological driving forces perceived by the founding fathers of GSM a decade ago, but it raises new technical issues, including tariffing and service-related problems such as portability of numbers between competing operators.

At present, European companies such as Finnish Nokia, German Siemens, and Ericsson of Sweden have the lion's share of deliveries of GSM base stations and digital switching equipment. The GSM switches can readily be interconnected with intelligent network (IN) nodes, so that new competitive service offerings can be extended into the fixed network. These include such features as storage of voice-mail messages to or from mobile subscribers who have temporarily logged out of the network and call forwarding.

Given their position in digital very-large-scale-integrated (VLSI) and radio-frequency (RF) microelectronics, Japanese and U.S. manufacturers such as Panasonic and Motorola could be poised to dominate the supply of the advanced digital terminals for GSM networks, a major consumer market with much competition. However, in the core area of total systems engineering and standardization, U.S. manufacturers and operators appear divided about the operational merits of various technologically advanced options for radio-channel access and digital modulation, in particular the choice between code division multiple access (CDMA) and time division multiple access (TDMA) transmission formats for cellular telephony (5). For the migration toward all-digital operation, the Federal Communication Commission (FCC) was looking for a new system that uses the same radio spectrum as the old analog system, but at least 10 times more efficient in spectrum use. A consensus on such a U.S.-wide standard has not been reached. Slowly, U.S. cellular network operators are implementing a narrowband TDMA version of AMPS, known as IS-54. At present, only small-scale experimental CDMA systems are in operation in several cities, based on the IS-95 standard (6).

As for the complementary personal communications services (PCS), new European standards such as DCS 1800 and the Digital European Cordless Telecommunications (DECT) in the 1800-MHz band includes many of the functionalities of the GSM architecture (2) and are intended to support wide-area (cellular) networks and low-cost pocket phones suitable for both residential (cordless) use and wireless access to local private exchanges (private automatic branch exchange, PABX), respectively. While GSM was originally intended and designed mainly as a vehicular system, it is now mostly marketed as a system for hand-held use and may temper the initial growth of PCS in Europe until the 900-MHz band becomes locally saturated. In the United States, the PCS standard based on GSM technology is known as PCS-1900. It presently offers the most attractive prices for handsets and network infrastructure due to the worldwide adoption of GSM.

The active presence of some of the most successful U.S. Bell Operating Companies (BOCs) in the consortia licensed by countries committed to the GSM standard can be taken as an American acknowledgment of the dominance of European manufacturers in the rapidly growing market for digital mobile network technologies. Conversely, the awards of such licenses in several countries of the European Union can also be seen as a European acceptance of the greater American service provision experience and marketing skills. The remarkable role of the BOCs abroad suggests that innovative technologies embedded in common standards are not enough to develop a modern mobile service market. This is confirmed by consideration of the evolution of another digital service in Europe, mobile data communications.

Mobile Data Networks

Until very recently, the only possible way in many countries to conduct data transmission over public mobile networks was to attach low-speed, voiceband modems to analog radiotelephone circuits. Generally, this results in a poor bit error rate due to the rapid signal fading and shadowing associated with mobile radio, the interference from adjacent cells, and the frequent handoffs between base stations inherent in systems with moving users. Improvement by suitable error-correcting codes or data-link protocols is possible (7), but the data throughput of the narrowband radio voice channel remains small.

With the advent of the digital GSM, there is a popular belief that mobile computer communications will become easier, cheaper, and better. While the circuit quality and data rate do indeed improve, a circuit-switched channel (supporting up to 9.6 kilobits per second [kb/s]) is not well suited to any bursty data source, even if using digital transmission. Furthermore, call setup procedures in GSM, requiring the exchange and sequential processing of about 24 control messages, are prohibitively long for some computer applications. With sufficient error control, digital voice circuits are useful for transfer of batch data such as scheduled up- or downloading of files. On the other hand, dialing up (and paying for!) a real-time, two-way circuit between end users is quite inefficient for the more frequent modes of mobile computer communications: electronic mail, interactive access to information services, computer messaging of the electronic data interchange (EDI) type, dispatch, and other types of fleet management and point-to-multipoint data "broadcasting." For the GSM packet data services, a number of applications for road transport informatics are envisioned, such as traffic information, route guidance, and fleet management. For such applications, the network features of classical packet switching appear desirable (i.e., nonblocking access for terminals and the ability to convert data rates and codes within the network). Wide-area flexibility and adaptability to serve computers and terminals with different functions and priorities may point toward dedicated mobile packet-radio networks, which can bill by traffic volume rather than by connect time. Such networks are also far more economic in the use of precious radio spectrum. This is a strong selling point in countries that charge operators for spectrum use or auction frequencies to the highest bidders.

To illustrate the possibilities with narrowband mobile data throughputs, imagine the transfer of the simple American route map shown in Fig. 3 to a mobile terminal. As demonstrated in Ref. 8, it is possible to encode this image and the associated text as a message string of some 1200 bytes from a graphic tablet or a modern pen computer using differential chain coding. This message occupies a nominal time of only half a second in a 19.2-kb/s packet data channel. This is considerably less than when dialing up a circuit-switched mobile telephone channel, especially if this were used to interconnect two standard

FIG. 3   EGA (extended graphics adapter) screen dump of route map encoded by differential chain coding into 1200-byte message based on Ref. 8. Note granularity due to individual picture elements in the screen.

facsimile terminals using run-length coding (8). Even if a digital, circuit-switched channel were used only to transfer the short 1200-byte message string, the time and the signaling overhead to build up the real-time circuit would likely result in a rather unattractive tariff. While a packet-radio protocol would, in practice, also cause some extra delay, especially during heavy traffic loads, the billing would still correspond to an effective data transfer of only 1200 bytes.

Dedicated mobile data networks operate in several European countries. It is significant to note here that Deutsche Bundespost Telekom in the spring of 1993 announced that its experimental pilot mobile data network, Modacom, would be available for full operation in most of Germany by 1995. This wireless extension of Telekom's own public X.25 packet service to mobile computers such as laptops, notebooks, and palmtops suggests that (at least) one of the major GSM operators in Europe has doubts about the competitiveness of the planned data-transmission modes in the GSM network (4).

Well ahead of the German Modacom network, the Scandinavian PTT administrations in Sweden, Norway, and Finland introduced the Mobitex Radio Data Standard. In the 1980s, Mobitex was supporting only a low-speed packet service at 1200 bits per second (b/s) plus an emergency voice service, reflecting the initial use by police patrols, fire brigades, and public utility services to exchange brief command-and-control messages with their headquarters. A decade later, the widespread acceptance of portable computers and the advent of more demanding commercial operators, including RAM Mobile Data in the United States, have pushed the Mobitex data rate up—and the hybrid voice channel out. The 19.2-kb/s data rate defined in Ref. 9 is twice that foreseen in GSM (4). Mobitex-based networks have been introduced in the United Kingdom and the Netherlands by RAM and in France by France Telecom, assisted by Bell South.

Once again, we note the involvement of U.S. expertise in service provision, though the network technology and standards are purely European. Still, European telecommunications manufacturers have met more U.S. competition in the supply of mobile data networks than in digital cellular telephone systems. Thus, Motorola's DataTAC technology, based on the Radio Data (RD) Link Access Protocol (LAP) for the logical radio link connection, was adopted by Deutsche Telekom for Modacom in Germany, by Hutchinson (operating a mobile data service in the United Kingdom and Hong Kong), and by the ARDIS companies in the United States and Canada.

Mention should be made of the work by the European Telecommunication Standardization Institute (ETSI) on two Trans European Trunked Radio (TETRA) standards, one for pure packet data services to single or multiple destinations and another also supporting additional circuit-switched data and speech channels (similar to the initial Mobitex systems in Scandinavia mentioned above). As we note below, a hybrid system cannot be optimized in terms of capacity for both types of services; hybrid systems may nevertheless be required for certain applications (e.g., in the transport sector). The European frequency allocations for TETRA will be in the ultra-high-frequency (UHF) band (parts of 380–400 MHz, 410–430 MHz, 450–470 MHz, and/or 870–890/915–933 MHz) and may adopt 25-kilohertz (kHz) channels for coexistence with existing mobile services. TETRA is thus a typical narrowband data standard. European harmonization is deemed essential to achieve cross-border operation with this second-generation standard in the interest of international courier services, railroads, and road and river transport companies.

Another important development is the digitalization of the networks for radio and television broadcasting. Once deployed, these cannot only disseminate audio and visual material to the public at large, but also provide new (multimedia) services with conditional access. Digital transmission allows more flexible multiplexing of different traffic streams. Hence, broadcast networks are not restricted to pure radio and television services, but also allow encryption for various purposes or watermarking of information for antipiracy enforcement and copyright protection. Systems designed for digital audio broadcasting (DAB) in the European EUREKA program also appear suitable, in terms of data rates, for transmission of still pictures or data files. Elegant novel modulation techniques have been developed that allow significantly more programs to be transmitted within the same bandwidth. In particular, single-frequency networks, allowing master and relay transmitters to operate on the same carrier frequencies, will provide a spectrum efficiently that cannot be achieved with frequency modulation (FM) transmitters.

Most broadcast systems are restricted to one-way transmission, a major drawback for interactive services, but can be augmented by using other networks for the reverse link. Future interactive services to mobile users are often highly asymmetric in their communication requirements. The bulk of data is likely to travel toward the (mobile) users, with only request messages traveling in the reverse direction. A broadcast network may support the high data rates required in the downlink. In the context of standardization for future digital video broadcasting (DVB), digital terrestrial television broadcasting (dTTb) concepts are also soon to be standardized. However, successful market introduction of innovations in public broadcast standards requires major investments, including the eventual replacement of receiver sets in all homes. Accordingly, governments and broadcasters hesitate to introduce such systems. This results in a significant technology lag for the broadcast services, which currently occupy about one-half of all the radio spectrum below 1 gigahertz (GHz), and the use of outdated standards.

Wireless Office Systems

Obviously, the need to link computers in a wireless mode does not only exist outdoors for public networks, but is often driven from the local-area networks (LANs) at the customer premises. The European DECT standard has made instantaneous data-link throughputs up to 1 Mb/s available, although the limited access time (50 milliseconds [ms]) makes this more suited for wireless linkage of multiple users with considerably lower individual throughputs, for instance, to a private branch exchange (PBX) connected to the public switched telephone network (PSTN) and/or ISDN.

Meanwhile, substantial efforts are made in the United States to market proprietary solutions to access the "information superhighway" through wireless links. As an important step toward the development of wireless computing or multimedia networks, the Federal Communications Commission in 1985 issued rules permitting "intentional radiators" to use the industrial, scientific, and medical (ISM) bands (902–928, 2400–2483.5, 5725–5850 MHz) at power levels up to 1 watt without end-user licenses. Originally, these bands had been reserved for unwanted, but unavoidable, emissions from industrial and other processes, but they also supported a few (often military) communications users. The new rules led to the development of a large number of consumer and professional products.

The interest in using these bands has been stimulated by several factors that differ substantially for the European approach of conscientious, but time-consuming, standardization (10). Most importantly, there is almost a complete absence of user restrictions—no registration procedure, no qualification of end users, no restrictions as to where the products can be used. The absence of license fees also contributes to the economic attractiveness of products. The proximity of the 902–928-MHz ISM band to the U.S. 800-MHz AMPS cellular telephone band and the suitability of low-cost silicon VLSI chip implementations for this band allow the design of products with mass-produced components. The U.S. experience with wireless broadband networks in the office domain has given such systems as Motorola's 18-GHz ALTAIR (meeting the Institute of Electrical and Electronics Engineers [IEEE] 802.3 Ethernet Standard) and AT&T's WaveLAN (using CDMA in the ISM band at 2.4 GHz), a competitive edge against the Europeans when it comes to capacity.

A drawback of the ISM band is the lack of any protection against interference. In particular, microwave ovens limit the useful range of such communications devices. To ensure some coexistence between new communications users and users already occupying the band, spread-spectrum transmission is mandatory, except for extremely low-power applications. Spread spectrum offers some protection both for the licensed narrowband users of the bands (since the average spectral power density of the new users is low in their existing channels) and also for new users (since the processing gain of spread-spectrum systems mitigates interference from existing intentional and nonintentional radiators). Some critics argue that technically it is a harder problem to protect a wanted signal from only a few interferers than to separate it from many weak interferers. Spread-spectrum transmission typically spreads all signals over the entire ISM band, so it also makes it likely that more users interfere than in narrowband scenarios.

The enthusiastic American pursuit of product and market opportunities offered by the ISM rules is partly caused by this advent of a means by which users can build low-cost mobile data systems (10). Applications include wireless LANs, short-range links for advanced traveler information and management systems (e.g., electronic toll collection), garage door openers, home audio distribution, cordless phones, private point-to-point links, remote control, and wireless telemetric systems (e.g., electrical power consumption metering). A promising future application appears to be wireless access to the Internet or other multimedia computer networks. This appears to provide communications infrastructure that can change, both in terms of technology and in terms of the range of services provided, the current broadcast services much more radically than by upgrading their transmission system.

The present use of the ISM band is characterized by proprietary systems. Associated with this are (10)

1. Little standards activity, except for the wireless LAN standardization work in the European Telecommunication Standardization Institute (ETSI) and IEEE 802.11
2. Little or no emphasis on interoperability
3. Great diversity in products and in the traffic they generate
4. Interference-limited (as opposed to noise-limited) system design
5. Little focus on coexistence, except as in Item 4 above

One interpretation of the industrial interest in these bands, voiced by both established computer manufacturers and small new companies, is that the ISM rules have permitted rapacious entrepreneurs to exploit a short-term opportunity without any thought to the longer-term implications when these bands will become intolerably crowded and unusable. Another view is that the rules are a useful experiment in spectrum allocation—a recognition of the incredible waste of our spectral resource due to exclusive licensing—and an attempt to improve this situation through demand assignment. Another, somewhat less radical, approach to this problem is being pursued in the technical development of the dynamically assigned spectrum "etiquette" now pursued for the newly allocated bands for unlicensed PCS in the United States.

Packet Radio

Products using ISM bands often employ packet-oriented transmission schemes. The foundations of packet radio were laid by U.S. researchers, mostly sponsored by military agencies. Perhaps as a result, more emphasis was put on hostile interference and strategies for network survivability than on optimum self-interfering systems and the random signal fluctuations in mobile radio channels. The choices for spread-spectrum rules seem to have been influenced by experiences from military research. Terms like *packet radio* or *packet broadcasting* seldom refer to the typical propagation features of realistic radio media, but rather to the purely architectural or information-theoretical notion of maximum connectivity among all terminals in a multiuser network (11).

In the 1970s, little was published in open literature on interference-limited system design and communication over problematic channels. The experimental use of satellite links with their nearly perfect Gaussian-noise-limited (additive white Gaussian noise, AWGN) channels did not stimulate much consideration of real channel impairments—except imperfect (hard-limiting) satellite amplifiers and jamming by an adversary, when appropriate. If terrestrial networks were considered, these were often appropriate to a tactical battlefield scenario, with geographically distributed store-and-forward repeater nodes linked by random paths with fixed, but unknown, losses. The desired packet communication modes were generally of the multihop type, designed to maximize the progress of packets in particular directions.

As a consequence of this strong research tradition, many researchers still intuitively expect the significant propagation impairments of typical terrestrial UHF/VHF (very-high-frequency) mobile channels to reduce the moderate theoretical throughput of contention ("collision-type") protocols. For computer networks with cable links between terminals and hosts, uncoordinated transmissions indeed run the risk of conflicting with each other, which results in the loss of all messages involved. However, coinciding packets sent over a radio channel with very different ground-wave losses or instantaneous fading levels do not necessarily all annihilate each other, given capabilities of the receiver to capture a strong packet. Therefore, throughput expressions for "poor" mobile channels indicate a higher capacity than intuitively suggested by the classical studies of contention protocols in "ideal" AWGN channels. Intelligent processing of received signals, containing both wanted signals and interference with partly known properties, can further enhance the performance of wireless multiuser networks.

The FCC Part 15.247 approach is an exercise in interference-limited system design (12). One particular form of spread spectrum, frequency hopping, was first patented in an electronic warfare context to prevent target deception by interference to radio-guided torpedoes. The specialized military expertise only gradually becomes available for commercial use, so much may still be learned from the (often classified) archives concerned with electronic warfare. But, one lesson of this environment is clear: Optimum interference-limited system design requires gaining knowledge about "your" interferers and exploiting it. Most often, this knowledge gives a statistical description of the probably biased behavior of interference signals. This is in contrast to the approach in noise-limited systems, in which Gaussian noise is known to be the utmost unpredictable type of signal. In a cooperative, interference-limited environment, a priori knowledge of the other party's behavior can be used to the mutual benefit of both interferer and victim. The subject is one of great (but still mostly unexplored) research interest. Successful system designs in the interference-limited environment now heavily rely on diversity (i.e., the receiver attempts to observe the transmit signal in as many ways as possible). Such multiple observations can be made, for instance, with difference in time, frequency, or location of the antenna.

With many technical problems at transmission and protocol levels still unresolved, the availability of ISM bands for communications applications has provided valuable insight into the needs of end users, especially at the lower end of the acceptable cost curve. Personal mobility with infrastructure access seems to be emerging as a common theme. This is illustrated by wireless headphones, controllers, and speakers in the home. Of course, cordless telephones are a prime example, but the new FCC rules permit building them with enhanced security, digital transmission, and much longer range. In the middle of the cost curve, we find the wireless PBX, again an example of access to an infrastructure. In systems in which performance is a primary consideration, the wireless LAN is an example of an application with both intrasystem and infrastructure access.

The ability to conceive radio-based communications systems is no longer constrained mainly by technology, but by our understanding of the needs of users. This understanding is still very imperfect and incomplete. This effectively precludes our ability to execute a formal, thorough design process and may also bear on our ability to formulate valid standards. The exciting, disorderly, energetic exploitation of the FCC Part 15 rules for the ISM bands contributes to a growth of understanding and to our future ability to offer more cooperative communications systems with the necessary capabilities. This approach is typical for the U.S.-dominated computer industry and differs radically from the orderly top down policy approach in the European evolution.

Differences Between Packet Data and Digital Voice Systems

Despite the many development programs currently under way for universal, integrated, and ubiquitous systems, the ongoing introduction of separate digital networks for mobile telephony and mobile data applications will not come as a surprise to experienced network engineers. In the past, optimum use of classical hard-wired transmission resources motivated separate communications networks and signaling protocols for telephone and computer traffic, adapted to the different statistical characteristics of the corresponding information sources. Real-time, blocking-type circuits for telephone conversations pose other network requirements than does the delayed, but nonblocking exchange of bursty data in computer sessions (e.g., using the X.25 Protocol). When the physical transmission media contribute additional random fluctuations (as in mobile radio due to signal fading and mutual interference) or when the channel capacity is too precious to allow the overhead of ISDN-type service integration (as in mobile radio due to spectrum shortage), further distinctions arise between voice and data networks.

The concept of the asynchronous transfer mode (ATM) could become a viable compromise between circuit- and packet-oriented network design. It supports teletraffic at widely ranging rates and burstiness over a guided communications infrastructure. A similar concept for wireless (or "unguided") communication has not yet been developed, although many research programs, including the European Advanced Communications Technologies and Services (ACTS) program, aim at demonstrating wireless access to multimedia services in the period 1996–2000.

It is the link outage probability, as determined by unacceptable co-channel interference between cells, that determines the optimum design and real-time capacity of optimum cellular telephone networks. Although digital modulation schemes can be made more robust to interference than analog modulation, they still require considerable spacing between cells using the same radio frequencies in order to avoid unacceptable degradation of voice circuits too often. (A notable exception is the spread-spectrum modulation as used in CDMA cellular systems, which trade carrier bandwidth for processing gain to tolerate higher co-channel interference levels. The tradeoff is the stricter requirement for power control in mobile spread-spectrum telephone systems in order to reap the full network capacity and spectrum efficiency in the face of the fluctuating interference in the mobile channel (6).)

It cannot be assumed that such cellular structures optimized for nonblocking voice circuits are also optimum for a mobile network of virtual circuits among users who are prepared to accept delays of their data packets. Users of packet protocols are quite willing to accept a significant risk of harmful collisions between coinciding packets in return for not having to schedule their bursty accesses very strictly or to defer transmission for a long time. Packet protocols are designed to repair any harmful interference between packets simply by retransmission. This contrasts with classical fixed and mobile telephone circuit requirements, which contain substantial a priori guarantees against occasional circuit outages, including those due to harmful interference.

When computer users are prepared to gamble against the risk of mutual conflicts in order to reap the benefit of lower average delay (13), they can also adopt a more tolerant attitude to the additional random vagaries of the mobile radio channel. Moreover, such users prepared to accept occasional mutual conflicts between their accesses may find it counterproductive to invoke the strict power control assumed in mobile telephone systems: If access powers are very unequal among competing signals, there is a better chance that at least one competitor wins the contest for the receiver than in the event of perfectly balanced signals, which will annihilate each other in a collision. Much research (e.g., Refs. 14 and 15) has shown that this effect causes mobile packet data networks to have a higher throughput than predicted by conventional models used for wireline networks.

Simple Multiple-Access Methods for Mobile Data Networks

Mobile transmitters sending bursty traffic in the form of data packets to a common base-station receiver can, in general, best use some kind of random access. The classical ALOHA protocol, according to which each mobile terminal is free to offer bursty packets to the channel in accordance with the simple flow diagram in Fig. 4, belongs in this category. It is well known that the (ideal) channel throughput can be doubled if active terminals are prepared to synchronize their packet transmissions into common time slots such that the risk of partial packet overlap is avoided (16). With high traffic loads, however, both unslotted and slotted ALOHA protocols become inefficient since the free competition between all transmitters exposes most of the offered data traffic to collisions and, hence, multiple retransmissions and increasing delays.

To reduce this risk, a transmitter can follow a more cautious strategy. By

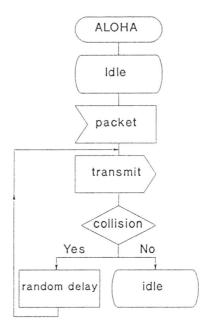

FIG. 4   Flow chart for mobile data terminal using non-slotted ALOHA.

first listening either to the common radio channel or to the return channel from the base station, a transmitter with a data packet can attempt to determine whether the shared radio facilities are already busy. The terminal approach based on the former listening method, known as Carrier Sense Multiple Access (CSMA), is illustrated in Fig. 5 in its simplest versions. In a realistic mobile channel, the various CSMA protocols may fail to detect ongoing radio transmissions of packets subject to deep fading on the listening path (17). Therefore, mobile radio CSMA proves less efficient than in classical hard-wired and satellite networks, in which contending terminals are not "hidden" from each other by individually different radio propagation effects. In such circumstances, mobile data terminals can better listen to the common base station, which broadcasts a "busy" signal to acknowledge an incoming transmission and inhibit prospective competitors and/or broadcasts an "idle" signal to invite transmissions (18–20). The throughput of most of these random-access protocols can also be improved by using them only as signaling preambles in order to schedule larger blocks of reserved (noncompetitive) airtime for a certain terminal. Obviously, this increases the risk of longer delays for all nonscheduled terminals.

In principle, the simplest random-access protocols are inherently unstable, given the standard assumptions of infinitely many users offering random Poisson-distributed traffic. In practice, however, realistic ALOHA models based on finite populations of competing mobile terminals and proper propagation characteristics of the shared mobile channel predict much better stability properties (14,21). Moreover, a number of simple, but effective, algorithms exist that

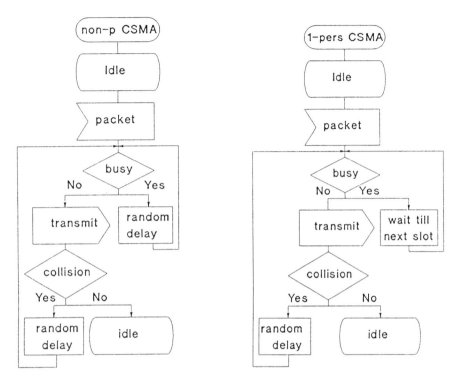

FIG. 5 Flow chart for mobile data terminal using non-persistent, non-slotted (left) or slotted (right) Carrier Sense Multiple Access (CSMA).

ensure stability even for the theoretical case of an infinitely large population of users. The throughput of mobile ALOHA networks depends on the ability of the receiver to discriminate between multiple signals, each arriving with a different power, and on the propagation environment. As Fig. 6 reveals, the positive effect of the propagation environment can be substantial in a collision channel (22).

In a mobile telephone network, on the other hand, every effort should be made to avoid propagation-induced outages of circuits once these have been set up for real-time connections. It is a normal goal in both cellular engineering and adaptive power-control schemes (23) to keep the capacity-limiting carrier-to-interference (C/I) ratios equally high at all receivers throughout a mobile telephone network in order to secure the prescribed grade of service and voice circuit quality at high spectrum occupation. If this real-time and "first-attempt-right" requirement is relaxed, frequency reuse in a network can be made extremely dense compared to conventional cellular voice networks. Even when all cells use the same bandwidth, the mutual interference between cells can prove to be acceptable for robust packet-oriented access protocols. This avoids the need for assigning different channels to different cells. Hence, it allows each cell to exploit locally the full system bandwidth at times of heavy traffic loads.

These arguments suggest a novel communications architecture that benefits from the vagaries of the mobile channel to the maximum extent possible. To

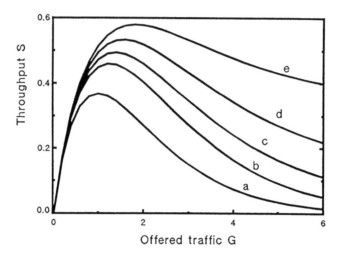

FIG. 6 Normalized throughput of single-cell mobile slotted ALOHA network, assuming statistical fading of all packets with spread σ and a base station with capture ratio $Z = 6$ dB. (Based on Ref. 22): *a*, "ideal channel"; *b*, Rayleigh fading for incoherent interference addition; *c*, shadowing channel ($\sigma = 6$ dB); *d*, Rayleigh fading and shadowing ($\sigma = 6$ dB) for incoherent interference addition; *e*, Rayleigh fading and shadowing ($\sigma = 6$ dB) for coherent interference addition.

this end, no clear boundaries are drawn between cell areas. The network simply consists of a number of cooperating base stations, all using the same radio channel. Terminals with a packet ready for transmission transmit on the common channel regardless of their location or other ongoing transmissions. For the case of base stations located at integer coordinates (i,j) with $i = 1, 2, \ldots$, and $j = 1, 2, \ldots$, Fig. 7 gives the probability of successful reception by at least one listening base station for a terminal location (x,y). A terminal will, on average, make a number of transmission attempts equal to 1 divided by its probability of receiver capture, which depends on its location. A considerable chance of capturing a base station from outside its proper cell is noted, especially when competition is not too heavy. The probabilities shown in Fig. 7 would represent an unacceptable probability of harmful interference between cells in a mobile telephone system, whereas a packet-switched system could work well under these circumstances. For bursty data traffic, such schemes are more efficient than conventional frequency reuse schemes assigning different frequencies to adjacent cells.

The system considered in Fig. 7 can exploit the fact that a message will be received at more base stations with different amounts of interference from other terminal transmissions. All base stations that receive a correct version of a message can forward this (by cable) to a coordinating station, which ensures that only one version of the packet is forwarded to the fixed network. Recently, it has been acknowledged that, even for wireless voice transmission, an ALOHA-type network access can be efficient (24,25).

On a downlink (i.e., in the reverse direction from base station to terminal) with data or multimedia services, queueing and retransmission delays are, again,

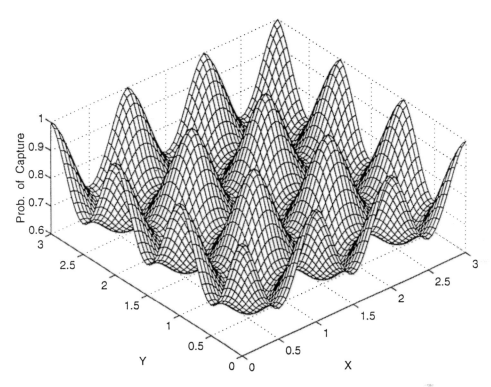

FIG. 7 Probability of immediate base-station capture by a mobile terminal using slotted ALOHA is a function of its position relative to a regular grid of base stations. Successful traffic equals 0.15 packets per slot per unit area.

more important performance measures than outage probability. Messages lost in a fade or due to accidental bursts of interference from other cells can simply be retransmitted, so the average outage probabilities may not be the appropriate design criterion. There is an inherent tolerance against interferers built into all contention-oriented, multiple-access schemes since any packet failures will ultimately be repaired by protocol measures (retransmissions). This allows considerably smaller optimal frequency reuse distances than in narrowband cellular telephone systems. Whether analog or digital, narrowband mobile telephony requires a considerable spacing between any two cells using the same frequency to ensure sufficient average circuit quality. On the other hand, using contiguous cells admits all base stations to the entire available bandwidth during bursty transmissions; this allows smaller delays of short packets even in the absence of spread-spectrum processing gains in the base-station receivers.

Radio Resource Management

The above considerations show that, if voice is to be integrated in a computer-oriented cellular network for more than occasional use, it thus would appear necessary to develop robust and delay-tolerant ways of packetizing it. As suggested above, the asynchronous transfer mode (ATM) experience may become relevant in this context. Conversely, if data packets are dispatched in a circuit-switched cellular network, inefficiencies of transmission and signaling resources would appear to result. When cost-based tariff regulations or plain competition apply, specialized mobile data networks therefore look more attractive to all users who attach only little value to the availability of voice transmission in the same network.

Driven by spectrum scarcity, radio resource management for bursty traffic is rapidly becoming a major research topic. The problem of managing information and data flows in radio and wireless optical networks appears to be significantly different from existing techniques for wired or "guided" communication. Decades of research on sharing communications resources among multiple users and services on wired networks has led to a wide variety of techniques for multiplexing, switching, and multiple access to communications resources such as coaxial cable local-area networks (LANs). The common goal of these schemes is the dynamic assignment of bandwidth during certain periods of time. Many of these multiple-access techniques were also adopted in radio data networks. However, it soon appeared that the performance of many random-access schemes substantially differs for guided (wired) and unguided (radio) channels, being highly dependent on the physical characteristics of the channel (14,15).

The aspects of spatially reusing scarce radio spectrum resources have mostly been addressed separately from allowing multiple users to share the same bandwidth–time resources. This is illustrated by the fact that most existing mobile data networks use a cellular frequency reuse pattern; within each cell, a random-access scheme is used independently of the traffic characteristics in other cells. The question of how to assign dynamically the totality of space–time–bandwidth resources in radio channels is not yet adequately answered, but is a topic of current research.

On a slow time scale, dynamic resource management is conducted, for instance, in the form of dynamic channel allocation (DCA). Instead of using a fixed reuse pattern (similar to that shown in Fig. 1, but with multiple colors in each cell), the system operator assigns frequencies or channels according to the instantaneous traffic demand. The European DECT system applies the DCA concept in a decentralized manner. Each terminal continuously searches for the best available channel and instructs base stations to communicate with it on this channel.

While some research programs now develop ATM-based wireless access techniques for multimedia traffic (thus including voice and data), other researchers challenge the need for transparent ATM links to and from the portable terminal. In their perception (e.g., see Ref. 26), the future terminal will likely evolve into a human interface, whereas all "intelligent" processing will be performed inside the network, in which processing power is more readily available, even for the most demanding tasks. Such system architectures are best supported by asymmetric wireless access links, supporting full-rate video toward the terminal, but with only low-rate control commands sent from the terminal.

Economic Issues

As stated in the section on the present status of European cellular telephony, one of the most interesting promises of modern wireless communications is a lower initial cost of connecting each subscriber to the public network, largely independent of distance. The access technologies employed in digital cellular radio personal communications services (PCS), such as the Digital European Cordless Telecommunications (DECT) and more full-fledged wireless office systems (WOSs) using broadband radio LANs, all offer interconnection with public networks at an investment that is mainly determined by the capacity required. This contrasts with the high initial investment of a hard-wired local loop (whether optical or not), the cost of which increases dramatically in rural areas, roughly in inverse proportion to the population density and required traffic capacity. Figure 8 illustrates these differences schematically (3).

It is the perverse cost structure of the traditional hard-wired access to public networks that caused "market failure" to occur in many low-traffic rural regions until now. This justified a local operating monopoly for public telephony in most European countries and the United States. However, the novel wireless technologies may help to avoid any need to cross-subsidize the local access from monopoly profits made elsewhere because costs proportional to area capacity remove the classical market-failure problem. In Fig. 8, wireless access to the public infrastructure would appear preferable to (installing new) twisted-pair loops left of point C1 and to (installing) optical fiber to the home left of point C2. In such circumstances, the economies of scale of the access network might become so marginal that local competition could be allowed under regulatory conditions to ensure fair spectrum allocation, consistent number planning, and equitable interconnectivity among the competitors. This is the case for the Pan-European GSM system, thanks to the policy directives from the European Commission in Brussels. Arguably, the recent acquisition of McCaw Cellular Communications by AT&T can be seen as the reentry of the divested long-lines company into local telephone operations—and hence a U.S. example of introduction of competition in a classical monopolistic field using wireless technology.

Networks for broadband interactive multimedia services are likely to face a more difficult market introduction. Most such services are not yet known to the public at large. In the past, it appeared difficult to introduce new technology (in new infrastructures) and new services at the same time. As wireless access does not have a cost structure that highly depends on spatial user densities, it may be feasible to start such services with a "thin" wireless infrastructure.

Market Value of the Spectrum

It is not yet evident on which basis equitable access by competitors to scarce frequency and number resources on the spectrum can best be granted. In the

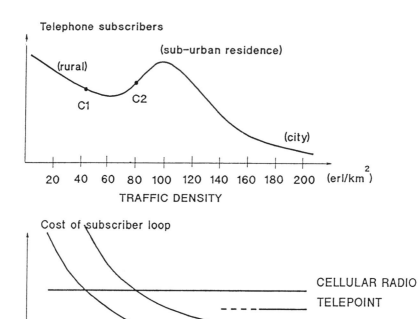

Telephone subscribers

(rural)

(sub-urban residence)

C1   C2

(city)

20 40 60 80 100 120 140 160 180 200 (erl/km²)

TRAFFIC DENSITY

Cost of subscriber loop

CELLULAR RADIO

TELEPOINT

OPTICAL FIBRE

TWISTED PAIR

WOS (?)

20 40 60 80 100 120 140 160 180 200 (erl/km²)

TRAFFIC DENSITY

FIG. 8   The economy of different access techniques to the public telephone network. At the top, subscriber distribution in a typical service area as a function of traffic density. The bottom shows relative cost structure of different subscriber access techniques in a typical service area. The cross-over points C indicate the transition at which cellular radio access techniques are cheaper than twisted-pair subscriber loop (C1) and optical fiber to the home (C2) (WOS = wireless office system).

United States, the FCC has experimented by replacing cumbersome administrative hearings with simple lotteries of frequency assignments (27). This resulted in rapid taking of "windfall profits" by fortunate winners, who simply sold their successful lots immediately after their award. This proves that frequencies have a substantial market value. When the public purse wishes to enjoy the profit of this, the government must design an auction system and the defined associated property rights very carefully to avoid being outsmarted by collusions of bidders. In the United States, PCS licensees have spent some $7.7 billion to buy 99 franchises in the 1900-MHz band in auctions organized by the FCC in early 1995. They now have to choose among the three available standards, PCS-1900 and IS-54 (TDMA) or IS-95 (CDMA).

It is always desirable to assign at least some fee for spectrum occupancy to encourage fairer and more efficient use of scarce resource, especially by avoiding extended "free parking" by inactive holders of frequency assignments. Even in the absence of competition, an incumbent holder of frequencies should be given sufficient incentive to vacate or share them for alternative use when not exploiting them fully. Examples may be broadcasters, outside active operation hours, that occupy spectrum with considerable value for public and private mobile communications. In the European teletext standard, any inactive video lines of a TV signal can be used to broadcast or download a substantial amount of data to arbitrary locations inside the coverage area of each TV transmitter network. This is commercially used in several European countries and by Luxembourg's commercial broadcasting satellites ASTRA. Narrowband public datacasting of traffic information to automobile radios is offered in several European countries and uses residual transmitting capacity on standard FM radio carriers in the broadcast band 88–108 MHz and the European Radio Data System (RDS) standard.

Conclusion: Different Cultural Paradigms

The development of novel wireless technologies in the past 15 years has spurred a tremendous drive toward new multiuser systems and applications on either side of the Atlantic Ocean. While the enabling technologies are largely the same, some clear differences in the approach and involvement of European and American regulatory and standardization bodies can be noted. The stronger European tradition of invoking public authorities in telecommunications leads to firmer and more widely accepted standards for new public wireless systems such as GSM, DECT, teletext, and RDS. These common standards are proving of great commercial value not only to European manufacturers and system integrators of mobile infrastructure, but also to terminal equipment manufacturers, service operators, and users worldwide.

In the United States, the government's role in steering communications system and technology research and development has traditionally been limited to the defense and aerospace sectors; the increasing competition between telecommunications operators seems to result in an overwhelming "smorgasbord" of alternative wireless technologies and voluntary standards (5). Frequently, these are based on military spin-offs such as CDMA and packet radio, with as yet no clear market winner in terms of successful public standards. On the other hand, the more competitive U.S. attitude to the use of computer and information technology, and the considerable financial resources of the BOCs, prove very powerful in the marketing of the new wireless services in the environment of the crumbling European telecommunications monopolies.

In terms of different research traditions in wireless communications, most foundations of packet radio and CDMA were certainly laid in the United States, as evidenced by Ref. 16. However, the shift from military to public multiuser networks required paradigm shifts in research focus that may perhaps have come easier in Europe. Thus, an earlier engineering attention appears in European system studies of the—often counterintuitive—influence of physical channel impairments on random-access methods, cellular engineering, and power control. Implications of other major paradigm shifts of information theory in the event of multiple self-interfering users, such as the nonapplicability of classical theorems in networks, have yet to be fully realized (11).

In more than one sense, all these changes and differences in European and American economic, regulatory, and research paradigms illustrate the fact that communications systems engineering is not merely a discipline related to the sciences of nature, but also to the protocols of culture.

Acknowledgment: The authors are grateful to Dr. David Bantz of the IBM Watson Research Center for his permission to include extended quotations from his invited presentation at the PIMRC '94 Conference.

Glossary

ALOHA. Algorithm that allows multiple terminals to share the same communications channel.

CIRCUIT SWITCHING. The allocation of network resources (link capacity, switches) for the entire duration of a communications session.

COMMISSION OF EUROPEAN COMMUNITIES (CEC). The daily government of the European Union in Brussels.

ERLANG (erl). A unit of telephone traffic.

FADING. Time variations of the signal strength received over a radio link.

INDUSTRIAL, SCIENTIFIC, AND MEDICAL (ISM). Bands of the radio spectrum.

INTELLIGENT NETWORK (IN). A secondary network used to create and deliver advanced services to subscribers to public telephone networks (fixed or mobile).

MEMORANDUM OF UNDERSTANDING (MoU). A pact between mobile operators.

MULTIPLE ACCESS. Method that allows multiple users to share the same communications channel.

PACKET. Message, or a piece of a message, treated as an independent segment of data by the network.

PACKET SWITCHING. The allocation of network resources by splitting the information flow into packets. These are sent from node to node in the network without prior reservations.

PAGING. Communications service that offers one-way transmission of short messages. Typically, a paging device (pager) produces an audible "bleep" when a message arrives.

VERY LARGE SCALE INTEGRATED (VLSI). Integration of microelectronic chips.

List of Acronyms

ATM   asynchronous transfer mode
AMPS   Advanced Mobile Phone System
BOC   Bell Operating Company
CDMA   code division multiple access

DCA   dynamic channel allocation
DECT   Digital European Cordless Telecommunications
EGA   extended graphics adapter
GSM   Global System for Mobile Communication
LAN   local area network
NMT   Nordic Mobile Telephone
PCS   personal communications services
RDS   radio data system
RF   radio frequency
TDMA   time division multiple access
WOS   wireless office system

References

1. Jakes, W. C. (ed.), *Microwave Mobile Communications*, Wiley, New York, 1974.
2. Rahnema, M., Overview of the GSM System and Protocol Architecture, *IEEE Commun.*, 92–100 (April 1993).
3. Arnbak, J. C., Economic and Policy Issues in the Regulation of Conditions for Subscriber Access and Market Entry to Telecommunications. In *Information Law Towards the 21st Century* (W. F. Korthals Altes et al., eds.), Kluwer, Deventer, 1992.
4. European Telecommunication Standards Institute (ETSI), *Bearer Services Supported by GSM PLMN*, Rec. GSM 02.02, January 1990.
5. Lynch, K., U.S. Seen Losing Cellular Advantage, *Commun. Weekly* (March 22, 1993).
6. Gilhousen, K. S., et al., On the Capacity of a Cellular CDMA System, *IEEE Trans. Vehicular Technol.*, VT-40(2):303–312 (May 1991); reprinted in Abramson, N. (ed.), *Multiple Access Communications – Foundations for Emerging Technologies* (Selected Reprint Volume), IEEE Press, New York, 1993.
7. Weissman, D., et al., Interoperable Wireless Data, *Commun.*, 68–77 (February 1993).
8. Arnbak, J. C., Bons, J. H., and Vieveen, J. W., Graphical Correspondence in Electronic-Mail Networks Using Personal Computers, *IEEE J. Sel. Areas Commun.*, J-SAC7(2):257–267 (February 1989).
9. Parsa, K., The Mobitex Packet-Switched Radio Data System, *Proc. 3rd IEEE PIMRC Symposium*, 534–538 (1992).
10. Bantz, D., The Use of the Industrial-Scientific-Medical Bands for Wireless Communication: An IBM Perspective, paper presented at *Proc. ICCC Symp. Wireless Computer Networks*, The Hague, September 19–23, 1994.
11. Cover, T. M., and Thomas, J. A., *Elements of Information Theory*, Wiley, New York, 1991.

    See, in particular, Chapter 14, "Network Information Theory."

12. Federal Communications Commission, FCC Regulations, Part 15. 247, ISM Rules, Federal Communications Commission, Washington, DC.
13. Massey, J. L., Some New Approaches to Random-Access Communications, *Proc. Performance '87*, 551–569 (1988); reprinted in Abramson, N. (ed.), *Multiple Ac-*

cess Communications—Foundations for Emerging Technologies (Selected Reprint Volume), IEEE Press, New York, 1993.

14. Linnartz, J. P. M. G., *Narrowband Land-Mobile Radio Networks*, Artech House, Boston, 1993.

15. Arnbak, J. C., and van Blitterswijk, W., Capacity of Slotted ALOHA in Rayleigh Fading Channel, *IEEE J. Sel. Areas Commun.*, JSAC-5:261–269 (February 1987).

16. Abramson, N. (ed.), *Multiple Access Communications—Foundations for Emerging Technologies* (Selected Reprint Volume), IEEE Press, New York, 1993.

17. Kleinrock, L., and Tobagi, F. A., Packet Switching in Radio Channels: Part 1—Carrier Sense Multiple Access Modes and Their Delay Throughput Characteristics, *IEEE Trans. Commun.*, COM-23(12):1400–1416 (December 1975).

18. Krebs, J., and Freeburg, T., U.S. Patent No. 4519068, Method and Apparatus for Communicating Variable Length Messages Between a Primary Station and Remote Stations at a Data Communications System, 1985.

19. Murase, A., and Imamura, K., Idle-Signal Casting Multiple Access with Collision Detection (ICMA-CD) for Land Mobile Radio, *IEEE Trans. Vehicular Technol.*, VT-36:45–50 (May 1987).

20. Andrisano, O., Grandi, G., and Raffaelli, C., Analytical Model for Busy Channel Multiple Access (BCMA) for Packet Radio Networks in a Local Environment, *IEEE Trans. Vehicular Technol.*, VT-39:299–307 (November 1990).

21. van der Plas, C., and Linnartz, J. P. M. G., Stability of Mobile Slotted ALOHA Network with Rayleigh Fading, Shadowing and Near-Far Effects, *IEEE Trans. Vehicular Technol.*, VT-39:359–366 (November 1990).

22. Safak, A., and Prasad, R., Effects of Correlated Shadowing Signals on Channel Reuse in Mobile Radio Systems, *IEEE Trans. Vehicular Technol.*, 40(4):708–713 (November 1991).

23. Zander, J., Distributed Cochannel Interference Control in Cellular Radio Systems, *IEEE Trans. Vehicular Technol.*, 41(3):305–311 (1992).

24. Zorzi, M., and Pupolin, S., Slotted ALOHA for High-Capacity Voice Cellular Communications, *IEEE Trans. Vehicular Technol.*, 43(4):1011–1021 (November 1994).

25. Kim, H. J., and Linnartz, J. P., Virtual Cell Concept: A New Communication Architecture with Multiple Access Ports, *44th IEEE Intl. Conf. Vehicular Technol.*, VTC-94:1055–1059 (June 1994).

26. Sheng, S., Chandrakasan, A., Broderson, R. B., A Portable Multimedia Terminal, *IEEE Commun.*, 30(12):64–75 (December 1992).

27. Mitchell, B., Allocating Spectrum for Cellular Telephones: U.S. Experience and Issues. In *Telecommunication New Signpost to Old Roads* (F. Klaver and P. Slaa, eds.), IOS Press, Amsterdam, 1992.

JENS CHRISTIAN ARNBAK
JEAN-PAUL M. G. LINNARTZ

Neural Networks and Their Application in Communications

Introduction

The human nervous system, as the most complex system of information processing, is the focus of intensive investigation by the life and systems scientists. Understanding of the functioning of this system will help life scientists to explain the causes and find therapeutic procedures for various neural disorders. On the other hand, study of the biological nervous system helps systems scientists and engineers to devise mechanisms that mimic specific features of the nervous system.

Devices and units realizing such mechanisms are called *artificial neural networks*, *neural systems*, *connectionist systems*, or *neurocomputers*. These are equivalent terms used by various authors in the field of neural networks to refer to such systems and mechanisms. Artificial neural networks are advantageous in some regards compared to conventional systems of information processing and control. For instance, in terms of parallelism, neural networks are far superior to serial computers.

The field of neural networks is based on mathematical modeling of the function of the mammalian nervous system in such a way that a specific feature of the nervous system is emulated by the model. Replication of the behavior of the nervous system with some degree of accuracy is therefore the ultimate goal of neural network scientists. This is pursued both at the single neural cell (neuronal) level, as well as at the systems level. It should be emphasized that successful utilization of artificial neural networks depends on the application needs versus the limitations in the capabilities of the utilized network and the technology used in the implementation of that network. In some applications, use of a specific neural network is justifiable; in others, their limitations outweigh their advantages.

Basic Neural Network Theory

Neural networks are highly connected sets of interacting devices or processors (*neurodes*) that produce and distribute outputs based on simple functions of their inputs (1). An artificial neural network is effectively a computing system made of a number of simple interconnected elements that process information by its dynamic state response to external inputs. They model real mammalian cerebral cortex, which consists of neurons with a cell and body transmitting activities to the dendrites of other neurons through interfaces called *synapses* (Fig. 1-*a*). A corresponding pictorial model showing interconnection of processing elements is given in Fig. 1-*b*.

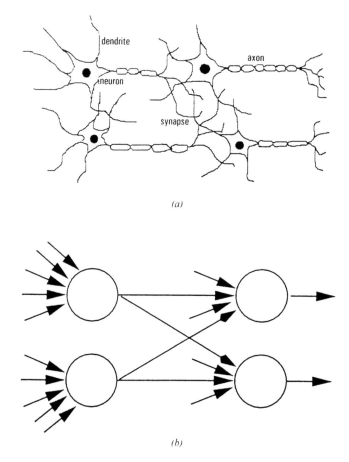

(a)

(b)

FIG. 1   Neural networks: *a*, interconnection of neurons in mammalian cerebral cortex; *b*, modeling of Fig. 1-*a* showing interconnection of processing elements.

The processing elements in a neural network are the counterparts of biological neurons that receive a number of input signals, $x_1, x_2, \ldots, x_n$. The sum of these input signals, weighted by a multiplier $w_1, w_2, \ldots, w_n$, respectively, is the effective input to the processing element as given by Eq. (1).

$$input = I = \sum_{i=1}^{n} w_i x_i \tag{1}$$

The output signal is the result of passing the effective input through a system function such as those shown in Fig. 2.

The form of the system function determines the type of neurode under study. A threshold T is specified for the overall input by the system function that limits the range of output. If the overall input is smaller than the threshold, the processing element will not have an appreciable output (does not fire). On the contrary, when input is larger than the threshold, the processing element will have a relatively large output (will fire). The output of this neurode will consti-

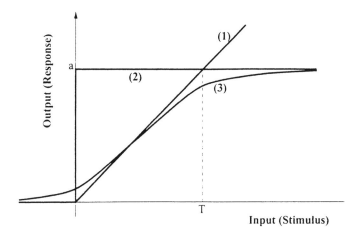

FIG. 2   System functions used in modeling the behavior of neurons.

tute the input to other neurodes. The weights w_i used for individual inputs can be positive or negative. In the first case, we have an *excitatory* input, whereas in the second case the input is *inhibitory*. The processing element modeling various components of the neuron is shown in Fig. 3.

From above, it is evident that the neural computers do not execute programs as much as they "behave" given a specific input. They are made to react, "self-organize," "learn," and "forget" (2). Unlike a serial computer, a neural computer is neither sequential nor necessarily deterministic. The neural network responds to the inputs presented to it in parallel. The result is the overall state of the network after it has reached equilibrium. The information is stored in a neural network in the way processing elements are connected and the weight of each input to the processing element. In this manner, neural networks have a potential to solve problems that cannot be solved with traditional computers. Such problems range from speech analysis and synthesis, pattern recognition, and control of movement, to operation and control of large-scale, wide-area telecommunications networks, to name a few.

A training process is involved in the creation of an artificial neural network for such tasks. In the following sections, various training procedures and learning algorithms are discussed.

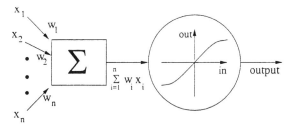

FIG. 3   Model of a neuron.

Training in Artificial Neural Networks

Training in artificial neural networks is performed when the response (outputs or reactions) to the environment (inputs or stimuli) is forced to change according to a specific rule. This is done by modification (recomputation) of the vector of weights used in conjunction with the inputs to produce an output. Most training algorithms currently discussed in the literature have evolved from the learning concepts introduced by D. O. Hebb (3). Hebb's learning rule is introduced in the section, "Hebbian Learning." Training procedures in common use today fall in three categories: supervised, unsupervised, and self-supervised training methods. In the following sections, these common training procedures are described.

Supervised Learning

A class of artificial neural networks is capable of being trained using known input-output data pairs. As such, the network will be exposed to some knowledge of what the response to a specific stimulus should be. This, in addition to the rule used for recomputation of the weight vectors, is known as *supervised training* and the result is *supervised learning*; it signifies the learning process in mammalian nervous system. The most common learning algorithm in this category is the least mean square law or delta rule (2). The network trained using the delta rule is called the Adaptive Linear Element Network (ADALINE) (2). The delta rule used in training ADALINE is given by

$$\Delta \mathbf{w} = \beta E \frac{\mathbf{x}}{|\mathbf{x}|^2} \qquad (2)$$

Here, $\Delta \mathbf{w}$ is the change in the weight vector and $\mathbf{x}$ is the input vector. E is the output error and β is the learning constant with values between 0 and 1. The learning constant β has to do with the stability of the training process and also the speed of convergence of the training. The system function assumed for ADALINE, as is also evident from the name, is linear, as given by Curve 1 in Fig. 2.

Repetition of the delta rule to calculate the weight vectors for a specific interconnection of several neurodes results in a network known as MADALINE, which is an acronym for Many ADALINEs. MADALINE is used in recognition of visual patterns without regard to their location in a scene (2). Generalization of the delta rule to the multilayer networks is known as the back-propagation method of supervised training (2). A two-layer network is shown in Fig. 4.

The column of circles on the left of Fig. 4 is not an independent layer since each circle acts only as a fan-out for each input. This layer is called the *input layer*. In the back-propagation method, an input from the training set is used to calculate activation from the input to the output layer starting with an initial distribution of weights. The computed output is then compared with the expected output pattern and errors are calculated for the output layer. The weight vectors are calculated using the delta rule, starting with the output layer and

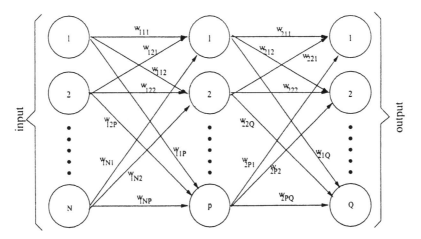

FIG. 4 A two-layer neural network.

proceeding through the network in a backward fashion. The error vector for the intervening layer, also called the *hidden layer*, is calculated as follows:

$$e_p = f'_p(I_p) \sum_{q=1}^{Q} w_{2pq} E_q \qquad (3)$$

where e_p is the pth component of the error vector for the hidden layer. I_p is the net input to the pth neurode of the hidden layer from the input layer, and $f'(I_p)$ is the derivative of the activation function of the pth neurode of the hidden layer for input I_p. Also, w_{2pq} is the weight corresponding to the output of the pth neurode of the hidden layer to the qth neurode in the output layer. The first index signifies the layer in the network. The activation function could be assumed as sigmoid, as shown by Curve 3 in Fig. 2, which is of the form of Eq. (4):

$$f(I) = \frac{a}{1 + \exp - (I + T)} \qquad (4)$$

Here, a is a constant multiplier, T is a threshold, and I is the overall input. This process is repeated for as many input patterns as necessary for the weights to converge to a stable set. The basis of the back-propagation method is the well-known gradient descent method in optimization. To ensure convergence, a generalized delta rule is used in back-propagation as given in Eq. (5):

$$\Delta \mathbf{w}_{pk} = \beta e_p \frac{\mathbf{x}_{pk}}{|\mathbf{x}_{pk}|^2} + \alpha \Delta \mathbf{w}_{p,k-1} \qquad (5)$$

Here, $\Delta \mathbf{w}_{pk}$ is the change in the weight vector for the pth neurode of the hidden layer for the kth input pattern and β is again the learning constant introduced in

Eq. (2). Also, e_p is the pth component of the error vector for the hidden layer and is calculated as in Eq. (3); $\mathbf{x}_{pk}$ is the vector of inputs to the pth neurode in the hidden layer due to the kth input to the network. The α is called the *momentum constant* that can prevent the gradient descent method getting caught in a local minimum. Back-propagation and the delta rule as supervised learning algorithms require a set of training data pairs. Moreover, they share the same pitfalls from which the gradient descent or hill-climbing algorithms suffer.

Unsupervised Learning

Learning in nature takes place for the most part without supervision. In other words, the output patterns are not known for specific inputs. Neural networks that are trained without supervision are capable of self-organization. Self-organization is the basis for learning by association, which is believed to be the main mechanism in the mammalian learning process. In the following section, the Hebbian learning rule as the most important training mechanism is introduced.

Hebbian Learning.   Hebb (3) proposed a model for unsupervised learning in which the synaptic strength (weight) between two connected neurons increases if both the source and destination neurons are activated (4). In this manner, the frequently used pathways in the network are strengthened and the phenomenon of habit through repetition is explained. A neural network using Hebbian learning will increase its weights according to the product of the outputs of the source and destination neurons, that is,

$$\Delta w_{ij} = \gamma y_i y_j \tag{6}$$

Here, Δw_{ij} is the change in weight and γ is the learning constant; y_i is the output of the ith neuron and input to the jth neuron. Also, y_j is the output of the jth neuron.

Kohonen's Network.   An example of the use of unsupervised training is the Kohonen's network (5), also called a self-organizing network. This network is not hierarchical and consists of a single layer of highly interconnected neurodes. The connections can also exist to the outside world.

If properly initialized and appropriately normalized, the weight vectors in Kohonen's network can self-organize according to the pattern of the input to the network. For each input, the neurodes compete for the privilege of learning (12). The neurode with the highest value of the net input is the winner and is allowed to modify its weight vector according to the following rule:

$$\Delta \mathbf{w} = \alpha(\mathbf{x} - \mathbf{w}) \tag{7}$$

Here again, $\Delta \mathbf{w}$ is the change in weight vector and $\mathbf{x}$ is the input vector. Kohonen's network follows the winner-takes-all philosophy in which the winner and only a small neighborhood of the winner are allowed to modify their weights and the only neurode that fires is the winner. When a neurode wins, it has an inhibiting effect on the firing of the other neurodes. This scheme of competition and inhibition is known as *on center/off surround*.

Kohonen's networks are capable of creating and preserving the topology or the statistical distribution of the inputs through adjustment of their weight vectors. Once the weight vectors are adjusted according to a given distribution, they will be retained so long as the inputs are drawn from the same population. The mammalian brain produces topology-preserving maps in the same manner. For example, the auditory portion of the cerebral cortex has tonotopic maps in which neurons have learned to be sensitive to different frequencies of sound in either increasing or decreasing order. Geotropic maps also develop in the brains of mice that have learned a maze.

More Advanced Neural Networks and Learning Rules. Grossberg has introduced models and training methods for neural networks that provide a mathematical basis for the explanation of several psychological effects. He introduced the concept of instar and outstar to model the fact that a neuron could be the source of many signals and also a receiver for many signals (2). His model also accounts for the fact that a neuron can have memory in the sense that the activity at any instant in time is a function of its activity in the past. Adaptive resonance theory (ART) and the subsequent models of the thought process are among the contributions of Grossberg and are highly recommended to those who are interested in the study of the neural networks (6).

Self-Supervised Training

In self-supervised training, the neural network generates an error signal internally. This error signal is then fed back to the network itself. A correct response is generated following a number of iterations (7).

Classification of Neural Networks

Neural networks can be classified according to their architecture, training law, or learning strategy and the system function used in their modeling.

Classification According to Architecture

In classification according to architecture, we have feedforward and feedback neural networks. In the first category, the output of each neurode is only connected to the input of the following neurode. Back-propagation can be used as the training method for this kind of network. In feedback neural networks, the output of some or all of the neurodes may be fed back to the input of the previous neurodes or even to the same neurode. A Kohonen's network is an example of a neural network with feedback.

Neural networks can have numerous layers. Each of these layers may include feedback. Multilayer neural networks with different characteristics can be constructed for different purposes. Theoretically speaking, a multilayer neural network can be constructed to emulate any behavior. The tradeoff, however, is the complexity of implementation and training.

If all the neurodes in the network have an input connection to all other neurodes in the network, the network is called *fully connected*. If a neural network is not fully connected, it is known to be *loosely connected*. A loosely connected network may include feedback.

Classification According to the Learning Rule

For classification according to the learning rule, training could be supervised, unsupervised, or self-supervised, as explained in the sections on training. Choice of system function also makes categories of neural networks with various distinct characteristics. For instance, the system function used for ADALINE is linear and the range of inputs can be $(-\infty, +\infty)$. For a class of neural networks called *perceptrons*, the system function is a step function, as shown by Curve 2 in Fig. 2. A system function such as Curve 3 in Fig. 2 represents neural networks that are closer to the behavior of the biological neurons. This function is known as a *sigmoidal function* and can be represented by Eq. (4); it is usually used in conjunction with the back-propagation learning rule. Hyperbolic tangent and arctangent functions are also used in literature to represent system functions that have symmetry at the origin and a bipolar value.

Performance of Neural Networks

The performance of a conventional computer is usually measured by its speed and memory (8). A brief discussion of the criteria used in measurement of the performance of neural networks is given below. Reference 9 should be consulted for an extensive discussion of this topic.

A neural network is treated as a state diagram in which threshold logic is carried out for the most part. The operation of the feedforward model is similar to the operation of a combinational circuit. The computation time is the time required for the signals to propagate and for the output to settle. The operation of the feedback model is closer to that of a sequential or asynchronous computer for which the system is initialized to a state and evolves in time to a final state.

The question of stability does not arise in feedforward networks because they are loop free. In feedback networks, the stability and convergence can be addressed by observing the change in the cost or energy function. The energy function is used in a neural network to optimize the weights as the state of the network (i.e., the weight vectors for various neurodes) converges to a final value in the training process.

The capacity of neural networks is measured in terms of the number of states that the network can distinguish. It can be proved that, for a feedback network with N neurodes, the capacity is proportional to N^3 (8).

Technology

Conventional computers are suitable for problems that can be solved using simple logic operations on a large scale and at great speed. Neural networks, on the other hand, are used to solve problems with a high degree of unpredictability. A prominent feature of neural networks is their massive connectivity. This is the limitation that controls the suitability of a given technology for an application using a specific architecture.

For networks with a small scale of connectivity, electronic systems can be used. Massive connectivity, on the other hand, is difficult in electronic systems due to the inherent interference and crosstalk between signal paths in those systems. In this regard, the field of optics has emerged as a promising technology to provide the massive connectivity needed in neural network implementations (9). A more promising approach in the near term is the use of hybrid optoelectronic processing elements. Gallium arsenide (GaAs) has been used for this purpose since it can be used to fabricate both fast electronic circuits and optical sources and detectors (9).

Metal oxide semiconductor (MOS) and complementary metal oxide semiconductor (CMOS) technologies have been used in the implementations of neural network integrated circuit (IC) chips. Mead has successfully designed analog very-large-scale-integrated (VLSI) chips implementing neural networks to help the blind, an optical motion sensor, a silicon retina, and an electronic cochlea (10).

Applications of Neural Networks in Communications

The past century has witnessed one of the most spectacular human inventions, the telephone. Von Neumann's machines, today's modern computers, on the other hand, appeared on a different horizon as the workhorses for massive processing of information. The marriage of computers and telecommunications created the web of computer networks that will record the epic saga of life and the strides of humans on this planet. Technology has evolved from circuit switching to packet switching and from frame relay to fast packet switching. Technology of the physical media has also evolved from copper wires to optical-fiber links, satellites, and mobile communications (11).

In the near future, Broadband Integrated Services Digital Networks (B-ISDNs) using the asynchronous transfer mode (ATM) at giga- or terabits per second will be used for the transport of information. This will include videoteleconferencing, video on demand, multimedia communications, high-definition TV, and personal communications services in addition to the future emerging services. In providing these services, challenging problems exist that are yet to be solved. These problems include routing of multimedia calls, traffic, congestion, and quality of service (QoS) control, to name a few.

Specification of multimedia traffic and management of resources based on the characteristics of the traffic are critical if conventional methods of network design are to be utilized. But, limitations of the conventional methods of network design, requiring accurate modeling of the stochastic processes that are often nonstationary, have rendered designs based on those methods inadequate. Adaptive approaches to the solution of such problems are therefore inevitable; neural networks offer excellent alternatives in this regard.

Along with strides in integration and transport of information, significant progress in sensing and transduction of information has also been made in recent years. Neural networks have proved their utility in most of these areas. In the following sections, the use of neural networks in the solution of a subset of such problems is briefly discussed and appropriate references in each case are cited for further studies. Numerous other applications of neural networks in telecommunications are not covered here. However, an extensive bibliography at the end of this article presents sources of study in corresponding subjects.

Neural Networks in Pattern Classification

Classical statistical pattern recognition is geared for non-real-time applications (12,13). Real-time problems in pattern classification include, but are not limited to, real-time speech recognition, vision, robotics, and so on. Neural networks provide the flexibility needed for such real-time applications due to their ability in massive parallel processing. Classifiers using neural networks include back-propagation classifiers, decision tree classifiers, Boltzmann's machines, feature map classifiers, learning vector quantizers, and modified nearest-neighbor classifiers. Detailed discussion of these classifiers is given by Lippmann (14). Different alternatives among these classifiers provide tradeoffs in memory and computation requirements, training, complexity, and ease of implementation and adaptation. In the following sections, we briefly introduce some of the work in pattern classification using neural networks.

Neural Networks in Recognition of Handwritten Digits

Automatic recognition of handwritten digits is of interest in many situations, such as postal sorting of mail. Although the real-time requirements of this application are not stringent, the speed with which handwritten patterns such as postal ZIP codes need to be recognized is critical. In the past, this problem has been approached using simple classification techniques applied to pixel images. These techniques, however, do not provide high recognition rates. High recognition rates can be achieved using predefined feature description of pixel images to train a neural network. Le Cun et al. have performed feature extraction using a neural network chip (15). They subsequently used the feature representation in a classifier consisting of a two-layer feedforward neural network trained with back-propagation to recognize handwritten digits.

To find the proper feature representation, Le Cun et al. have incorporated sufficient knowledge of the topology of the handwritten digits into the classifier so that it automatically generates an appropriate change of representation. They also have used some preprocessing of the data in terms of tilt control and line thinning and feature extraction using local template mapping. Their method has been used for recognition of postal ZIP codes using a general-purpose neural network chip used in a larger network. The weights were obtained by back-propagation learning and were implemented using commercial digital signal processing hardware; they were able to obtain throughput rates of more than 10 digits per second.

Neural Networks in Speech Recognition

Automatic recognition of speech has been the subject of attention for use in many applications. Patterns of audio signals corresponding with acoustic signals of various phonemes, letters, words, and sentences can be used to train neural networks. This kind of system has been successful in low-noise environments. For environments with a high-noise level, such systems demonstrate poor performance (16). A combination of acoustic (audio) signals along with visual speech signals to train neural networks can provide significant improvement in performance in noisy environments (17–19). Yuhas studied the automatic speech recognition from visual speech signals (20). Yuhas further studied the methods for combining the acoustic speech signals with visual speech signals to improve automatic speech recognition.

Neural Networks in Switching

Neural networks can provide a framework for designing massively parallel machines such as switching networks (21). A switch can be abstracted as a device that takes a set of N signal inputs and is able to reproduce them in any permuted order at the output. A basic switch is an $N \times N$ crossbar switch (Fig. 5), which is an example of a so-called nonblocking class.

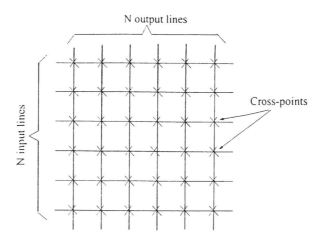

FIG. 5 An $N \times N$ crossbar switch.

The crosspoint count of a switch is often used as a measure of its complexity. To reduce the number of crosspoints, larger switches are built from stages of smaller crossbar switches. This may result in switches that are not nonblocking. However, they can be designed either with a small probability of blocking or such that any sequence of legal call requests can be connected through the switch as long as the routing of calls already in progress can be rearranged.

In the case of rearrangeable switches, the rearrangements need to be made with certainty and not rely on blocking probability. The questions are then how we should rearrange the calls when a new call request is made and which of the many available routes through a switch can accommodate the new call. The problems can be solved utilizing the parallelism of a neural network by matching the neural network topology to the problem. Details of this process are covered in Ref. 22, in which a winner-takes-all neural network is used. Generalization of the winner-takes-all network used in those references is known as K-winners-take-all network, which is covered in Ref. 23.

Neural Networks in Asynchronous Transfer Mode Transport

Introduction

In an ATM network, user information is transported by the payload of the 53-byte ATM cells. ATM cells belonging to a specific user are statistically multiplexed with the cells from other virtual connections. Statistical multiplexing, therefore, plays a pivotal role in ATM transport. In such networks, various multimedia calls (including data, voice, and video) arrive at random. Cells belonging to various connections also arrive at different switches randomly. Controls should be provided in such networks at various levels so that new calls with various traffic and QoS requirements can be set up without impairing the quality of service for the existing calls. Such controls should be provided proactively in the form of traffic control mechanisms, as well as reactively in the form of congestion controls.

In the proactive category, a functionality known as connection admission control (CAC) is provided to control admission of newly arriving calls. Once a call is admitted, traffic controls should be provided to prevent the users from overdriving the network. This is done in the form of usage parameter control and network parameter control (UPC/NPC) at the user–network interface (UNI) and the network-node interface (NNI) (24), respectively. In the reactive category, congestion-control mechanisms are needed to cure the congestion state and prevent the spread of congestion when it happens locally.

Due to the nonstationary behavior of the processes involved in the transport process, adaptive methods are needed for such controls. Utilization of neural networks for the solution of such problems is therefore the logical approach. In the following sections, applications of neural networks to the control of ATM networks are briefly discussed and references are cited for further studies.

Asynchronous Transfer Mode Network Control by Neural Networks

The early work in application of neural networks in the control of ATM networks was done by Hiramatsu (25). Hiramatsu applied an N input feedforward neural network to create a multiplexer for an ATM network. The input vector to the neural network has components equal to the number of cells that arrived in each of the last N time slots. The single output of the neural network would represent the dichotomy of exceeding a cell loss objective or not exceeding it. Back-propagation is used to train the network. This network is then used in the multiplexer to block calls for a period of time until the cell loss objective is met. It is shown that this network would successfully keep cell loss below the desired threshold by rejecting the excess calls. Milito subsequently applied a feedforward neural network for call admission in which a stochastic approximation along with back-propagation are used for training (26). Connection admission control (CAC), though, should be capable of admission of end-to-end calls by multilink routes.

CAC functionality restricts admission of new calls in order for the quality of service objectives of the existing calls to be maintained while the newly admitted call also meets its QoS objectives. An additional objective is to maximize the long-term throughput. This may imply selective rejection of narrowband calls in order to save bandwidth for wider bandwidth calls (27). The CAC is implemented by routing, link allocation, and link admission control. The routing function finds the best among a set of possible routes connecting an origin to a destination. A method for routing is proposed by Dziong and Mason based on dynamic programming techniques in which calls are assumed to have Poisson arrivals with exponentially distributed holding times (28). However, Poisson call arrivals and exponential holding times are only valid for interactive services. ATM traffic control should be such that non-Poisson arrivals can also be managed along with nonexponential holding time distributions, which can arise in the B-ISDNs.

At the link level, a control functionality should be provided to determine the set of feasible paths. The links should be admitted on the basis of long-term maximization of throughput also. Capacity change on a link needs an additional control when call-level traffic volume varies. This functionality should first attempt to find the feasible links and then select the link that maximizes long-term throughput.

Brandt et al. (29) have proposed a hybrid link admission control (LAC) based on analytical approximations and neural networks with data from the accurate fluid flow model (30). An analytical upper bound on the cell loss probability controls the admission at light loads on the link, while the neural network performs admission at higher loads. The input vector to the neural network characterizes the aggregate cell arrival rate obtained from the traffic descriptor of each call. Brandt et al.'s results show that the use of neural networks does indeed increase the utilization of resources.

Summary

In this article, the basic theory of neural networks is presented with some detail. Several applications of neural networks in electronic communications have also been briefly described and appropriate references are cited. References for many applications that are not covered in the main text of the article can be found in the bibliography.

Acknowledgment. The author is indebted to Dr. R. Milito for providing a major part of the bibliography.

Bibliography

General

Baram, Y., Nested Neural Networks and Their Codes, *Proc. 1990 IEEE Intl. Symp. Information Theory*, 9 (1990).

Chen, S., Gibson, G. J., and Cowan, C. F. N., Adaptive Channel Equalization Using a Polynomial-Perceptron Structure, *IEEE Proc.*, 137:257–264 (1990).

Farotimi, O., Dembo, A., and Kailath, T., Absolute Stability and Optimal Training for Dynamic Neural Networks. In *Conference Record. 23d Asilomar Conference on Signals, Systems and Computers* (R. R. Chen, ed.), Maple Press, San Jose, CA, 1989, pp. 133–137.

Franklin, J. A., Smith, M. D., and Yun, J. C., Learning Channel Allocation Strategies in Real Time, *IEEE Vehic. Technol. Conf. '92* (May 1992).

Hrycej, T., Self-Organization by Delta Rule, *Proc. Intl. Joint Conf. Neural Networks*, 307–312 (1990).

Kohonen, T., Raivio, K., Simula, O., and Henriksson, J., Performance Evaluation of Self-Organizing Map Based Neural Equalizers in Dynamic Discrete-Signal Detection, *Proc. Intl. Conf. Artificial Neural Networks*, 2:1677–1680 (June 1991).

Kohonen, T., Raivio, K., Simula, O., Venta, O., and Henriksson, J., Combining Linear Equalization and Self-Organizing Adaptation in Dynamic Discrete-Signal, *Proc. Intl. Joint Conf. Neural Networks*, 1:223–228 (June 1990).

Lee, T., and Peterson, A.-M., Adaptive Vector Quantization with a Structural Level Adaptable Neural Network, *Proc. IEEE Pacific Rim Conference on Communications, Computers and Signal Processing*, 517–520 (1989).

Takefuji, Y., and Lee, K. C., An Artificial Hysteresis Binary Neuron: A Model Suppressing the Oscillatory Behavior of Neural Dynamics, *Biological Cybernetics*, 64: 353–356 (1991).

Tank, D. W., and Hopfield, J. J., Simple "Neural" Optimization Networks: An A/D Converter, Signal Decision Circuit, and a Linear Programming Circuit, *IEEE Trans. Circuits and Systems*, CAS-33 (May 1986).

Whittle, P., The Achievement of Memory by an Antiphon Structure. In *Developments in Neural Computing. Proceedings of a Meeting on Neural Computing, 1989, IOP and London Math Society*, Adam Hilger, Bristol, UK, pp. 119–124.

Network Control and Management

Ansari, N., and Liu, D., The Performance Evaluation of a New Neural Network Based Traffic Management Scheme for a Satellite Communication Network, IEEE Globecom '91, Phoenix, AZ, December 1991.

Chen, X., and Leslie, I., A Neural Network Approach Towards Adaptive Congestion Control in Broadband ATM Networks, IEEE Globecom '91, Phoenix, AZ, December 1991.

Erfani, S., Malek, M., and Mehrotra, P., Potential Applications of Neural Networks in Network Management, Proc. 36th Midwest Symp. Circuits and Systems, Detroit, MI, August 16–18, 1993.

Funabiki, N., Takefuji, Y., and Lee, K. C., A Neural Network Model for Traffic Controls in Multistage Interconnection Networks, *Proc. Intl. Joint Conf. Neural Networks*, A898 (July 1991).

Goodman, R. M., Miller, J., and Latin, H., NETREX: A Real Time Network Management Expert System, IEEE Globecom Workshop on the Application of Emerging Technologies in Network Operation and Management, FL, December 1988.

Hiramatsu, A., ATM Communications Network Control by Neural Networks, *IEEE Trans. Neural Networks*, 1(1) (March 1990).

Hiramatsu, A., Integration of ATM Call Admission Control and Link Capacity Control by Distributed Neural Networks, *IEEE J. Sel. Areas Commun.*, 9(7) (September 1991).

IEEE Contr. Syst. Magazine, Special Section on Neural Networks for Systems and Control (April 1988).

Khasnabish, B., and Ahmadi, M., Congestion Avoidance in Large Supra-High-Speed Packet Switching Networks Using Neural Nets, IEEE Globecom '91, Phoenix, AZ, December 1991.

Milito, R. A., Guyon, I., and Solla, S., Neural Network Implementation of Admission Control. In *Advances in Neural Information Processing Systems 3*, Morgan Kauffmann, 1991.

Morris, R. J., and Samadi, B., Neural Network Control of Communications Systems, *IEEE Trans. Neural Networks*, 5(4) (July 1994).

Ogier, R., and Beyer, D., Neural Network Solution to the Link Scheduling Problem Using Convex Relaxation, IEEE Globecom '90, San Diego, December 1990.

Takahashi, T., and Hiramatsu, A., Integrated ATM Traffic Control by Distributed Neural Networks, ISS '90, Stockholm, Sweden, May 1990.

Vakil, F., and Lazar, A. A., Flow Control Protocols for Integrated Networks with Partially Observed Voice Traffic, *IEEE Trans. Auto. Contr.* (January 1987).

Mobile Communications, Cellular, and Satellite

Andersson, G., and Andersson, H., *Generation of Soft Information in a Frequency-Hopping HF Radio System Using Neural Networks* (in Swedish), Linkoping Institute of Technology, Linkoping, Sweden, 1991.

Ansari, N., Managing the Traffic of a Satellite Communication Network by Neural Networks. In *Dynamic, Genetic and Chaotic Programming of the 6th Generation series* (B. Soucek and the IRIS Group, eds.), John Wiley and Sons, New York, 1992, pp. 339–352.

Ansari, N., and Chen, Y., Configuring Maps for a Satellite Communication Network by Self-Organization, J. Neural Network Computing, 2(4):11–17 (Spring 1991).

Ansari, N., and Chen, Y., A Neural Network Model to Configure Maps for a Satellite Communication Network, *1990 IEEE Global Telecom. Conf.*, 1042–1046 (1990).

Ansari, N., and Liu, D., The Performance Evaluation of a New Neural Network Based Traffic Management Scheme for a Satellite Communication Network, *Proc. 1991 IEEE Globecom*, 110–114 (1991).

Chan, P. T. H., Palaniswami, M., and Everitt, D., Dynamic Channel Assignment for

Cellular Mobile Radio System Using Feedforward Neural Networks, *Proc. Intl. Joint Conf. Neural Networks*, 1242–1247 (November 1991).

Chan, P. T. H., Palaniswami, M., and Everitt, D., Dynamic Channel Assignment for Cellular Mobile Radio System Using Self-Organizing Neural Networks, *6th Australian Teletraffic Seminar*, 89–96 (November 1991).

Kanz, D., Channel Assignment for Cellular Radio Using Neural Networks, *IEEE Trans. Vehic. Technol.*, 40(1):188–193 (February 1991).

Munoz-Rodriguez, D., Moreno-Cadenas, J. A., et al., Neural Supported Hand Off Methodology in Micro Cellular Systems, *IEEE Vehic. Technol. Conf. '92*, May 1992.

Nakano, K., Sengoku, M., Shinoda, S., Yamaguchi, Y., and Abe, T., Channel Assignment in Cellular Mobile Communication Systems Using Neural Networks, Singapore Intl. Conf. Commun. Systems, 531–534 (November 1990).

Sivarajan, K. N., Spectrum Efficient Frequency Assignment for Cellular Radio, doctoral dissertation, California Institute of Technology, Pasadena, June 1990.

Routing

Ali, M., and Kamoun, F., A Neural Network Approach to the Maximum Flow Problem, IEEE Globecom '91, Phoenix, AZ, December 1991.

Chang, F., and Wu, L., An Optimal Adaptive Routing Algorithm, *IEEE Trans. Auto. Contr.* (August 1986).

Goudreau, M. W., and Giles, C. L., Neural Network Routing for Multiple Stage Interconnection Networks, Proc. Intl. Joint Conf. Neural Networks, A885 (July 1991).

Goudreau, M. W., and Giles, C. L., Neural Network Routing for Random Multiple Stage Interconnection Networks. In *Advances in Neural Information Processing Systems 4* (J. E. Moody, S. J. Hanson and R. P. Lippmann, eds.), Morgan Kauffmann, San Mateo, CA, 1992, pp. 722–729.

Jensen, J. E., Eshara, M. A., and Barash, S. C., Neural Network Controller for Adaptive Routing in Survivable Communications Networks, *Proc. Intl. Joint Conf. Neural Networks*, 2:29–36 (1990).

Kamoun, F., and Ali, M., A Neural Network Shortest Path Algorithm for Optimum Routing in Packed-Switched Communications Networks, *IEEE Globecom '91*, Phoenix, AZ, December 1991.

McDonald, K., Martinez, T. R., and Campbell, D. M., A Connectionist Method for Adaptive Real Time Network Routing, *Proc. 4th Intl. Symp. Artificial Intelligence*, 371–377 (1991).

Melsa, P. J. W., Kenney, J. B., and Rohrs, C. E., A Neural Network Solution for Call Routing with Preferential Call Placement, *IEEE Globecom '90*, 1377–1382 (December 1990).

Melsa, P. J. W., Kenney, J. B., and Rohrs, C. E., A Neural Network Solution for Routing in Three Stage Interconnection Networks, *Proc. 1990 Intl. Symp. Circuits and Systems*, 483–486xi (May 1990).

Rauch, H. E., and Winarske, T., Neural Networks for Routing Communication Traffic, *IEEE Control Systems Magazine*, 8(2):26–31 (April 1988).

Zhang, L., and Thomopoulos, S. C. A., Neural Network Implementation of the Shortest Path Algorithm for Traffic Routing in Communication Networks, *Proc. Intl. Joint Conf. Neural Networks*, 591 (June 1989).

Switching

Brown, T. X., and Liu, K.-H., Neural Network Design of a Banyan Network Controller, *IEEE J. Sel. Areas Commun.*, 8(8):1428–1438 (October 1990).

Hakim, N. Z., and Meadows, H. E., A Neural Network Approach to the Setup of the Benes Switch, *Proc. Infocom '90*, 397–402 (1990).

Marrakchi, A. M., and Troudet, T., A Neural Net Arbitrator for Large Crossbar Packet-Switches, *IEEE Trans. Circuits and Systems*, 36(7):1039–1041 (July 1989).

Troudet, T. P., and Walters, S. M., Neural Network Architecture for Crossbar Switch Control, *IEEE Trans. Circuits and Systems*, 38:42–56 (1991).

Classification

Lancini, R., Perego, F., and Tubaro, S., Some Experiments on Vector Quantization Using Neural Networks, *IEEE Globecom '91*, Phoenix, AZ, December 1991.

Error Control and Coding

Alston, M. D., and Chau, P. M., A Decoder for Block-Coded Forward Error Correcting Systems, *Proc. Intl. Joint Conf. Neural Networks*, 2:302–305 (January 1990).

Bruck, J., and Blaum, M., Neural Networks, Error-Correcting Codes, and Polynomials over the Binary n-Cube, *IEEE Trans. Inform. Theory*, 35(5):976–987 (1989).

Caid, W. R., and Means, R. W., Neural Network Error Correcting Decoders for Block and Convolutional Codes, *IEEE Globecom '90*, 1028–1031 (1990).

Chen, C., and Chen, T., Preliminary Study of the Local Maximum Problem of the Energy Function for the Neural Network in Decoding of Binary Block Codes, *Intl. Symp. Inform. Theory and Its Applications, ISITA '90*, 727–729 (1990).

Hussain, M., and Bedi, J. S., Decoding a Class of Non Binary Codes Using Neural Networks, *33rd Midwest Symp. Circuits and Systems*, 1990.

Hussain, M., and Bedi, J. S., Decoding Scheme for Constant Weight Codes for Optical and Spread Spectrum Applications, *Elect. Letters*, 27(10):839–842 (1991).

Hussain, M., and Bedi, J. S., Performance Evaluation of Different Neural Network Training Algorithms in Error Control Coding, *SPIE '91, Appl. Artif. Intell. and Neural Networks*, 697–707 (1991).

Hussain, M., and Bedi, J. S., Reed–Solomon Encoder/Decoder Application Using a Neural Network, *SPIE '91, Appl. Artif. Intell. and Neural Networks*, 463–471 (1991).

Jeffries, C., Code Recognition with Neural Network Dynamical Systems, *Soc. for Ind. and Applied Math Review*, 32(14):636–651 (1990).

Jeffries, C., and Protzel, P., High Order Neural Models for Error Correcting Code, *Proc. SPIE – The Intl. Soc. of Optical Eng.*, 510–517 (1990).

Petsche, T., and Dickinson, B. W., A Trellis-Structured Neural Network, *Neural Inform. Proc. Systems*, 592–601 (1988).

Santamaria, M. E., Lagunas, M. A., and Cabrera, M., Neural Nets Filters: Integrated Coding and Signaling in Communication Systems, *Proc. MELCON '89*, 532–535 (1989).

Schnell, M., *Multilayer Perceptrons and Their Application to Decoding Block Codes*, German Aerospace Research Establishment, Inst. for Comm. Tech., Oberpfaffenhofen, Germany, 1990.

Sourlas, N., Spin-Glass Models as Error-Correcting Codes, *Nature*, 339:693–695 (1989).

Takefuji, Y., Hollis, P., Foo, Y. P., and Cho, Y. B., Error Correcting System Based on Neural Circuits, *Proc. IEEE 1st Intl. Conf. Neural Networks*, 3:293–300 (1987).

Yang, J., Chen, C., and Lee, J., Neural Networks for Maximum Likelihood Error Correcting Systems, *Proc. Intl. Joint Conf. Neural Networks*, 1:493–498 (1990).

Yuan, J., Bhargava, V. K., and Wang, Q., An Error Correcting Neural Network, *Proc. IEEE Pacific Rim Conf. Commun., Computers and Signal Proc.*, 530–533 (1989).

Yuan, J., Bhargava, V., and Wang, Q., Maximum Likelihood Decoding Using Neural Nets, *J. Inst. Elect. and Telecom. Engineers*, 36(5–6):367–376 (September–December 1990).

Yuan, J., and Chen, C. S., Correlation Decoding of the (24,12) Golay Code Using Neural Networks, *IEE Proc. Commun., Speech, and Vision*, 138:517–524 (1991).

Yuan, J., and Chen, C. S., Neural Net Decoders for Some Block Codes, *IEE Proc. Commun., Speech, and Vision*, 309–314 (1990).

Zeng, G., Hush, D., and Ahmed, N., An Application of Neural Net in Decoding Error-Correcting Codes, *Proc. 1989 IEEE Intl. Symp. Circuits and Systems*, 782–785 (1989).

Zetterberg, L. H., and Zhang, Q., Signal Detection Using Neural Networks and Error Correcting Codes, TRITA-TTT, No. 9114, KTH, Stockholm, 1991.

References

1. Posner, E. C., Guest Editorial, *IEEE Commun. Mag.*, 27(14) (November 1986).
2. Caudill, M., Neural Network Primer, Parts I–V, *AI Expert*, (December 1987–November 1988).
3. Hebb, D. O., *Organization of Behavior*, Science Editions, New York, 1961.
4. Wasserman, P. D., *Neural Computing, Theory and Practice*, Van Nostrand, New York, 1989.
5. Kohonen, T., Self-Organization and Associative Memory, Springer-Verlag, New York, 1984.
6. Grossberg, S., *Studies of Mind and Brain*, Reidel Press, 1982.
7. Defense Advanced Research Project Agency (DARPA), *Neural Network Study*, Armed Forces Communications and Electronics Association (AFCEA), 1988.
8. Abu-Mostafa, Y., Information Theory, Complexity, and Neural Networks, *IEEE Commun. Mag.*, 27(14) (1986).
9. Psaltis, D., et al., Optoelectronic Implementation of Neural Networks, *IEEE Commun. Mag.*, 27(14) (November 1986).
10. Mead, C., *Analog VLSI and Neural Systems*, Addison-Wesley, Reading, MA, 1989.
11. Habib, I. W., Neurocomputing in High-Speed Networks, guest editorial, *IEEE Commun. Mag.*, 33(10) (October 1995).
12. Duda, R. O., and Hart, P. E., *Pattern Classification and Scene Analysis*, John Wiley and Sons, New York, 1973.
13. Fukunaga, K., *Introduction to Statistical Pattern Recognition*, Academic Press, New York, 1972.
14. Lippmann, R. P., Pattern Classification Using Neural Network, *IEEE Commun. Mag.*, 27(14): (November 1986).
15. Le Cun, Y. L., et al., Handwritten Digit Recognition: Application of Neural Network Chips and Automatic Learning, *IEEE Commun. Mag.*, 27(14): (November 1986).
16. Allen, J., A Perspective on Man-Machine Communication by Speech, special issue on man-machine speech commun., *Proc. IEEE*, 73:1539–1550 (1985).
17. McGurk, H., and MacDonald, J., Hearing Lips and Seeing Voices, *Nature*, 264: 746–748 (1976).

18. McGurk, H., and MacDonald, J., Visual Influences on Speech Processes, *Perception and Psychophysics*, 24:253–257 (1978).
19. Kuhl, P. K., and Meltzoff, A. N., The Bimodal Perceptrons of Speech in Infancy, *Science*, 218:1138–1141 (1982).
20. Yuhas, B. P., Integration of Acoustic and Visual Speech Signals Using Neural Networks, *IEEE Commun. Mag.*, 27(14) (November 1986).
21. Brown, T. X., Neural Networks for Switching, *IEEE Commun.*, 27(14) (November 1986).
22. Marakchi, A., and Troudet, T., A Neural Network Arbitrator for Large Crossbar Packet Switches, *IEEE Trans. Circuits and Systems*, 36(7):1039–1041 (July 1989).
23. Majani, E., et al., On the K-Winners-Take-ALL Network. In *Advances in Neural Information Processing Systems I* (D. Turetzky, ed.), Morgan Kaufman, San Mateo, CA, 1989, pp. 634–642.
24. *Traffic Control and Congestion Control in B-ISDNs*, ITU-T Recommendation I.371, November 1995.
25. Hiramatsu, A., Integration of ATM Call Admission Control and Link Capacity Control by Distributed Neural Networks, *IEEE Trans. Commun.*, 35:1347–1356 (1987).
26. Milito, R. A., Neural Network Implementation of Admission Control, *TIMS XXX – SOBRAPO XXIII*, Rio de Janiero, Brazil, July 1991.
27. Nordstrom, E., et al., Neural Networks for Adaptive Traffic Control in ATM Networks, *IEEE Commun. Mag.*, 33(10) (October 1995).
28. Dziong, Z., and Mason, L., An Analysis of Near Optimal Call Admission Control and Routing Model for Multiservice Loss Networks, *Proc. Infocom '92*, Florence, Italy, 1992.
29. Brandt, H., et al., A Hybrid Neural Network Approach to ATM Admission Control, *Proc. Intl. Switching Symp. (ISS '95)*, Berlin, April 1995.
30. Ritter, M., and Tran-Gia, P., Performance Evaluation and Design of Multiservice Networks, European Cooperation in the Field of Scientific and Technical Research, COST 224, Final Report, Luxembourg, 1992.

MAHMOOD R. NOORCHASHM

The 1992 Cable Television Act

Introduction

The regulatory status of cable television changes on almost a daily basis, on almost every legal level. Federal, state, and local governments enact new statutes, regulations, and ordinances; cities, states, and the Federal Communications Commission (FCC) impose new or different requirements; and the courts review or enforce all of these. At the time of this writing, the FCC had several dozen rulemaking proceedings in different stages of development, state and local authorities were adopting rules and setting subscriber rates, and many of these matters were subject to review in the courts.

To address these issues in any real detail would be an invitation to obsolescence. Instead, this article attempts to identify the main themes in the last four decades of cable regulation, outline the general provisions of the 1992 Cable Act, and tentatively project some possible legal and regulatory developments, from the vantage point of 1995.

To begin, it may be useful to identify some of the major players in the cable regulation game. Unlike most other electronic media, cable is subject to regulation by a variety of different entities on the federal, state, and local levels. The most significant authorities on a long-term basis naturally are Congress, the courts, and the administration. Their roles change in scope and importance from time to time. Congress generally was quiescent in cable regulation until the early 1980s, when it became involved in a series of issues leading to the 1984 Cable Act—and then to its more comprehensive successor, the 1992 Cable Act. The courts played their traditional role of reviewing both congressional legislation and administrative rules; with the advent of the relatively detailed 1992 Cable Act, however, the number of appeals increased tremendously. The administration has become increasingly involved as of late, such as when President Bush unsuccessfully vetoed the 1992 Cable Act or when President Clinton proposed the National Information Infrastructure (NII) as an "electronic superhighway."

On the federal level, however, the Federal Communications Commission (FCC or commission) has been by far the most significant long-term policy maker. It began the cable regulatory process in the 1960s (see "Early Regulatory and Legal Initiatives" section), and has continued to promulgate rules both on its own motion and pursuant to the 1992 Cable Act. To the limited extent that any real enforcement of federal cable policy exists, its primary responsibility has fallen to the commission. On a more limited level, the Department of Commerce's National Telecommunications and Information Administration (NTIA) has been involved in long-range planning—most notably in development of the NII and encouragement of minority ownership.[1]

This discussion does not include consideration of the Telecommunications Act of 1996, which affects several provisions of the 1992 Cable Act. Most notably, Sections 651–653 allow telco entry into cable, by mergers, acquisitions, and facilities construction.

411

About a dozen states have cable regulatory agencies, which vary tremendously in the nature of their powers. For example, the New York State Commission on Cable Television usually just advises local governments on issues in franchising cable systems[2]; only if a local government and cable operator are unable to resolve basic issues does the state commission use its preemptive powers to grant or deny a "certification." By contrast, in New Jersey, only the Office of Cable Television of the Department of Public Utility Commissioners can authorize a company to operate a cable system.[3] In reality, however, the difference between the New York and New Jersey approaches may not be great; New Jersey follows the local government's approach unless it has major legal deficiencies. Although there was a brief trend during the early 1980s for state governments to become involved in the franchising process, most states have eschewed or withdrawn from any intensive involvement in cable franchising or other regulation.

Historically, local governments have played the major role in cable regulation—including sensitive issues such as franchise renewal or rate regulation. Although the 1984 Cable Act cut back local governments' roles in both of these areas, the 1992 Cable Act largely restored it. Realistically, however, most municipalities are too involved with more politically pressing matters to take a substantial interest in cable issues; nevertheless, in some cases they become heavily involved—and have substantial powers under the 1992 Cable Act to enforce their policies, particularly through rate regulation.[4]

Cable regulation thus comes from a variety of potential sources, often with differing mandates. For example, under the 1992 Cable Act, the FCC has the ultimate authority to take over subscriber rate decisions from local governments[5]; but, the agency is far more concerned with vertical integration of cable programmers and systems or compatibility of cable converters and other consumer electronics equipment.

Private entities naturally play major roles in the regulatory process by taking positions for or against any given rule. As with most regulated firms, the primary issues involve entry of and competition by other firms. With cable, several other industries are relevant.

First, broadcasters have a natural interest in insuring not only that cable operators carry their signals, but also that cable systems do not create too much competition with their own advertisers, such as through new satellite networks (such as the Cable News Network [CNN] or ESPN). Broadcasters traditionally have lobbied for "mandatory carriage" by cable systems and against offering of attractive programming by satellite networks and other nonbroadcast program providers (see the "Must Carry" section).

Second, and perhaps most important in the 1990s, local-exchange telephone companies have had increasing—although undefined—interests in offering traditional types of cable programming. As discussed below, in the "Cable/Telco Relationships" section, the "cable/telco" confrontation largely concerns local-exchange telephone companies' (LECs') ability to control the content of traditional common-carrier channels.

Neither the cable operators nor the LECs seem very sure of the directions in which they are headed. For example, some cable operators are interested in the United Kingdom's experience of offering competitive LEC service by cable;

others believe that it is a fruitless approach. Similarly, some LECs are eager to provide program content, while others believe that it has no long-range value. In any event, no single long-range strategy has emerged.

Other interested parties include intellectual property interests (e.g., studios that produce both motion pictures and television programming). Cable has opened up a new market for them in relatively inexpensive programming, which in turn can be resold to nonnetwork stations and foreign broadcasters. Depending on how cable or other industries (e.g., the LECs) develop, the producers' ability to sell more types of programming may change dramatically.

Another sector is banking and financial services. Many loans to cable operators—and other new media—are or may have been a bit speculative in nature. As the economics of cable and other industries stabilize, however, banks and venture capital firms find investments in these industries attractive. Indeed, some of the 1980s major cable television acquisitions took place only because "junk bonds" were easily available. Even if they do not provide financing, these firms play a major role in potential mergers and acquisitions. For example, in the abortive Bell Atlantic/TCI merger—the largest potential transaction during the mid-1990s—investment and legal advisors made handsome profits from their advice.

Cable television regulation always has involved clashes between a number of industries—primarily broadcasting and production companies. Over the years, cable has become increasingly involved with a number of other industries, as both partners and competitors—sometimes in both roles. For example, the LECs offer both signal carriage and potential competition. All these interests naturally have had and will have an impact in forging cable policy.

History and Development of the Cable Industry

Early Developments

Video distribution through coaxial cable is hardly a "new technology." Engineers demonstrated video transmission through coax to the Federal Radio Commission in the early 1930s, but the cable industry did not even begin to develop until the early 1950s since few people had television receivers. Although sets were available on the retail market and more than 40 experimental stations were on the air by 1940, World War II delayed the introduction of all new electronic media—including broadcast television and FM (frequency modulation) radio.

It is impossible to identify the "first" cable television operation; at least a dozen entrepreneurs—most of them still active in the business—claim that status. Initial operations began during the 1950s and were technologically and financially unsophisticated. The early entrepreneurs usually served communities located in "white" areas inaccessible to broadcast stations because of terrain, most commonly valleys, hence the dozens of cable systems named "Valley Cablevision" or the like. Operators erected a high-gain receiving antenna on the tallest piece of terrain and then ran a low-capacity cable with a few amplifiers to the nearest town—often lashing the cable to trees when utility poles were unavailable. These systems generally carried only three to seven channels. Distribution amplifiers were primitive, unreliable, and expensive; set-top converters were not available until the early 1970s.

The only available signals were local television stations since microwave relays did not develop until the 1960s and satellite dishes until the 1970s. Even the economic underpinnings of the early cable industry were less than clear; some entrepreneurs were department or hardware stores, which initially got into the cable business in order to give subscribers a reason to buy television sets—much as had happened in the early days of radio. Within a few years, however, operators universally recognized that the real profits were in subscription fees. They thus began to search for ways to expand their systems' number and type of services.

Until the end of the 1970s, cable thus was severely limited in its offerings and had comparatively few subscribers—about a million by the early 1960s. The almost simultaneous development of satellite programming services (e.g., Home Box Office [HBO], CNN, ESPN) and high-capacity cable, amplifiers, and converters in the late 1970s created a new market for cable systems and allowed them to attract subscribers even in areas with good broadcast television reception.

Technological and Economic Overview

Cable television uses essentially the same transmission technology as broadcast television, except that cable distributes signals through a coaxial cable rather than over the air. The main limitation on a cable system's bandwidth is not so much the cable as the associated electronics—distribution amplifiers and the like. By the mid-1990s, a realistic channel limit was about 70 channels. Cable operators claimed that they would upgrade their systems to 500 or more channels by introducing digital compression, fiber optics, and switching "nodes." Since this involved rebuilding almost every part of a cable system, however, most operators were slow to implement these plans. Moreover, local governments' and the FCC's limitations on rates (see the section "The 1984 Cable Act") may have limited cable operators' access to capital to fund these upgrades. This situation may change if the country ultimately converts to a system of fiber-optic broadband networks, which may be able to carry thousands of channels on an "electronic superhighway." However, this does not seem likely in the near future.

At present, a cable system receives signals from a variety of sources (e.g., local television stations, satellites, microwave relays, videocassettes) at a central "head-end" processing unit. Equipment at the head-end descrambles, amplifies, and changes the frequencies of signals in preparation for their distribution.

A cable distribution system uses *tree and branch architecture*, in which signals move "downstream" from the head-end to subscribers through a series of cables, known as trunk, feeder, and drop cables in descending order of magnitude. A cable system thus resembles other traditional distribution plants, such as those for water, electricity, and natural gas. Perhaps most important, cable is not a "switched" system, like a telephone local-exchange carrier. A cable system cannot connect one subscriber with another. A few cable systems have limited "two-way" or "interactive" capability that allows subscribers to send signals back "upstream" to the head-end. Although this technology has existed for a decade, it is expensive and unreliable. "Pay-per-view" ordering of individual movies and sports events generally depends on the telephone as the viewer's return link to the head-end. Although the cable industry ultimately may make upstream transmission work, it has not accomplished this so far. The introduction of fiber optics into the home may make switched video possible in the long term.

Early Regulatory and Legal Initiatives

Local Regulation

The first cable regulatory bodies were local governments. Virtually all municipalities have inherent jurisdiction over cable since systems must use public rights-of-way—either directly by installing their own poles and conduits or indirectly by using existing utility facilities. As early as the 1960s, state courts had upheld local governmental authority over cable.[6]

Until the restrictions in the 1984 Cable Act, there were virtually no limitations on the concessions a local franchising authority could extract from a cable operator, depending on its bargaining power. In a few cases, to be sure, this resulted in payoffs and other corrupt practices. More significant, however, municipalities negotiated deals that were not only unprofitable for cable operators, but also unworkable in the long run for local governments.

In some cases, this took the form of extremely high franchise fees. Although the FCC purported to limit payments to 5% of gross revenues,[7] in fact franchise fees often approached 10% because of provisions in "side agreements" with cable operators, "contributions" to various local activities, and the like. In some cases, cable operators agreed to buy unused school buildings, provide senior citizen bus service, or fund drug rehabilitation centers.

Alternatively, some franchises required cable operators to build high-tech facilities with "bells and whistles," even though there was no real need for them. Towns of 50,000 suddenly found themselves with extensive local production studios, "institutional networks" connecting all government buildings, and the like. The costs of these facilities not only burdened cable operators, but also ultimately were passed along to rate payers. As discussed below, they were a major cause for passage of the 1984 Cable Act.

State Regulation

About a dozen state governments became involved with cable regulation in one way or another. From the very beginning, there was no doubt that cable fell within a state's general powers to promote the public health, safety, and welfare.[8]

A few state governments have totally or partially preempted local jurisdiction by creating state-level regulatory authorities. Some states simply have taken over all nonfederal cable regulation; the Connecticut Public Utilities Commission issues certificates of public convenience and necessity directly to cable operators—municipalities have no role whatsoever.[9] Other states tend to play more of an advisory role; the New York State Commission on Cable Television offers detailed advice to local authorities, subjects franchises to a general review, and promulgates procedural and minimal substantive rules.[10]

Some states, such as Connecticut, just legislatively added cable to the jurisdiction of existing public utilities commissions. This approach has the advantage of vesting jurisdiction in an existing agency with regulatory experience; it has the disadvantage of giving the same agency jurisdiction over competing media — cable and local telephone exchange companies. Other states, like New York, have created totally new agencies to deal solely with cable; the problem with this type of single-industry agency is that it may become a bit too solicitous of the firms that it regulates, thus causing conflicts with other agencies.

Federal Regulation

By the late 1950s, broadcasters had become sufficiently concerned about the potential development of cable — even though it had less than a million subscribers — to request the FCC to regulate it. The commission's initial reaction was negative. In a 1959 case and rulemaking,[11] the FCC held that cable did not yet represent a significant threat to "free" broadcast television. In a decision that would come back to haunt it, the commission held that it might not have jurisdiction over cable in any event since it was neither a common carrier nor a broadcaster.

As the cable industry began to grow during the early 1960s, the FCC revised this approach rather quickly. In a 1962 decision, it refused to allow a microwave common carrier to provide distant broadcast signals to a cable system unless the cable operator carried all local signals — the precursor to the "must carry" rules.[12] Also in the 1966 *Second Report and Order*,[13] the commission imposed the first real regulation on cable.

The 1966 rules dealt almost exclusively with carriage of broadcast signals. First, the "must carry" rules required cable operators to offer their subscribers all stations within a defined distance of a system's community (e.g., coverage by a predicted Grade A or predicted Grade B contour).[14] This was designed to insure that broadcasters did not lose their traditional viewers by cable operators' not carrying their channels. Whether the rule was necessary at that time is questionable; cable systems carried virtually any available signal because microwave and satellite programming were not yet available.

Second, the "may carry" signals prohibited cable operators from carrying "distant" signals (i.e., those from beyond a local television market) without a prior FCC finding that carriage would not hurt local stations by "fragmenting" their audiences. This rule required an extremely lengthy hearing[15] — only one of which ever was completed. In effect, it thus amounted to a "freeze" on importation of new distant signals by cable systems.

Finally, the rule prohibited cable systems from carrying a network television program on a more distant station if a more local station offered it simultaneously.[16] Here again, the goal was to protect local broadcasters by insuring that viewers did not inadvertently tune into a network program on a nonlocal station — hence depriving the local station of ratings and advertising revenues.

After all the sturm und drang, the Supreme Court had no difficulty in upholding the FCC's jurisdiction over cable—at least as to signal carriage. In the unanimous 1968 case of *United States versus Southwestern Cable Company*,[17] the Court held that cable might be neither a common carrier nor a broadcaster, but that the FCC had jurisdiction over it for actions "reasonably ancillary" to its authority over broadcasting.[18] In highly ambiguous language, the Court held that

> It is enough to emphasize that the authority which we recognize today under § 152(a) is restricted to that reasonably ancillary to the effective performance of the Commission's various responsibilities for the regulation of television broadcasting. The Commission may, for these purposes, issue "such rules and regulations and prescribe such restrictions and conditions, not inconsistent with law," as "public convenience, interest, or necessity requires." We express no views as to the Commission's authority, if any, to regulate CATV under any other circumstances or for any other purposes.

The breadth of the *Southwestern* holding never has been tested; it may apply only to protection of broadcasters. The point seems largely moot, however, since the 1992 Cable Act provides a new and independent source of jurisdiction for the FCC beyond the vague reasonably ancillary standard. Whether reasonably ancillary jurisdiction still exists is less than clear and not likely to be tested in the foreseeable future.

The 1966 rules—particularly the distant signal freeze—created a situation in which the cable industry could not grow much, even beyond its rather limited beginnings. After the *Southwestern* case established that the FCC had some jurisdiction over cable, the industry began to negotiate a way for cable to enter the video media market with broadcast stations and program producers. In this process, they had the active encouragement of then FCC Chairman Dean Burch, whose staff even participated in the negotiations. Over a period of several years, the industry representatives worked out a "consensus agreement" that eventually became the FCC's 1972 *Cable Television Report and Order* (*CTRO*).[19]

The *CTRO* basically retained the must carry and nonduplication provisions, while relaxing the distant signal provisions and adopting new access channel requirements.[20] In the end, few of these rules lasted very long. As discussed below in the "Must Carry" section, the courts invalidated the must carry rules until the 1992 Cable Act legislatively reinstated them.[21] Similarly, the FCC quickly decided that the restrictions on distant signals were unnecessary and repealed them in 1980.[22]

Perhaps more important, all parties were increasingly dissatisfied with the interaction between local governments and cable operators. Large multiple system operators (MSOs) of cable felt that they could not live with the franchise agreements that they had negotiated only a few years before, and local governments became increasingly fearful that they might lose through legislation or litigation what they had gained through franchise negotiation. The result was the 1984 Cable Act.

The 1984 Cable Act

Because of its concerns about the potential dangers of the franchise renewal process, in the early 1980s the cable industry began a multipronged attack on local governmental powers. First, and most expeditiously, it convinced the FCC to preempt many aspects of local cable regulation—regulation of all rates except those for broadcast signals.[23] The FCC took comparatively little urging to go down the preemption path; since the commission already had embarked on an extensive program of federal cable deregulation, preemption of local powers—the only remaining regulation—de facto amounted to deregulation.

Second, the industry lobbied extensively for federal statutory limitations on a variety of local government powers—perhaps most important, franchise renewals. Initially, the industry's efforts took the form of a Senate bill, S. 66,[24] which was largely deregulatory and did little more than preempt state and local regulation. The National Cable Television Association (NCTA) apparently realized that it could not move legislation through Congress unilaterally; it thus initiated negotiations in 1983 with the cities, represented by the National League of Cities (NLC).

At first, the NLC was relatively uninterested in negotiating since the NCTA seemed to have little bargaining power. In the summer of 1984, the cities' position changed dramatically in the wake of *Capital Cities Cable, Incorporated, versus Crisp*.[25] In a much misunderstood opinion, the Court there invalidated on preemption grounds an Oklahoma statute that prohibited cable systems from carrying liquor advertisements. The Court based its holding on the effect of federal copyright legislation as well as FCC signal carriage rules, but many observers—particularly local governmental interests—interpreted the decision as giving the FCC carte blanche to continue preempting municipal jurisdiction over cable. The NLC thus found new reason to negotiate with the cable industry; it, the NCTA, and then-Congressman Timothy Wirth (chair of the House Telecommunications Subcommittee) launched into a series of negotiations that continued to the very day that the legislation passed the House as the Cable Communications Policy Act of 1984.[26]

One of the capstone—and most hotly contested—provisions of the act was the renewal provision, which imposed both substantive and procedural limitations on a local government's ability to withhold renewal of a franchise. The act provided that a city must renew if a franchise could show that

> (A) the cable operator has substantially complied with the material terms of the existing franchise and with applicable law;
> (B) the quality of the operator's service including signal quality, response to consumer complaints, and billing practices, but without regard to the mix, quality, or level of cable services or other services provided over the system, has been reasonable in light of community needs;
> (C) the operator has the financial, legal, and technical ability to provide the services, facilities, and equipment as set forth in the operator's proposal; and
> (D) the operator's proposal is reasonable to meet the future cable-related community needs and interests, taking into account the cost of meeting such needs and interests.[27]

These four tests still are quite significant since the 1992 Cable Act took them over intact, but their meaning is unclear since there have been only a few court decisions to date. Nevertheless, the statute's plain language and legislative history indicate that its purpose was substantially to foreclose local governments' discretion in denying franchise renewals. Unless the four criteria above are interpreted very expansively, any vaguely competent cable operator should qualify for renewal.

Moreover, the statute's procedure also makes it less likely that a city will throw a franchise into jeopardy. In order to conduct franchise denial proceedings, a city must take the following steps: (1) begin reviewing a franchisee's performance during a "window" 30 to 36 months before franchise termination, (2) receive a renewal proposal from the incumbent cable operator, and (3) "commence an administrative proceeding after providing prompt public notice."[28] This type of endeavor naturally requires a city to exercise a considerable amount of planning and foresight—something for which municipal officials are not renowned. In addition, the formalities of a renewal proceeding may be enough to deter any but the most serious city officials. Finally, the law allows judicial review of a city's action in federal district court, with de novo consideration of a city's "compliance with the procedural requirements of this section."[29]

The 1984 act also severely restricted local governments' powers to regulate rates. Except when the FCC found no "effective competition" for cable, municipalities could not regulate cable rates at all. In no event could they regulate the price of "premium" or "pay" tiers. Since the commission initially decided that effective competition constituted predicted Grade B contours from three broadcast stations, 97% of the country's households were deemed to have effective competition. (The commission eventually modified its definition of effective competition, and the 1992 act has made the test considerably more stringent,[30] with the result that only about 50% of households are in areas with effective competition.[31])

The quid pro quo for the partial guarantee of franchise renewals and immunity from rate regulation was that local governments could charge a franchise fee of up to 5% of gross revenues—also carried forward by the 1992 Act.[32] The 5% statutory cap is considerably more stringent than the FCC's prior restrictions since it excludes a variety of other payments, such as the side agreements, contributions, and the like previously allowed by the FCC.[33] It also limits payments for local access and similar programming activities to capital contributions; a cable operator may agree to build an access production facility, but not to pay its ongoing salary and other expenses.

The 1984 act thus represented a basic tradeoff between the cable industry and local government: a guaranteed renewal in return for a guaranteed 5% franchise fee. The National Cable Television Association urged its members not to raise their subscription rates, even though they were allowed by the new statute. Many of them did so, however, thus setting the stage for the 1992 Cable Act—potentially the most severe regulatory regime under which the cable industry ever has operated.

Prelude to the 1992 Cable Act

For the first few years after enactment of the 1984 Cable Act, most operators largely conducted business as usual. Although their rates were frozen by the legislation until the end of 1986, few systems initiated large increases after the thaw. Many operators were aware of the importance of not looking like profiteers in their new free market environment.

Some cable systems used their new rate freedom, however, to increase rates significantly. According to the Government Accounting Office, basic cable service rates increased more than 30% per year shortly after 1986.

Not unjustifiably, some cable operators argued that their basic rates had been artificially depressed after a decade of municipal rate regulation, and that they thus needed major increases. These claims met with little sympathy, however, because rates had been stable for such a long period of time.

The debate might have ended there, but in early 1990 a junior senator named Al Gore initiated hearings as to cable rate increases. Without much difficulty, he was able to identify a number of systems that had increased their rates by 100% or more.

A general cry went up for "reregulation," and bills were introduced in both the House and Senate. Senator Gore's bill passed the Senate by an easy margin, only to face what had been promised as a guaranteed veto by President Bush. For the first time in the Reagan–Bush administrations, however, Congress overrode the veto on October 5, 1992.

The 1992 Cable Act

Background and Rationale

Congress passed the Cable Television and Consumer Protection and Competition Act of 1992 (1992 Cable Act) in the midst of outcries from consumers who had seen their cable rates rise at nearly three times the rate of inflation (as measured by the consumer price index) since the 1984 Cable Act deregulated cable rates. Broadcasters also had initiated an intense public campaign to convince regulators that cable operators had attained sufficient market power to control the information received by a large segment of the population. These forces brought increased scrutiny to the market structure and practices of the cable industry.

Congress attributed rising cable rates to a lack of competition for cable. Due to restrictive local franchise requirements and the immense cost of building competitive cable systems, nearly all cable operators are monopolies within their geographic markets. Therefore, Congress felt it imperative to encourage alternative forms of competition such as satellite technologies.

Since the development of competition in the local cable market necessarily would take some time, Congress felt that it was necessary to control cable rates in the meantime. The 1992 Cable Act charged the Federal Communications Commission with the responsibility of developing and administering price controls.

In addition to cable rates, Congress also was concerned that cable operators had developed sufficient market power to control the content of the programming carried on their cable systems. Cable originally served mainly as a conduit by which subscribers received clearer signals for broadcast television, but cable companies also had become premier producers and distributors of programming. Cable services receive a significant share of the audience. More than 60% of households subscribe to cable service.[33] While the networks commanded 90% of the market 15 years ago, they currently are viewed by just over 60% of the nation's viewers. Regulators considered the ascendance of cable companies a threat to the economic viability of "free" broadcast television. Broadcast television stations rely exclusively on advertising for revenue. Since networks owned and operated by the cable companies were now competing for the same advertising revenue, Congress feared that cable operators would have an incentive to drop broadcast stations from their systems.

Signal Carriage Provisions

Must Carry

The FCC first placed mandatory carriage obligations on cable systems in 1962. As amended in 1966, the rules only applied to those cable systems that carried at least one broadcast signal.[34] Therefore, a cable system could avoid the requirement by simply not carrying any broadcast signals. This loophole, which ran counter to the intent of the rules, was eliminated by new must carry rules in 1984.[35] Under the 1984 must carry rules, all cable systems were required to carry the signals of local broadcasters. Cable operators viewed the new regulations as compelling speech in violation of the First Amendment, and the must carry rules soon faced their first constitutional challenge.

The must carry rules first were struck down as an unconstitutional infringement on the speech of cable operators in *Quincy Cable TV versus Federal Communications Commission*.[36] The D.C. Circuit determined that the must carry rules did not further a substantial governmental interest and restricted speech protected under the First Amendment to an extent greater than was essential to further that interest.[37] The court followed the Supreme Court's decision in *United States versus O'Brien*,[38] which established the "intermediate scrutiny test" used when a regulation is "content neutral."[39] Since the *Quincy* court found that the must carry rules were instituted to protect "free, locally controlled" broadcast television from the "potentially monolithic national" cable television industry and did not turn on any specific content carried, the rules were deemed to be content neutral.[40]

The *Quincy* court did not rule that must carry provisions were unconstitutional per se. The court did find that the government failed to show that the rules furthered a substantial government interest, however, or did so in a way that was no greater than necessary to further that interest. The court held that the FCC had not provided adequate proof that the economic health of local broadcast television was threatened by cable since the FCC had not presented empirical evidence sufficient to back up its conclusions.[41] Without such evidence, the government could not maintain that the must carry rules served any substantial government interest.

Further, the court found that the 1984 version of the must carry rules was "grossly" overinclusive. Since the rules protect all local broadcasters regardless of the amount of local service, the rules were deemed to be a greater intrusion into the First Amendment right of cable operators than was necessary.[42] In addition, the court took notice of the fact that broadcast stations did not have to show that they individually were economically threatened.[43]

In response to the *Quincy* decision, the FCC modified the must carry rules. Attempting to tailor the rules more narrowly, the FCC imposed the must carry requirement on cable for a five-year period to allow the public to become familiar with a "switching box." The switching box would allow the consumer to switch between broadcasting and cable channels. The new rules, however, were also struck down as unconstitutional in *Century Communications Corporation versus Federal Communications Commission.*[44]

In *Century*, the court again found that the government had failed to show that the must carry rules furthered a substantial government interest. Relying principally on the same evidence as it had two years earlier in *Quincy*, the government failed to produce empirical evidence that the broadcasting was indeed threatened by the cable industry. The court characterized the FCC's assertions as "dubious" and "more fanciful than real."[45]

The court also questioned the five-year period in which must carry would be imposed. The court concluded that this was substantially longer than subscribers needed to learn to use a relatively simple device.[46] Therefore, the rules were not sufficiently narrowly tailored to address the government interest involved.

The present must carry rules were implemented in the 1992 Cable Act. The rules are coupled with a "retransmission consent" option for the broadcasters. Under the 1992 Cable Act, every three years broadcasters have the option either to negotiate with the cable operators for a price to carry their signals (retransmission consent) or simply to require the cable operator to carry the signals without a fee (must carry).

To satisfy the narrowly tailored prong of the *O'Brien* test, Congress's new rules vary from their predecessors in that the present version does not require that all local broadcast stations be carried by cable operators. Two types of broadcast stations are afforded separate consideration under the rules: noncommercial educational (NCE) and commercial stations. The number of each type of station that must be carried by cable operators depends on the system's number of channels. Cable systems with 12 or fewer usable activated channels must carry 1 local NCE station and at least 3 local commercial television stations, leaving up to 8 channels to the discretion of the cable operator.[47] Cable systems with 13 or more usable activated channels must carry up to 3 local NCE stations, but are not required to carry any NCE stations that substantially duplicate programming already being broadcast by another qualified NCE station.[48] In addition, such cable systems must carry local commercial television stations up to one-third of the usable activated channels, but need not carry a station that substantially duplicates the programming of an already-carried station. This will leave at least six, and in most cases substantially more, channels to the discretion of the cable operator.[49]

The 1992 Cable Act gives the cable operator a choice of which signals it will carry to meet the must carry obligations when there are more qualified stations than the number of signals it is obligated to carry. This provides the cable operator with a degree of editorial discretion absent in the former versions of the rule.

Congressional drafters of the 1992 Cable Act were well aware that the new must carry rules were likely to be brought before the courts like their predecessors. Therefore, the act contained a clause allowing appeal of a decision by a federal district court directly to the Supreme Court of the United States, bypassing the court of appeals.[50]

As anticipated, the must carry rules were quickly brought before the courts in *Turner Broadcasting versus Federal Communications Commission*.[51] A three-judge panel granted summary judgment for the government, concluding that the must carry rules were essentially economic regulation designed to remedy market dysfunction caused by the imbalance of power between cable operators and broadcasters. Under the intermediate scrutiny test, the court concluded that the rules did not violate the First Amendment; preserving local broadcasting was deemed to be a substantial government interest, and the rules were sufficiently narrowly tailored to meet that purpose.

The Supreme Court had denied certiorari in both *Quincy* and *Century*. The Supreme Court thus decided for the first time what level of scrutiny to apply to regulations that curtailed the speech of cable television operators.

The Supreme Court 5–4 refused to apply the rational basis scrutiny test used for broadcast regulation. The Court noted that cable television did not suffer the same physical limitations as broadcasting, especially since technology such as digital compression and fiber optics potentially would provide hundreds of alternative avenues of expression for cable operators. Turning to the familiar *O'Brien* test, the Court reasoned that the must carry rules were intended to protect the financial viability of broadcasters, not to protect any particular content of speech. The Court concluded that the government had not satisfied its burden of proof, however, when it asserted that the rules were in fact necessary to achieve that purpose. Concluding that Congress had relied on incomplete and unconvincing evidence, the Court ruled that the District Court for the District of Columbia erred when it granted summary judgment in favor of the government and remanded the case to the lower court for further consideration.[52]

Retransmission Consent

Under Section 6 of the 1992 Cable Act, no cable system operator or other multichannel video programming distributor may retransmit the signal of a broadcast station without the station's consent unless the station has elected to be treated as a must carry station under Section 614 of the Communications Act.[53]

Broadcasters had to decide either to forego compensation to take advantage of the must carry requirements or to seek compensation for cable carriage and take the risk that the local cable system either would not carry them or would pay little for transmission consent. A broadcaster generally may not change its election during the prescribed three-year period. If a broadcaster elects for retransmission consent and is unable to negotiate an agreement, it faces the prospect of not being carried by a cable system for three years.

If a broadcast station does not make an election, it is deemed to have elected must carry status. The FCC reached this result by concluding that, if a station did not even bother to make an election, it would not be likely to expend the effort to negotiate a retransmission consent agreement. In this event, the public could lose access to a broadcast station.

Congress treated retransmission consent as a new right. Consequently, the retransmission consent right exists regardless of whether the owners of copyrights in programming have consented to cable retransmission. In other words, retransmission consent is different from the bundle of rights held by copyright owners under federal copyright law, as discussed in the sections on intellectual property.

The first round of negotiations between broadcasters and cable operators in 1993 was bitterly fought in the press. Cable operators contended that they would refuse to pay for broadcast channels that they always had transmitted for no cost, and broadcasters vowed that they would not consent unless adequately compensated. But, each medium was intricately dependent on the other. Cable operators knew that many customers ordered cable to receive clear broadcast signals, and broadcasters knew that withholding consent would lead to smaller viewership and advertising revenue.

Initiating a trend, the Fox Network negotiated a retransmission pact with TCI. TCI agreed to compensate Fox for carrying a new cable network channel if Fox's broadcast channel were provided at no cost. Though this provision circumvented the direct payment contemplated by the drafters of the retransmission provision, it allowed for compensation. NBC (National Broadcasting Co.) and ABC (American Broadcasting Co.) soon entered into similar agreements with various cable operators. But networks and group owners without cable programming, including CBS (Columbia Broadcasting System), found themselves with little bargaining power and settled for no payment in order to be carried.

The next retransmission election will occur in 1996 and promises interesting results. With the birth of the direct broadcast satellite (DBS) market, in future retransmission consent negotiations broadcasters might have other media distribute their programming, and unless changed by Congress, broadcasters will be allowed to sign exclusive agreements with cable operators or other carriers. This, no doubt, will provide some added leverage to the broadcaster in future retransmission consent negotiations.

Rate Regulation

Procedural Issues

The 1992 Cable Act reintroduced rate regulation, abolished only eight years earlier in the 1984 Cable Act.[54] Before the 1984 Cable Act was passed, rate regulation was largely a function of the local governments.[55] In contrast, though Section 3 of the 1992 Cable Act allows local governments to regulate subscriber rates, ultimate authority for policy functions rests with the FCC.[56]

Section 3 required the FCC to determine criteria for identifying unreasonable rates. Factors that must be considered in establishing the criteria for reasonable rates were rates of cable systems subject to effective competition, comparison of cable rates over time to the consumer price index, and capital and operating costs and revenue of the cable operator.[57]

Substantive Standards

The provisions concerning rates, like the 1992 Cable Act generally, were largely a response to the monopoly structure of the cable market. Under Section 3 of the Act,[58] only the rates of cable operators not subject to "effective competition" are subject to regulation. Under the act's definition, however, effective competition exists if any of the following standards are met:

1. Fewer than 30% of the households in the system's franchise area subscribe to the operator's service;
2. The franchise area is (1) served by at least two "multichannel video programming distributors," each of which offers comparable programming to at least 50% of a franchise area's households; and (2) more than 15% of the franchise area's households subscribe to service offered by the distributor other than the largest one in the area; or
3. The area's franchising authority operates a multichannel video programming service available to at least 50% of the households in the area.

New entrants potentially could satisfy the second of the standards. New DBS providers cover 50% of a franchise area's households; a satellite's footprint probably would cover the entire franchise area. But, capturing 15% of the households will be no small task. Roughly 25% of all cable customers would have to switch services, which is not likely unless initial costs are lowered. Satellite receivers initially cost $700, many times the cost of cable installation. Alternatively, cable and DBS companies may offer services jointly to provide a large number of channels without expensive rebuild costs.

Section 3(b)(7) of the 1992 Cable Act requires that a cable operator provide some channels on a separate, low-cost basic service tier. Only the subscription to the basic service tier may be a condition for provision of service—the "anti-buy-through" requirement. Programming included in this provision includes:

1. All must carry broadcast stations;
2. All other broadcast stations that the cable operator carries other than distant signals (including superstations outside their local market);
3. All public, educational, and governmental (PEG) access programming required by the cable system's franchise.[59]

The purpose of this requirement is to assure that cable operators do not shift programming from the basic tier to a more expensive tier in order to force customers to pay premium charges for additional channels they now receive at no extra charge, thus circumventing the purpose of the rules. The FCC has allowed cable operators to shift a limited number of program services from basic tiers into unregulated "enhanced tiers." The commission generally has focused on whether an operator was moving new or long-established services.

Access Channels

Access channels allow third-party programmers to require cable systems to carry their material. They include both nonprofit (e.g., churches, fraternal associations) and for-profit entities. Local franchises usually require one free channel each for public, educational, and governmental purposes. Leased access is a quasi-carrier service for commercial entities for which cable operators may charge a "reasonable" fee. The FCC cannot require PEG channels.[60] The 1992 act, however, requires cable operators to offer commercial "leased access" channels.

Leased Access

Section 2 of the 1984 Cable Act, as incorporated in the 1992 Act, requires cable operators to set aside a certain number of cable channels for use by unaffiliated programmers.[61] The number of channels to be set aside varies with a system's number of channels. Cable systems with between 36 and 54 channels must make available 10% of their capacity (either 3 or 4 channels), while cable operators with more than 55 channels must make available 15% of their capacity. Cable operators with between 36 and 100 channels are specifically exempted from these requirements, however, when otherwise available channels are required for use by federal law or regulation (e.g., channels preempted for aviation use by the Federal Aviation Administration). Further, cable systems with less than 36 activated channels need not set aside channels for leased access unless the cable system is required to do so under a franchise agreement in effect on enactment of the 1984 Cable Act. Unless provided under the terms of a franchise agreement, a cable operator can elude the requirements of Section 2 completely simply by carrying 35 or fewer channels.

These rules are separate and apart from the must carry rules. Unlike the must carry channels, a cable operator has an absolute right to compensation for use of its signal capacity—hence the term *leased access*. In order to insure that cable operators provide unaffiliated programmers with meaningful access, Section 2 gives the FCC power to establish maximum rates and reasonable terms and conditions for the use of such channels, including permissible billing and collection procedures. Cable operators nevertheless may charge reasonable fees for use of leased access channels.

Since the purpose of Section 2 is to promote diversity of programming, cable operators are specifically forbidden to exercise any editorial control of the programming carried on these channels.[62] But the Act allows cable operators to prohibit programming "that the cable operator reasonably believes describes or depicts sexual or excretory activities of organs in a patently offensive manner as measured by contemporary community standards."[63]

Public, Educational, and Governmental Access

To further encourage diversity in programming carried on cable television, Section 2 of the 1984 Cable Act authorizes local governments to establish requirements that a cable system offer channels for public, educational, or governmental use.[64] This requirement differs substantially from that of the federally imposed leased access requirement because the franchise authority cannot unilaterally impose PEG access requirements. PEG access must be negotiated by the parties, and franchise terms depend on their bargaining power.

As with the leased access, however, the cable operator may not exercise any editorial control of the content on any PEG channel. Again, a small exception applies: A cable operator may exercise editorial discretion when programming contains "obscene material, sexually explicit conduct, or material soliciting or promoting unlawful conduct."[65]

Franchise Fees

As discussed in the section "Local Regulation," franchise authorities have extracted very large fees from cable operators, often to pad local budgets. Congress recognized that if cable operators were required to pay excessive franchise fees, however, consumers eventually would have to pay the cost through higher monthly cable bills. Finding this result to be against the public interest, franchise fees were limited to 5% of gross revenues, in addition to all general fees and taxes charged by the government.

Capital costs incurred in providing PEG channels, however, are not included in the calculation of franchise fees under a franchise agreement made after the 1984 Cable Act; but operating expenses (e.g., staff salaries, rent equipment maintenance) count against the 5% franchise cap. For franchises in effect before the 1984 act, any payment made in support of PEG channels, whether or not related to the actual cost of making the channel available, also is excluded from the calculation of the franchise fee. While franchise fees are limited, local governments still can extract considerable resources from cable operators. The Cable Act provides an incentive for local governments not to pinch cable operators unreasonably. Cable operators may itemize on subscribers' bills the amounts attributed to franchise fees, other fees or taxes, and costs associated with PEG channels—which may generate consumer reactions.

Franchise Renewals

Since large amounts of capital are necessary to build the infrastructure of a cable system, potential operators understandably want long-term franchise agreements to recoup their investments. Contracts for 15 years generally have become the standard. For the same reasons, if a cable operator at the end of its term was uncertain whether it would be granted renewal, the operator would have no incentive to upgrade or even maintain essential facilities. It thus would appear essential to begin franchise renewal proceedings far enough in advance of franchise termination to prevent such neglect by the cable operator. Not coincidentally, the provisions in the Cable Act attempt to remedy these problems.

Under the renewal provisions of the 1984 Cable Act,[66] the franchising authority may conduct a proceeding as to whether a franchise renewal is in the public interest. As discussed in the section on the 1984 Cable Act, to do so, it must initiate such a proceeding within a 6-month window that begins 36 months before expiration of the franchise and ends 30 months before the end of the franchise. The cable operator may submit a written request thereafter; if it does, the franchising authority must commence a renewal proceeding within 6 months of the request.

Consumer Protection

The perceived need for consumer protection laws may best be explained by the traditional monopoly status of most cable operators within their franchise areas. Without competition driving cable operators, regulators fear that subscribers simply will have to put up with cable operators' standards. In the absence of competition, cable operators presumably would provide just enough service to avoid the risk that a consumer would decide not to subscribe to cable.

Cable operators' monopoly over multichannel video systems may change in the long term, with the potential entry of other delivery systems. For example, by the mid-1990s in the United States, multichannel offerings were available in some areas from a mix of media (e.g., DBS, MMDS ["wireless cable"], and LECs). The success or coexistence of these media probably will take a generation to be decided.

Consumer protection standards are set at the federal, state, and local levels. Section 8 of the 1992 Cable Act requires the FCC to establish standards regarding customer-service requirements.[67] The commission must determine the amount of time it should take to get cable equipment installed or repaired, the maximum acceptable length for outages, the minimum length of office hours, and the standards for governing bills and refunds.

Section 8 specifically allows local franchise authorities to negotiate with cable companies for more stringent consumer service standards than are required by the FCC. In addition, both the states and the franchise authorities may enact further consumer protection laws to the extent that they are not preempted by any section of the Communications Act.

State and local consumer protection agencies generally focus on a few major issues. Their primary concerns tend to be (1) response time for installation and repair calls, (2) credits for system outages, and (3) treatment of subscribers by cable customer-service representatives. These issues resulted in a combination of administrative rules and industry guidelines.

Ownership Limitations

Section 28 of the 1992 Cable Act authorizes the FCC to restrict the amount of cable subscribers served by any particular operator and the number of channels occupied by a video programmer affiliated with the cable operator.[68] The act also limits the ability of cable operators to enter other markets. Under Section 28, the same party may not operate a cable system in the same market in which it owns a television broadcast station, a multichannel multipoint distribution service, or satellite master antenna service separate from the franchised cable service.

Section 28 also prevents LECs, other than those in rural areas not served by cable operators, from offering video programming directly to subscribers. Several Regional Bell Operating Companies (RBOCs) successfully challenged the provision in federal court, and some were upheld in the courts of appeals.[69]

In *Chesapeake and Potomac*,[70] the Fourth Circuit Court of Appeals determined that the purpose of the restrictions was to prevent telephone companies from cross-subsidizing their entry into the video programming market with their revenues from telephone rate payers. The court found this purpose to be unrelated to the regulation of the content of the programs that would have been carried over telephone lines. In terms of content-neutral regulation of speech, rules must involve only proper time, place, and manner restrictions under the *O'Brien* test.[71] Under this standard, the rules must further a substantial government purpose and restrict speech no more than necessary to achieve that purpose.

The court had little difficulty in determining that prevention of cross-subsidization to the detriment of local telephone rate payers constituted a substantial government purpose, but the court found that depriving a telephone company of all editorial discretion in providing programming to customers restricted its speech more than was necessary to accomplish the purpose enunciated by the government. The court thus struck down the rules as violative of the free speech rights of telephone companies.

Required Sale of Programming to Other Multichannel Media

To gain the advantages associated with competition, the Congress decided to promote diversity of programming distribution providers, particularly for those in rural areas who could not receive cable and thus relied on other multichannel distributors to receive their programming. Congress had become increasingly concerned that cable operators had achieved sufficient market power to keep new entrants out of the distribution market. Cable companies increasingly were vertically integrated with both production and distribution. Even when a cable operator did not actually own the source of programming, regulators hypothesized that cable operators would use their monolithic market power to forge exclusive agreements that would deny new competitors (e.g., DBS, MMDS) the same programming.

Section 19 of the 1992 Cable Act prohibits a cable operator from preventing other multichannel video programming distributors from obtaining programming for their subscribers.[72] The act charges the FCC with the responsibility of adopting and enforcing regulations to stop cable operators from exerting their market power to the detriment of competition.

Requirements of Compatibility with Consumer Electronics

Manufacturers of consumer electronics continually develop new products with innovative features. For example, new videocassette recorders (VCRs) allow owners to view a program on one channel while simultaneously recording a program on another. Another feature allows an owner to program the VCR to tape two consecutive programs on separate channels. Cable technology to prevent theft of services (such as scrambling, encoding, and encryption), however, can interfere with the use of these features.

In order to encourage manufacturers of consumer electronics to develop innovative features that are desired by consumers, Congress passed Section 17 of the 1992 Act.[73] This directs the FCC to develop standards to allow consumers to use innovative consumer electronic equipment in conjunction with cable television service, while still allowing the cable industry to protect their signals from theft.

Section 17 requires the FCC to develop standards to make television sets and VCRs "cable compatible." To protect consumers, the section also requires the FCC to develop rules that cable operators must notify subscribers if they may be unable to use some features on their television sets or their VCRs. In addition, cable operators are to notify subscribers that other commercially available devices, such as remote controls, are available from other sources and the operators may not prevent their cable boxes from operating with such units.[74]

Theft of Services

The cable industry loses large revenues to unauthorized interception of signals. "Cable piracy" is apparently quite common, and Congress has adopted stiff penalties for theft of cable signals.

Section 28 of the 1992 Cable Act makes it a crime to intercept or receive service without the authorization of the cable operator.[75] The act also makes it a crime to assist in the unauthorized reception of such service, which includes the manufacturing or distributing of equipment intended for that purpose.

The act authorizes fines of up to $1000 or imprisonment for up to six months for willful violations. If the manufacturers or distributors of illegal equipment violate the provision for the purpose of "commercial advantage or financial gain," they may be fined up to $50,000 and imprisoned up to two years for a first offense and fined $100,000 and imprisoned up to five years for any subsequent offense. A plaintiff cable operator may recover damages and attorneys' fees or choose statutory damages, which range from $250 to $50,000, depending on the severity of the offense.

Privacy

Cable television companies often obtain sensitive personal information about their customers in the course of doing business. For example, in order to receive pay per view, a customer must order a specific program. Cable companies keep this information for billing purposes. Cable companies also can monitor the channels specific customers are viewing. Such information is used for both maintenance and marketing purposes.

Section 20 of the 1992 Cable Act protects the privacy of cable customers by limiting the permissible uses of information received by the cable companies.[76] Congress attempted to protect the privacy of the customers while still allowing the cable companies to make use of the information for legitimate purposes.

The first requirement of Section 20 is that the cable company must disclose to its customers the nature of the information the cable companies collect and the purpose for which it is collected. The cable companies also must provide notice regarding the nature, frequency, and purpose of any disclosure and, in addition, to whom such information will be made available. This notice must be furnished when entering into an agreement to provide cable service and at least once a year thereafter.

This limitation applies only to "personally identifiable information." Thus, aggregate or other general information may be used at the discretion of the cable operator. Even personally identifiable information may be used by the cable operator in order to collect information deemed necessary to render cable service or to prevent unauthorized reception of service. In addition, the cable operators also may provide such information pursuant to a court order.

Intellectual Property

Background

Generally, in order for works to be copyrightable under the 1976 Copyright Act, they must be "fixed in any tangible medium of expression . . . from which they can be perceived, reproduced, or otherwise communicated, either directly or with the aid of a machine or device."[77] If sufficiently fixed, video programs may be classified as "motion pictures and other audiovisual works," the latter of which are defined as

> Works that consist of a series of related images which are intrinsically intended to be shown by the use of machines or devices such as projectors, viewers, or electronic equipment, together with accompanying sounds, if any, such as films or tapes, in which the works are embodied.[78]

As with regulation, the copyright history of cable television began in the courts. The Supreme Court consistently held that transmission of distant signals had no copyright significance.[79] The Court viewed cable systems as passive intermediaries that "simply carry, without editing whatever programs they receive."[80] Cable systems thus could carry broadcast programming without incurring any copyright liability. In effect, the Court declared the problem to be too complicated for judicial policy making and left the issue to a legislative resolution. This came as part of the general Copyright Revision Act of 1976.

1976 Copyright Revision Act

Compulsory License

Section 111 of the new law created a compulsory copyright scheme under which cable systems may retransmit broadcast programs in return for statutorily defined royalty fees. The act also created a new administrative agency, the Copyright Royalty Tribunal (CRT), to administer the compulsory copyright.[81] The CRT collects fees from cable operators under Section 111 and passes them out to producers and other copyright owners—under formulas that seen to satisfy no one.

Section 111's basic thrust is to give cable systems compulsory licenses for all local signals and to require systems, except in specifically exempted situations, to pay for the use of distant signals. The availability of a compulsory license depends on whether carriage of a particular signal is "permissible under the rules, regulations, or authorizations of the Federal Communications Commission."[82] Section 111 thus uses the FCC's rules to define the scope of a compulsory copyright.

Tracking the commission's must carry and may carry rules, Section 111 basically sets up two primary categories of payments. First, a cable system must pay a flat fee of about 1% of its gross revenues to carry local television signals. If must carry rules were not in effect—a situation certainly not foreseen by the act's drafters—a system arguably might be able to avoid this payment by carrying no local signals.

Second, cable operators must pay additional royalties for every "distant signal equivalent" (DSE) carried. The act defines a DSE as "any nonnetwork television programming carried by a cable system in whole or in part beyond the service area of the primary transmitter of such programming," that is, a distant independent station.[83] But, public television stations count for only one-fourth of a DSE, and "specialty," "late night," and partially carried stations are even less of a DSE.

Calculating a system's royalties involves some complexities. Under the copyright act, the definition of a local signal depends on the old must carry rules; a signal is local if a cable system would be required to carry it. Section 111(f) defines the "local service area of a primary transmitter" as

> The area in which such station is entitled to insist upon its signal being retransmitted by a cable system pursuant to the rules . . . of the Federal Communications Commission in effect on April 15, 1976.

A signal's classification as distant or local—and a potential fee difference of several thousand percent—thus depends on application of the old must carry rules.

Suffice it to say, these are somewhat arcane in nature. Under the old rules, must carry status depended not only on the size of the market in which a cable system was located, but also on the type of station seeking carriage. For example, in the top 50 television markets, a cable system had to carry

(1) Television broadcast stations within whose specified zone [a 35-mile radius from a designated reference point in the station's city of license] is located, in whole or in part.

(2) Noncommercial educational television broadcast stations within whose Grade B contours the community of the system is located, in whole or in part;

(3) Television translator stations, with 100 watts or higher power, licensed to the community of the system;

(4) Television broadcast stations licensed to other designated communities of the same major television market (Example: Cincinnati, Ohio–Newport, Kentucky television market);

(5) Commercial television broadcast stations that are significantly viewed in the community of the system.[84]

To characterize Section 111's implementation as idiosyncratic would be an understatement. Nevertheless, none of the industries involved seem interested in abandoning the basic notion of a compulsory copyright; the transactions cost of an open marketplace might be staggering — as experience with "retransmission consent" has indicated (see the section on retransmission consent).

Calculation of Payments

Initially, the fee for DSEs was set on a sliding scale and was relatively low — that is, 0.675% of gross revenues for the first DSE, 0.425% for the second, third, and fourth, and 0.2% for each thereafter. Precisely because Congress foresaw that the commission might modify its signal carriage rules, however, the statute gives the Copyright Royalty Tribunal the right to increase compulsory copyright fees to reflect changes in the FCC's rules. After the commission's deletion of its distant signal and syndicated exclusivity restrictions, the tribunal increased royalty fees significantly in two ways. First, it imposed a flat — as opposed to sliding — fee of 3.75% of gross revenues for each DSE. Second, it created a sliding scale surcharge of about 1% per DSE to compensate for the loss of syndicated exclusivity.

Section 111 limits these payments in three major ways. First, a system pays a percentage of "the gross receipts from subscribers . . . for the basic service of providing" broadcast signals. This provision reduces royalty payments of larger cable systems by excluding revenues from pay television programming — which may constitute half of a system's gross receipts.

Second, Section 111 explicitly provides a bargain basement rate for small cable systems. For example, a cable system with about 1,000 subscribers would pay very little for carriage of local signals alone. Some systems thus will pay only nominal royalties.

Finally, cable systems pay on the basis not of individual distant signals, but rather of distant signal equivalents. In recognition of the fact that educational stations do not attract large audiences, Section 111 treats them as only one-fourth of a distant signal equivalent. Moreover, the previously discussed bits and pieces of available distant signal programming (i.e., "specialty" and "late-night" programs) count toward a distant signal equivalent only in fractional amounts. A cable system thus could carry two distant independent signals and four educational signals for the price of three DSEs.

Future Developments

Cable/Telco Relationships

Status Quo with Enhanced Capacity and Interaction

In the unlikely event that cable/telco cross-ownership rules and line of business restrictions endure, each will try to enlarge its market as much as possible. In the case of the cable industry, fiber optics and digital compression will increase the capacity of cable systems. Enhanced capacity will create the ability to provide "interactive" services. A typical use of this enhanced capacity will be the ability of cable subscribers to order programs or goods they choose. Thus, programming available to subscribers will be coextensive with the operators' "video archives."

Cable operators also may provide interactive games, lessons, and the like. Since cable networks do not include switches, however, cable operators cannot connect customers to each other. Thus, as a practical matter, customers can only interact with the cable operator, though that activity can take several forms.

Cooperation with the Local-Exchange Carriers

In 1993, Bell Atlantic, the largest LEC, and TCI, the largest cable operator, announced a plan to merge into one company providing both telecommunications and video programming. While some marveled at the interactive capabilities such a system could produce, others were concerned about the prospect of one company controlling all the information coming in and out of businesses and homes. Claiming that intervening cable rate regulation had made the deal unprofitable, Bell Atlantic and TCI called off the merger only a few months later. Other announcements of planned mergers between RBOCs and large MSOs followed, but also failed to consummate.

Despite this awesome potential, the LECs and cable operators apparently have second guessed the wisdom of merger. Several factors have contributed to this determination. First, the initial appeal of combining the two industries was based on the fact that each group had a valuable resource the other did not: LECs owned switches and the MSOs would provide video programming. However, the infrastructure alone necessary to combine the two networks would cost billions of dollars. Weighed against the large costs are uncertain demands for impressive, yet nonessential, services. Instead, cable companies have apparently chosen new allies and shifted their focus to competing with the LECs. However, changes in technology, market conditions, and regulatory environment could well rekindle opportunistic alliances between the two industries.

Competition with the Local-Exchange Carriers

Prior to 1984, AT&T provided most of the customers in the United States with both their long-distance and local services. As a result of the AT&T divestiture in 1984 (otherwise known as the *Modification of Final Judgment* or *MFJ*), AT&T was broken up into the present long-distance carrier and seven local-exchange carriers, known as the Regional Bell Operating Companies (RBOCs or Baby Bells). Though AT&T has encountered significant competition from the likes of MCI and U.S. Sprint, the RBOCs have remained monopolies with control of over 80% of the local-exchange market.

The cable companies, like the LECs, have wires going directly to most of the homes in their service areas. Having replaced much of their coaxial copper with fiber optics, cable companies have now positioned themselves to offer telecommunications services. However, cable companies face both technical and legal hurdles. Technically, cable companies now have "interactive" capability. That is, the consumer can both receive messages from and deliver messages to the cable operator. But, cable cannot connect end users to each other. This necessitates the use of a switch to route the call. In addition, the cross-ownership provisions of the Communications Act forbid the provision of telephone service by cable operators within the area of their cable franchise.[85] Thus, cable companies currently have invested heavily in the competitive access provider (CAP) industry.

Competitive access providers provide various services in competition with the LECs. For example, CAPs provide "private-line" service between two locations for the same customer (e.g., various hospital databases). A customer can avoid using the local telephone company for high-traffic data needs and receive a discount. In addition, CAPs provide "special access" by which a customer can place a long-distance call and "bypass" the local-exchange carrier, thus saving money on the local portion of the call. Cable companies have an incentive to invest in CAPs because they are able to use the conduits already laid by the cable companies, thus providing an extra source of revenue for the cable operator.

CAPs currently provide local transport in several states, but have limited authority to provide switched local-exchange service. Bills debated in both chambers of Congress in 1994 would have opened up the local-exchange markets to competition. Unless Congress ultimately acts to open up the local-exchange markets to competition, competitors will have to rely on each state to do so independently.

Relationship to the National Information Infrastructure

In practice, the future relationships between the cable and telephone industries largely will define the National Information Infrastructure. Although the Clinton administration attempted to impose its vision on the NII, in fact the nation will accept whatever network the cable, telephone, and other electronic industries end up building.

The LECs' entry into broadband networks and content control may not consciously be driven by the NII's strictures. The resulting infrastructure, however, will be the NII.

Potential Legislative Responses

Bills to Date

In 1994, both the House and the Senate debated bills that would have allowed cable operators, LECs, and long-distance telephone companies to enter each other's line of business. Though the measure passed in the House by a vote of 430 to 5, the bill was stalled in the Senate because members could not agree on the appropriate safeguards necessary to insure that the RBOCs would not use their tremendous market power to leverage unfair advantages in adjacent markets, particularly long distance. Generally, Democrats favored a plan that would have kept the RBOCs at bay until there was effective competition in the local marketplace, while Republicans advocated allowing the RBOCs to enter other market segments at a "date certain."

Access/Common Carriage Requirements

As cable companies attempt to enter the local exchange market, whether in their own right or through a CAP, they must have access to the essential facilities of the RBOCs. First, the cable operators must be interconnected to the LECs' network. Otherwise, customers of the new cable competitor would not be able to send or receive calls from LEC customers.

Universal Service

The concept behind universal service requirements is to provide as many people as possible access to core services, such as local telecommunications service. As common carriers, LECs must serve all customers who request service. However, some customers are much more expensive to serve than others. For example, local-loop infrastructure is more expensive in sparsely populated rural areas where more wire must be laid to accommodate each line. If rates were to reflect average cost, telecommunications services would become considerably more expensive for rural residents, who arguably rely more heavily on telecommunications service. In order to keep prices at levels at which a majority of citizens can afford service, telephone companies historically have subsidized internally high-cost rural customers with revenues from low-cost urban rate payers.

External subsidies also exist. For example, high-cost funds such as the Universal Service Fund (USF) have been established to promote affordable telecommunications service in rural and low-income areas. The Universal Service Fund, administered by the National Exchange Carrier Association (NECA), supports the cost of building and maintaining local infrastructure. For a variety of reasons, the fund is currently paid by large long-distance carriers based on their total number of presubscribed lines and then distributed to individual LECs to support high-cost local loops for which the cost exceeds the national average by a specified percentage.

As cable operators and other participants enter the sphere of telecommunications, universal service support mechanisms will need to be modified. For example, since only the LECs receive USF, how will an entrant compete in high-cost areas against an incumbent who is subsidized? Restructuring of the distribution mechanism thus will be necessary to insure that new competitors have the incentive to serve high-cost areas.

Conclusion

Cable has emerged from a literally backwoods industry into one of today's major forces in U.S. telecommunications policy. A host of questions await resolution—probably not in the near future. Will cable upgrade to more than 500 channels? Will it incorporate switched technology? Will it make peace with the LECs? Will it play a politically acceptable role with the NII? If not, what is its future?

Fortunately or unfortunately, no answers exist.

Acknowledgment. The authors wish to thank Ms. Catherine Lenti for her assistance in preparing this piece.

Notes

1. E.g., National Telecommunications and Information Administration, *National Telecommunications Infrastructure: Agenda for Action* (1993).
2. N.Y. Exec. L. section 818 et seq. (McKinney 1982).
3. N.J. Stat. Ann. Tit. 48:17–18 (1993).
4. 47 U.S.C. section 543(b) (Supp. V 1993).
5. 47 U.S.C. section 543(c) (Supp. V 1993).
6. See, e.g., *City of New York versus Comtel, Inc.*, 25 N.Y. 2d 922, 252 N.E. 2d 285, 304 N.Y.S. 2d (1969).
7. 47 C.F.R. section 76.32 (1972).
8. *TV Pix versus Taylor*, 396 U.S.556 (1970) (upholding Nevada state-level regulation of cable television).
9. Conn. Stat. Ann. Ch. 289 (1988).
10. N.Y. Exec. L. section 818 et seq. (McKinney 1982).
11. *Frontier Broadcasting*, 24 FCC 251 (1959); *First Report and Order*, 26 FCC 403 (1959).

12. *Carter Mountain Transmission Corp* v. *FCC*, 321 F. 2d 359 (D.C. Cir.), cert. denied 375 U.S. 951 (1962).

13. 2 FCC 2d 725 (1966). A prior 1965 rulemaking was virtually identical in its substantive requirements, but applied only to microwave-served cable systems because of the commission's doubts as to its jurisdiction.

14. These are measurements of signal strength. Throughout the commission's wrestling with the must carry rules, until the 1992 Cable Act it generally required cable operators to carry weaker signals from public than private stations (e.g., predicted Grade B versus predicted Grade A). In fact, a predicted Grade B signal usually does not even provide a usable signal. Under the 1992 Cable Act, as discussed here, the commission has adopted a completely different—and perhaps more troublesome—standard.

15. 47 C.F.R. section 74.1107 (1972).

16. 47 C.F.R. section 74.1103 (1972).

17. 392 U.S. 157, 177 (1968).

18. One of the few other major cases concerning the FCC's reasonably ancillary jurisdiction involved the validity of the commission's "access" rules, which the Court held to be *ultra vires*. The court indicated that the FCC's jurisdiction might include promotion of activities unrelated to broadcasting. *FCC versus Midwest Video Corp.*, 440 U.S. 689 (1979).

19. 36 FCC 2d 143 (1972).

20. The access channel requirements were invalidated by the Supreme Court in the *Midwest Video II* case, see Note 18.

21. 47 U.S.C. sections 614, 615 (Supp. V 1993). The Supreme Court narrowly upheld the constitutionality of the rules. *Turner Broadcasting Co. versus United States*, U.S. (1994).

22. See *Malrite T.V. of New York, Inc. versus FCC*, 652 F. 2d 1140 (2d Cir. 1981), cert. denied sub nom. *National Football League versus FCC*, 454 U.S. 1143 (1982).

23. *Community Cable*, FCC 2d (1983).

24. 98th Cong., 1st sess. (1983).

25. 462 U.S. 691 (1984). The court there invalidated Oklahoma's prohibitions on the advertisement of liquor on either broadcast or cable television. In doing so, the court relied largely on the preemptive effect of the federal cable compulsory copyright laws, which obviously would have restricted the decision's preemptive impact heavily in other areas of cable regulation.

26. H.R. 4103, 98th Cong., 2d sess. (1984). This was codified primarily in Title V of the Communications Act of 1934; parts of the 1984 Cable Act have been retained by the 1992 Cable Act.

27. The 1992 Cable Act retained this provision as 47 U.S.C. section 546(c)(1) (Supp. V 1993).

28. 47 U.S.C. section 546(c) (Supp. V 1993).

29. 47 U.S.C. section 546(E)(2) (Supp. V 1993).

30. 47 U.S.C. section 543(1) (Supp. V 1993).

31. 47 U.S.C. section 542(b) (Supp. V 1993).

32. 47 C.F.R. section 76.31(a) (1973).

33. 1991 S. Rep. No. 92 at 3.

34. See, generally, *Quincy*, 768 F.2d 1434, 1439, *infra*, for a brief history of the must carry rules.

35. 47 C.F.R. section 76.5 (1984).

36. 768 F.2d 1434 (D.C. Cir. 1985), cert. denied 476 U.S. 1169 (1986).

37. Note 36 at 1451.

38. 391 U.S. 367 (1968).

39. If the court found that the must carry regulations were content based, the rules

would have been subject to a tougher strict scrutiny standard. Content neutrality relates to whether a rule imposes requirements based on the content of speech.

40. 768 F.2d at 1451.

41. Note 40 at 1434.

42. Note 40 at 1460.

43. Note 40 at 1461.

44. 835 F.2d 292 (D.C. Cir. 1987).

45. Note 44 at 300.

46. Note 44 at 301, 302.

47. *Cable Television Services; Must-Carry and Retransmission Consent Provisions*, 57 FR 56298 (1992).

48. See Note 47.

49. See Note 47.

50. 47 U.S.C. section 555(c)(1) (Supp. V 1993).

51. 819 F. Supp. 32 (D.C. D. C. 1993).

52. *Turner Broadcasting versus FCC*, 114 S. Ct. 2445 (1994).

53. 47 U.S.C. section 325(b) (Supp. V 1993).

54. 47 U.S.C. section 543 (Supp. V 1993).

55. Though the local governments were initially more active in the regulation of rates, the FCC maintained the ability to preempt local rate regulation, a power it considered to be important in order to assure that overly burdensome local rate restrictions did not stifle the development of nascent portions of the cable industry. In the same year the 1984 Cable Act was passed, the Supreme Court affirmed the right of the FCC to preempt inconsistent local regulation. *Capital Cities Cable, Inc. versus Crisp*, 467 U.S. 691 (1984).

56. 47 U.S.C. section 543(a)(3) (Supp. V 1993).

57. 47 U.S.C. section 543(c) (Supp. V 1993).

58. 47 U.S.C. section 543 (Supp. V 1993).

59. 47 U.S.C. section 543(b)(8) (Supp. V 1993).

60. See discussion of *Midwest Video*, Note 18.

61. 47 U.S.C. 532.

62. 47 U.S.C. section 532(c)(2) (Supp. V 1993).

63. 47 U.S.C section 532(h) (Supp. V 1993). One court has ruled that Congress may not constitutionally authorize a cable operator to ban indecent material from leased access channels. *Alliance for Community Media versus FCC*, 10 F. 3d 812 (Ct. App. D.C. Cir. 1993).

64. 47 U.S.C. section 531 (Supp. V 1993).

65. 47 U.S.C. section 532(h) (Supp. V 1993).

66. 47 U.S.C. section 546 (Supp. V 1993).

67. 47 U.S.C. section 552 (1992).

68. 47 U.S.C. section 533 (1992).

69. See, e.g., *Chesapeake and Potomac Telephone Company versus United States*, 42 F. 2d 181 (4th Cir. 1994), affirming 830 F. Supp. 909 (Va. 1993).

70. See Note 69.

71. This standard was used to examine the constitutionality of the must carry rules, discussed in a separate section above.

72. 47 U.S.C. section 548 (1992) (Supp. V 1993).

73. 47 U.S.C. section 544a (Supp. V 1993).

74. 47 U.S.C. section 624(c) (Supp. V 1993).

75. 47 U.S.C. section 553 (Supp. V 1993).

76. 47 U.S.C. section 551 (Supp. V 1993).

77. 17 U.S.C. section 102 (1986).

78. 17 U.S.C. section 101 (1986).

79. *Fortnightly Corporation versus United Artists Television*, 392 U.S. 390 (1968); *Teleprompter Corporation versus CBS*, 415 U.S. 394 (1974).
80. Note 79, 392 U.S. at 400.
81. 17 U.S.C. section 801(a) (1986).
82. 17 U.S.C. section 111(c)(1) (1986).
83. 17 U.S.C. section 111(f) (1986).
84. 47 C.F.R. section 76.61(a) (1986).
85. However, as discussed in text, this provision has been challenged successfully in the courts. In addition, Congress has considered repeal of the ban.

MICHAEL BOTEIN
L. FREDRIK CEDERQVIST

Nonionizing Electromagnetic Wave Energy Exposures of Portable Cellular Telephones

Introduction

Recently, there has been speculation that the use of portable cellular telephones may stimulate the growth of human brain tumors in the region of the head in proximity to the antenna (1). This speculation has triggered some lawsuits and the advent of a variety of devices, called *antenna shields*, that are claimed to decrease substantially the exposure of the users of cellular telephones to the radio-frequency (RF) electromagnetic (EM) energy from the antenna without excessively degrading the coverage performance of the telephone.

This article does not address in detail the issue of brain carcinogenesis of RF electromagnetic energy; it deals essentially with the dosimetric aspects of the exposure of users of cellular telephones. In addition, it presents an experimental analysis of the performance of some antenna shields currently available on the market.

Methodology and Measurement Apparatus

Significant Exposure Parameters

The exposure of users of cellular telephones can be quantified in terms of incident electric and magnetic fields or the specific absorption rate (SAR), which measures the rate at which electromagnetic energy is absorbed by lossy dielectric media with nonmagnetic dissipative properties (2). Previous research showed that the incident electric field is poorly correlated with the electromagnetic energy that actually penetrates deeply into the tissue (3). With present knowledge, brain tissue absorbs RF electromagnetic energy essentially through joule heating associated with rotational motion collisions of large and small polar biomolecules and by ohmic losses of freely moving ions in the cellular fluids (4).

Recently, some traces of biomagnetite have been found in the human brain (5), but their biological import in terms of RF electromagnetic energy absorption in the frequency band of cellular telephones (824–849 megahertz [MHz]) remains unclear. In the absence of known relevant magnetic absorption phenomena of the organs residing in the human head, the obvious and best choice for characterizing the exposure of the user of cellular telephones is the specific absorption rate or SAR.

As shown in the following sections, evaluation of the magnetic field in the

© 1995 IEEE. Reprinted, with permission, from *IEEE Transactions on Vehicular Technology*, Vol. 44, No. 3 (August 1995).

close vicinity of the cellular telephone is an excellent diagnostic tool to analyze the RF currents on the cellular telephone antenna and radio case. Since, close to RF sources, the SAR patterns are well correlated with the near magnetic field distribution, it is important during the design phase of a cellular telephone to establish that no undesired RF current bunching effects, due to stray phenomena, are accumulating at locations close to the head of the user.

The electric and magnetic field probes used in the measurements of SAR and the near magnetic field have been already described in the literature (6,7). The measurement apparatus and the calibration of the probes are discussed in the next sections. The temperature increase method to measure SAR cannot be used in conjunction with portable cellular telephones because of the low RF power (0.6 watt [W]) emitted by these devices (8). The temperature increase method has been used to calibrate the SAR-sensing probes.

Phantom Model

The phantom model, the simulated human head to be used in the SAR measurements, has been selected with several criteria in mind. The SAR tests are not aimed to determine with great accuracy the RF absorption pattern in a specific anatomy or a set of specific anatomies. The tests are aimed to measure the highest peak SAR value that could happen in the head of a user. The SAR tests were set up as a pass or fail test for new models of cellular telephones. The pass or fail level is the peak SAR value established by the *NCRP* (*National Council on Radiation Protection*) *Report 86* for the general public exposure, 1.6 milliwatts per gram (mW/g) (9). The NCRP 86 exposure limits represented, at the time of the introduction of the portable cellular telephones in 1984, the most stringent U.S. national guidelines for the exposure of humans to RF energy in the frequency band of 800–900 MHz. The 1986 SAR limits of NCRP in the frequency band of cellular communications have been also adopted by the more recent (1992) human safety standard of ANSI (American National Standards Institute), the ANSI C95.1 – 1992. Given the facts that (1) a specific anatomy is not of great relevance, (2) it is important to simulate the maximum coupling of the RF energy between the telephone and the phantom human in normal and unusual conditions of use, one must decide which are the target organs of greatest interest to the effect of peak SAR measurements.

From simple anatomical considerations, it is obvious that the cranial skin; the skull bones in the parietal, temporal, and occipital regions; the brain mass in the same areas; the acoustic organs; and the lower portion of the face from the ear to the mouth are the regions most exposed to RF energy emitted by the cellular telephone. The posterior portion of the eyes, the tissues of the neck and upper torso, and the hand holding the cellular telephone are also exposed to RF energy emanating from the radio.

Let us briefly analyze the significance of the exposure of the organs just listed. The skin (relative dielectric constant $\epsilon_r = 51$, conductivity $\sigma = 1.4$ S/m [siemens/meter] at 800–900 MHz) is a water-rich organ that readily absorbs RF energy in the cellular telephone band (10), but it is also a very rugged organ constantly exposed to a variety of environmental agents. The same can be said about the hand. The cranial bones are organs from a delicate tissue and their exposure must be carefully considered. Wet bones have a lower dielectric constant and conductivity than skin; they have similar dielectric properties to those of the subcutaneous fat also present below the skin in the human head. The relative dielectric constant and conductivity of wet bone are $\epsilon_r = 5\text{-}6$ and $\sigma = 70\text{-}140$ mS/m, respectively, in the 800–900-MHz band (10).

The dielectric characteristics of live bone tissue have not yet been fully quantified experimentally. This topic is discussed further in the discussion section. The thickness of the bone and subcutaneous fat layer varies between 0.6 and 1 centimeter (cm) over the adult skull, with the maximum at the squamous portion (rocca squamosa) of the temporal bone close to the ear.

Brain is a tissue also rich in water, but it has also a substantial fatty content, with the exception of the cerebrospinal fluid, which is contained in the ventricles, deep in the cerebral mass. When a portable cellular telephone is in the typical use position, the nearest brain tissue is matter of relatively uniform dielectric characteristics with macroscopic values of dielectric constant and conductivity $\epsilon_r = 41$ and $\sigma = 1.1\text{-}1.2$ S/m in the frequency band of interest.

The eye is another organ of great importance that is proximal to the radio case of cellular telephones. The posterior chamber of the eye is a small globe of tissue rich in water, the vitreous humor, with dielectric properties similar to those of skin. The facial region between the ear and the mouth is made up of muscles in addition to the skin and subcutaneous fat. The skeletal muscles involved are the masseter, the zygomaticus major and minor, and the risorius, which form the fleshy strip of facial anatomy between the ear and the mouth. Muscle tissue has a high water content and readily absorbs RF energy in the frequency band of cellular telephones. The dielectric properties of muscle tissue are very similar to those of skin (10).

The tissues in the neck should also be given consideration. In addition to the cervical spine, containing a part of the central nervous system, the main contents of the neck are the muscular masses that move the head, the tongue, the larynx, the pharynx, the trachea, and the important thyroid and thymus glands. These organs have very complicated geometries and widely different dielectric properties, ranging from air $\epsilon_r = 1$, $\sigma = 0$ S/m to skeletal muscle with $\epsilon_r = 51$, $\sigma = 1.4$ S/m in the frequency band of cellular telephones.

The organ closest to the radio case and the base of the antenna is actually the external ear (pinna), where one can reasonably expect the highest values of exposure because of the possible proximity to the strongest RF currents. The external ear is a cartilaginous organ covered by a thin layer of skin, with an appendage, the lobule, made up of essentially fatty tissue. The external ears are rugged organs. They keep the upper part of the radio case and the base of the antenna from direct contact with the skin in the temporal and occipital regions of the head.

Given the complexity of the structure of the portion of the human head closest to the cellular telephone, a phantom model could be made to approximate the human anatomy to a sufficient accuracy only with great efforts (11). The cranial skin and the ears are not easily representable through phantom models. The subcutaneous fat and the skull can be properly modeled by simulated tissue (3). The same considerations hold for the muscles of the face and the posterior chamber of the eye (12).

Since we are concerned mainly with E fields generated by H fields (2) in materials without magnetic dissipative properties, the deposition of RF energy in bone and fatty tissue near RF currents can be approximated by using simulated brain material in their place. This substitution of phantom tissues, in addition to resulting in a larger simulation of the RF energy absorbed by the organs in the skull beneath the skin, also substantially simplifies the phantom model.

The phantom used in the RF deposition measurements for cellular telephones can be a shell of fiberglass about 1.5-millimeters (mm) thick, filled with a material simulating brain tissue as given in Ref. 11. Simple and relatively easy to reproduce, it has the significant benefit of giving the worst case SAR absorbed by brain tissue and the skull bones, a necessary feature for safety measurement purposes. It has the disadvantages of not simulating correctly the RF deposition at the surface skin, the earlobes, the facial muscles, and the posterior chamber of the eye. As mentioned above, the earlobes and the skin are rugged organs exposed to external physical agents. The facial muscles and the eye, given their greater distance from the highest RF currents, are not the primary exposed organs of the head, so their accurate simulation, at least in an initial phase, is not as critical as for the brain and the skull. Their exposure can be more accurately simulated with a head phantom having a "cranial" chamber filled with simulated brain tissue and a "facial" chamber filled with simulated muscle tissue. This further refinement of the phantom model is left for future work, if necessary.

The phantom used in the experimental evaluation of the worst-case exposure of the users of cellular telephones is a fiberglass enclosure 1.5-mm thick, shaped like a human head, and filled with a mixture simulating the dielectric characteristics of brain in the 800–900-MHz band. To maximize the coupling between the simulated user and the telephone, a large model of the human head has been used. The maximum width of the cranial model is 17 centimeters (cm), the cephalic index is 0.7, and the crown circumference of the cranial model is 61 cm.

The distal part (from the telephone) of the phantom enclosure has been removed from the model to permit scanning the probe in the region of the simulated anatomy close to the RF sources. A close-up picture of the phantom model is given in Fig. 1.

Probe Positioning System and Measurement Instrumentation

In the near field of portable communications antennas in the 800-MHz band, the positioning of the probes must be performed with sufficient accuracy to obtain repeatable measurements in the presence of rapid spatial attenuation phenomena and variations of the field distribution (13). The smallest physical dimension of the antennas involved in the class of cellular telephones under investigation is of the order of 1 mm, so a 0.5-mm positioning accuracy in three dimensions is required for SAR measurement repeatability because the antenna can be arbitrarily close to the target tissue.

The accurate positioning of the E- and H-field sensors has been accom-

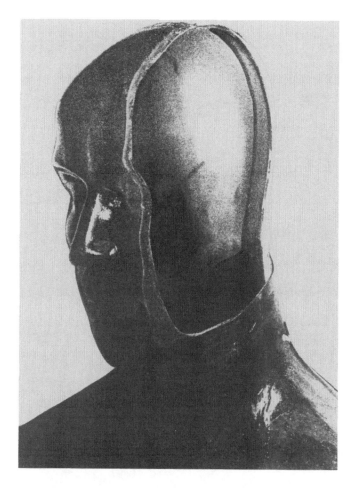

FIG. 1 Phantom model used in experimental evaluation.

plished by using a robot: the MicroSmooth Model 660 (IDX Robotics, Inc., Corvallis, OR). The robot can be taught to position the probe sensor following a specific pattern of points. In a first sweep, the sensor is positioned as close as possible to the brain's surface, with the sensor enclosure touching the inside of the fiberglass shell of the phantom. The SAR is measured on a grid of points 1.0 cm apart, which covers the curved surface of the head nearest to the portable cellular telephone.

The E-field probes are extremely fragile. To ensure their protection, they have been enclosed in a hollow plastic protective cylinder of 5-mm outer diameter and 0.5-mm thickness. The external surface of the cylindrical cap protrudes 0.41 cm beyond the dipole sensors of the E-field probes. The closest distance that the sensor can be placed to the phantom shell is therefore 0.41 cm. The SAR falloff in this 0.41-cm gap is estimated by using experimental data, as discussed below.

Figure 2 shows the robot in a typical measurement session. Figure 3 gives a

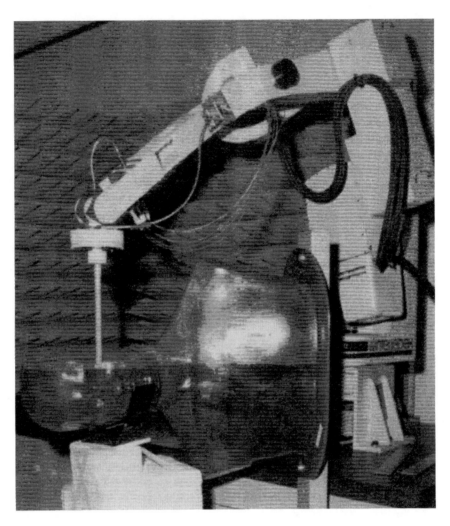

FIG. 2   Robotic positioning system for a typical measurement session.

close-up view of the phantom with the cellular telephone. The position of the cellular telephone is dictated by the human anatomy. The earpiece is located at the ear of the phantom and retained with a Styrofoam fixture using light holding pressure. If the earpiece cup is located over the inner earlobe for optimum or near-optimum sound coupling, the cellular telephone radio case has been shaped to generally follow the human face's anatomy and to place the microphone by the mouth of the user.

The telephone can be tilted away from the face within an angular range of 10° to 15° without losing the acoustic coupling between the microphone and the mouth (Fig. 4). The telephone position can be altered also by sliding the earpiece over the earlobe or by tilting the radio case so that the microphone is substantially above or below the mouth (Fig. 5). As shown below, even substantial

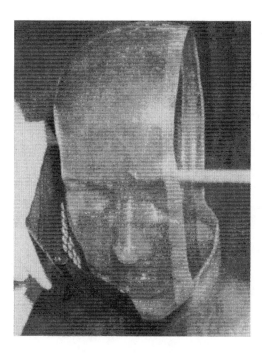

FIG. 3   Close-up view of phantom with cellular phone.

perturbations of the telephone position with respect to the one with the best acoustic coupling only marginally affect, or actually reduce, the peak SAR in the user's head.

From Fig. 3, one can see that the telephone is tested without a hand holding it. The human hand is an anatomical and geometrical complex structure that can assume a practically infinite number of configurations over the radio case. As discussed in the next section, the effect of the hand is generally to reduce the peak SAR in the head. Because of the almost irreproducible SAR effects of the hand over the radio case, it was decided to perform the measurements of the peak SAR in the head without the presence of the hand holding the telephone, thus introducing a bias for higher peak SAR values. This bias for the worst case is acceptable in safety measurement.

The origin of the coordinates of the scanning grid is located inside the phantom head at the intersection of the antenna axis and the top surface of the telephone. The coordinate system is oriented so that the positive x axis is in the direction of the antenna and the positive y axis penetrates the brain mass in the direction of the local normal to the cranial surface; the positive z axis forms an orthogonal, left-handed coordinate system with the positive x and y axes. These x and z coordinates are curvilinear and follow the contour of the inside of the phantom cranium. The phantom surface is considered locally planar, at least over a 0.5-cm incremental distance. Figure 6 gives a sketch of the coordinate system used during the measurement sessions.

The coarse scanning on a 1-cm grid locates the region or regions of local

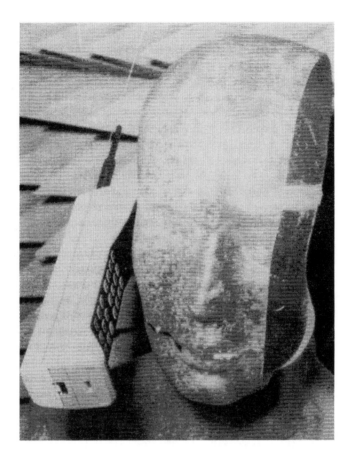

FIG. 4   Cellular phone tilted away from the face.

SAR maxima at the phantom surface. Since the phantom surface is substantially planar in the sections close to the antenna feed point and the radio case, the peak value of SAR is located at the surface of the simulated tissue mass because of (1) the space attenuation of the magnetic field and (2) the absence of possible lens effects. The latter condition was checked by searching for focusing phenomena in the parietal and occipital regions of the brain, but none were found.

The region or regions of locally maximized SAR are further probed by scanning the surface of the brain over a square grid with 1-mm increments. The absolute peak SAR near the surface of the phantom brain is located in this fashion within a square with 2.5-mm sides. The robot then proceeds with a y scan (the y axis penetrates locally the brain mass) at 0.5-cm intervals over a 5-cm length. A plot of the falloff of the SAR versus the depth is obtained. From this plot, the peak SAR at the surface of the phantom is evaluated using a 3-point interpolation method (14). In this fashion, the SAR falloff in the 0.41-cm gap between the sensors and the phantom brain surface is estimated by using experimentally determined attenuation values. The calibration of the

FIG. 5 Cellular phone positioned below the mouth.

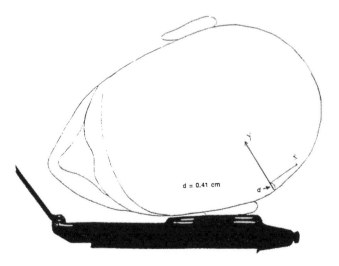

FIG. 6 Phantom coordinate system used during the measurement sessions.

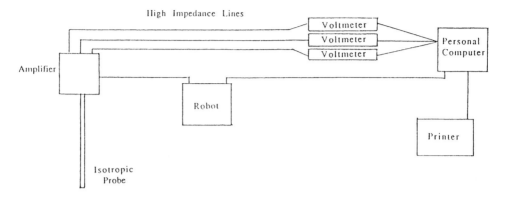

FIG. 7   Computer-based data-collection system.

probes, described in the next section, shows that no significant experimental error is introduced using this method to evaluate the peak SAR in locally planar phantoms.

A direct current (DC) electrical signal proportional to the $|E|^2$ or $|H|^2$ detected by the sensors forms the input of a DC amplifier designed for this special application. The electronic devices, one channel for each sensor, are contained in a metal box 5.2 cm $\times$ 4.8 cm $\times$ 5.8 cm located 30 cm from the sensors of the probing devices. The gain of the three amplifiers is variable over the range 10 to 13 dB and is set once for each channel of the probe as described in the following section.

The circuitry of the three DC amplifiers has been carefully isolated by metal shielding, so its performance is not affected by the presence of incident electromagnetic fields.

The output of each DC signal amplifier is connected to a voltmeter, one for each channel, by the means of 30 feet of high-impedance lines. The DC resistance of this wire is relatively high, 4–5 k Ω/foot (ft); the RF resistance is extremely high because the wire is made of carbon composite, which is a dielectric at the 800–900-MHz frequency band.

The DC voltage is measured by a millivoltmeter, the HP 3478A, the output of which is monitored by a GPIB interface connected to an IBM-compatible personal computer (PC). A sketch of the electronics is given in Fig. 7. The computer is programmed to have the robot move the probes through a predetermined path and perform the electromagnetic field sensing. The probe is driven to the appropriate location; after waiting one (1) second (s) to let all the mechanical vibrations in the mixture and the robot arm die out, the outputs of the sensors are read by the voltmeters and summed by the computer. The total voltage is proportional to $|E|^2$ or $|H|^2$, depending on the probe used. The typical measurement session for determining the peak SAR of a cellular telephone model takes approximately 20 minutes after the equipment has been set up and the robot has been "trained" to scan a specific grid.

Probe Calibration

The free-space calibration of the E-field and H-field probes can be performed using a transverse electromagnetic (TEM) cell manufactured by IFI (Instruments for Industry, Farmingdale, NY) operating at the frequency band of cellular telephones (15,16). The SAR calibration of the E-field probe requires a procedure involving temperature increase measurements.

The Vitek Electrothermia Monitor Model 101 was used to determine temperature during the RF exposure of a planar phantom. This thermometer has an RF-transparent sensor system that is nominally nonperturbing to the incident RF field. The temperature increase at the feed point of a balanced dipole operating at 840 MHz was measured at the inner surface of a planar phantom filled with brain-simulating mixture. Figure 8 shows a picture of the planar phantom used in the calibration measurements; note that the probe is positioned by inserting it into the mixture from the free surface side and that the dipole antenna is very close to the surface of the phantom. The initial temperature increase slopes and the relative SAR values for various radiated power levels are plotted in Fig. 9. The exposure time of the phantom for each of the points in Fig. 9 was 30 s. The following simple equation relates SAR to the initial temperature slope:

$$\text{SAR } \Delta t = c\Delta T \tag{1}$$

In Eq. (1), Δt is the exposure time (30 seconds), c is the thermal capacity of the simulated tissue ($c = 4.0$ joules/°C/gram [g]) and ΔT is the temperature increase due to the RF exposure. Clearly, the SAR is proportional to $\Delta T/\Delta t$ or the initial rate of tissue heating before thermal diffusion takes place (17).

From Eq. (1), it is possible to quantify the field in tissue using (17)

$$\text{SAR } = |E|^2\sigma/\rho \tag{2}$$

where σ is the simulated tissue dielectric loss and ρ its density; in this case, $\rho = 1.25$ g/cubic centimeter (cm^3).

The E-field probe tip is positioned at the same location as the thermal sensor. Since the E-field sensors are located at a 0.41-cm distance from the surface of the phantom, this field attenuation in the corresponding tissue layer is computed in the following fashion. The probe is scanned along the normal to the plane phantom at the point of thermal measurement. A plot of the field attenuation is obtained, as in Fig. 10, in which it is also shown the result of the extrapolation method used to compute the $|E|^2$ field at the surface of the phantom, the location of the maximum SAR. This method has given highly repeatable results. The linearity of the incremental temperature readings of the Vitek thermometer is excellent, as checked with a thermocouple standard. The experimental error in the determination of the SAR using the E-field probe method is less than 10% if an antenna radiating more than 5 W is used in determining the temperature increase during a 30-second exposure.

FIG. 8   Planar phantom used for calibration of instruments.

Experimental Results

All measurements reported in the experimental results sections have been performed with 0.6-W RF (maximum setting) power output from the cellular telephones tested.

Tested Telephone Models

There are two classes of portable cellular telephones evaluated in terms of the SAR exposure of the user. The first class has been commercially available since

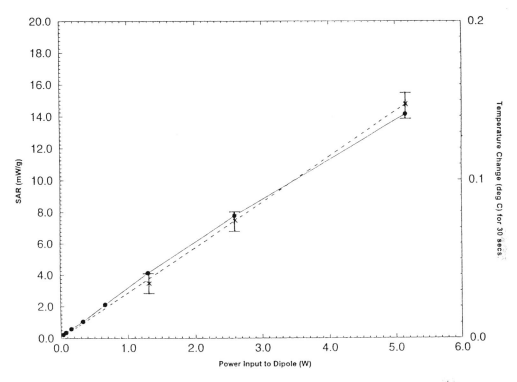

FIG. 9   Initial temperature increase and specific absorption rate (SAR) calibration curves.

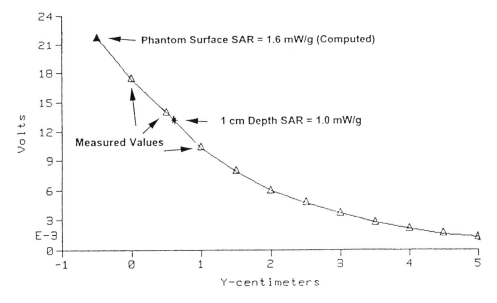

FIG. 10   Field attenuation versus depth.

1984. These telephones, since they are derivatives of the old model, are called Dynatac (Motorola, Schaumberg, IL) or "classic." A typical sample of the classic telephones is shown in Fig. 11. The classic telephones have a case dimension (front-to-back distance) that varies between 6.0 cm and 8.0 cm. All these telephones have a shortened sleeve dipole that is 14 cm long (Fig. 11).

The other type of cellular portable devices has been commercially available since 1989. These are called Microtac personal telephones (Motorola), also known as "flip" telephones. The flip telephone radio case (without the battery), depending on the model, has a dimension (front-to-back distance) that varies between 2.5 cm and 2.0 cm, the most recent models being the thinnest. All flip telephone models have a 2-hour (2-h) battery, 1-cm thick, and an 8-h battery, 2-cm thick. The type of battery has practically little effect on the SAR measurement results.

FIG. 11   "Classic" cellular phone model.

The flip telephones have a dual-antenna system. In Fig. 12, the same telephone is shown with its antenna extended and collapsed. In the dome above the telephone case, there is a small (1.75-cm long) helical wire that forms the RF radiator with the metal in the radio case when the primary antenna is collapsed. The primary antenna is a half-wave dipole, 12-cm long, and is end fed by the helical wire when pulled out of the radio case (18). Typical free-space magnetic field measurement results for the flip telephones are shown in Fig. 13 (antenna collapsed) and Fig. 14 (antenna extended). The free-field patterns have been recorded by moving the sensor at a constant 5-mm distance from the radio case and the antenna. The salient features of the patterns in Figs. 13 and 14 are the peaks of the H field at the feed point of the helical wire and the substantial decrease of the H-field intensity over the radio case when the primary antenna is extended.

Figure 15 shows the results of the $|H|^2$ field measurements recorded by moving the H probe a 5-mm distance from the radio case and the antenna of a typical classic cellular telephone. It is worth noting that the center-fed sleeve dipole does effectively choke the RF current on the radio case, which is not an integral part of the radiating mechanism as in the case of the flip telephone. This is rigorously true only in the free-space conditions.

Results of the Specific Absorption Rate Measurements

Dozens of portable cellular telephones have been tested since the introduction of these devices in 1984. The results are best summarized in tabular form, as in Tables 1 and 2. The tested telephone models are kept in an archive and are retested whenever any component of the SAR measurement system is upgraded or modified. All new portable telephone models are tested for compliance with the SAR limits of NCRP 86 as part of the new model shipping acceptance.

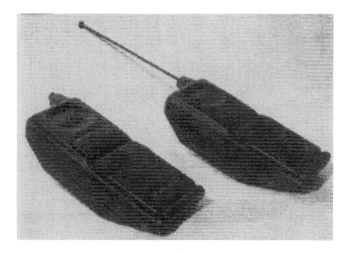

FIG. 12   "Flip" cellular phone model.

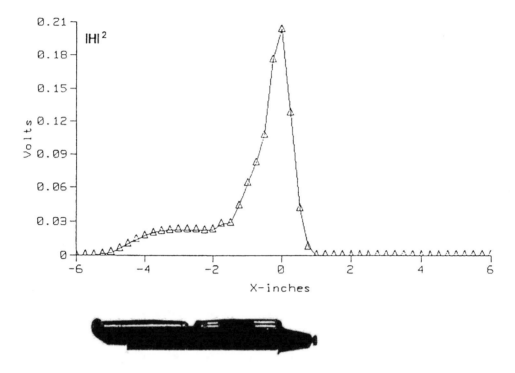

FIG. 13   H-field scan (antenna collapsed) of "flip" phone.

   Table 1 gives the maximum and the minimum values of measured SAR averaged over 1 g of tissue for the classic portable telephones. The columns of Table 1 refer to various services that are offered in cellular telephony in the United States and in Europe. FM CW identifies the frequency modulation of the RF continuous wave almost universally used at the present time (May 1995) by cellular telephones. GSM (Global System for Mobile Communication) is the new digital service of the Pan European cellular mobile telephones; the RF signal of the subscriber unit is pulsed at 217 Hz; each channel can service eight users. The average RF power driving the pulse is 2 W; the time average is 0.25 W. The NADC (North American Digital Cellular, or IS-54) is a multiple-access cellular service, three users to a channel; the subscriber unit transmits an RF signal pulsed at 50 Hz. The average RF power during the pulse is 0.6 W, with the time average being 0.2 W. The NADC service was introduced in 1993.
   Table 2 compiles the measured peak SAR values for the flip telephones. The data in Tables 1 and 2 have been collected without a human hand holding the telephone.
   As one can see, there is substantial dispersion of the experimental SAR values within each telephone model of the various transmission strategies. The

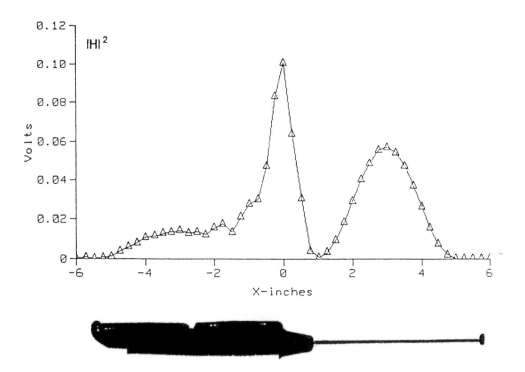

FIG. 14   H-field scan (antenna extended) of "flip" phone.

dispersion is due to various positions of the radio case, to different sizes of the radio case, and to different packaging of the electronic subassemblies within the radio case.

It is essential, at this point, to locate the position of the peak SAR values within the phantom. The worst-case maps of the measured SAR values at 0.41 cm inside the simulated tissue, using an FM CW flip telephone with the antenna collapsed and extended, is shown in Figs. 16 and 17, respectively. Although the peak SAR values in Fig. 16 exceed 1.6 mW/g, the 1-gram average of SAR is 1.6 mW/g or less. Figure 18 shows a similar map for the classic phones. The values given in Figs. 16–18 are the detected SAR levels without averaging them over 1 g of tissue shaped like a cube.

Figure 19 plots the SAR pattern at the surface of the simulated brain tissue in the plane parallel to the antenna axis and containing the point of peak SAR. In Fig. 19 is also shown an outline of the phantom contour. The origin of the x axis is the bottom of the touchtone pad; the x axis that follows the inward surface of the human head has been rectified for ease of representation. Note that the peak SAR is in the temporal area of the phantom user. Analogous information is shown in Fig. 20 for a GSM classic portable cellular phone.

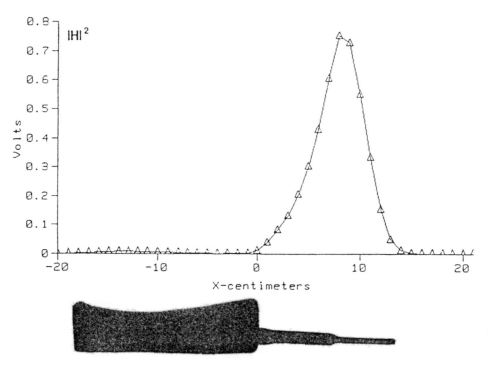

FIG. 15 H-field scan of "classic" phone.

From the results shown in Figs. 19 and 20, one can see that the exposure of the face of the phantom is well diffused over the entire area of the radio case and the base of the antenna. The exposure values are peaked at essentially very few locations (one or two).

Figure 21 shows the region of the human head where the peak SAR values have been detected for all the portable telephones tested using the phantom described in the phantom model section. Note that no SAR peak was found past the earlobe in the temporal area. Most peak SARs are detected in the area of the display and of the touchtone pad, located just below the plastic case of the

TABLE 1 Measured Peak Specific Absorption Rate Values from Classic Portable Cellular Telephones

Antenna	FM CW (0.6 W)	GSM (2 W)	NADC (0.6 W)
Dipole	0.2–0.4	0.09–0.1	0.07–0.09

All SAR values are in watts per kilogram (W/kg) averaged over 1 g of tissue shaped like a cube. For the GSM and NADC telephones, the peak SAR value is also time averaged over the frame of the signal. The range of values is due to different models and operating positions. No hand holds the phones.

TABLE 2 Measured Peak Specific Absorption Rate Values from Flip Portable Cellular Telephones

Antenna	FM CW (0.6 W)	GSM (2 W)	NADC (0.6 W)
Collapsed	0.9–1.6	0.2–0.3	0.2–0.8
Extended	0.6–0.8	0.1–0.2	0.1–0.4

All SAR values are in W/kg averaged over 1 g of tissue shaped like a cube. For the GSM and NADC telepones, the peak SAR value is also time averaged over the frame of the signal. The range of values is due to different models and operating positions. No hand holds the phones.

telephone, which is touching the face of the phantom in the area close to the ear. The antenna, except for its feed point at the top of the telephone case, causes no sharp SAR peaks in the occipital or parietal area of the head.

The effects of a hand holding the telephone have been difficult to quantify reliably. The results of the experiments are briefly summarized because of their lack of repeatability. If the radio case is lightly held by the fingertips, practically

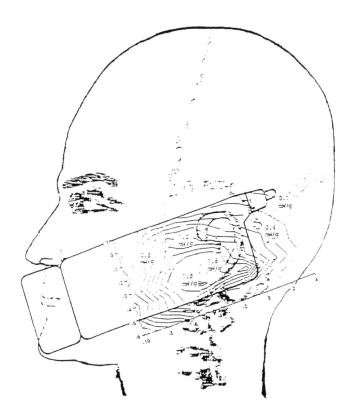

FIG. 16 Specific absorption rate map of flip phone with collapsed antenna.

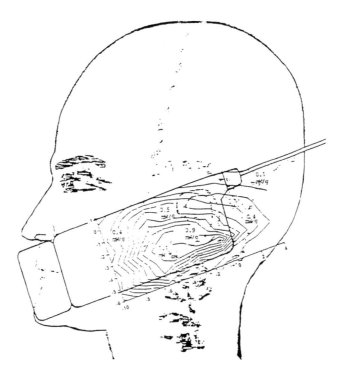

FIG. 17   Specific absorption rate map of flip phone with extended antenna.

no effect on its SAR patterns and their peaks has been detected. In general, the SAR patterns of the classic telephones are affected minimally by the location or the position of the palm of the holding hand. Flip telephone SAR performance is affected by the touch of the palm of the hand over the radio case and its location. Measurements show that peak SAR values are 10–30% lower if the palm of the hand is in direct contact with the radio case; the higher value is detected if the hand is close to or on top of the base of the antenna.

Antenna Shields

Two types of antenna shields have been tested. One is called Cell Shield (manu-factured by Dynaspek Inc.) (19). The device is a hemicylinder 15-cm long, with a radius of 2 cm. Its base makes electrical contact with the ground of the antenna. The device fits only the classic type of cellular telephones. Installed following the directions supplied, the shield is placed between the antenna and the head of the user.

The other device, called Cell-u-Shield (Cellu-Shield, Palm Beach, FL), also fits only the classic telephones; it is a metal wedge about 15-cm long and 2-cm wide on the side that must be shrink-wrapped around the antenna. The metal wedge also must be located between the head and the antenna; this device does

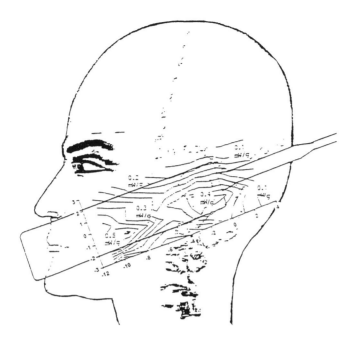

FIG. 18   Specific absorption rate map of classic phone.

not contact the antenna ground and is electrically floating. Both devices perform an electromagnetic shielding function at a substantial cost to the gain of the antenna. Table 3 shows the results of the SAR and gain measurements performed using various models of the classic telephones with and without the two shields.

The peak SAR value measurements have been performed following the methods described in methodology and measurement apparatus sections of this article. The average gain was measured by integrating the contour plots of the radiation patterns taken over an angular section 360° azimuth and ±18° elevation.

Both devices shield the users not by redirecting the RF energy in space as claimed in Ref. 19, but by reflecting it back into the telephone, where it is dissipated in the filter and the heat sink of the final RF amplifier of the radio. During the tests, the telephone RF amplifiers become extremely hot, beyond their normal temperature range. There are two serious disadvantages for the users of these shields. One is the obvious loss of coverage. With a gain loss of about 11 decibels (dB), the user can expect a drop in range of a factor of 2 to 3 in an urban environment (20), and a consequent loss of area coverage by a factor of 4 to 9. In these conditions, on average, the telephone can be used to carry on a conversation only 10% to 25% of the time compared to full coverage. The other big disadvantage is that the RF power output of the telephone transmitter will be kept nearly always at the maximum level because of low signal strength at the base station (21), thus causing overheating and loss of useful life of the telephone.

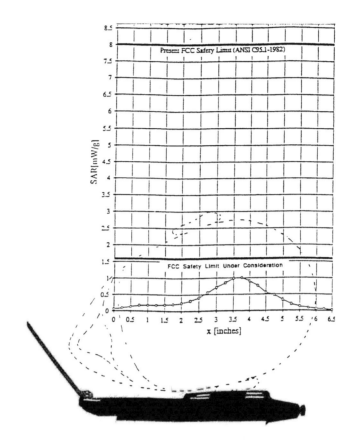

FIG. 19   Plot of peak specific absorption rate distribution of flip phone with antenna collapsed (FCC = Federal Communications Commission; ANSI = American National Standards Institute).

Another type of device that purports to protect the user of portable cellular telephones from absorbing RF energy is the Ireland antenna. This antenna is marketed by Universion Systems, Incorporated, of Miami, Florida, and fits only the classic type of telephones. The literature of this product states: "The unique technology in the reactance cancellator eliminates the ground return system, which means that the human body does not absorb the radiation from the antenna."

The Ireland antenna is a half-wave whip base loaded with a matching coil. This type of radiating element is analogous to the "long" antenna offered with the classic telephones. Figure 22 plots the near H-field values recorded at 0.5-cm distance from the classic telephone with the antenna shipped by the manufacturer, while Fig. 23 shows the same results using the Ireland antenna.

As one can see, the RF currents on the radio case side placed against the head of the user are no different for the two antennas.

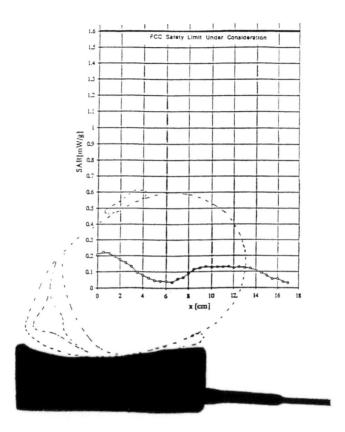

FIG. 20 Plot of peak specific absorption rate distribution of Global System for Mobile Communication cellular phone.

Discussion

The experimental results given in the sections above clearly do not refer to a human. The phantom is a control model, not a real model. Yet, the SAR values shown in the previous sections can be interpreted to provide the range of the highest expected values of the exposure of the brain of humans using cellular telephones. Although fatty tissues do not absorb RF strongly because of their hydrophobic properties, living bones are perfused by fluids rich in water and cellular nutrients that cause the relative dielectric and the RF attenuation constant to be not much lower than those of brain tissue. In various attempts to treat brain tumors by RF diathermy, it was found that it was very difficult to heat the brain tissue of anesthetized animals (particularly swine) at 915 MHz because of the RF absorption of the bones of their head (22). Although reliable data are not available for the electrical characteristics of living bone in the 800–900 MHz frequency band, it is reasonable for the following considerations to use the dissipation factor of brain tissue in characterizing the RF attenuation in

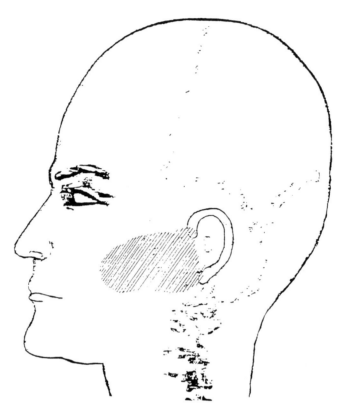

FIG. 21   Area of detected peak specific absorption rates.

the skull bones adjacent to the cellular telephone as shown in Fig. 21. Recent measurements show that this hypothesis is essentially correct (23).

Using the space attenuation and the RF dissipation factor of simulated brain tissue in assessing the field decrease in the bones, in the surface skin, and in the subcutaneous fat of the temple (which are 0.7–1-cm thick), one can see from Fig. 10 that the exposure at 0.7–1 cm inside the simulated tissue is attenuated by about a factor of 1.6 with respect to the peak value, which is always at the surface of the planar model. If we apply the same attenuation factor of 1.6 to the SAR values in Tables 1 and 2, one finds that the FM CW cellular telephones

TABLE 3   Antenna Shield's Performance

Parameter	With/Without Cell Shield	With/Without Cell-u-Shield
Average gain (dBi)	−12.5/−1.1	−11.7/−1.1
Peak SAR value (W/kg)	0.008/0.38	0.06/0.38

dBi = decibels above isotropic

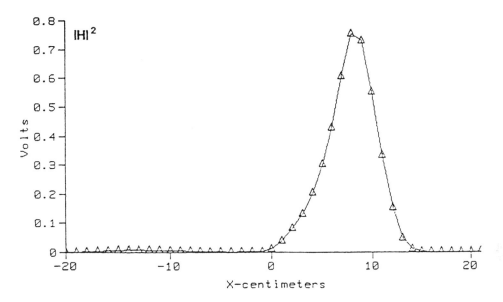

FIG. 22   H-field scan of classic phone with standard antenna.

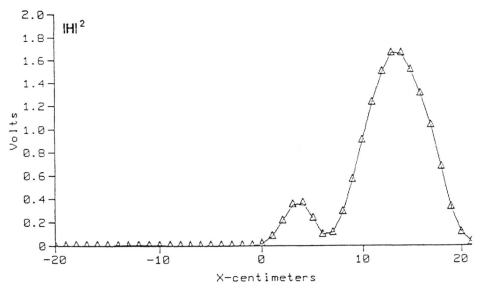

FIG. 23   H-field scan of classic phone with Ireland antenna.

expose brain tissue to localized peak SAR values in the range 0.1 to 1.0 W/kg. At these levels, the exposure can be considered athermal, that is, causing no macroscopic temperature increase above the physiological range of the target tissue.

Brain tissue is metabolically very active. The brain absorbs approximately one-fifth to one-sixth of the blood flow output of the heart (24); the basal metabolic rate of human brain tissue is in the range of 8 to 10 W/kg, that is, a full order of magnitude higher than the peak localized SAR of the exposure. In addition, the peak exposure of the brain is always at the surface of the organ, the location of the pia mater and the outer cerebral cortex. The pia mater is a vascular membrane covering the surface of the cerebral cortex as a rich capillary bed; the outer cerebral cortex is also rich in capillaries.

The local anatomy and physiology of the brain exposed to the highest values of electromagnetic energy absorption rate in simulated brain tissue from portable cellular telephones tend to exclude the possibility of a macrothermal exposure of the human brain.

Microthermal effects of the type discussed in Ref. 25 cannot be excluded at this time, given the gross characteristics of the phantom and the probes used in this study. However, a runaway temperature increase in brain tissue at the microscopic level is difficult to conceive at the exposure levels presented here. At the present time, there are neither scientific evidence nor credible theoretical models of the interaction between EM fields and living tissue that indicate an adverse biological effect at athermal levels of exposure to constant amplitude, 800–900-MHz waves with incoherent phase (26,27). These are the characteristics of the FM signals emitted by cellular telephones.

Conclusion

A practical method and control model to evaluate the RF exposure of humans using portable cellular telephones has been presented. The model tends to over-state the exposure of cerebral tissue compared to more realistic anatomical models.

Although the model and RF sensors can indicate only the gross features of the RF absorption by the telephone users, the results collected so far suggest that adverse biological effects from thermal damage are highly unlikely or outright impossible. Given the physical characteristics of the RF waves used in analog cellular telephony (FM), there are no other known mechanisms of adverse bioeffects that can be attributed to the cellular telephone usage on the basis of existing scientific knowledge.

Future work will endeavor both to refine the anatomical model and to reduce the size of the RF sensors so that microthermal effects (if any) at tissue boundaries can be experimentally detected. At this time, there is a considerable theoretical effort to predict the exposure of a user of cellular telephones by computer simulation (28,29). This theoretical effort will facilitate future experimental investigations using more realistic phantoms. The results given in Ref. 28 show

SAR levels lower than those reported here. The reader should not be surprised because the model presented in this article gives an upper bound of the exposure of the users of cellular telephones.

Although research to date suggests that pulsed RF signals such as those of GSM and NADC, which are below or substantially below the NCRP protection guidelines, are also biologically neutral, the data do not yet permit so ready a conclusion to that effect as is the case for FM signals. Biological research should be and is being conducted to test further the hypothesis that the new cellular telephone RF signals are as biologically neutral as are the FM systems they are replacing.

Acknowledgments: The authors wish to thank W. Ross Adey, M.D., of the Loma Linda Medical Center, Loma Linda, California, for many years of guidance in the area of the biological effects of RF electromagnetic fields and Howard Bassen of the Food and Drug Administration, Center for Devices and Radiological Health, Rockville, Maryland, for his contribution to experimental RF dosimetry. Also, the accurate typing of Kaye DeMuro is gratefully acknowledged.

References

1. Fischetti, M., The Cellular Phone Scare, *IEEE Spectrum*, 43–47 (June 1993).
2. Kuster, N., and Balzano, Q., Energy Absorption Mechanism by Biological Bodies in the Near Field of Dipole Antennas Above 300 MHz, *IEEE Trans. Vehic. Technol.*, 41(1):17–23 (February 1992).
3. Balzano, Q., et al., Energy Deposition in Simulated Human Operators of 800 MHz Portable Transmitters, *IEEE Trans. Vehic. Technol.*, VT-27(4):174–181 (November 1978).
4. Pethig, R., *Dielectric and Electronic Properties of Biological Materials*, John Wiley and Sons, New York, 1979.
5. Kirschvink, J. L., et al., Magnetite Biomineralization in the Human Brain, *Proc. Natl. Acad. Science USA*, 89:7683–7687 (1992).
6. Bassen, H., Internal Dosimetry and External Microwave Field Measurements Using Miniature Electric Field Probes, *Proc. Conf. Biological Effects and Measurements of Radio Frequency/Microwave, 136–151 (February 16–18, 1977)*.
7. Balzano, Q., and Siwiak, K., The Near Field of Annular Antennas, *IEEE Trans. Vehic. Technol.*, VT-36(4):173–183 (November 1987).
8. Guy, A. W., and Chou, C. K., Specific Absorption Rate of Energy in Man Models Exposed to Cellular UHF Mobile Antenna Fields, *IEEE Trans. Microwave Theory and Techniques*, MTT-34(6):671–680 (June 1986).
9. National Council on Radiation Protection and Measurements, *NCRP Report No. 86*, NCRP, Bethesda, MD, 1986.
10. Johnson, C. C., and Guy, A. W., Nonionizing Electromagnetic Wave Effects in Biological Materials and Systems, *Proc. IEEE*, 60:692–718 (June 1972).
11. Hartsgrove, G., et al., Simulated Biological Materials for Electromagnetic Radiation Absorption Studies, *Bioelectromagnetics*, 8(1):29–36 (1987).
12. Cleveland, R. F., and Athey, T. W., Specific Absorption Rate (SAR) in Models of the Human Head Exposed to Hand-Held UHF Portable Radios, *Bioelectromagnetics*, 10(2):173–186 (1989).
13. Balzano, Q., et al., The Near Field of Omnidirectional Helical Antennas, *IEEE Trans. Vehic. Technol.*, VT-31(4):173–185 (November 1982).

14. Abramowitz, M., and Stegun, I. (eds.), *Handbook of Mathematical Functions*, Dover, New York, May 1968.
15. Crawford, M. L., Generation of Standard EM Fields Using TEM Transmission Cells, *IEEE Trans. Electromagnetic Compatibility*, EMC-10(4):189–195 (November 1974).
16. American National Standards Institute, *Recommended Practice for the Measurement of Potentially Hazardous Electromagnetic Fields—RF and Microwaves*, ANSI C95.3—1991, ANSI.
17. Johnson, C. C., and Guy, A. W., Nonionizing Electromagnetic Wave Effects in Biological Materials and Systems, *Proc. IEEE*, 60:702 (June 1972).
18. Balzano, Q., Land Mobile Antenna Systems (J. James and K. Fujimoto, eds.), Artech House, London, 1994.
19. *Mobile Products News*, 9(6):53–54 (June 1993).
20. Olt, G. D., and Plitkins, A., Urban Path-Loss Characteristics at 820 MHz, *IEEE Trans. Vehic. Technol.*, VT-27(4):189–197 (November 1978).
21. Cellular Telecommunication Industry Association, *Primer on Cellular Operations*, FDA presentation, February 19, 1993.
22. Sutton, C., private communication, November 1979.
23. Gabriel, C., University Microwave, London, private communication, September 1993.
24. Guiton, A. C., *Text Book of Medical Physiology*, 5th ed., W. B. Saunders, Philadelphia, PA, 1976.
25. Adey, W. R., Tissue Interactions with Nonionizing Electromagnetic Fields, *Physiol. Rev.*, 61:435–514 (1981).
26. Illinger, K. H., Spectroscopic Properties of *In Vivo* Biological Systems, *Bioelectromagnetics*, 3(1):9–16 (1982).
27. Adey, W. R., Mechanism Mediating Athermal Bioeffects of Nonionizing Electromagnetic Fields, invited paper, 1993 National URSI Conf., Kleinheubach, Germany.
28. Gandhi, O., Electromagnetic Absorption in the Human Head and Neck for Some Cellular Telephones, private communication, December 2, 1993.
29. Kuster, N., Multiple Multipole Method Applied to an Exposure Safety Study. In *ACES Special Issues on Bioelectromagnetic Computations*, Vol. 7 (A. Fleming and K. H. Joyner, eds.), Applied Computational Electromagnetics Society, 1992, pp. 43–60.

QUIRINO BALZANO
OSCAR GARAY
TOM MANNING

North American 800 Services

Introduction

Virtually everywhere we look today, there are 800 numbers. In fact, they are so common in our everyday lives that we rarely give them a second thought. We encounter them when we are watching television, listening to the radio, reading a newspaper or magazine, or see them on billboards when we are driving. Most of us frequently use 800 numbers, but seldom give them much thought. It is simply a means to get to a business, just as a road is a means to drive to a store. We dial the 800 number, and it delivers our call to its intended destination. We expect the 800 number to be provided; we expect it to work.

But, the commonality of 800 services belies their importance to businesses and society in general. In essence, since their inception, they have become a critical and essential method of conducting business, regardless of whether it is business to business or consumer to business. The average consumer does not understand that behind that simple 10-digit number is a wide array of complex technologies and an equally complex set of business and managerial issues. Beyond the average consumer, many businesses still do not comprehend the power or importance of 800 services.

One needs only to examine a few statistics to understand the impact of 800 services on the business world. Every year, Americans make more than 10 billion 800 calls. But, the sheer volume of calls does not tell the entire story of how hard 800 services work for business. Of customer service 800 numbers, 14% operate 24 hours a day, 7 days a week. An additional 14% above that figure operate during weekends. Fully 25% operate Monday through Friday with extended hours of service beyond the standard 8 working hours.

These statistics only reveal the scope of 800 services. What they do not reveal is their rapid growth rate. In a 1992 study conducted by the Society of Consumer Affairs Professionals in Business (SOCAP), 63% of the respondents used 800 services as opposed to 50% in 1988 and 38% in 1983. In addition, 10% of the respondents indicated that, while they did not currently use 800 services, there were definite plans to add these services in the future.

Not only has the use of 800 services increased, but usage per 800 number has increased. The median number of calls handled by an 800 number has increased 250%, from 600 calls per week on 1988 to 1500 calls per week in 1992. This is an increase of more than 350%. Now consider that the productivity of 800 service representatives has doubled over the past four years. Combine this with the fact that the median number of full-time professionals used to staff 800 numbers has increased 50% over the past four years.

What has sparked such changes in the uses of such services? Changes in the economy and society in general are at least part of the answer. For instance, there has been much speculation about how the American economy is making a transition from a manufacturing base to one that is service oriented. In this new service-based economy, there is much pressure to deliver products faster, less expensively, and with better service. Also, in a free-market economy in which the consumer has many choices and prices may be similar, what makes the consumer buy from one company in lieu of another? This is a question that invites many answers, and probably none of them would be simple. But, one important piece of the puzzle is the use of 800 services. Modern businesses not only need to provide 800 service, but they also constantly need to seek new methods to utilize them better. Businesses explore innovative advertising techniques or new 800 features that may allow them to service their customers better. Conversely, the providers of 800 services (i.e., long-distance carriers) constantly struggle to find new and innovative ways to provide 800 services to business via new technology or more attractive pricing.

Businesses that understand the importance of 800 services and apply their efforts toward both providing them and enhancing them are often rewarded. Studies have shown that businesses that provide 800 services experience increased customer satisfaction and loyalty, positive word of mouth, an enhanced corporate image, decreased service costs, enhanced product design and quality, decreased customer-caused problems, and new and untapped sources of revenue.

Customer satisfaction and loyalty are enhanced because 800 services increase customer contact. Any business is going to have its fair share of problems, and there has to be a methodology for addressing those problems. How quickly or efficiently a business addresses problems has an impact on customers' perceptions. Since an 800 number increases the chances of customer contact, there is more opportunity to solve more problems in a timely manner.

The 800 services also help to generate positive word of mouth—and the power of word of mouth should never be discounted. Studies show that customers tell twice as many people of negative experiences as opposed to positive. Since 800 services have the ability reach a large number of customers in a short period of time, there is a great potential to satisfy a large volume of customers, thereby diminishing the possibility of negative word of mouth.

What exactly is an 800 number? What are the principles behind it and how does it work? Who are the organizations that provide these 800 services? Are there different types of 800 numbers? If there are, do they all have the same features? What happens when the existing pool of 800 numbers is exhausted and new toll-free numbers are implemented, such as 888 numbers? The answers to these questions are not simple. There is actually considerable diversity among 800 services. The physical nature of the connection, the structure of the billing, and geographical coverage and advanced technical features can vary considerably from service to service. It is essential to understand the differences so that intelligent decisions can be made best to service both the customer and the subscriber.

An 800 number is a unique type of telephone number. It is 10 digits long, just as any normal long-distance number is, but the caller always dials "800" in lieu of an area code. The 800 is not really an area code, but a code that indicates to the local telephone company the call is of a special nature. The 800 is universally accepted in North America as a designation of a free telephone call, which contributes to its appeal as outlined in the beginning of this article. Toll-free services are also prevalent in other countries; however, they use dialing plans other than the North American Numbering Plan (NANP) and digits other than 800. The 800 services are also known as inbound, "toll free," or INWATS (inward wide-area telephone service). The WATS was originally a family of long-distance services (both inbound and outbound) that were developed by AT&T. In fact, 800 services were originally introduced exclusively as INWATS services. The official designation was changed to 800 services because the Federal Communications Commission (FCC) was examining the name and nature of WATS services and was considering ruling that WATS services were discriminatory. Today, all of these terms are commonly used to designate 800 services. In the future, these terms will include the new toll-free code 888, which is scheduled to go into effect in the spring of 1996.

When an 800 number is dialed, the call is always free to the caller and the long-distance charges are billed to the subscriber of the 800 service. The 800 number is provided as a service or convenience to a customer (or at least to a potential customer). The subscriber of the 800 service is charged based on three factors: usage, the fixed cost of the trunk lines, and possibly a monthly service charge. Providers of 800 services (normally a long-distance carrier — but not always) mix and match these factors to produce more attractive 800 programs and pricing. The combination of these factors changes often due to the competitive nature of the long-distance business. Many subscribers have their 800 services installed at large centers that are designed to service large volumes of calls. This is a concept known as a *call center*. Call centers have become a standard method of servicing customers in the modern business world, and 800 services normally comprise a large portion of the call center budget.

The concept of 800 services was originally developed by AT&T under the old monopoly system. These services have been in existence since the 1960s and have been used almost exclusively by businesses. In recent years, however, some innovative new applications have yielded personal 800 services for residential use.

The North American Numbering Plan

Before examining how 800 services work, it is essential to understand the telephone numbering plan that services North America. This numbering plan provides a standard numbering scheme, and strict guidelines have been set for carriers concerning the transport of local and long-distance calling. Understanding this system is necessary because it helps to understand how basic 800 services work; this information also sets the stage for understanding how advanced 800 features work.

NANP is a uniform plan that was originally developed by AT&T. The plan is, as of this writing, administered by Bellcore (Bell Communications Research, an organization that provides research and other services of common interest to local carriers). It is based on the premise that all telephone numbers in the public network will be 7 digits for local calling and 10 digits for long distance. This provides a logical foundation for telecommunications in the United States, Canada, and certain sections of Mexico.

Under the format of this plan, telephone numbers are identified in three ways: a three-digit area code, a three-digit exchange — or central office (CO) — code, and a four-digit subscriber code.

The first three digits are commonly known as an *area code*, but the technical designation is NPA, which stands for Numbering Plan Area. Within any area code, no two seven-digit local telephone numbers may be duplicated. There are currently over 200 area codes in the United States, Canada, Bermuda, the Caribbean, northwestern Mexico, Alaska, and Hawaii. Until recently, the format of the area code was to provide a number range of between 2 and 9 for the first digit, a 1 or 0 for the second digit, and the numbers 2 through 9 for the third digit. This format has changed because the demand for numbers is rising and the number of available area codes is limited. Now, it is not unusual to see almost any combination of digits used as an area code.

The second 3 digits of a 10-digit number is commonly known as the *exchange*. The technical designation is NXX. The area code, combined with the exchange is often referred to as NPA NXX. The NXX identifies the CO of the local-exchange carrier (LEC), the carrier responsible for transporting local telephone traffic. The N represents any digit from 2 to 9. The X can be any digit. However, just as area code formats are undergoing changes, so is the local-exchange format. Many COs, formally bound by the NNX format, are now changing to an NXX format. This change has resulted in many exchanges looking like area codes.

The remaining four digits of a standard telephone number are known as the subscriber code. Whereas the area code and exchange service specific geographic areas, these four digits are assigned specifically to a business or individual subscriber. The subscriber code identifies a trunk and/or trunk group. The trunk is an important concept in the design and performance of 800 services. The trunk line is a channel that carries a telephone conversation and possibly data that are used to process the call. Trunks can be either analog or digital (see the section "Methods of 800 Access" section).

Carriers

Carriers of telephone service can be divided into two categories: the LEC and the interexchange carrier (IXC). Both types of carrier play crucial roles in the 800 network architecture, although IXCs are the true providers of 800 services.

An LEC has the authority to provide local telephone service (i.e., dial tone), carry local telephone traffic, provide the physical interface to either a residence or business (i.e., the cable), and provide some services that are associated with local telephone service. The LEC is allowed to carry telephone traffic within the local access and transport area (LATA). A LATA is a geographical area; there are 161 LATAs within the United States. The geographical coverage can actually be quite large, and so the term *local* can be deceiving. For example, almost the entire southwestern portion of Pennsylvania is a single LATA. A person may call within their own city and may not be charged for the call, which may be included in the monthly telephone service agreement. But, the same person might call a point 60 miles away; the call is still carried by Bell Atlantic, the LEC, but it is charged at a long-distance rate. If, however, the call would cross the LATA, the call is carried by an IXC.

An LEC is commonly a subsidiary of a regional holding company (RHC). When the AT&T monopoly was broken up in January 1984 (commonly known as the divestiture), AT&T was no longer permitted to carry local telephone traffic. The old Bell system was divided into seven RHCs: Ameritech, Bell Atlantic, New England Telephone, NYNEX, Southern Bell, Southwestern Bell, and U.S. West. Although LECs are commonly Bell companies, there are also many independent telephone companies (ITCs). ITCs can actually be very large, providing advanced state-of-the-art communications capabilities, or they can be very small and of limited capabilities.

True long-distance (i.e., inter-LATA) traffic is the domain of the IXC. AT&T, MCI, and U.S. Sprint are all examples of IXCs, and all long-distance companies that have their own networks are referred to as such. Prior to divestiture, carriers other than AT&T were commonly referred to as other common carriers (OCCs), but this is an archaic term and is seldom used today.

The largest IXC is AT&T, followed by MCI and U.S. Sprint. In the early days of divestiture, AT&T maintained a distinct advantage in network quality and reliability. Other carriers, such as MCI and U.S. Sprint, marketed primarily on the basis of substantial savings over AT&T. After a decade of competition, MCI and U.S. Sprint have achieved high-quality networks, and the price difference among all of the carriers has diminished significantly.

Other providers of 800 service are smaller regional carriers (still known as IXCs), aggregators, and resellers. Resellers have been common for years in the long-distance industry. The premise is simple. A company subscribes to special long-distance discount plans that are normally only available to large corporations. They resell the service to smaller companies, charging a percentage or service fee, and realize a profit. Many resellers purchase long-distance service from a variety of IXCs and may often change the IXC they use for long-distance service, depending on the type of discounts they can negotiate. A subscriber of reseller services may be serviced by any of a number of IXCs. A key point to consider when differentiating between a reseller and an aggregator is that the reseller takes on full billing responsibility to the subscriber.

A new type of long-distance provider that has emerged in recent years is the aggregator. The difference between the reseller that sells a variety of services and the aggregator is that the aggregator acts as a sales agent for a specific IXC (i.e., they only sell one carrier's service). The aggregator subscribes to a master 800 agreement provided by a single IXC. The aggregator makes a profit by charging a fixed monthly charge, a percentage of savings, or additional costs per minute. The customer may be billed either directly from the aggregator or the IXC.

Portability

Since the divestiture of AT&T, the FCC has striven for even competition in the world of telecommunications. The long-distance industry in general has fared well for the consumer due to this competition with advancing technologies and lower prices. But, the one area that lagged behind for many years was that of 800 services. AT&T had developed the concept of 800 services and retained a lion's share of the business since, for a very long time, there was simply no other alternative. Also, the network hierarchy that supported 800 services was limited, and there was actually very little that could be done to change the fact that AT&T had a large advantage in this part of the long-distance business.

The concept of portability was mandated by the FCC in 1991 and put into effect in May 1993. The concept was to make 800 services "portable." For instance, prior to May 1993, if a subscriber wished to change carriers of their 800 service, they also had to change their 800 number. Companies were often reluctant to do this. They may have been using the same number for many years and feared that a change would confuse their customers or disrupt service. Or, they may have had a vanity number (one that spells a name or concept) that they found appealing for their marketing efforts. Obviously, the situation weighed heavily in favor of AT&T. While IXCs such as MCI and U.S. Sprint were able to provide 800 service, it was determined by the FCC that this situation was unfair.

Today, any IXC can carry the traffic of any 800 number. The 800 services are now truly portable. Under this concept, the subscriber to the 800 service actually has ownership of the 800 number. This is in stark contrast to the preportability era when the carrier had ownership of the 800 number. Now, a company may change long-distance carriers for a specific 800 number as often as they wish. They may even, and for a variety of reasons, designate more than one carrier for a single 800 number. For instance, calls on the East Coast may be carried by U.S. Sprint, and calls on the West Coast may be carried by MCI. Or, it is possible to designate carriers by the time of day or specific days of the week.

Portability did not come easily. Prior to 1993, the LECs were only required to screen the first six digits of a telephone number in order to determine how it would be routed. The 800 was constant and indicated a special type of call. The NXX told the LEC which IXC would carry the call, and the database was constant. AT&T had a range of NXXs that were exclusive to their network, as did other carriers such as MCI and U.S. Sprint.

In order to offer the concept of portability, the LECs were required to invest heavily in new equipment, and the entire architecture that supports 800 services had to be changed. Because of the sheer scope of the changes, many users of 800 services expressed concern for how aggressive the schedule was. Their concerns were that the LECs would not be able to make the changes in time, that calls would be lost, and that call setup time would be affected. The implementation was actually postponed due to pressure by an ad hoc 800 users' group.

But, when the concept was implemented in May 1993, it proved to be a resounding success. There were no noticeable changes for the general public, and 800 users (especially the larger ones) breathed a sigh of relief. Now that true competition exists in the world of 800 services, users of 800 service have realized many benefits, such as the ability to change carriers without changing 800 numbers and the ability to use multiple carriers on a single 800 number. But the IXCs that provide 800 services, with the playing field being even, are now in a fiercely competitive situation. There will be much pressure to provide new and innovative products and technology at lower prices. The clear winners in this situation are 800 subscribers, who will realize the benefits of this competition.

How the 800 Services Work

When a customer dials an 800 number, it normally only takes about four seconds from the time dialing is finished until the time the call is delivered to the subscriber's telephone system. In that short period of time, a large volume of information is passed and processed among various types of computer and communications equipment and among various organizations. Processing an 800 call encompasses six concepts: a method of accessing the public network, the use of switching equipment, signaling methods, database translation, the actual transport of the call, and the actual delivery of the call to a telephone system. These concepts become clearer as the 800 network architecture is examined more closely.

When an 800 number is dialed (Fig. 1), each digit is identified and provides information that is processed by the LEC. In almost all cases, the standard 10-digit number is prefaced by the number "1." The 1 indicates to the LEC that the call is long distance. The 800 itself is not an area code, but a code that indicates that special processing will be required.

All residential telephone calls go through the LEC CO. The call is identified as an 800 number by the service switching point (SSP). Call information (not the call itself) is then sent to the signal transfer point (STP). The STP is actually

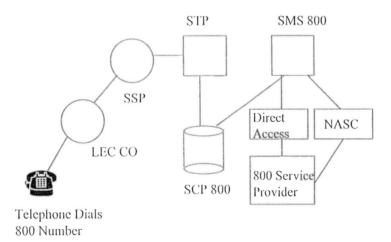

FIG. 1 The process of dialing an 800 number (LEC CO = local-exchange central office; NASC = National Administration and Service Center; SCP = service control point; SMS = service management system; SSP = service switching point).

a packet switch (a method of transmitting data, i.e., a packet of data). The STP asks a database called a service control point (SCP) where to send the call. The SCP is provided by the LEC and is capable of storing millions of records. Most LECs provide two SCPs and sometimes more, depending on the volume of calls they must handle.

The SCP identifies the IXC that is responsible for transmission of the 800 number. The SCP is the heart of the 800 routing system because, in addition to identifying the 800 carrier, it provides specific instructions for routing the call. Once the IXC has been identified, the call is then sent to the appropriate IXC point of presence (POP). The POP is the IXC equivalent of a CO and sometimes also is referred to as the technical operating center (TOC).

The SCP database is constantly updated by pulling data from the national central database, known as the services management system (SMS/800). The update happens every 15 minutes or on a special request basis. The SMS/800 is actually a large IBM mainframe computer that is, as of this writing, administered by Bellcore. The SCP is actually a smaller version of the SMS/800.

The specific organization under Bellcore responsible for administering 800 number assignments and the SMS/800 database is the National Administration and Service Center (NASC). The RHCs requested in the late 1980s that an independent third party administer the NASC. The FCC agreed with this and, in 1992, Lockheed Information Management Systems was selected to maintain the NASC. Ultimately, Lockheed will replace Bellcore.

With even a cursory glance at the 800 network architecture, it is obvious that a tremendous amount of information must be passed very quickly in order for this system to operate efficiently. The system that allows this is Signaling System No. 7 (SS7), a digital, high-speed, out-of-band method of transmission. SS7 uses common channel signaling (CCS). In this out-of-band method of transmission, the voice conversation and signaling information are transmitted on two distinct channels. This is the signaling method that allowed portability to become a success because it serves two basic requirements very well. First, the transmission is very fast; second, the out-of-band signaling allows for an excellent method for database lookup.

The final leg of the 800 call is when the IXC has been identified, the call is transmitted through the appropriate network, and then delivered to the 800 subscribers' trunk lines. Once again, this entire process takes approximately four seconds. The concern of subscribers was that this elaborate system would cause significant increases in call setup time or postdial delay (i.e., the time that it takes to set up an 800 call before the transmission actually begins).

This is the system that allows for the concept of portability. The design of the system is to query the network for a carrier. The system works efficiently and allows 800 users to change carriers for the same 800 number or even use multiple carriers. But, how are carriers assigned and who accesses the SMS/800 database to make changes? Portability demanded that a system be developed, hence the concept of the RespOrg (responsible organization) was developed; this is any organization that has been authorized to access the SMS/800 database. A RespOrg serves three functions: reserving 800 numbers, assigning 800 numbers, and making changes to an 800 number.

A number may be reserved for a period of 60 days, at the end of which the

RespOrg no longer has the rights to the number and it is then added back to the available pool of unused 800 numbers. A RespOrg also assigns numbers to a customer or client. This assignment is often registered with national directories. The RespOrg may also make changes to a number, even if it stays with the same client. For instance, the customer may wish to have a different IXC carry its East Coast calls than its West Coast calls. Or, the same company may wish to have its evening traffic carried by an alternate IXC in order to achieve a more favorable rate. The customer may use a single RespOrg for all of these changes.

What is the RespOrg? Technically, it can be any type of organization, but, typically, a RespOrg will be a long-distance carrier. A fee is charged to access the SMS/800, so the number of organizations willing to be a RespOrg is usually limited to those that have a vested interest in administering 800 services. The largest carriers (e.g., AT&T, MCI, and U.S. Sprint) have dedicated data links to the SMS/800. Smaller carriers or resellers may only have dial-up access.

Technical Aspects of 800 Access Methods

A subscriber to 800 services must select a way to connect the subscriber's telephone equipment with the public network in order to receive 800 calls. This is actually the last stage of processing an 800 call: delivering it to a subscriber's telephone system. The call is delivered to the subscriber's telephone system via the telephone trunk line, which is also the access method or the local loop. The trunk line affects 800 services in a number of ways: cost, versatility, capacity, and sometimes features.

Methods of access can be divided into three categories: switched access, dedicated access, and special access. These three types of local loops are commonly provided by the LEC, even though the 800 customer may have placed a single order for 800 service with the IXC. The IXC actually subscribes to a type of local loop and then passes on the charges incurred from the LEC to the 800 subscriber. Although it is possible, it would be rare that an IXC would provide 100% connectivity and transmission facilities for a single 800 service.

Provisioning the LEC local loop by the IXC is actually easier for all parties involved. The LEC usually already has cable facilities built into a facility, and there is normally a minimum amount of physical work that needs to be done. But, it is also perfectly legal to use a competitive access provider (CAP). These companies have been flourishing in recent years by providing private cabling for a number of applications, including alternate access to IXC POPs. They have found popularity because of discount rates over standard LEC offerings. The CAPs can be viewed in the same manner as the OCC was 10 years ago. They offer a discount over standard LEC cable services, but their coverage is limited. As these fledgling companies increase in size, it is quite possible that they will be formidable competitors to the LECs. Already, due to the severely competitive nature of the long-distance business, the IXCs have been increasing their use of CAPs. It is only logical to assume this trend will continue in the future.

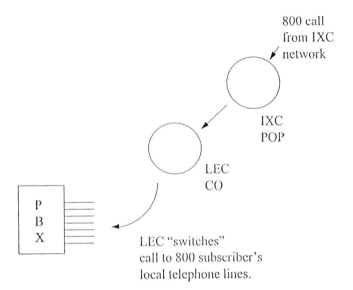

FIG. 2   The basic premise of switched access for providing 800 services (IXC = interexchange carrier; PBX = private branch exchange; POP = point of presence).

Switched Access

Switched access has become a popular method of providing 800 services. The basic premise (Fig. 2) is that the IXC uses a basic local telephone CO line as a target for the 800 number. The LEC "switches" the 800 number over to the local line, hence the term *switched access*. The CO line may be a stand-alone line or the first member of a hunt group. The CO lines are commonly provided in hunt groups. The premise is that if the first trunk line is busy, the next call will hunt over to the next available line (see Fig. 3). Products provided by the major IXCs are AT&T's Readyline, MCI's Common Business Line, and U.S. Sprint's Fonline 800.

The switched access 800 products are usually designed with the smaller business in mind. The beauty of the switched access 800 services is that they are so simple to install and use. For instance, consider a small business that feels it may be time to add 800 service. They are uncertain if 800 services are needed, but would like to try them. To add a dedicated 800 number would require the addition of telephone equipment, an expense that would mount to several thousand dollars. This is something they may be reluctant to do. However, if

FIG. 3   The principle of hunting. First the call is delivered to the telephone or trunk line. Second, if busy, the call will hunt over to the next available telephone or trunk line.

the 800 service proves to be effective, they will have no problem investing money.

The switched access 800 service is an elegant solution to this problem. An 800 number is ordered and assigned to the company's listed directory number (LDN). Any call dialed on the new 800 number will be directed to the same trunk group that supports the company's main number that rings into the operator's console. The 800 service has been added and no expensive modifications were required.

But, switched access 800 service is also a double-edged sword. There are operation difficulties presented with these types of services. Because the CO line is a two-way line for both inbound and outbound services, managing 800 services on them becomes much more difficult.

Dedicated Access

Dedicated access for 800 services is the use of trunking that is exclusive to the 800 service (i.e., it is inbound only). This method of access is often commonly referred to as a dedicated access line (DAL). The DAL is often thought of as dedicated analog trunk lines; however, it is actually a generic term that may refer to either an analog or digital connection. All of the major IXCs have standard offerings in the dedicated arena. AT&T has Masterline, MCI offers Dedicated Termination, and U.S. Sprint markets Ultra 800 (which is more commonly installed with a DS-1 [Digital Signal Level 1] circuit).

Dedicated access via an analog offering was once the only method of providing 800 services, but the use of these types of services has been diminishing in recent years. This is due to longer lead times until installation, being more expensive than switched access, and the proliferation of inexpensive, high-capacity digital services. There are still many dedicated access 800 services installed, but the installation of newer 800 services has favored switched access or special access.

Dedicated access, either analog or digital, can be thought of as a direct connection to the IXC POP. In many cases, however, the cable will actually be provided by an LEC, and the connection will go through the CO wiring center and then to the IXC POP. The IXC may also bypass the LEC and use cable that is provided by a CAP. In either case, the subscriber cannot seize a local dial tone or accept local calls on the DAL.

Special Access

Special access 800 services have gained wide acceptance in recent years, especially with larger users of 800 services. Some of the better known products are AT&T's MEGACOM 800, MCI's Prism 800, and U.S. Sprint's Ultra 800. These products have evolved as a result of the wide availability and low cost of high-speed digital lines in recent years.

The special access 800 service is offered with a very low usage charge, but with a higher monthly service charge than either switched access or dedicated access. The concept of the special access is to provide a high-capacity digital link to the 800 subscriber's telephone equipment. The most common method is that of the DS-1 circuit, which is more commonly known as a T1 circuit. The DS-1 circuits offer 24 channels for transmission (24 simultaneous conversations). Special access 800 services can also be provided with a DS-3 circuit, which offers 672 channels for transmission for very large applications. Analog access is also possible, although it is rare. Once again, the special access can be thought of as a direct connection to the IXC POP.

Digital Technology

There are two reasons why digital technology bears closer examination. First, the world of telecommunications is rapidly making a change to an infrastructure that is completely digital. This swift metamorphosis is taking place with transmission capabilities, network switching systems, and customer premises equipment (CPE). It is therefore essential to understand some fundamental principles behind digital technology since these changes also have an impact on the use of 800 services.

Until very recently, the field of telecommunications was supported almost exclusively by analog technology. *Analog* comes from the word *analogous*, which means similar to. In the world of telecommunications, it means that the signal being transmitted is analogous to the original signal. A good visual representation of an analog signal is seen through an oscilloscope. If a microphone were plugged into this device, the output of the oscilloscope is an electronic graph of a human voice on a small cathode-ray tube (CRT). This graph is actually waves that represent the various frequencies of the voice, called *sine waves*. In the old analog world, a call initiated on an 800 number, through the public network, and to a company's telephone system, would never show a variance of the original sine wave (at least if there were no interference).

However well the analog telecommunications infrastructure worked, it did offer some problems. It was very susceptible to outside interference, picking up electrical interference from a variety of sources. This noise was often amplified along with the original signal. Digital service is not prone to such outside interferences and so offers many advantages in clarity of transmission and capacity.

Digital transmission is vastly different from its counterpart. Analog signals are transformed into bits of information that can be processed and transmitted in a number of ways. Digital technology uses much of the same language that is used in the computer world. Computers are digital and need to use information that is divided into functional units. The smallest unit is the bit. Bits are then formed together to make computer words, or bytes. Computers can only use binary numbers, so only a 1 or 0 can be used as a bit of information in a byte.

In order to change an analog signal into a digital signal, the sine wave is sampled by a coder/decoder, which is commonly called a CODEC. This is a common device in modern telephone equipment that samples the analog sine wave at a standard rate of 8000 times per second, a process known as pulse code modulation (PCM). The sampling rate is then turned into an 8-bit byte for transmission. The transmission rate of 64,000 bits per second (b/s) is equal to 8000 bits multiplied by an 8-bit byte rate. This rate is often referred to as 64 kb/s (kilobits per second).

The digital circuit used by the special access 800 services is the DS-1 circuit. DS stands for Digital Signal. The most common name is actually T1, but this is an erroneous designation because it is vendor specific (i.e., AT&T) and only refers to digital transmission over copper lines.

The fundamental building block of the DS-1 is the 64-kb/s transmission channel. A DS-1 circuit has 24 channels, with a bandwidth (i.e., capacity) of 64 kb/s for each channel. The individual channel is also known as the DS-0 (Digital Signal Level 0). The transmission method is via a serial bit stream using a concept known as time division multiplexing (TDM) (Fig. 4). A sample of all 24 channels is known as a frame (24 multiplied by 8 equals 192). There is also a bit that is used for synchronization called a framing bit. This makes the frame 193 bits. Each of these frames is transmitted at a rate of 8000 per second. If you multiply the frame by the transmission rate, it yields the total bandwidth of 1.54 million bits of data per second or 1.54 Mb/s (megabits per second). It should be noted that transmission on a DS-1 circuit can be either voice or data.

This format is not restricted to the 1.54-Mb/s bandwidth or 24 channels. Transmission methods with higher bandwidths are available, and their use is becoming more common every day. Using the same fundamental building blocks as the DS-1, the following formats may also be used for special access 800 services; however, it should be noted that, typically, only DS-1 or DS-3 circuits are used:

DS-1	1.546 Mb/s	24 channels
DS-1C	3.152 Mb/s	48 channels
DS-2	6.312 Mb/s	96 channels
DS-3	44.736 Mb/s	672 channels
DS-4	274.176 Mb/s	4032 channels

Digital circuits may be interfaced directly into telephone equipment if the switch is digital. If it is an analog switch, then a device called a channel bank is required. The channel bank allows the digital-to-analog conversion to take place and vice versa.

DS-1 circuits are normally provided through the LEC. The circuit is passed

Frame 1   Frame 2   Frame 3   Frame 4   Frame 5

FIG. 4 Time division multiplexing 24 channels on a DS-1 (Digital Signal Level 1) circuit. Each frame contains bits of information that can be telephone conversations. Information is broken into this framing concept and reassembled at the distant end.

though the LEC CO and then to the IXC POP. It is also common to contract for private cable via a CAP that connects directly to the IXC POP. DS-1 transmission is available on a variety of formats. The most common is copper cable, but it is not uncommon to find fiber-optic cable and microwave transmission used for this.

Integrated Services Digital Network

Integrated Services Digital Network (ISDN) is a concept that has been around for nearly a decade, but has only recently been finding widespread acceptance. This was due to slowly developing standards, limited applications, and expensive network and equipment costs. These obstacles are being overcome, and the advantages of ISDN are becoming apparent to the business world. This concept also has a special appeal for the users of 800 services that will become apparent as the concept is further explained.

As stated above, the world of telecommunications is migrating from an analog infrastructure to one that is digital. The ISDN provides a standard method of digital transmission for the world of telephony. The differences between a basic analog telephone line and an ISDN line provide insight into how ISDN works and why it is a superior method of transmission over analog technology.

Aside from the obvious advantages of digital versus analog technology, ISDN provides a number of channels for transmission and signaling, which adds great versatility. With any telephone line, there is a need to send signaling information through the network so that the LEC and the IXC will know how to process the call. On a standard analog line, there is only one channel for both signaling and transmission. Often, once a telephone number is dialed, the caller will hear a pause and then digits being transmitted through the public network. The caller is actually hearing part of the processing of the call, known as the call setup. Once the call is established, then the same channel is used for transmission (i.e., the telephone conversation). This example demonstrates a fundamental difference between analog and digital transmission methods. Using the same channel for signaling is known as in-band signaling. Using a separate channel is known as out-of-band signaling.

The ISDN provides for a number of channels, known as bearer (B) and data (D) channels. Call setup information is sent through the data channel, while the actual transmission is sent through the bearer channel. The B channel has a standard bandwidth of 64 kb/s, while the bandwidth of the D channel varies depending on what type of ISDN service is being used.

There are different types of ISDN service. The most fundamental type is the Basic Rate Interface (BRI), which is also known as 2 B + D. BRI can be thought of as a basic telephone line. ISDN was developed to be a service that encompassed both voice and data communications, hence the two B channels. If a residential customer were to subscribe to BRI, a typical setup may be a telephone with a data port (normally an RS-232 connection). The telephone is plugged into a data port on a personal computer (PC) that is equipped with a special ISDN interface card. The two B channels each offer 64 K of bandwidth, with one channel for voice and one for data, although it is not absolutely necessary to use the bandwidth in this manner. The D channel offers 16 K for out-of-band signaling. This is where information is often passed and received and can be used for a variety of purposes.

The BRI is obviously geared for small applications. A larger model is required for businesses if they are to use ISDN as their trunk access. Primary Rate Interface (PRI) provides a standard configuration of 23 B + D. Once again, the standard B channel offers a bandwidth of 64 K, but the D channel is 64 K as compared to the 16 K of the BRI.

Why is ISDN important to 800 services? If you compare the bandwidth of a DS-1, you will see it is the same as PRI. One difference is the out-of-band signaling. The D channel is used to send call setup information through the network, but also to send information such as the caller's telephone number. This concept is known as automatic number identification (ANI). Under this system, the telephone number of the caller is forwarded to a telephone system, identified (by either the telephone system ANI routing tables or a computer database—a concept known as computer-to-telephone integration [CTI]), and routed to a telephone agent, who may be able to service the call properly. This is a versatile and powerful form of transmission because it reduces call setup time (remember the example of listening to the digits of a dialed number as they are passed through a network) and provides a medium for sending data through a network, independent of the actual telephone conversation.

The Economics of 800 Services

The modern call center has two components that can comprise up to 90% of its budget: personnel and 800 services. This, of course, varies with the equipment used and the application. For instance, in the case of CTI, the equipment portion of the budget would be considerably larger. Regardless of the application, 800 services are a large portion of the call center budget, and it is necessary to understand how these services are priced. Understanding the components of 800 pricing allows a call center manager to make adjustments to various factors in order to seek the most favorable costs.

There are three components to 800 pricing: usage, service charges, and access charges. Each component bears closer examination. There are several factors that affect how usage is charged. Usage is actually a rate that is priced on a number of parameters. Three factors are typically used to charge rates: service areas (i.e., the distance of the call), cost per timed increment (e.g., a per-minute rate), and the time of the call.

The IXC, or any regulated telephone company, files their rates in a tariff with the FCC or, in the case of an LEC, with the state public utilities commission (PUC). A tariff is an officially approved document that specifies the rate structure of a long-distance service and is available to the public. The purpose of the tariff is to document the rates so that members of the public know they are receiving pricing that is offered to all customers at a rate that has been deemed fair and reasonable. Tariffs are not legally binding. While they do provide a reference point, carriers often violate their own tariffs and offer discount rates to be more competitive, especially to their larger customers.

The service area is actually a geographical area and may also be known as a range or band. Rates that are charged strictly by geographic origination and termination are known as distance sensitive. Service areas are designated by all IXCs with numbers, 1 through 6. Service Area 1 is the area closest to the point of termination (i.e., where the access lines are terminated) and Service Area 6 is the area farthest from that point. For instance, Service Area 1 might be an area that encompasses the states that immediately surround a call center. In reality, the service area is a mileage band. As more service areas are added (2 through 6), sections of coverage are added until the concept encompasses the entire country. If a call center was located in New Jersey and a customer called from eastern Pennsylvania, it would be a call from Service Area 1. If a customer called the same 800 number from California, it would be from Service Area 5. The reader may also encounter an older, and seldom used, term, banded INWATS. Before divestiture, AT&T used the bands to designate service areas. Bands were also designated as 1 through 6, similar to today's service areas. A key difference between modern 800 services and the old banded concept is that the older products were marketed with geographical restrictions. For instance, if a user had a Band 1 800 number, it only offered service to the surrounding states. If a call center were located in Pennsylvania, the coverage would only be for states such as Ohio, New York, West Virginia, and the like. If a potential customer from California tried to call, the customer would hear a message saying that the customer was outside the calling area. It is obvious that this system was complex, cumbersome, and restrictive. Today's 800 services do not have geographic restrictions and offer service to the entire country unless limitations are specifically requested by the 800 subscriber. The 800 services that have service areas, but no limitations, are often referred to as virtual or seamless 800 services.

Service areas are priced by assigning a timed rate to each area. A rate can be expressed in increments of seconds, minutes, or hours. The rate may not be a true indication of what is actually billed. If a carrier publishes a rate of $13.36 per hour, it reflects what one solid hour of talk time would be. There is also a billing increment that needs to considered. Over the years, IXCs have used a variety of billing increments to charge their customers. For instance, the first minute may be rounded, and all subsequent talk time is billed in 6-second increments. So, if a customer calls an 800 number but only talks for 30 seconds, the subscriber is billed for a full minute. All subsequent talk is billed in 6-second increments. Most IXCs charge in 1-second increments, but there are smaller carriers, resellers, and aggregators that use billing increments (e.g., 6 seconds, 30 seconds, or 60 seconds) to gain profit.

It is also common to see rates reduced by hours of usage. These types of rates are known as tapered. For instance, an IXC may charge 23 cents a minute up to 5 hours worth of usage. Once the 5-hour mark has been met, the rate reduces to 22 cents per minute. The IXC typically sets up blocks of hours that trigger new discounts, up to a maximum at which rates remain constant. A common way of evaluating 800 services is to determine the cost per minute. This is achieved by adding the total costs that an 800 service incurs. The total cost equals usage charges, service charges, and access charges. The total cost is then divided by the total number of minutes of usage. This yields the cost per minute. In the case of tapered rates, the more the 800 service is used, the lower the per-minute rate becomes. Since there will inevitably be variations from one IXC to another in terms of rates and service charges, cost per minute is a good barometer for comparing 800 carriers and products.

Rates are determined by distance and time increments. In addition, they are also based on the time at which a call was placed. The 800 charges, and long-distance charges in general, are divided into three rate periods: the day rate, the evening rate, and the night and weekend rate. Day rates are the most expensive, and normally range from 8 A.M. to 5 P.M. Evening rates generally offer a discount over day rates, and the rate period is normally from 5 P.M. to 11 P.M. The night and weekend rate is the least expensive and is defined from 11 P.M. to 8 A.M. during the weekdays and during the entire day on Saturdays and Sundays. Some IXC rate structures may charge the evening rate at certain times during the weekend (e.g., Sunday evening). It may appear that 800 rates are offered by competing IXCs with so many variables that an apples-to-apples comparison would be almost impossible. Fortunately, the structure of pricing is very similar from one IXC to another, and it is actually fairly easy to make comparisons. For instance, notice the 800 rates in Table 1. These rates will change and may be obsolete by the time this article goes to press, but they are perfectly adequate for demonstration purposes.

Notice that the special access products are very similar in structure. A quick analysis could be made by multiplying usage (minutes or hours) by rates and then adding service and access charges. Actually, though, most IXCs have software that will analyze long-distance bills and will perform this service free of charge.

There are several other factors that affect 800 bills. A service charge is a common part of 800 services. This is a flat rate that is charged on a monthly basis. There is also normally an installation charge, but with the competitive nature of the long-distance business, it is not unusual to have installation charges waived. Long-distance companies (both IXCs and resellers) are constantly offering specials, and it is also not uncommon to find a limited amount of free usage being given away if a customer changes carriers.

The rates that have been discussed so far are standard tariffed rates; large users of 800 services have an additional option. IXCs offer special agreements for the larger users. Special agreements may be a standard offering for lower rates, or the agreement can be a result of special negotiations that yield customer-specific rates. In return for offering these low rates, the IXC will want a commitment (in dollar volume, length of contract, or both) from the 800 subscriber. If the subscriber wishes to terminate service prematurely, there is normally a termination liability, which may be the balance of the minimum monthly dollar amount for the remainder of the contract.

The final portion of pricing 800 services is the access charges. These charges include installation costs and a monthly recurring charge. The monthly charge

TABLE 1   Rates for 800 Services (in dollars)

Service	Per Hour of Use				
	Business Day Rate	Evening Rate	Night/ Weekend Rate		
AT&T Readyline and Masterline* (switched and dedicated)					
Service Area 1	14.34	11.82	9.52		
Service Area 2	14.84	12.23	9.84		
Service Area 3	15.09	12.42	10.01		
Service Area 4	15.57	12.82	10.32		
Service Area 5	15.81	13.02	10.48		
Service Area 6	17.28	14.22	11.44		
AT&T MEGACOM 800 Service* (special access— DS-1 or DAL)					
Service Area 1	9.75	7.97	6.65		
Service Area 2	10.28	8.39	6.99		
Service Area 3	10.55	8.62	7.18		
Service Area 4	11.09	9.07	7.55		
Service Area 5	11.35	9.30	7.75		
Service Area 6	12.99	10.60	8.85		
	Per Minute Rate				
MCI Switched WATS†					
Range 1	0.2288	0.1907	0.1501		
Range 2	0.2370	0.1978	0.1557		
Range 3	0.2410	0.2008	0.1583		
Range 4	0.2489	0.2077	0.1636		
Range 5	0.2534	0.2112	0.1666		
Range 6	0.2777	0.2318	0.1829		
	Peak	Off-Peak			
MCI Dedicated Termination† (special access—DS-1 or DAL)					
Range 1 (0–292 miles)	0.1538	0.1230			
Range 2 (292+ miles)	0.1856	0.1485			
	0 to 4.99 Hours	5 to 24.99 Hours	25 to 74.99 Hours	75 to 149.99 Hours	Over 150 Hours
Fonline business day rates§ (switched access)					
Band 1	0.0270	0.0247	0.0235	0.0231	0.0224
Band 2	0.0287	0.0267	0.0245	0.0242	0.0236
Band 3	0.0287	0.0267	0.0245	0.0242	0.0236
Band 4	0.0287	0.0267	0.0245	0.0242	0.0236
Band 5	0.0287	0.0267	0.0245	0.0242	0.0236

TABLE 1 Continued

Service	Per Minute Rate		
	Business Day Rate	Evening Rate	Night/ Weekend Rate
Sprint Ultra 800‡ (special access – DS-1 or DAL)			
Service Area 1	0.1560	0.1210	0.1210
Service Area 2	0.1880	0.1460	0.1460
Service Area 3	0.1880	0.1460	0.1460
Service Area 4	0.1880	0.1460	0.1460
Service Area 5	0.1880	0.1460	0.1460

*Rates effective January 1, 1993.
†Rates effective March 1994.
‡Evening and night/weekend rates are also offered at a lower rate and applied to the same format.
§Rates effective January 1, 1995.

may be a flat rate that is charged by the LEC to all business customers or may be charged on a mileage basis if the customer is located far from the CO. The IXCs will provide the trunk access as a natural part of their 800 service. However, the subscriber is free to provide his own access, either by installing cable or by negotiating an individual contract with either the LEC or a CAP.

Equipment may also have an effect on the use of different access methods and their associated cost. The access methods of 800 services are also generically known as the trunk lines. Trunk lines are attached to the subscribers' CPE. There are three basic types of telephone systems classified as CPE that interface to 800 services: key systems, private branch exchanges (PBXs), and automatic call distributors (ACDs). A fourth type of system, Centrex, is not classified as CPE. This system is actually for the use of the LEC CO. The CO software is programmed to provide telephone system functionality for the subscriber.

The trunk line, whether it is a CO, DAL, or DS-1 circuit, has to interface to a specific type of circuit card in the telephone switch. The CO and DALs are analog circuits and, since most modern telephone systems are digital, they will actually convert the analog signal to one that is digital. Each type of circuit card has a capacity for a specific number of circuits. For instance, a CO card may have a capacity of 4 on 1 (4 possible circuits on 1 CO card), 8 on 1, or 16 on 1. If it is determined that more circuits are required and the capacity of a card has been reached, then it will be necessary to add another. Circuit cards can be expensive, ranging from $1000 to $3000, depending on the type of system and card.

Switched access 800 services simply use the LEC-provided CO line. Prices for CO lines vary in different areas of the country, but the monthly charge is usually fairly reasonable. Rates are as low as $15 per month and as high as $40. A CO line can usually be installed within a week or two after the order is placed, and installation charges can vary, but are still fairly reasonable. It would be safe to assume that, in most areas, the charges will range from $100 to $120.

The DAL line is actually a dedicated circuit that is only used for inbound service. The monthly charge for this type of circuit can range from $50 to $150. It should be noted at this point that both a CO and a DAL carry one telephone conversation per circuit. A DS-1 is a single circuit capable of carrying 24 voice conversations. This is a key issue when choosing an access method, weighing both capacity and economics.

The DS-1 circuit actually presents two possible scenarios for equipment. First, older analog telephone systems require some method of converting the 24 channels of the digital DS-1 format to 24 analog signals that it can process. This is accomplished with the use of a channel bank. The channel bank also is a device that converts the 24 channels into a tie line interface. A tie line is a special type of circuit that is used to link two types of systems. So, the channel bank would perform the digital-to-analog conversion and then interface to the telephone system via 24 tie lines. A tie line is a special type of dedicated circuit (also known as a private line—a family of dedicated circuits that have various purposes) that is used to "tie" two telephone systems together. In this case, the channel bank ties the DS-1 circuit to the telephone system. Channel banks can be either purchased or leased, and the subscriber may either procure a channel bank or the IXC may have a leasing option. Purchasing the channel bank is normally the most cost-effective solution since leasing will incur interest payments and an eventual buyout.

The second type of DS-1 interface is the direct DS-1 card. Most modern telephone systems, whether ACD or standard PBX, are digital and offer a DS-1 interface card. The DS-1 simply interfaces directly into the system, allowing for the standard 24 channels.

The remaining piece of equipment that is necessary for both channel bank and direct DS-1 interface is the channel service unit (CSU). This device actually synchronizes the frames of the DS-1 circuit. There are several protocols for DS-1 framing, such as superframe (SF) or extended superframe (ESF); the CSU synchronizes these various methods of framing a DS-1 circuit.

In order to justify the implementation of a DS-1 circuit, the subscriber will need to justify the expense of additional equipment or a telephone system upgrade. The justification for making such a change is to weigh the savings that will be realized by changing to a special access 800 service. If, for instance, there is an up front cost of $3000 to purchase a DS-1 card and the projected monthly savings on 800 costs is $500, the payback period will be six months. Even if the equipment is not purchased outright and the subscriber decides to lease and incur a monthly charge, there may be justification for upgrading the 800 service. If the monthly lease payment is $250 over a five-year period, the monthly savings will still be $250 per month or $15,000 total for the duration of the lease.

New Numbers—Diminishing Reserves

The supply of existing 800 numbers is limited, and the demand for new numbers is steadily increasing. How serious is the problem? There are some industry analysts who felt it was entirely possible that the business world could run out of 800 numbers by the end of 1995 or early 1996, which would have caused a real dilemma for the business world.

The reason for this dilemma is very simple: it is a case of supply and demand. The supply is limited, and the demand is growing at a tremendous rate. For example, AT&T alone has experienced a 49% growth rate in the past year on the 800 side of their business. A quick look at the installed base of 800 numbers clearly indicates how the use of such services is accelerating. In 1993, 2.4 million 800 numbers were being used. In 1995, there were 5.5 million numbers being used, a 100% increase over two years. Compare this increased usage with the maximum number of usable 800 numbers (7.6 million), and the serious nature of the problem becomes apparent. Currently, it is estimated that 80% of all the available 800 numbers either are being used or are reserved, and the consumption of available numbers is progressing at a rate of 4% per month.

Why the dramatic acceleration in 800 usage? There are actually a number of reasons. In recent years, 800 services have been marketed with relatively low costs, ease of use, and ease of installation. This, in turn, has attracted small businesses and residential customers, relatively new players in the 800 game. Larger users have also been hoarding 800 numbers, keeping large banks of numbers in reserve for future use. An expanding economy and one that is more service oriented have also contributed to the problem since 800 numbers play a critical role in a service-based economy. In addition, RespOrgs (responsible organizations) may reserve numbers up to a number that equals 15% of their installed base. Once a number is reserved, it is taken out of circulation for a period of 60 days. The problem has been further exacerbated by 800 customers who do not disconnect an 800 number, even though it may no longer be used. Many companies simply place the unused 800 number into an internal company pool for future use.

The problem is considered to be so critical that an emergency meeting of the Industry Numbering Committee (INC)—made up of carriers, RHCs, and 800 customers—was called in early May 1995. The INC addressed a number of issues that might serve to gain efficiencies within the current system and release numbers that were previously not available.

One of the most glaring problems is the system by which 800 numbers are reserved, a system that has often been considered to be too liberal. RespOrgs actually have free rein to reserve any 800 number as long as they do not exceed the allowable 15%. While it may be difficult to police why a number is reserved (i.e., a customer has specifically made a request that a number is reserved), the problem has been addressed by reducing the allowable reservation percentage to 8%.

The INC also agreed to modify the aging process for previously used 800 numbers, reducing the aging time to four months. Previously, when an 800 number was released back into the general pool, the number was not made available for six months. The reasoning behind this methodology was to allow the previous user to change advertising or listings so that a new user would not incur difficulties with the previous user's customer base.

In addition to modifying existing policies, the INC has agreed to release numbers that have been withheld from public use for a variety of reasons. For instance, 1-800-555-0100 through 1-800-555-0199 have been reserved for the entertainment industry. Now, these numbers will be released for public use with the exception of 1-800-555-0199, which will be the sole number used for movies or television shows.

Also, N11 codes will be opened. In the North American Numbering Plan, N is a designation of a digit, 2 through 8. The N11 range has been in reserve for emergency applications, but now 211 though 811 will be made available for general business use. Additional numbers were also found in the 250 exchange that had been reserved for testing. The range of 1-800-250-0000 through 1-800-250-0099 will still be retained for testing; however, 1-800-250-1000 through 1-800-250-9999 will made available for general public use.

Additional numbers had also been reserved for future use with Caribbean countries. The INC drafted a letter to the post, telegraph, and telecommunications (PTT) administrations of these countries requesting that the reserved bank of numbers be released for general public use.

Regardless of what actions are taken, it is inevitable that the existing range of 800 numbers will be exhausted. This has necessitated the creation of a new toll-free exchange 888, which went into effect in early 1996. The committee is also considering additional toll-free exchanges, 877 and 866. When these new toll-free exchanges are put into place, new issues will surface regarding toll-free services. First, there may be confusion on the part of the general public. Second, a brand protection policy will need to be considered for redundant and vanity numbers that are duplicated across the 800 and 888 exchanges. Some interesting problems may even surface for competitors (i.e., the same vanity number for 888 vs. 800). Will the actions of the INC solve the current problems? Only time will tell; however, a whole new set of problems may surface, making the toll-free arena even more confusing than it is today.

List of Acronyms

ACD	automatic call distribution
ACS	automatic call sequencer
ANI	automatic number identification
ATB	all trunks busy
ATM	asynchronous transfer mode
Bellcore	Bell Communications Research
BRI	Basic Rate Interface
BTN	billing telephone number
CAP	competitive access provider
CCS	common channel signaling
CDR	call detail recording
Centrex	Central Exchange
CLID	calling line identification
CO	central office
CODEC	coder/decoder

CPE	customer premises equipment
CRT	cathode-ray tube
CSU	channel service unit
CTI	computer-to-telephone integration
DAL	dedicated access line
DNIS	Dialed Number Identification Service
DS-0	Digital Signal Level 0
DS-1	Digital Signal Level 1
DS-3	Digital Signal Level 3
ESF	extended superframe
FCC	Federal Communications Commission
FX	foreign exchange
INC	Industry Numbering Committee
ISDN	Integrated Services Digital Network
ITC	independent telephone company
IVR	interactive voice response
IXC	interexchange carrier
kb/s	kilobits (1000 bits) per second
LATA	local access and transport area
LDN	listed directory number
LEC	local-exchange carrier
LED	light-emitting diode
Mb/s	megabits (million bits) per second
MIS	management information system
MOH	music on hold
NANP	North American Numbering Plan
NASC	National Administration and Service Center
NCD	network call distributor
NPA	Numbering Plan Area
OCC	other common carrier
PBX	private branch exchange
PCM	pulse code modulation
PDRC	Primary Disaster Recovery Center
POP	point of presence
PRI	Primary Rate Interface
PUC	public utility commission
RespOrg	responsible organization
RHC	regional holding company
RNA	ring no answer
SCP	service control point
SDRC	secondary disaster recovery point
SF	superframe
SLO	service-level objective
SMDR	station message detail recording
SMS/800	service management system
SS7	Signaling System No. 7
SSP	service switching point
STP	signal transfer point

TDM	time division multiplexing
TOC	technical operating center
TOPMS	Telemarketing Operations Performance Management System
UCD	uniform call distribution
VRU	voice recognition unit
WATS	wide-area telephone service
WTN	working telephone number

Bibliography

Aiuto, J., and Matinez, L., 800 Network Control with the Accumaster Services Workstation (ASW), AT&T Inbound Services Management Group, April 1994.

Aggregator Warning, *Teleconnect* (June 1990).

Baer, S. M., Bellcore's Vision for Multimedia-Based Communications Services, *Telecommun.* (September 1993).

Barca, M., Help Your Agents Serve More Callers, More Effectively, AT&T Inbound Services Management Group, April 1994.

Bergman, D., The Future of Intelligent, Networked ACDs, *Business Commun. Rev.* (May 1991).

Bodin, M., 800 Portability—The Database, *Call Center* (August 1993).

Bodin, M., The Technology of Tomorrow, Today, *Call Center* (September 1993).

Bodin, M., Welcome to the New World of 800, *Call Center* (November 1993).

Braca, M., Help Your Agents Serve More Callers, More Efficiently, AT&T Inbound Services Management Group, April 1994.

Brice, K., Tools for Effective Customer Service, AT&T Inbound Services Management Group, April 18, 1994.

Bozeman, M., End Users and Portability: The Summer of Their Discontent? *Voice Processing* (October 1993).

Briere, D., AT&T Tariff FCC No. 12 Custom-Designed Integrated Services: Overview, Data Pro Communications Series: Communications Networking Services, October 1993.

Bulluck, C., Francus, J., and Everett, W., Interactive Voice Response: Technology and Applications Using AT&T Infoworx, AT&T Inbound Services Management Group, April 1994.

Cleveland, B., Advanced Network Services for ACDs: Today's Realities, *Business Commun. Rev.* (April 1990).

Coleman, R., ISDN Issues, Data Pro: Managing Voice Networks, February 1992.

Crowe, T. K., 800 Portability: End-Users Beware, *Voice Processing* (September 1993).

Dawson, K., 800 Portability Is Here, *Call Center* (June 1993).

Dawson, K., 800 Services—Look Hard Before You Leap, *Call Center* (February 1994).

Dawson, K., ISDN Equipment—Tapping into New Services, *Call Center* (July 1993).

Dawson, K., The Next Phase of 800, *Call Center* (February 1993).

Dawson, K., Open Architecture—New Ways to Make Your Call Center More Efficient, *Call Center* (September 1992).

Deixler, L., T-1's Coming of Age, *Teleconnect* (April 1993).

Domurath, F., and Schreibstein, S., Juggling Resources Without Dropping Any Calls, AT&T Inbound Services Management Group, April 1994.

Donachy, L., Frankel, R. J., Kranz, L., and Shapiro, B., Call Redirection Overview, AT&T Inbound Services Management Group, April 1994.

Don't Spell It Out, *Call Center* (April 1994).

800 Boosts Responses, *Call Center* (December 1993).

800 Digits, *Teleconnect* (September 1990).

800 Survey: Question to Ask Before You Switch, *Call Center* (February 1993).

800 Tips, *Teleconnect* (September 1990).

Erickson, M., and Watt, K., Protect the Lifeline of Your Business by Planning for the Unexpected, AT&T Inbound Services Management Group, April 1994.

Flanagan, D., Grieco, N., Schmidt, J. R., and Weber, B., Business Solutions Through ISDN, AT&T Inbound Services Management Group, April 1994.

Frydman, I., Managing a Call Center, Data Pro Managing Voice Networks, March 1993.

Fulgham, G., and Giacoia, D., Fine Tune Your Marketing to Better Serve Your Customer, AT&T Inbound Services Management Group, April 1994.

Gable, R. A., Disaster Planning—Planning 800 Disaster Rerouting in Distributed Call Centers, *Business Commun. Rev.* (June 1993).

Gable, R. A., *Inbound Call Centers: Design, Implementation, and Management*, Artech House, Boston, MA, 1993.

Gable, R. A., IXC Management Tools for 800 Services, *Business Commun. Rev.* (August 1994).

Gable, R. A., Market Trends: Number Portability—What Happened? *Business Commun. Rev.* (October 1993).

Haley, C., Hance, P., and Rowe, P., 800 Information Management with the Accumaster Services Personal Computer Software (ASPC), AT&T Inbound Services Management Group, April 1994.

Hirsch, P., and Montgomery, K. J., 800 Services Network Reliability and Your Business, AT&T Inbound Services Management Group, April 1994.

Identifying, Developing, and Implementing an IVR Application, National Rolm Users Group, April 28, 1994.

Impellizeri, J., and Stropes, B., Help Your 800 Network Run More Smoothly, More Profitably, AT&T Inbound Services Management Group, April 1994.

Jepperson, R. M., Common Channel Signaling System Number 7, Data Pro Communications Series: Broadband Networking, March 1992.

Johnson, J. T., Users Rate Long-Distance Carriers, Data Pro Communications Series: Communications Networking Services, February 1993.

Kennedy, M. D., Improving Customer Service Through Advanced Networking Capabilities, *Telecommun.* (November 1992).

Leibowitz, E., The ANI Phenomenon, *Teleconnect* (November 1991).

Leibowitz, E., Rochelle's Caller ID Quest, *Teleconnect* (April 1994).

Luhmann, R., Class Service Universal ANI? *Teleconnect* (September 1990).

Newton, H., *800 Pocket Guide*, Telecom Library, New York, 1992.

Newton, H., 800 Portability—Does It Really Make a Difference? *Call Center* (February 1993).

Newton, H., How 800 Services Work, *Teleconnect* (June 1990).

Newton, H., Your Very Own STP, *Teleconnect* (August 1992).

Newton, H., A Summary of Erector Set Telecom, *Teleconnect* (May 1993).

A 1992 Profile of 800 Numbers for Customer Service, Society of Consumer Affairs Professionals in Business, June 1992.

Questions to Ask Before You Switch, *Call Center* (February 1993).

Rendleman, J., Carriers Offer Few New "800" Features, *Commun. Week* (February 28, 1994).

Skvarla, C. A., 800 Number Portability, Data Pro: Managing Voice Networks, March 1993.

Skvarla, C. A., Sprint 800 Services, Data Pro Communications Series: Communications Networking Services, May 1993.

Skvarla, C. A., and Womack, A., AT&T 800 Services Family, Data Pro Communications Services: Communications Networking Services, January 1994.

Sprint Opens Its Network to Developers and Large Users, *Call Center* (May 1993).

Toth, V. J., Progress Report: 800 Database Access, *Business Commun. Rev.* (September 1992).

Vanity Numbers, *Call Center* (February 1994).

Waite, A., *The Inbound Call Center – How to Buy and Install Automatic Call Distributors*, Telecom Library, New York, 1989.

Wexler, J., AT&T Advances on the 800 Front, *Network World* (May 2, 1994).

Womack, A. Y., MCI Communications Corp. 800 Service, Data Pro Communications Series: Communications Networking Services, August 1993.

Womack, A. Y., Status of the North American Numbering Plan, Data Pro: Managing Voice Networks, August 1993.

ROBERT A. GABLE

NSFNET

NSFNET is a program developed by the National Science Foundation (NSF) to provide electronic network access and support for the research and education community of the United States. NSF is funded by the U.S. government to support research and education in science and engineering within the United States. A key element and the major component of the NSFNET program until 1995 was the NSFNET backbone network service, a high-speed network linking 19 sites within the continental United States to which midlevel regional networks were connected. Beginning in April 1995, this infrastructure was replaced by another, consisting of a series of network access points (NAPs) linked by network service providers (NSPs) from the private sector, which in turn linked with the midlevel networks to provide a national network infrastructure. This new architecture is being funded on a five-year decreasing scale by the NSFNET program. Following is a brief history of the NSFNET program, an outline of its status in 1995, and a summary of future directions for the program.

The history of networking in the United States has as its starting point the establishment of ARPANET (Advanced Research Projects Agency Network) in 1969 as an experimental datagram network designed to facilitate communications among defense research contractors within universities and national laboratories (1). From then, ARPANET grew steadily in size until by the mid-1980s it had clearly demonstrated the importance and utility of networking and had stimulated the appearance of other major networks within the community such as BITNET (Because It's Time Network), CSNET (Computer Science Network), Usenet, and MFEnet, the precursor to ESnet (Energy Sciences network). Particularly noteworthy was the adoption in 1983 of the TCP/IP (Transmission Control Protocol/Internet Protocol) nonproprietary switching protocol (2), the protocol that eventually became the de facto standard in the research network world.

In 1982, the report of the NSF panel, Large Scale Computing in Science and Engineering, chaired by Peter D. Lax, called for the "establishment of a system of effective computer networks that joins government, industrial and university scientists and engineers" (3). The existence and success of ARPANET made the implementation of this recommendation a practical possibility. The Lax report led to the establishment of five NSF supercomputer centers in 1986 and to the development of NSFNET to provide links to them.

The first NSFNET backbone network was developed in 1986 with 5 nodes, 56-kilobit-per-second (kb/s) lines, PDP-11/70 "fuzzball switches," and used the TCP/IP protocol suite for switching developed by ARPANET. (The PDP-11/70s were computers manufactured by Digital Equipment Corporation.) In that same year, the first midlevel regional networks (SDSCnet, SURAnet, JVNCnet, and NYSERNET came into being and became the first networks funded by the NSFNET Connections Program. Subsequently, several other midlevel networks came into being with the assistance of the NSFNET Connections Program.

By 1987, the NSFNET backbone network had become severely congested. A new NSF solicitation led to an award to Merit, Incorporated, a consortium of Michigan universities, with IBM and MCI as industrial partners to establish the new phase of NSFNET. This new phase in 1988 took the form of a T1 (1.5-megabits-per-second [Mb/s]) backbone with 13 nodes across the United States and still using the TCP/IP protocol suite.

In 1989, the backbone was reconfigured to triple connectivity at each node, providing improved redundancy and throughput. Then, in 1990, three more nodes were added and the T3 (45-Mb/s) upgrade began. The upgrade was completed in 1991 and three further T3 nodes were added, bringing the total to 19.

Unfortunately, "teething problems" persisted in the T3 network, which led to the maintenance of an independent T1 "safety net" to ensure continuance of service. During this period, the NSFNET International Connections Manager (ICM) program was started with an award to U.S. Sprint. Under this program, NSF collaborated with the National Aeronautics and Space Administration (NASA), Defense Advanced Research Projects Agency (DARPA), and U.K. agencies to establish the U.K. "fat pipe" at 512 kb/s across the Atlantic. Meanwhile, FIX-East and FIX-West (FIX means federal Internet exchange) were established as managed interagency interconnects allowing NSFNET to connect smoothly with other U.S. government agency networks.

In 1992, the T3 backbone service became stable, enabling the cessation of the T1 service. The NSFNET program broadened to provide increased support for network services through the funding of the InterNIC awards to Network Solutions, Incorporated, for network registration services, to General Atomics for network information services, and to AT&T for network database services. In the same vein, an award was made to MCNC for the Clearinghouse for Networked Information Discovery and Retrieval (CNIDR) to provide a central focus and forum for NIDR tools.

Throughout its history, the NSFNET model was based on a three-level architecture consisting of the high-speed backbone linking midlevel networks, which in turn connected campus networks at individual institutions. Figure 1 illustrates the structure of the T3 backbone, the nodes by which it was connected to the midlevel networks, and the geographical areas of the midlevel networks. These networks used the TCP/IP protocol suite, which eventually became the standard for the worldwide Internet to which they are connected.

The backbone network and services such as the InterNIC were funded by the NSFNET program. Partial support was also provided to the various midlevel networks and educational and research institutions to extend network connectivity throughout that community. By 1992, it was clear that this had become a stable commodity service, much of which could be left in the hands of the private sector to provide. Consequently, after extensive consultation with the network community, a solicitation for bids to switch to a new architecture was issued in 1993. The solicitation called for proposals to implement and operate a series of network access points within the United States to which national service providers would connect. The solicitation also called for the provision of funding for the midlevel networks to negotiate contracts with the NSPs to connect them to the NAPs, the establishment of a routing arbiter (RA) to do research in network routing and to provide equitable routing within the new architecture, and for the establishment of a very-high-speed backbone network service (vBNS) linking the supercomputer centers and to be accessed via the NAPs. The

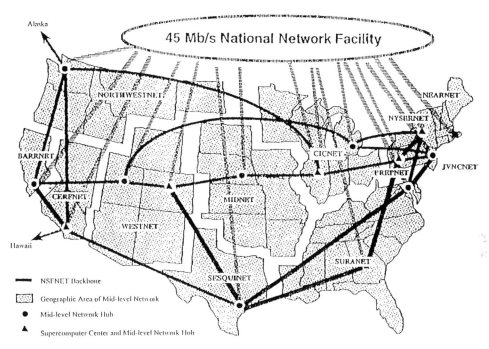

FIG. 1 The former NSFNET backbone.

switchover to the NAPs, the NSPs, and the RA took place over several months prior to full implementation in April 1995. The vBNS was implemented later in 1995.

Looking to the future, funding for the NAP-based architecture will decrease incrementally each year to zero after five years. The private sector will take over the funding of the NAPs through the fees charged for access to them. However, the NSFNET program will continue to fund the vBNS as part of the NSF's efforts to keep advancing the leading edge of network technology. It will also continue to fund the InterNIC, CNIDR, and related network services and will continue to expand connectivity within the United States through the Connections Program. The NSFNET backbone that played a critical role in the advancement of networking within the United States is now part of the very rich history of networking, but the NSFNET program will continue to fund projects aimed at pushing forward the frontiers of network service, as well as advancing the level of connectivity within the research and education community.

The NSFNET program is administered through contracts awarded by the NSF's Division of Networking and Communications Research and Infrastructure through a competitive bidding process. Generally, bids for the operation of major facilities such as the NAPs are generated in response to the issuance of a solicitation for proposals. However, unsolicited proposals are welcomed for funding to establish connections to one of the midlevel networks or the vBNS or to fund networking research. Multiyear contracts generally are subject to a periodic review by panels of disinterested experts to ensure compliance with the conditions of the awards.

References

1. McKenzie, A. A., and Walden, D. C., ARPANET, the Defense Data Network, and Internet. In *The Froehlich/Kent Encyclopedia of Telecommunications*, Vol. 1 (F. E. Froehlich and A. Kent, eds.), Marcel Dekker, New York, 1990, pp. 341–376.
2. Haring, D. R., Internetworking with Transmission Control Protocol/Internet Protocol. In *The Froehlich/Kent Encyclopedia of Telecommunications*, Vol. 9 (F. E. Froehlich and A. Kent, eds.), Marcel Dekker, New York, 1995, pp. 459–490.

ROGER TAYLOR

NYNEX

Overview

NYNEX Corporation provides a full range of communications services in the northeastern United States and select markets around the world, including the United Kingdom, Thailand, Gibraltar, Greece, Indonesia, the Philippines, Poland, Slovakia, and the Czech Republic. The corporation has expertise in telecommunications, wireless communications, cable television, and entertainment and information services.

NYNEX has four main business groups: NYNEX Telecommunications, NYNEX Worldwide Communications and Media Group, NYNEX Asia Communications, and financial subsidiaries.

NYNEX Telecommunications Group is NYNEX's largest business group. The group includes New York Telephone Company, New England Telephone and Telegraph Company, and their subsidiaries, which do business under the NYNEX brand. Telecommunications operations provided NYNEX with approximately 87% of its operating revenues in 1994. Approximately 87% of those telecommunications revenues derived from operations in New York and Massachusetts.

NYNEX's Worldwide Communications and Media Group includes the corporation's interests in wireless communications (Bell Atlantic NYNEX Mobile Communications, PCS PrimeCo, Tomcom, STET Hellas), wholly owned subsidiaries and interests in cable TV and entertainment and information services (NYNEX CableComms, NYNEX Entertainment and Information Services, NYNEX Information Resources Company, TELE-TV), and subsidiaries and interests in worldwide communications operations (FLAG [Fiberoptic Link Around the Globe], Gibraltar NYNEX Communications).

NYNEX Asia Communications includes joint ventures that are constructing and operating telecommunications and cable television networks in Thailand, Indonesia, and the Philippines.

NYNEX's financial subsidiaries include NYNEX Credit Company, NYNEX Capital Funding Company, and NYNEX Trade Finance Company.

The communications industry is going through significant changes prompted by rapid new developments in technology and regulation and the "convergence" of the telecommunications, cable television, computing, entertainment, and information industries. The changes were recognized in passage of the Telecommunications Act of 1996, which is intended to stimulate competition in local and long distance telecommunications and cable television.

In 1993, a comprehensive analysis of its operations and work processes, NYNEX made a strategic decision to reengineer itself to improve customer service while implementing workforce reductions and producing cost savings necessary to operate in an increasingly competitive environment. Through this repositioning, the "New NYNEX" is building on its heritage as a local telecommunications provider in the northeastern United States to transform itself into

499

a full-range provider of customer-focused communications, entertainment, and information services, both in the Northeast and around the globe.

NYNEX has stated that its long-term goal is to capture 10% of a $300–$350 billion global communications market by controlling costs, growing revenues, expanding selectively, and exceeding expectations—and giving customers the combined benefits of merging telecommunications, entertainment, and information in a new, digital world.

On April 22, 1996, NYNEX and Bell Atlantic announced a definitive agreement for a merger of equals. The companies said their goal was to close the merger within 12 months. The new company will be called Bell Atlantic.

Corporate Highlights (as of December 31, 1995)

Headquarters	New York City
Net income (before changes)	$1.4 billion
Revenues (not including cellular joint venture)	$13.1 billion
Employees	65,776
Share owners	864,833
Number of shares outstanding	426.5 million

1995 Telecommunications Group Highlights

Total customer access lines	17.1 million
Customer access lines served by digital offices	15.5 million
Percentage of digital central offices	87.0%
Customer access lines served by digital offices	90.6%
Fiber-optic (conductor) miles	1.3 million

NYNEX continues to invest for the future, upgrading outside plants and installing advanced operating systems to reduce costs and improve service. In 1995, NYNEX Telecommunications invested some $2.5 billion in network expansion and improvements. NYNEX is one of the leading regional Bell companies in terms of the percentage of lines served by digital switching, with more than 90%.

Worldwide Communications and Media Group Highlights

Bell Atlantic NYNEX Mobile communications (BANM), a jointly managed partnership formed in July 1995, serves the largest single cellular service territory in the United States, stretching from Maine to South Carolina and covering a total population of 53 million people. BANM ended with 3.4 million customers—adding more than 1 million customers during the year—and achieved revenues of $1.7 billion through innovative, value-priced offerings such as Talk-*Along*[sm] and Mobile Reach® roaming.

Bell Atlantic NYNEX Mobile communications is a partner with AirTouch and US WEST in a premier alliance, PrimeCo Personal Communications, L.P., to provide nationally branded, innovative, and easy-to-use wireless communications services. The alliance owns licenses for personal communications services

(PCS) in 11 additional major markets and reaches 24 out of the top 25 cellular markets nationally, with more than 160 million potential customers. The cellular properties of the partners will be managed and owned as separate entities. PrimeCo PC conducted its first test calls last year in Texas.

NYNEX CableComms holds 16 cable franchises including some 2.7 million homes in the United Kingdom, making it one of the world's largest combined telecommunications and cable television operators. CableComms more than doubled its revenues in 1995. An initial public offering in 1995 valued Cable-Comms at approximately $2 billion.

TELE-TV, a joint venture of NYNEX, Bell Atlantic, and Pacific Telesis Group, is working to deliver the next generation of home entertainment, information, and interactive services. TELE-TV is developing a portfolio of nationally branded programming and services and the technology needed to deliver them over the three companies' wireline and wireless networks. Howard Stringer, former president of the CBS Broadcast Group, is TELE-TV's chairman and CEO.

NYNEX and Bell Atlantic led the way in recognizing the potential of digital wireless cable technology. Together they have invested $100 million in CAI Wireless Systems, Inc., an Albany, N.Y.,–based wireless cable company, to speed the delivery of competitive video services throughout the Northeast. NYNEX plans to begin deployment of wireless cable later this year and throughout 1997 as it continues to invest in its landline telephony network.

In 1995, TELE-TV announced the selection of Thomson Consumer Electronics to produce up to 3 million TV set-tops and systems for the delivery of wireless video entertainment. The equipment will be used to provide consumers with a high-quality service that includes local broadcast, basic and premium cable channels, and limited interactive services.

NYNEX Information Resources Company's Big Yellow$^{(sm)}$ (http://www. bigyellow.com) is one of the largest advertiser-supported sites on the Internet, with content far exceeding that of any other online shopping directory. Big Yellow includes the names, addresses, and phone numbers of more than 16 million businesses nationwide, and more than 9,000 local national advertisements.

NYNEX published telephone directories in the Czech Republic, Gibraltar, Poland, and Slovakia.

NYNEX is managing sponsor of FLAG (Fiberoptic Link Around the Globe), which is building the world's longest undersea fiber optic cable. FLAG will link Japan, China, Hong Kong, India, and Thailand with landing points in the Middle East and Europe. NYNEX's co-sponsors in FLAG are from Japan, Thailand, Hong Kong, Saudi Arabia, and the United States.

NYNEX Asia Communications Group Highlights

NYNEX formed NYNEX Asia Communications Group in September 1995 to focus its resources and build on its success in the world's highest-potential telecommunications growth region.

Through TelecomAsia, its joint venture with the Charoen Pokpkhand (CP) Group, one of Thailand's largest conglomerates, NYNEX is constructing and operating a 2.6 million-line telecommunications network—larger than Manhattan's. UTV, a TelecomAsia subsidiary, also is providing cable TV services in Bangkok over its fiber backbone network.

In Indonesia, NYNEX is the strategic operating partner in Excelcomindo, a joint venture that will bring digital cellular telephone service to 195 million potential customers in the world's fourth most populous nation. Excelcomindo holds one of three nationwide licenses for GSM digital cellular service, and plans to begin offering digital mobile telephone service in mid-1996 and reach cities throughout the country by 1998. NYNEX owns 23% of the venture and provides technical, customer service, marketing, and sales management expertise.

In the Philippines, NYNEX and its partners in Bayan Telecommunications Holdings Corp. hold licenses to install a minimum of 300,000 telephone lines in metropolitan Manila and the Philippines provinces.

Thailand is helping to plan and design telephone and cable television networks for the budding communications market of Southeast Asia.

NYNEX Network Systems Company and China's Ministry of Electronics Industry jointly will study the feasibility of interconnecting various existing networks in China and the technologies required to deploy advanced telecommunications services in selected Chinese markets.

Technology and Product Highlights

NYNEX invested more than $150 million in research and development in 1995. New products and services accounted for more than 70% of the growth in NYNEX local service revenues in 1995.

Hundreds of thousands of New York customers have signed up for NYNEX's new optional calling plans, NYNEX Business Link, Unlimited Calling, and Cents Per Minute, since they were introduced in September 1995.

NYNEX has more than 90,000 lines in service of ISDN—Integrated Services Digital Network—which gives customers high-speed, high-capacity connections to the Internet. In 1995, NYNEX announced two contracts with equipment suppliers that will enable installation of up to one million lines of ISDN by the year 2000, and agreements with equipment suppliers and other telephone companies to work toward a common set of standards that will radically simplify the process of ordering ISDN.

Bell Atlantic NYNEX Mobile's innovative, value-priced offerings include Talk*Along*(sm), a PCS-like product appealing to a new class of wireless customers, and MobileReach® roaming, which takes advantage of BANM's service territory. Reaching from Maine to South Carolina, BANM's service territory is the largest single cellular service territory in the United States.

NYNEX VoiceDialing(sm) service—the first of its kind—is now licensed by other regional Bell operating companies for use in their voice dialing products.

NYNEX VoiceDialing service enables customers to place calls by simply speaking into the phone.

NYNEX introduced NYNEX Enterprise Services, a family of private-line services with flexible network options on demand, in April 1993. These services allow NYNEX to reconfigure a business customer's network within one hour of a customer's request. NYNEX was the first communications company to offer such services. At the end of 1995, NYNEX had sold more than 46,000 Enterprise circuits.

NYNEX and the Securities Industry Association are joining in an effort under which NYNEX is taking a leadership role in working with other foreign and domestic telephone companies to ensure that compatible services are available to SIA members in New York and other financial capitals around the world. This will facilitate the creation of a global telecommunications network that provides state-of-the-art voice and data services.

NYNEX will develop with IBM an integrated and transparent network computing platform to provide a full set of value-added services targeted at the business market. The alliance and services are part of *iMPOWER*, NYNEX's strategic vision for business. As part of the *iMPOWER* concept, NYNEX also will deploy a new information infrastructure called the NYNEX Business Network Architecture, which will be based on ATM (asynchronous transfer mode) technology, and carry voice, data, image, and video to the desktop.

In 1995, Bell Atlantic NYNEX Mobile invested more than $700 million in its cellular network, adding cell sites, installing indoor microcells, and continuing its conversion to digital technology. BANM plans to spend another $700 million in 1996 on construction.

Trials and Alliances

NYNEX was the first of the former Bell System companies to offer "video dial tone" to customers on a trial basis. NYNEX fiber optic cables carried video signals for five cable TV providers for some 2,500 residents in three apartment buildings on the east side of Manhattan, who received between 80 and 150 channels of cable programming. Fifty of the trial participants also received interactive on-demand service, which let them view movies, sports, and other programs over the NYNEX network whenever they wanted. A digital trial high-rise apartment building is planned when the technology becomes available.

NYNEX is working with the City University of New York to trial a new distance learning service and video conferencing capability for administrators. The system, featuring leading-edge wideband technology, also is being used by the New York City Department of Corrections and Department of Health.

NYNEX is participating in the first U.S. health communications network, linking independent rural hospitals with medical specialists in urban areas. The network connects 11 hospitals in the Buffalo, N.Y., area through more than 100 miles of fiber optic lines.

NYNEX has invested $1.2 billion in Viacom to form a strategic alliance.

NYNEX Chairman and Chief Executive Officer Ivan G. Seidenberg and NYNEX Vice Chairman–Finance and Business Development Frederic V. Salerno have seats on Viacom's board of directors.

Corporate Philanthropy

In 1995 NYNEX awarded $19.5 million in grants and matching gifts to non-profit organizations. NYNEX's grant-making philosophy is guided by three priorities: to improve the quality of education with an emphasis on creating a better learning environment for economically disadvantaged children; to promote diversity by sponsoring cultural and community partnerships; and to promote long-term wellness by supporting health and human services programs.

NYNEX Corporate Officers

Ivan G. Seidenberg, chairman and chief executive officer
Frederic V. Salerno, vice chairman and chief financial officer
Morrison Des. Webb, executive vice president, general counsel, and secretary
Robert T. Anderson, vice president, business development
Jeffrey A. Bowden, vice president, Strategy and Corporate Assurance
John M. Clarke, vice president, Law
Saul Fisher, vice president, Law
Mel Meskin, vice president, Financial Operations and comptroller
Patrick F. X. Mulhearn, vice president, Public Relations
Donald J. Sacco, vice president, Human Resources
Thomas J. Tauke, vice president, Government Affairs
Colson P. Turner, vice president and treasurer

Officers of Principal Operating Groups

Richard A. Jalkut, president and group executive, Telecommunications
Donald B. Reed, president and group executive, External Affairs and Corporate Communications
Richard W. Blackburn, president and group executive, Worldwide Communications & Media Group
Arnold J. Eckelman, executive vice president and group executive, New York City
Joseph C. Farina, acting vice president and group executive, Quality Assurance and Operations Support
Roy Berger, acting president, Entertainment and Information Services

John Diercksen, acting vice president and treasurer, Telecommunications
Stanley Fink, senior vice president, Regulatory and Government Affairs

Bell Atlantic NYNEX Mobile Communications
Dennis F. Strigl, president and chief executive officer

NYNEX Asia Communications
Arthur J. Troy, president

NYNEX Asset Management Company
Candace Cox, president

NYNEX CableComms Group
John F. Killian, president and chief executive officer

NYNEX Capital Funding Company
Colson P. Turner, president

NYNEX Credit Company
Richard E. Lucey, president

NYNEX Global Systems Company
Daniel C. Petri, president

NYNEX Information Resources Company
Matthew J. Stover, president and chief executive officer

NYNEX Network Systems Company
Robert T. Anderson, president

NYNEX Science & Technology
Casimir S. Skrzypczak, president

NYNEX Trade Finance Company
Richard W. Frankenheimer, president

DAVID FRAIL